Make: Electronics

Make: Electronics

Learning by Discovery

Charles Platt

with photographs and illustrations by the author

Sebastopol, CA

Make: Electronics

by Charles Platt

Printed in Canada.

Published by Maker Media, Inc., 1005 Gravenstein Highway North, Sebastopol, CA 95472.

Maker Media books may be purchased for educational, business, or sales promotional use. Online editions are also available for most titles (*my.safaribooksonline.com*). For more information, contact O'Reilly Media's corporate/institutional sales department: 800-998-9938 or *corporate@oreilly.com*.

Editors: Dale Dougherty and Brian Jepson

Development Editor: Gareth Branwyn

Production Editor: Rachel Monaghan

Technical Editor: Andrew "Bunnie" Huang

Copyeditor: Nancy Kotary

Proofreader: Nancy Reinhardt

Indexer: Julie Hawks

Cover Designer: Mark Paglietti

Interior Designer: Ron Bilodeau

Illustrator/Photographer: Charles Platt

Cover Photographer: Marc de Vinck

Print History:

December 2009: First Edition.

ISBN: 978-0-596-15374-8

[TI] [2014-02-07]

For my dearest Erico

Contents

Preface

How to Have Fun with This Book

Everyone uses electronic devices, but most of us don't really know what goes on inside them.

Of course, you may feel that you don't need to know. If you can drive a car without understanding the workings of an internal combustion engine, presumably you can use an iPod without knowing anything about integrated circuits. However, understanding some basics about electricity and electronics can be worthwhile for three reasons:

- By learning how technology works, you become better able to control your world instead of being controlled by it. When you run into problems, you can solve them instead of feeling frustrated by them.
- Learning about electronics can be fun—so long as you approach the process in the right way. The tools are relatively cheap, you can do all the work on a tabletop, and it doesn't consume a lot of time (unless you want it to).
- Knowledge of electronics can enhance your value as an employee or perhaps even lead to a whole new career.

Learning by Discovery

Most introductory guides begin with definitions and facts, and gradually get to the point where you can follow instructions to build a simple circuit.

This book works the other way around. I want you to start putting components together right away. After you see what happens, you'll figure out what's going on. I believe this process of *learning by discovery* creates a more powerful and lasting experience.

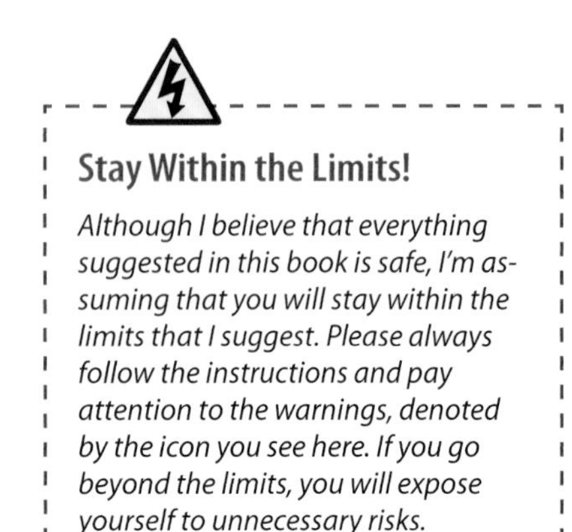

Stay Within the Limits!

Although I believe that everything suggested in this book is safe, I'm assuming that you will stay within the limits that I suggest. Please always follow the instructions and pay attention to the warnings, denoted by the icon you see here. If you go beyond the limits, you will expose yourself to unnecessary risks.

Learning by discovery occurs in serious research, when scientists notice an unusual phenomenon that cannot be explained by current theory, and they start to investigate it in an effort to explain it. This may ultimately lead to a better understanding of the world.

We're going to be doing the same thing, although obviously on a much less ambitious level.

Along the way, you will make some mistakes. This is good. Mistakes are the best of all learning processes. I want you to burn things out and mess things up, because this is how you learn the limits of components and materials. Since we'll be using low voltages, there'll be no chance of electrocution, and so long as you limit the flow of current in the ways I'll describe, there will be no risk of burning your fingers or starting fires.

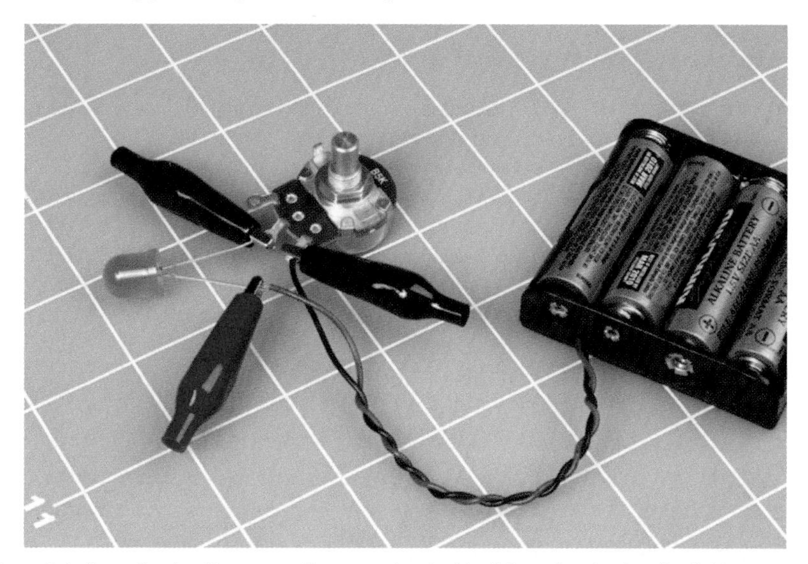

Figure P-1. *Learning by discovery allows you to start building simple circuits right away, using a handful of cheap components, a few batteries, and some alligator clips.*

How Hard Will It Be?

I assume that you're beginning with no prior knowledge of electronics. So, the first few experiments will be ultra-simple, and you won't even use solder or prototyping boards to build a circuit. You'll be holding wires together with alligator clips.

Very quickly, though, you'll be experimenting with transistors, and by the end of Chapter 2, you will have a working circuit that has useful applications.

I don't believe that hobby electronics has to be difficult to understand. Of course, if you want to study electronics more formally and do your own circuit design, this can be challenging. But in this book, the tools and supplies will be inexpensive, the objectives will be clearly defined, and the only math you'll need will be addition, subtraction, multiplication, division, and the ability to move decimal points from one position to another.

Moving Through This Book

Basically there are two ways to present information in a book of this kind: in tutorials and in reference sections. I'm going to use both of these methods. You'll find the tutorials in sections headed as follows:

- Shopping Lists
- Using Tools
- Experiments

You'll find reference sections under the following headings:

- Fundamentals
- Theory
- Background

How you use the sections is up to you. You can skip many of the reference sections and come back to them later. But if you skip many of the tutorials, this book won't be of much use to you. Learning by discovery means that you absolutely, positively have to do some hands-on work, and this in turn means that you have to buy some basic components and play with them. You will gain very little by merely imagining that you are doing this.

It's easy and inexpensive to buy what you need. In almost any urban or suburban area in the United States, chances are you live near a store that sells electronic components and some basic tools to work with them. I am referring, of course, to RadioShack franchises. Some Shacks have more components than others. Many may carry the Maker Shed kits that were developed to go with this book.

You can also visit auto supply stores such as AutoZone and Pep Boys for basics such as hookup wire, fuses, and switches, while stores such as Ace Hardware, Home Depot, and Lowe's will sell you tools.

If you prefer to buy via mail order, you can easily find everything you need by searching online. In each section of the book, I'll include the URLs of the most popular supply sources, and you'll find a complete list of URLs in the appendix.

Fundamentals

Mail-ordering components and tools

Here are the primary mail-order sources that I use myself online. I leave it to you to shop around for bargains, if you are so inclined.

http://www.radioshack.com
: RadioShack, a.k.a. The Shack. For tools and components. The site is easy and convenient, and some of the tools are exactly what you need.

http://www.mouser.com
: Mouser Electronics.

http://www.digikey.com
: Digi-Key Corporation.

http://www.newark.com
Newark.

Mouser, Digi-Key, and Newark are all good sources for components, usually requiring no minimum quantities.

http://www.allelectronics.com
All Electronics Corporation. A narrower range of components, but specifically aimed at the hobbyist, with kits available.

http://www.ebay.com
You can find surplus parts and bargains here, but you may have to try several eBay Stores to get what you want. Those based in Hong Kong are often very cheap, and I've found that they are reliable.

http://www.mcmaster.com
McMaster-Carr. Especially useful for high-quality tools.

Lowe's and Home Depot also allow you to shop online.

Figure P-2. *You'll find no shortage of parts, tools, kits, and gadgets online.*

Companion Kits

Maker Shed (*www.makershed.com*) offers a number of *Make: Electronics* companion kits, both toolkits and bundles of the various components used in the book's experiments. This is a simple, convenient, and cost-effective way of getting all the tools and materials you need to do the projects in this book.

Comments and Questions

Please address comments and questions concerning this book to the publisher:

Maker Media, Inc.
1005 Gravenstein Highway North
Sebastopol, CA 95472
800-998-9938 (in the United States or Canada)
707-829-0515 (international or local)
707-829-0104 (fax)

We have a web page for this book, where we list errata, examples, larger versions of the book's figures, and any additional information. You can access this page at:

http://oreilly.com/catalog/9780596153748

To comment or ask technical questions about this book, send email to:

bookquestions@oreilly.com

For more information about our publications, events, and products, see our website at:

http://makermedia.com

Safari® Books Online

Safari Books Online is an on-demand digital library that lets you easily search over 7,500 technology and creative reference books and videos to find the answers you need quickly.

With a subscription, you can read any page and watch any video from our library online. Read books on your cell phone and mobile devices. Access new titles before they are available for print, and get exclusive access to manuscripts in development and post feedback for the authors. Copy and paste code samples, organize your favorites, download chapters, bookmark key sections, create notes, print out pages, and benefit from tons of other time-saving features.

O'Reilly Media has uploaded this book to the Safari Books Online service. To have full digital access to this book and others on similar topics from O'Reilly and other publishers, sign up for free at *http://my.safaribooksonline.com*.

1 Experiencing Electricity

I want you to get a taste for electricity—literally!—in the first experiment. This first chapter of the book will show you:

- How to understand and measure electricity and resistance
- How to handle and connect components without overloading, damaging, or destroying them

Even if you have some prior knowledge of electronics, you should try these experiments before you venture on to the rest of the book.

IN THIS CHAPTER

Maker Shed (www.makershed.com) has put together a series of Make: Electronics *companion kits. These include all of the tools and components used in book's experiments. This is a quick, simple, and cost-effective way of getting everything you need to complete the projects in this book.*

Shopping List: Experiments 1 Through 5

If you want to limit your number of shopping trips or online purchases, look ahead in the book for additional shopping lists, and combine them to make one bulk purchase.

In this first chapter, I will give you part numbers and sources for every tool and component that we'll be using. Subsequently, I won't expect you to need such specific information, because you will have gained experience searching for items on your own.

Tools

Small pliers

RadioShack 4.5-inch mini long-nose pliers, part number 64-0062, or Xcelite 4-inch mini long-nose pliers, model L4G.

Or similar. See Figures 1-1 through 1-3. Look for these tools in hardware stores and the sources listed in the preface. The brand is unimportant. After you use them for a while, you'll develop your own preferences. In particular, you have to decide whether you like spring-loaded handles. If you decide you don't, you'll need a second pair of pliers to pull the springs out of the first.

Wire cutters

RadioShack 4.5-inch mini diagonal cutters, part number 64-0061, or Stanley 7-inch model 84-108.

Or similar. Use them for cutting copper wire, not harder metals (Figure 1-4).

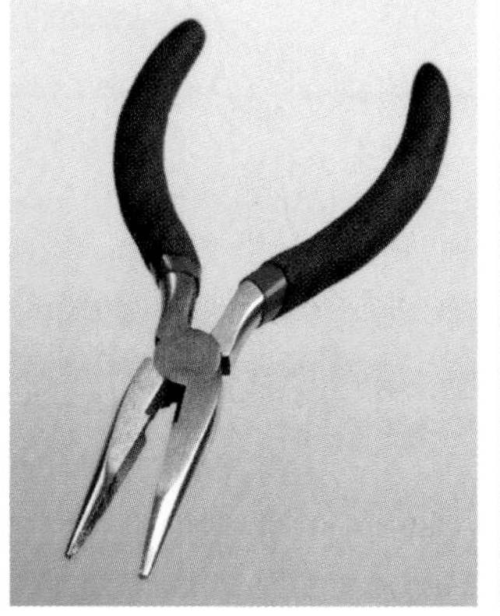

Figure 1-1. *Generic long-nosed pliers are your most fundamental tool for gripping, bending, and picking things up after you drop them.*

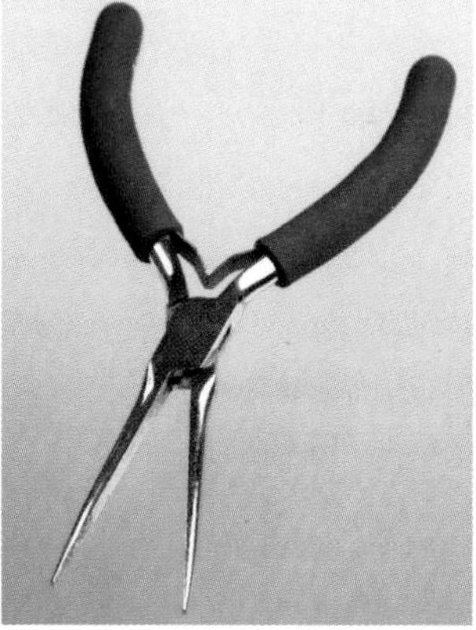

Figure 1-2. *Longer-nosed pliers: these are useful for reaching into tiny spaces.*

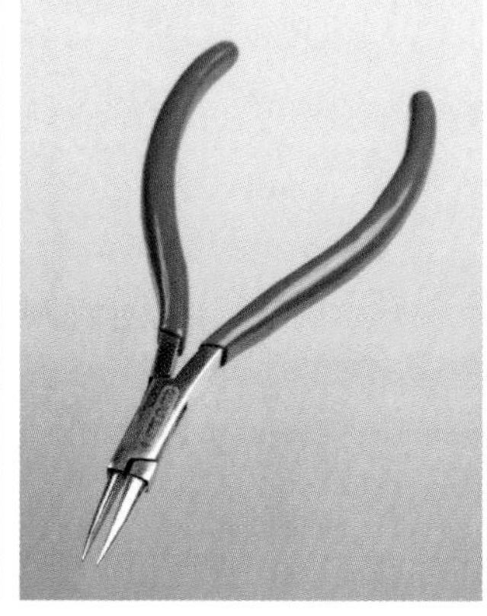

Figure 1-3. *Sharp-pointed pliers are designed for making jewelry, but are also useful for grabbing tiny components.*

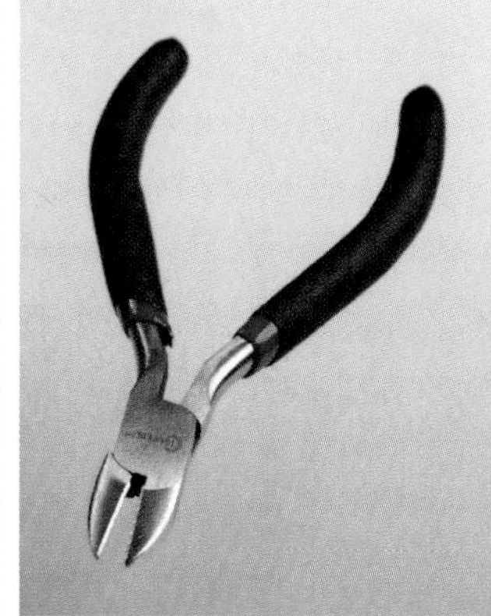

Figure 1-4. *Wire cutters, sometimes known as side cutters, are essential.*

Multimeter

Extech model EX410 or BK Precision model 2704-B or Amprobe model 5XP-A.

Or similar. Because electricity is invisible, we need a tool to visualize the pressure and flow, and a meter is the only way. A cheap meter will be sufficient for your initial experiments. If you buy online, try to check customer reviews, because reliability may be a problem for cheap meters. You can shop around for retailers offering the best price. Don't forget to search on eBay.

The meter must be digital—don't get the old-fashioned analog kind with a needle that moves across a set of printed scales. This book assumes that you are looking at a digital display.

I suggest that you do not buy an autoranging meter. "Autoranging" sounds useful—for example, when you want to check a 9-volt battery, the meter figures out for itself that you are not trying to measure hundreds of volts, nor fractions of a volt. The trouble is that this can trick you into making errors. What if the battery is almost dead? Then you may be measuring a fraction of a volt without realizing it. The only indication will be an easily overlooked "m" for "millivolts" beside the large numerals of the meter display.

On a manual-ranging meter, you select the range, and if the source that you are measuring is outside of that range, the meter tells you that you made an error. I prefer this. I also get impatient with the time it takes for the autoranging feature to figure out the appropriate range each time I make a measurement. But it's a matter of personal preference. See Figures 1-5 through 1-7 for some examples of multimeters.

Figure 1-5. *You can see by the wear and tear that this is my own favorite meter. It has all the necessary basic features and can also measure capacitance (the F section, for Farads). It can also check transistors. You have to choose the ranges manually.*

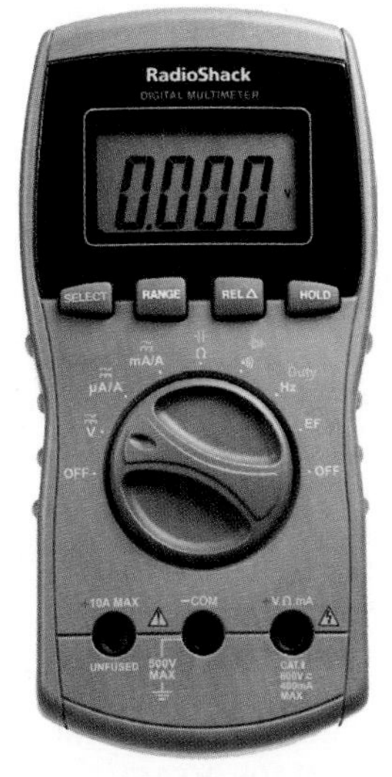

Figure 1-6. *Mid-priced RadioShack meter, which has the basic features; however, the dual purpose for each dial position, selected with the SELECT button, may be confusing. This is an autoranging meter.*

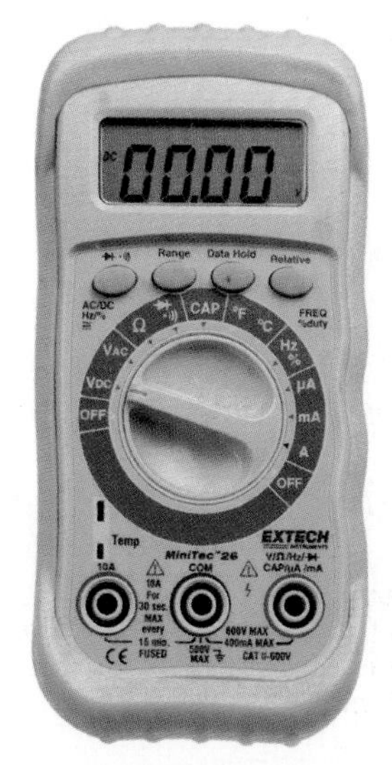

Figure 1-7. *An autoranging meter from Extech offers basic functions, plus a temperature probe, which may be useful to check whether components such as power supplies are running unduly hot.*

Supplies

Batteries

9-volt battery. Quantity: 1.

AA batteries, 1.5 volts each. Quantity: 6.

The batteries should be disposable alkaline, the cheapest available, because we may destroy some of them. You should *absolutely not* use rechargeable batteries in Experiments 1 and 2.

Battery holders and connectors

Snap connector for 9-volt battery, with wires attached (Figure 1-8). Quantity: 1. RadioShack part number 270-325 or similar. Any snap connector that has wires attached will do.

Battery holder for single AA cell, with wires attached (Figure 1-9). Quantity: 1. RadioShack part number 270-401 or Mouser.com catalog number 12BH311A-GR, or similar; any single-battery holder that has thin wires attached will do.

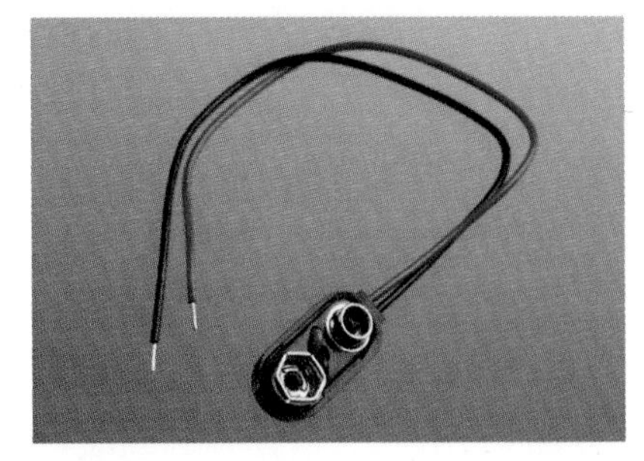

Figure 1-8. *Snap connector for a 9-volt battery.*

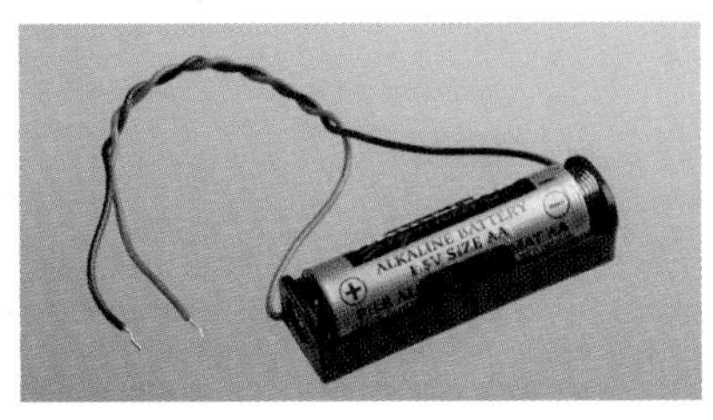

Figure 1-9. *Single AA-sized battery carrier with wires.*

Battery holder for four AA cells, with wires attached (Figure 1-10). Quantity: 1. All Electronics catalog number BH-342 or RadioShack part 270-391 or similar. Also, one battery carrier to hold two AA cells, from the same sources.

Figure 1-10. *Battery carrier for four AA cells, to be installed in series, delivering 6 volts.*

Alligator clips

Vinyl-insulated. Quantity: at least 6. All Electronics catalog number ALG-28 or RadioShack part number 270-1545 or similar (Figure 1-11).

Figure 1-11. *Alligator clips inside vinyl sheaths, which reduce the chance of accidental short circuits.*

Components

You may not know what some of these items are, or what they do. Just look for the part numbers and descriptions, and match them with the photographs shown here. Very quickly, in the learning by discovery process, all will be revealed.

Fuses

Automotive-style, mini-blade type, 3 amps. Quantity: 3. RadioShack part number 270-1089, or Bussmann part ATM-3, available from automotive parts suppliers such as AutoZone (Figure 1-12).

Or similar. A blade-type fuse is easier to grip with alligator clips than a round cartridge fuse.

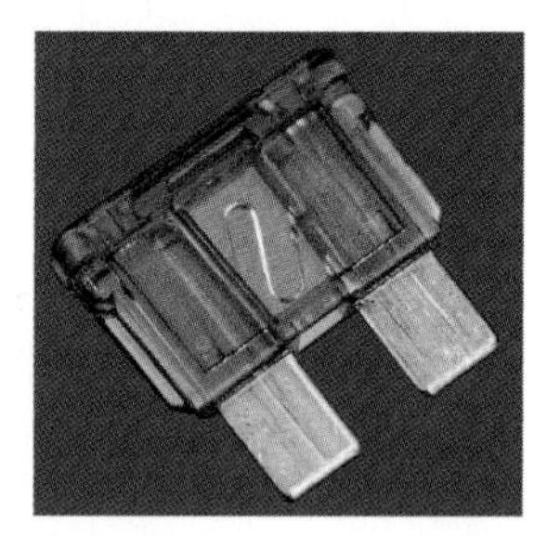

Figure 1-12. *A 3-amp fuse intended primarily for automotive use, shown here larger than actual size.*

Potentiometers

Panel-mount, single-turn, 2K linear, 0.1 watt minimum. Quantity: 2. Alpha part RV170F-10-15R1-B23 or BI Technologies part P160KNPD-2QC25B2K, from Mouser.com or other component suppliers (Figure 1-13).

Or similar. The "watt" rating tells you how much power this component can handle. You don't need more than 0.5 watts.

Figure 1-13. *Potentiometers come in many shapes and sizes, with different lengths of shafts intended for different types of knobs. For our purposes, any style will do, but the larger-sized ones are easier to play with.*

Resistors

Assortment 1/4-watt minimum, various values but must include 470 ohms, 1K, and 2K or 2.2K. Quantity: at least 100. RadioShack part number 271-312.

Or search eBay for "resistor assorted."

Light-emitting diodes (LEDs)

Any size or color (Figures 1-14 and 1-15). Quantity: 10. RadioShack part number 276-1622 or All Spectrum Electronics part K/LED1 from Mouser.com.

Or similar. Just about any LEDs will do for these first experiments.

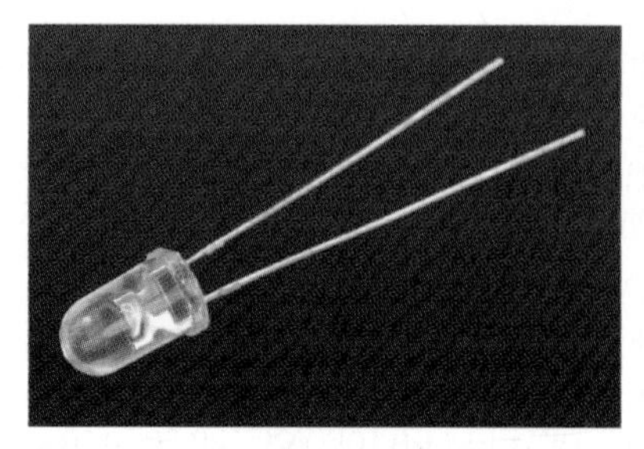

Figure 1-14. *Typical 5-mm diameter light-emitting diode (LED).*

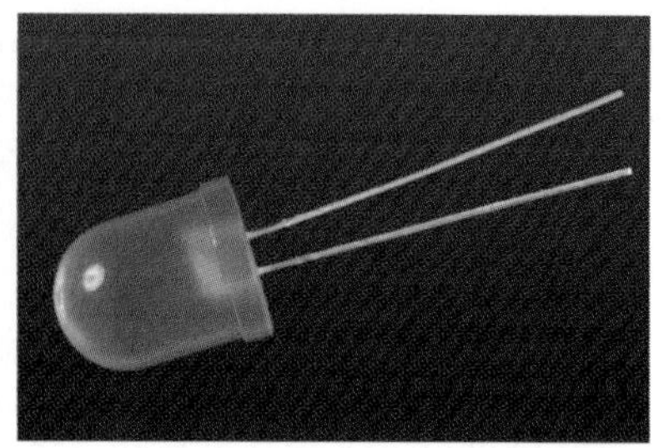

Figure 1-15. *Jumbo-sized LED (1 cm diameter) is not necessarily brighter or more expensive. For most of the experiments in this book, buy whatever LEDs you like the look of.*

Experiment 1: Taste the Power!

Can you taste electricity? Maybe not, but it feels as if you can.

You will need:

- 9-volt battery
- Snap connector for battery terminals
- Multimeter

No More Than 9 Volts

A 9-volt battery won't hurt you. But do not try this experiment with a higher-voltage battery or a larger battery that can deliver more current. Also, if you have metal braces on your teeth, be very careful not to touch them with the battery.

Procedure

Moisten your tongue and touch the tip of it to the metal terminals of a 9-volt battery. The sudden sharp tingle that you feel is caused by electricity flowing from one terminal of the battery (Figure 1-16), through the moisture on and in your tongue, to the other terminal. Because the skin of your tongue is very thin (it's actually a mucus membrane) and the nerves are close to the surface, you can feel the electricity very easily.

Figure 1-16. *Step 1 in the process of learning by discovery: the 9-volt tongue test.*

Now stick out your tongue, dry the tip of it very thoroughly with a tissue, and repeat the experiment without allowing your tongue to become moist again. You should feel less of a tingle.

What's happening here? We're going to need a meter to find out.

Tools

Setting up your meter

Check the instructions that came with the meter to find out whether you have to install a battery in it, or whether a battery is preinstalled.

Most meters have removable wires, known as *leads* (pronounced "leeds"). Most meters also have three sockets on the front, the leftmost one usually being reserved to measure high electrical currents (flows of electricity). We can ignore that one for now.

The leads will probably be black and red. The black wire plugs into a socket labeled "COM" or "Common." Plug the red one into the socket labeled "V" or "volts." See Figures 1-17 through 1-20.

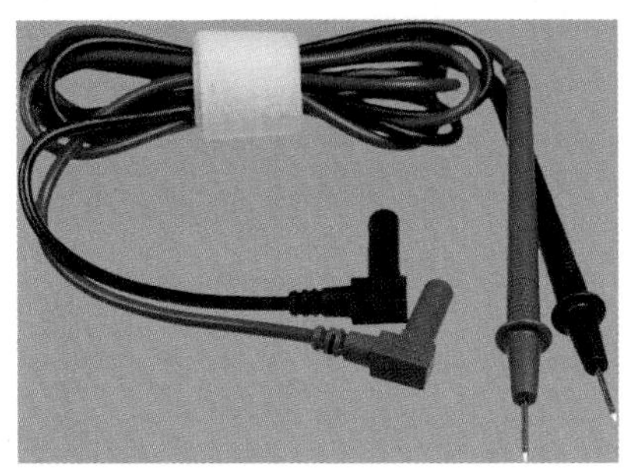

Figure 1-17. *The black lead plugs into the Common (COM) socket, and the red lead plugs into the red socket that's almost always on the righthand side of a multimeter.*

The other ends of the leads terminate in metal spikes known as *probes*, which you will be touching to components when you want to make electrical measurements. The probes detect electricity; they don't emit it in significant quantities. Therefore, they cannot hurt you unless you poke yourself with their sharp ends.

If your meter doesn't do autoranging, each position on the dial will have a number beside it. This number means "no higher than." For instance if you want to check a 6-volt battery, and one position on the voltage section of the dial is numbered 2 and the next position is numbered 20, position 2 means "no higher than 2 volts." You have to go to the next position, which means "no higher than 20 volts."

If you make a mistake and try to measure something inappropriate, the meter will show you an error message such as "E" or "L." Turn the dial and try again.

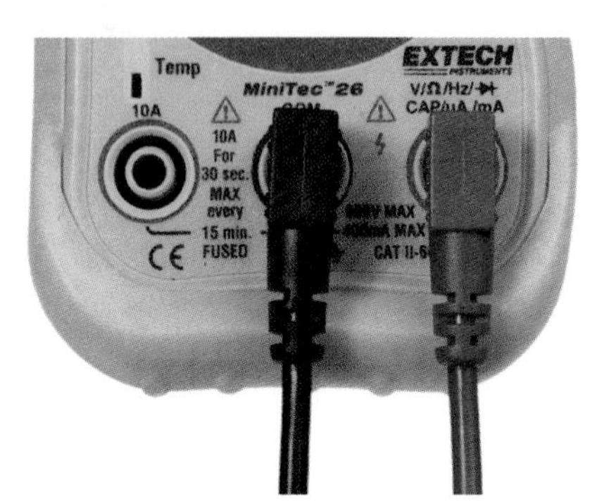

Figure 1-18

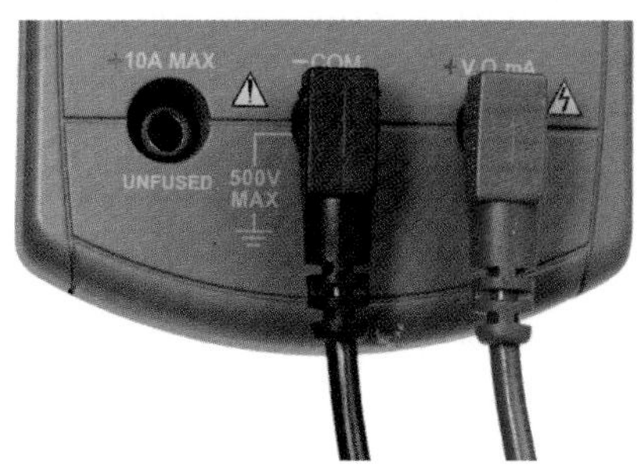

Figure 1-19

Figure 1-20. *To measure resistance and voltage, plug the black lead into the Common socket and the red lead into the Volts socket. Almost all meters have a separate socket where you must plug the red lead when you measure large currents in amps, but we'll be dealing with this later.*

FUNDAMENTALS

Ohms

We measure distance in miles or kilometers, mass in pounds or kilograms, temperature in Fahrenheit or Centigrade—and electrical resistance in ohms. The ohm is an international unit.

The Greek omega symbol (Ω) is used to indicate ohms, as shown in Figures 1-21 and 1-22. Letter K (or alternatively, KΩ) means a kilohm, which is 1,000 ohms. Letter M (or MΩ) means a megohm, which is 1,000,000 ohms.

Number of ohms	Usually expressed as	Abbreviated as
1,000 ohms	1 kilohm	1KΩ or 1K
10,000 ohms	10 kilohms	10KΩ or 10K
100,000 ohms	100 kilohms	100KΩ or 100K
1,000,000 ohms	1 megohm	1MΩ or 1M
10,000,000 ohms	10 megohms	10MΩ or 10M

A material that has very high resistance to electricity is known as an *insulator*. Most plastics, including the colored sheaths around wires, are insulators.

A material with very low resistance is a *conductor*. Metals such as copper, aluminum, silver, and gold are excellent conductors.

Figure 1-21. *The omega symbol is used internationally to indicate resistance on ohms.*

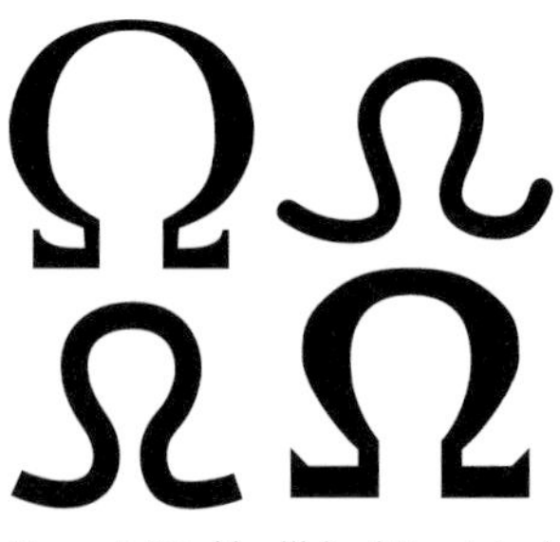

Figure 1-22. *You'll find it printed or written in a wide variety of styles.*

Procedure

We're going to use the meter to discover the electrical resistance of your tongue. First, set your meter to measure resistance. If it has autoranging, look to see whether it is displaying a K, meaning kilohms, or M, meaning megohms. If you have to set the range manually, begin with no less than 100,000 ohms (100K). See Figures 1-23 through 1-25.

Touch the probes to your tongue, about an inch apart. Note the reading, which should be around 50K. Now put aside the probes, stick out your tongue, and use a tissue to dry it very carefully and thoroughly. Without allowing your tongue to become moist again, repeat the test, and the reading should be higher. Finally, press the probes against the skin of your hand or arm: you may get no reading at all, until you moisten your skin.

When your skin is moist (for instance, if you perspire), its electrical resistance decreases. This principle is used in lie detectors, because someone who knowingly tells a lie, under conditions of stress, may tend to perspire.

A 9-volt battery contains chemicals that liberate electrons (particles of electricity), which want to flow from one terminal to the other as a result of a chemical reaction inside it. Think of the cells inside a battery as being like two water tanks—one of them full, the other empty. If they are connected with a pipe, water flows between them until their levels are equal. Figure 1-26 may help you visualize this. Similarly, when you open up an electrical pathway between the two sides of a battery, electrons flow between them, even if the pathway consists only of the moisture on your tongue.

Electrons flow more easily through some substances (such as a moist tongue) than others (such as a dry tongue).

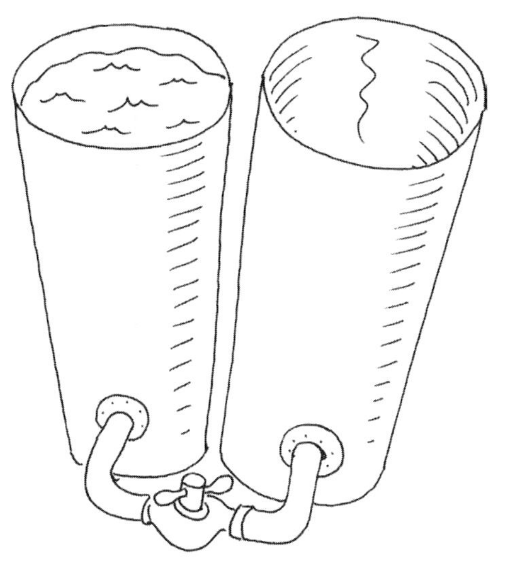

Figure 1-26. *Think of the cells in a battery as being like two cylinders: one full of water, the other empty. Open a connection between the cylinders, and the water will flow until the levels are equal on both sides. The less resistance in the connection, the faster the flow will be.*

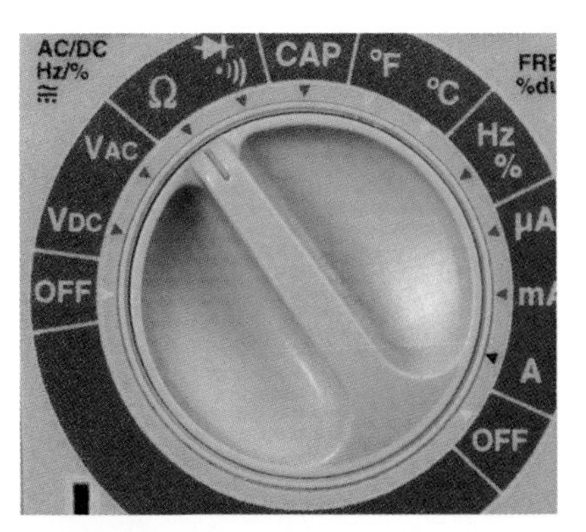

Figure 1-23

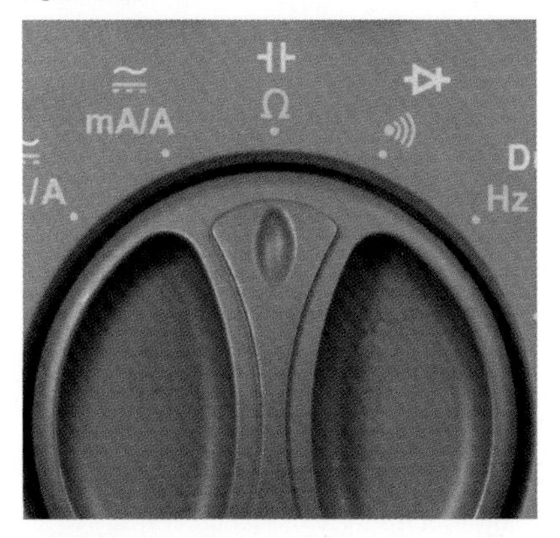

Figure 1-24

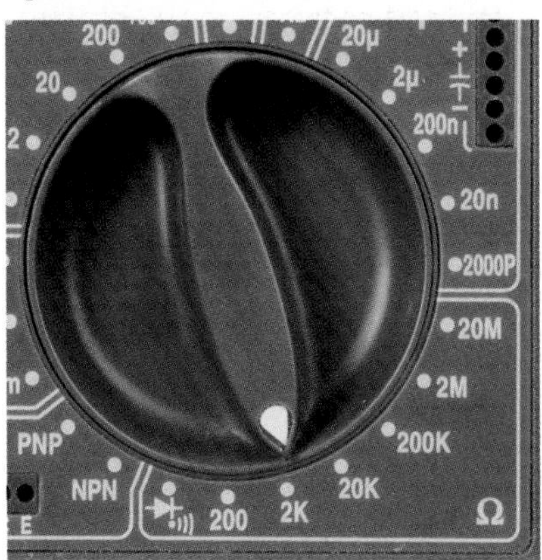

Figure 1-25. *To measure ohms, turn the dial to the ohm (omega) symbol. On an autoranging meter, you can then press the Range button repeatedly to display different ranges of resistance, or simply touch the probes to a resistance and wait for the meter to choose a range automatically. A manual meter requires you to select the range with the dial (you should set it to 100K or higher, to measure skin resistance). If you don't get a meaningful reading, try a different range.*

BACKGROUND

The man who discovered resistance

Georg Simon Ohm, pictured in Figure 1-27, was born in Bavaria in 1787 and worked in obscurity for much of his life, studying the nature of electricity using metal wire that he had to make for himself (you couldn't truck on down to Home Depot for a spool of hookup wire back in the early 1800s).

Despite his limited resources and inadequate mathematical abilities, Ohm was able to demonstrate in 1827 that the electrical resistance of a conductor such as copper varied in inverse proportion with its area of cross-section, and the current flowing through it is proportional to the voltage applied to it, as long as temperature is held constant. Fourteen years later, the Royal Society in London finally recognized the significance of his contribution and awarded him the Copley Medal. Today, his discovery is known as Ohm's Law.

Figure 1-27. *Georg Simon Ohm, after being honored for his pioneering work, most of which he pursued in relative obscurity.*

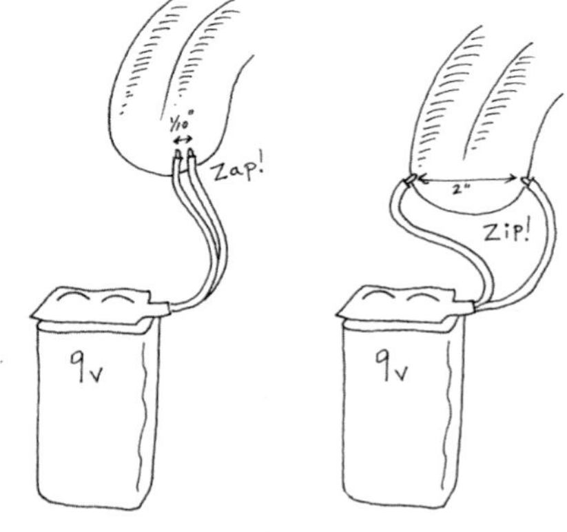

Figure 1-28. *Modifying the tongue test to show that a shorter distance, with lower resistance, allows greater flow of electricity, and a bigger zap.*

Further Investigation

Attach the snap-on terminal cap (shown earlier in Figure 1-8) to the 9-volt battery. Take the two wires that are attached to the cap and hold them so that the bare ends are just a few millimeters apart. Touch them to your tongue. Now separate the ends of the wires by a couple of inches, and touch them to your tongue again. (See Figure 1-28.) Notice any difference?

Use your meter to measure the electrical resistance of your tongue, this time varying the distance between the two probes. When electricity travels through a shorter distance, it encounters less total resistance. As a result, the current (the flow of electricity per second) increases. You can try a similar experiment on your arm, as shown in Figure 1-29.

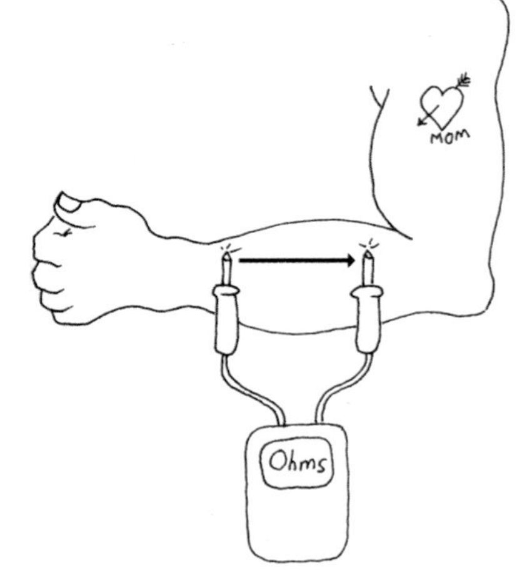

Figure 1-29. *Moisten your skin before trying to measure its resistance. You should find that the resistance goes up as you move the meter probes farther apart. The resistance is proportional to the distance.*

Use your meter to test the electrical resistance of water. Dissolve some salt in the water, and test it again. Now try measuring the resistance of distilled water (in a clean glass).

The world around you is full of materials that conduct electricity with varying amounts of resistance.

Cleanup and Recycling

Your battery should not have been damaged or significantly discharged by this experiment. You'll be able to use it again.

Remember to switch off your meter before putting it away.

Experiment 2: Let's Abuse a Battery!

To get a better feeling for electrical power, you're going to do what most books tell you not to do. You're going to short out a battery. A short circuit is a direct connection between the two sides of a power source.

Short Circuits

Short circuits can be dangerous. Do not short out a power outlet in your home: there'll be a loud bang, a bright flash, and the wire or tool that you use will be partially melted, while flying particles of melted metal can burn you or blind you.

If you short out a car battery, the flow of current is so huge that the battery might even explode, drenching you in acid (Figure 1-30).

Lithium batteries are also dangerous. Never short-circuit a lithium battery: it can catch fire and burn you (Figure 1-31).

Use only an alkaline battery in this experiment, and only a single AA cell (Figure 1-32). You should also wear safety glasses in case you happen to have a defective battery.

Figure 1-30. *Anyone who has dropped an adjustable wrench across the bare terminals of a car battery will tell you that short circuits can be dramatic at a "mere" 12 volts, if the battery is big enough.*

Figure 1-31. *The low internal resistance of lithium batteries (which are often used in laptop computers) allows high currents to flow, with unexpected results. Never fool around with lithium batteries!*

You will need:

- 1.5-volt AA battery
- Single-battery carrier
- 3-amp fuse
- Safety glasses (regular eyeglasses or sunglasses will do)
- Alligator clip (small or large)

Procedure

Use an alkaline battery. Do not use any kind of rechargeable battery.

Put the battery into a battery holder that's designed for a single battery and has two thin insulated wires emerging from it, as shown in Figure 1-32. Do not use any other kind of battery holder.

Use an alligator clip to connect the bare ends of the wires, as shown in Figure 1-32. There will be no spark, because you are using only 1.5 volts. Wait one minute, and you'll find that the wires are getting hot. Wait another minute, and the battery, too, will be hot.

Figure 1-32. *Shorting out an alkaline battery can be safe if you follow the directions precisely. Even so, the battery is liable to become too hot to touch comfortably. Don't try this with any type of rechargeable battery.*

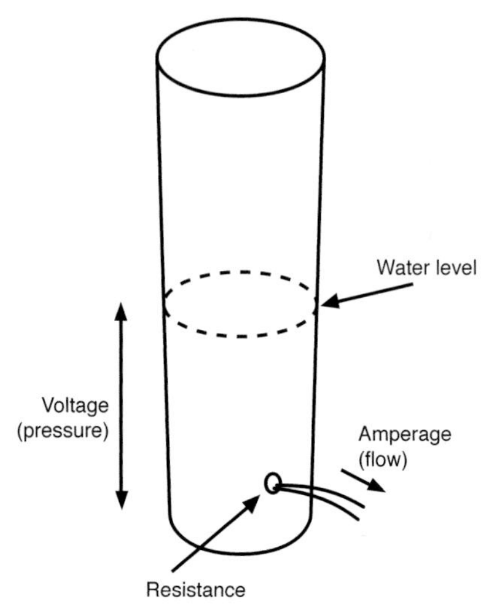

Figure 1-33. *Think of voltage as pressure, and amperes as flow.*

The heat is caused by electricity flowing through the wires and through the electrolyte (the conductive fluid) inside the battery. If you've ever used a hand pump to force air into a bicycle tire, you know that the pump gets warm. Electricity behaves in much the same way. You can imagine the electricity being composed of particles (electrons) that make the wire hot as they push through it. This isn't a perfect analogy, but it's close enough for our purposes.

Chemical reactions inside the battery create electrical pressure. The correct name for this pressure is *voltage*, which is measured in volts and is named after Alessandro Volta, an electrical pioneer.

Going back to the water analogy: the height of the water in a tank is proportionate to the pressure of the water, and comparable to voltage. Figure 1-33 may help you to visualize this.

But volts are only half of the story. When electrons flow through a wire, the flow is known as *amperage*, named after yet another electrical pioneer, André-Marie Ampère. The flow is also generally known as current. It's the current—the amperage—that generates the heat.

BACKGROUND

Why didn't your tongue get hot?

When you touched the 9-volt battery to your tongue, you felt a tingle, but no perceptible heat. When you shorted out a battery, you generated a noticeable amount of heat, even though you used a lower voltage. How can we explain this?

The electrical resistance of your tongue is very high, which reduces the flow of electrons. The resistance of a wire is very low, so if there's only a wire connecting the two terminals of the battery, more current will pass through it, creating more heat. If all other factors remain constant:

- Lower resistance allows more current to flow (Figure 1-34).
- The heat generated by electricity is proportional to the amount of electricity (the current) that flows.

Here are some other basic concepts:

- The flow of electricity per second is measured in amperes, or amps.
- The pressure of electricity, measured in volts, causes the flow.
- The resistance to the flow is measured in ohms.
- A higher resistance restricts the current.
- A higher voltage overcomes resistance and increases the current.

Figure 1-34. *Larger resistance results in smaller flow—but if you increase the pressure, it may overcome the resistance and increase the flow.*

If you're wondering exactly how much current flows between the terminals of a battery when you short it out, that's a difficult question to answer. If you try to use your multimeter to measure it, you're liable to blow the fuse inside the meter. Still, you can use your very own 3-amp fuse, which we can sacrifice because it didn't cost very much.

First inspect the fuse very carefully, using a magnifying glass if you have one. You should see a tiny S-shape in the transparent window at the center of the fuse. That S is a thin section of metal that melts easily.

Remove the battery that you short-circuited. It is no longer useful for anything, and should be recycled if possible. Put a fresh battery into the battery carrier, connect the fuse as shown in Figure 1-35, and take another look. You should see a break in the center of the S shape, where the metal melted almost instantly. Figure 1-36 shows the fuse before you connected it, and Figure 1-37 depicts a blown fuse. This is how a fuse works: it melts to protect the rest of the circuit. That tiny break inside the fuse stops any more current from flowing.

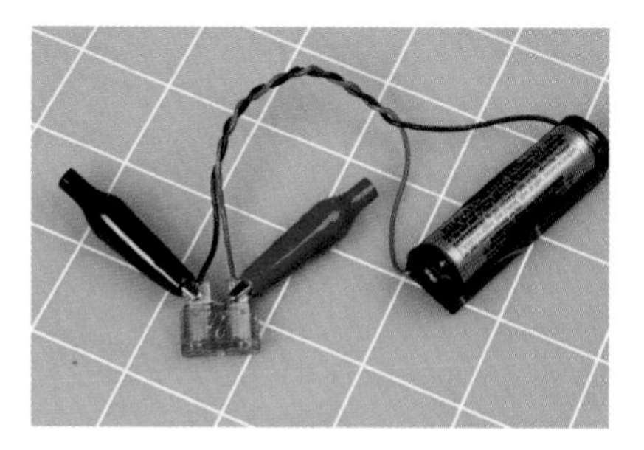

Figure 1-35. *When you attach both wires to the fuse, the little S-shaped element inside will melt almost instantly.*

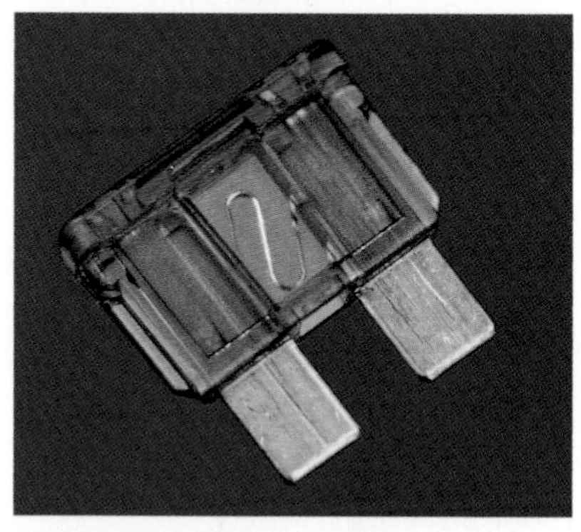

Figure 1-36. *A 3-amp fuse, before its element was melted by a single 1.5-volt battery.*

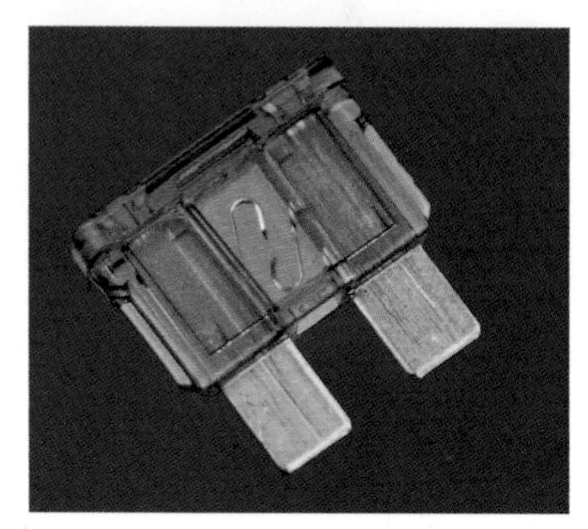

Figure 1-37. *The same fuse after being melted by electric current.*

FUNDAMENTALS

Volt basics

Electrical pressure is measured in volts. The volt is an international unit. A millivolt is 1/1,000 of a volt.

Number of volts	Usually expressed as	Abbreviated as
0.001 volts	1 millivolt	1 mV
0.01 volts	10 millivolts	10 mV
0.1 volts	100 millivolts	100 mV
1 volt	1,000 millivolts	1V

Ampere basics

We measure electrical flow in amperes, or amps. The ampere is an international unit, often referred to as an "amp." A milliamp is 1/1,000 of an ampere.

Number of amperes	Usually expressed as	Abbreviated as
0.001 amps	1 milliamp	1 mA
0.01 amps	10 milliamps	10 mA
0.1 amps	100 milliamps	100 mA
1 amp	1,000 milliamps	1A

BACKGROUND

Inventor of the battery

Alessandro Volta (Figure 1-38) was born in Italy in 1745, long before science was broken up into specialties. After studying chemistry (he discovered methane in 1776), he became a professor of physics and became interested in the so-called galvanic response, whereby a frog's leg will twitch in response to a jolt of static electricity.

Using a wine glass full of salt water, Volta demonstrated that the chemical reaction between two electrodes, one made of copper, the other of zinc, will generate a steady electric current. In 1800, he refined his apparatus by stacking plates of copper and zinc, separated by cardboard soaked in salt and water. This "voltaic pile" was the first electric battery.

Figure 1-38. *Alessandro Volta discovered that chemical reactions can create electricity.*

FUNDAMENTALS

Direct and alternating current

The flow of current that you get from a battery is known as *direct current*, or DC. Like the flow of water from a faucet, it is a steady stream, in one direction.

The flow of current that you get from the "hot" wire in a power outlet in your home is very different. It changes from positive to negative 60 times each second (in many foreign countries and Europe, 50 times per second). This is known as *alternating current*, or AC, which is more like the pulsatile flow you get from a power washer.

Alternating current is essential for some purposes, such as cranking up voltage so that electricity can be distributed over long distances. AC is also useful in motors and domestic appliances. The parts of an American power outlet are shown in Figure 1-39. A few other nations, such as Japan, also use American-style outlets.

For most of this book I'm going to be talking about DC, for two reasons: first, most simple electronic circuits are powered with DC, and second, the way it behaves is much easier to understand.

I won't bother to mention repeatedly that I'm dealing with DC. Just assume that everything is DC unless otherwise noted.

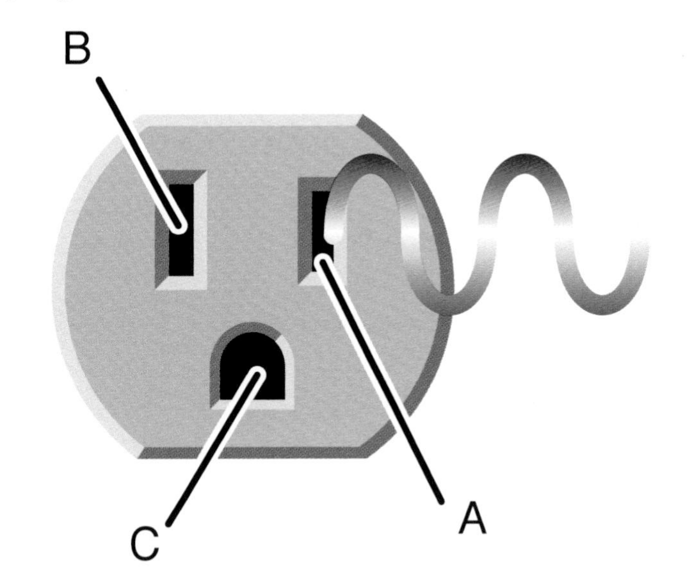

Figure 1-39. *This style of power outlet is found in North America, South America, Japan, and some other nations. European outlets look different, but the principle remains the same. Socket A is the "live" side of the outlet, supplying voltage that alternates between positive and negative, relative to socket B, which is called the "neutral" side. If an appliance develops a fault such as an internal loose wire, it should protect you by sinking the voltage through socket C, the ground.*

Cleanup and Recycling

The first AA battery that you shorted out is probably damaged beyond repair. You should dispose of it. Putting batteries in the trash is not a great idea, because they contain heavy metals that should be kept out of the ecosystem. Your state or town may include batteries in a local recycling scheme. (California requires that almost all batteries be recycled.) You'll have to check your local regulations for details.

The blown fuse is of no further use, and can be thrown away.

The second battery, which was protected by the fuse, should still be OK. The battery holder also can be reused later.

Experiment 3: Your First Circuit

Now it's time to make electricity do something that's at least slightly useful. For this purpose, you'll use components known as resistors, and a light-emitting diode, or LED.

You will need:

- 1.5-volt AA batteries. Quantity: 4.
- Four-battery holder. Quantity: 1.
- Resistors: 470Ω, 1K, and either 2K or 2.2K (the 2.2K value happens to be more common than 2K, but either will do in this experiment). Quantity: 1 of each resistor.
- An LED, any type. Quantity: 1.
- Alligator clips. Quantity: 3.

Setup

It's time to get acquainted with the most fundamental component we'll be using in electronic circuits: the humble resistor. As its name implies, it resists the flow of electricity. As you might expect, the value is measured in ohms.

If you bought a bargain-basement assortment package of resistors, you may find nothing that tells you their values. That's OK; we can find out easily enough. In fact, even if they are clearly labeled, I want you to check their values yourself. You can do it in two ways:

- Use your multimeter. This is excellent practice in learning to interpret the numbers that it displays.
- Learn the color codes that are printed on most resistors. See the following section, "Fundamentals: Decoding resistors," for instructions.

After you check them, it's a good idea to sort them into labeled compartments in a little plastic parts box. Personally, I like the boxes sold at the Michaels chain of crafts stores, but you can find them from many sources.

BACKGROUND

Father of electromagnetism

Born in 1775 in France, André-Marie Ampère (Figure 1-40) was a mathematical prodigy who became a science teacher, despite being largely self-educated in his father's library. His best-known work was to derive a theory of electromagnetism in 1820, describing the way that an electric current generates a magnetic field. He also built the first instrument to measure the flow of electricity (now known as a *galvanometer*), and discovered the element fluorine.

Figure 1-40. *Andre-Marie Ampere found that an electric current running through a wire creates a magnetic field around it. He used this principle to make the first reliable measurements of what came to be known as amperage.*

FUNDAMENTALS

Decoding resistors

Some resistors have their value clearly stated on them in microscopic print that you can read with a magnifying glass. Most, however, are color-coded with stripes. The code works like this: first, ignore the color of the body of the resistor. Second, look for a silver or gold stripe. If you find it, turn the resistor so that the stripe is on the righthand side. Silver means that the value of the resistor is accurate within 10%, while gold means that the value is accurate within 5%. If you don't find a silver or gold stripe, turn the resistor so that the stripes are clustered at the left end. You should now find yourself looking at three colored stripes on the left. Some resistors have more stripes, but we'll deal with those in a moment. See Figures 1-41 and 1-42.

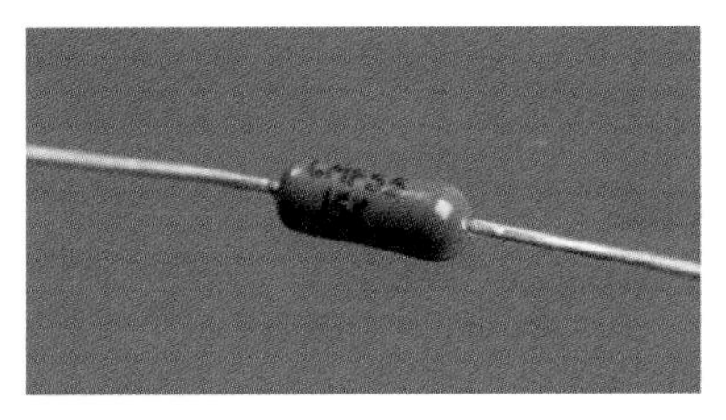

Figure 1-41. *Some modern resistors have their values printed on them, although you may need a magnifier to read them. This 15K resistor is less than half an inch long.*

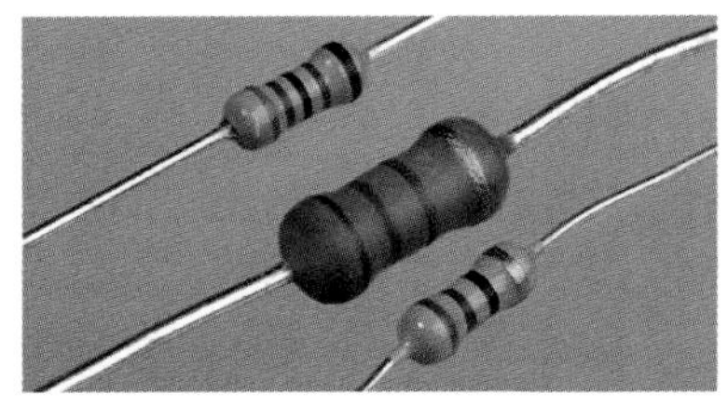

Figure 1-42. *From top to bottom, these resistor values are 56,000 ohms (56K), 5,600 ohms (5.6K), and 560 ohms. The size tells you how much power the resistor can handle; it has nothing to do with the resistance. The smaller components are rated at 1/4 watt; the larger one in the center can handle 1 watt of power.*

Starting from the left, the first and second stripes are coded according to this table:

Black	0
Brown	1
Red	2
Orange	3
Yellow	4
Green	5
Blue	6
Violet	7
Gray	8
White	9

The third stripe has a different meaning: It tells you how many zeros to add, like this:

Black	-	No zeros
Brown	0	1 zero
Red	00	2 zeros
Orange	000	3 zeros
Yellow	0000	4 zeros
Green	00000	5 zeros
Blue	000000	6 zeros
Violet	0000000	7 zeros
Gray	00000000	8 zeros
White	000000000	9 zeros

FUNDAMENTALS

Decoding resistors (continued)

Note that the color-coding is consistent, so that green, for instance, means either a value of 5 (for the first two stripes) or 5 zeros (for the third stripe). Also, the sequence of colors is the same as their sequence in a rainbow.

So, a resistor colored brown-red-green would have a value of 1-2 and five zeros, making 1,200,000 ohms, or 1.2MΩ. A resistor colored orange-orange-orange would have a value of 3-3 and three zeros, making 33,000 ohms, or 33KΩ. A resistor colored brown-black-red would have a value of 1-0 and two additional zeros, or 1KΩ. Figure 1-43 shows some other examples.

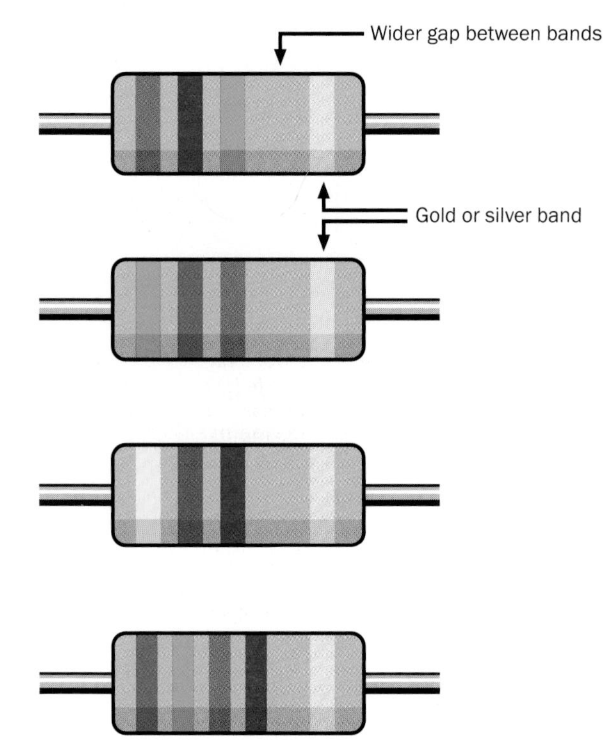

Figure 1-43. *To read the value of a resistor, first turn it so that the silver or gold stripe is on the right, or the other stripes are clustered on the left. From top to bottom: The first resistor has a value of 1-2 and five zeros, or 1,200,000, which is 1.2MΩ. The second is 5-6 and one zero, or 560Ω. The third is 4-7 and two zeros, or 4,700, which is 4.7KΩ. The last is 6-5-1 and two zeros, or 65,100Ω, which is 65.1KΩ.*

If you run across a resistor with four stripes instead of three, the first *three* stripes are digits and the *fourth* stripe is the number of zeros. The third numeric stripe allows the resistor to be calibrated to a finer tolerance.

Confusing? Absolutely. That's why it's easier to use your meter to check the values. Just be aware that the meter reading may be slightly different from the claimed value of the resistor. This can happen because your meter isn't absolutely accurate, or because the resistor is not absolutely accurate, or both. As long as you're within 5% of the claimed value, it doesn't matter for our purposes.

Lighting an LED

Now take a look at one of your LEDs. An old-fashioned lightbulb wastes a lot of power by converting it into heat. LEDs are much smarter: they convert almost all their power into light, and they last almost indefinitely—as long as you treat them right!

An LED is quite fussy about the amount of power it gets, and the way it gets it. Always follow these rules:

- The *longer* wire protruding from the LED must receive a *more positive* voltage than the shorter wire.
- The voltage difference between the long wire and the short wire must not exceed the limit stated by the manufacturer.
- The current passing through the LED must not exceed the limit stated by the manufacturer.

What happens if you break these rules? Well, we're going to find out!

Make sure you are using fresh batteries. You can check by setting your multimeter to measure volts DC, and touching the probes to the terminals of each battery. You should find that each of them generates a pressure of at least 1.5 volts. If they read slightly higher than this, it's normal. A battery starts out above its rated voltage, and delivers progressively less as you use it. Batteries also lose some voltage while they are sitting on the shelf doing nothing.

Load your battery holder (taking care that the batteries are the right way around, with the negative ends pressing against the springs in the carrier). Use your meter to check the voltage on the wires coming out of the battery carrier. You should have at least 6 volts.

Now select a 2KΩ resistor. Remember, "2KΩ" means "2,000 ohms." If it has colored stripes, they should be red-black-red, meaning 2-0 and two more zeros. Because 2.2K resistors are more common than 2K resistors, you can substitute one of them if necessary. It will be colored red-red-red.

Wire it into the circuit as shown in Figures 1-44 and 1-45, making the connections with alligator clips. You should see the LED glow very dimly.

Now swap out your 2K resistor and substitute a 1K resistor, which will have brown-black-red stripes, meaning 1-0 and two more zeros. The LED should glow more brightly.

Swap out the 1K resistor and substitute a 470Ω resistor, which will have yellow-violet-brown stripes, meaning 4-7 and one more zero. The LED should be brighter still.

This may seem very elementary, but it makes an important point. The resistor blocks a percentage of the voltage in the circuit. Think of it as being like a kink or constriction in a flexible hose. A higher-value resistor blocks more voltage, leaving less for the LED.

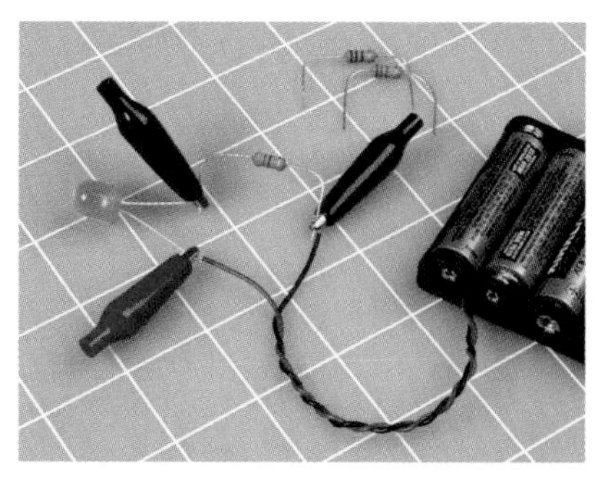

Figure 1-44. *The setup for Experiment 3, showing resistors of 470Ω, 1KΩ, and 2KΩ. Apply alligator clips where shown, to make a secure contact, and try each of the resistors one at a time at the same point in the circuit, while watching the LED.*

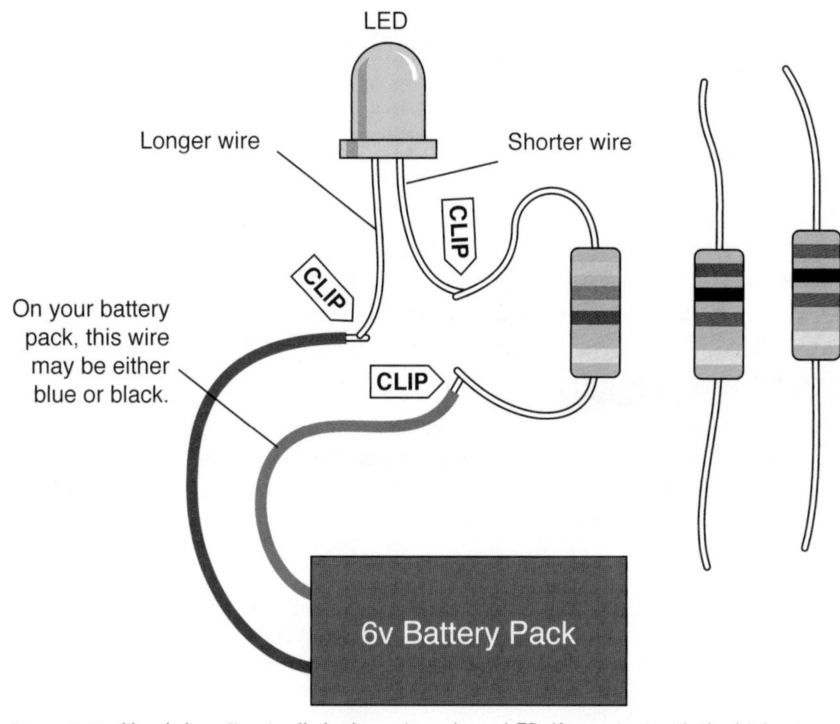

Figure 1-45. *Here's how it actually looks, using a large LED. If you start with the highest value resistor, the LED will glow very dimly as you complete the circuit. The resistor drops most of the voltage, leaving the LED with insufficient current to make it shine brightly.*

Cleanup and Recycling

We'll use the batteries and the LED in the next experiment. The resistors can be reused in the future.

Experiment 4: Varying the Voltage

Potentiometers come in various shapes and sizes, but they all do the same thing: they allow you to vary voltage and current by varying resistance. This experiment will enable you to learn more about voltage, amperage, and the relationship between them. You'll also learn how to read a manufacturer's data sheet.

You will need the same batteries, battery carrier, alligator clips, and LED from the last experiment, plus:

- Potentiometer, 2KΩ linear. Quantity: 2. (See Figure 1-46.) Full-sized potentiometers that look like this are becoming less common, as miniature versions are taking their place. I'd like you to use a large one, though, because it's so much easier to work with.
- One extra LED.
- Multimeter.

Look Inside Your Potentiometer

The first thing I want you to do is find out how a potentiometer works. This means you'll have to open it, which is why your shopping list required you to buy two of them, in case you can't put the first one back together again.

Most potentiometers are held together with little metal tabs. You should be able to grab hold of the tabs with your wire cutters or pliers, and bend them up and outward. If you do this, the potentiometer should open up as shown in Figures 1-47 and 1-48.

Figure 1-46

Figure 1-47

Figure 1-48. *To open the potentiometer, first pry up the four little metal tabs around the edge (you can see one sticking out at the left and another one sticking out at the right in Figure 1-47). Inside is a coil of wire around a flat plastic band, and a pair of springy contacts (the wiper), which conduct electricity to or from any point in the coil when you turn the shaft.*

Depending whether you have a really cheap potentiometer or a slightly more high-class version, you may find a circular track of conductive plastic or a loop of coiled wire. Either way, the principle is the same. The wire or the plastic possesses some resistance (a total of 2K in this instance), and as you turn the shaft of the potentiometer, a wiper rubs against the resistance, giving you a shortcut to any point from the center terminal.

You can try to put it back together, but if it doesn't work, use your backup potentiometer instead.

To test your potentiometer, set your meter to measure resistance (ohms) and touch the probes while turning the potentiometer shaft to and fro, as shown in Figure 1-49.

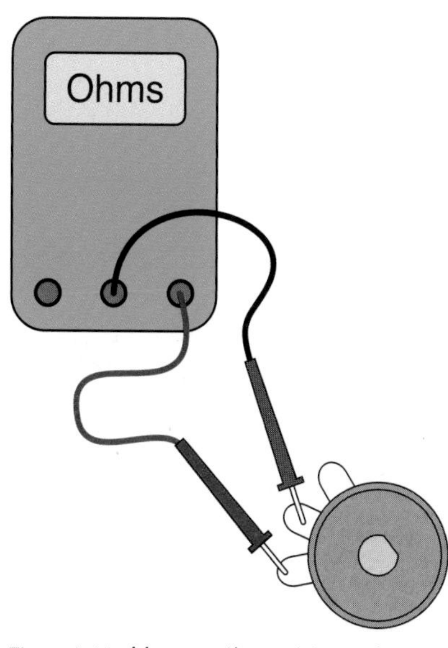

Figure 1-49. *Measure the resistance between these two terminals of the potentiometer while you turn its shaft to and fro.*

Dimming Your LED

Begin with the potentiometer turned all the way counterclockwise, otherwise you'll burn out the LED before we even get started. (A very, very small number of potentiometers increase and decrease resistance in the opposite way to which I'm describing here, but as long as your potentiometer looks like the one in Figure 1-48 after you open it up, my description should be accurate.)

Now connect everything as shown in Figures 1-50 and 1-51, taking care that you don't allow the metal parts of any of the alligator clips to touch each other. Now turn up the potentiometer *very* slowly. You'll notice the LED glowing brighter, and brighter, and brighter—until, oops, it goes dark. You see how easy it is to destroy modern electronics? Throw away that LED. It will never glow again. Substitute a new LED, and we'll be more careful this time.

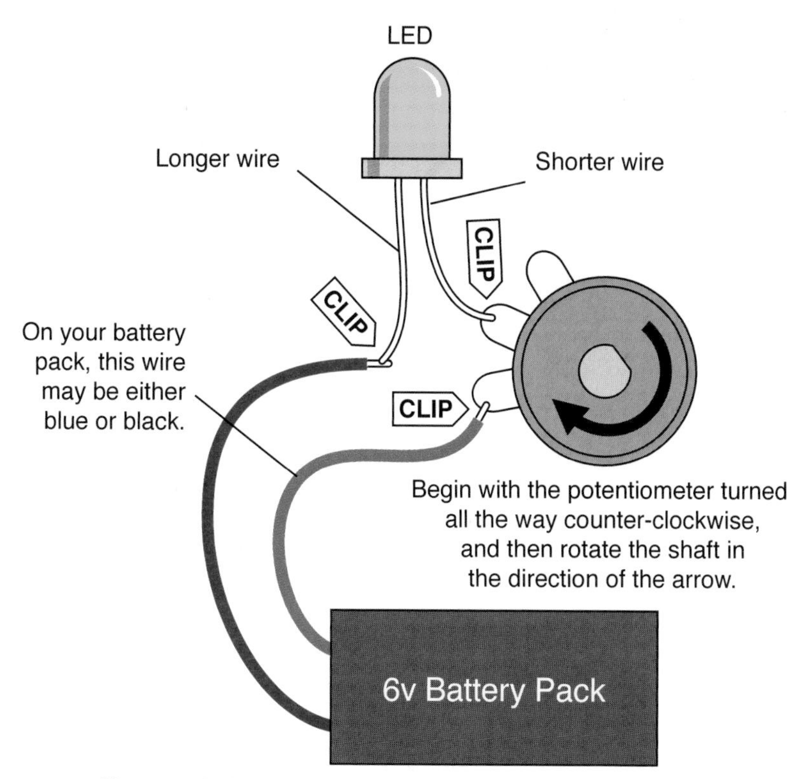

Figure 1-50. *The setup for Experiment 4. Rotating the shaft of the 2K potentiometer varies its resistance from 0 to 2,000Ω. This resistance protects the LED from the full 6 volts of the battery.*

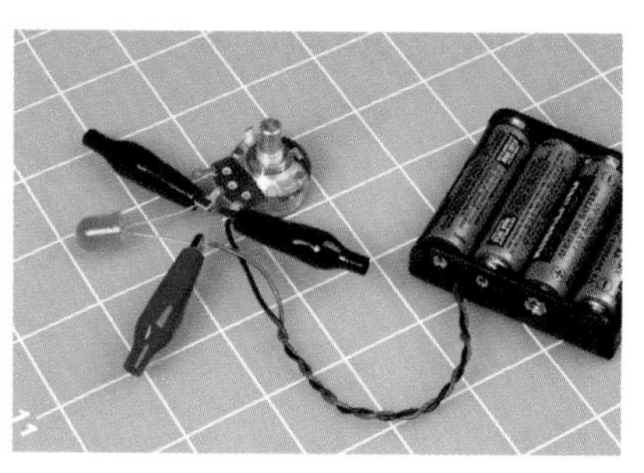

Figure 1-51. *The LED in this photo is dark because I turned the potentiometer up just a little bit too far.*

Figure 1-52

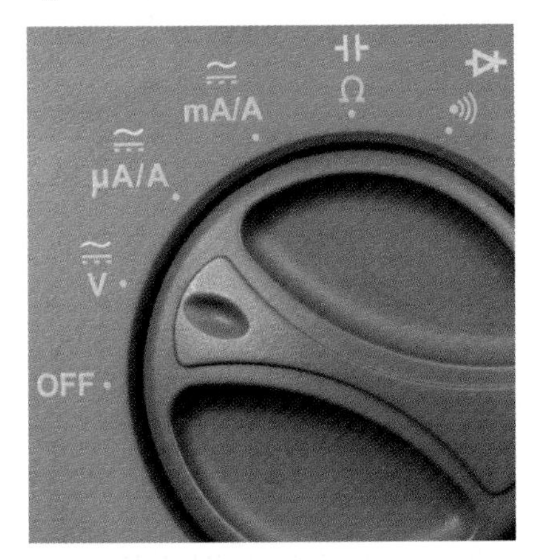

Figure 1-53

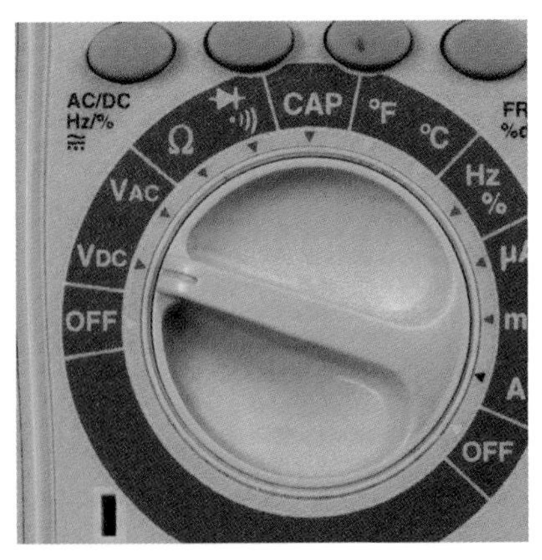

Figure 1-54. *Each meter has a different way to measure volts DC. The manually adjusted meter (top) requires you to move a slider switch to "DC" and then choose the highest voltage you want to measure: In this case, the selected voltage is 20 (because 2 would be too low). Using the autoranging RadioShack meter, you set it to "V" and the meter will figure out which range to use.*

While the batteries are connected to the circuit, set your meter to measure *volts DC* as shown in Figures 1-52 through 1-54. Now touch the probes either side of the LED. Try to hold the probes in place while you turn the potentiometer up a little, and down a little. You should see the voltage pressure around the LED changing accordingly. We call this the *potential difference* between the two wires of the LED.

If you were using a miniature old-fashioned lightbulb instead of an LED, you'd see the potential difference varying much more, because a lightbulb behaves like a "pure" resistor, whereas an LED self-adjusts to some extent, modifying its resistance as the voltage pressure changes.

Now touch the probes to the two terminals of the potentiometer that we're using, so that you can measure the potential difference between them. The potentiometer and the LED share the total available voltage, so when the potential difference (the voltage drop) around the potentiometer goes up, the potential difference around the LED goes down, and vice versa. See Figures 1-55 through 1-57. A few things to keep in mind:

- If you add the voltage drops across the devices in the circuit, the total is the same as the voltage supplied by the batteries.
- You measure voltage relatively, between two points in a circuit.
- Apply your meter like a stethoscope, without disturbing or breaking the connections in the circuit.

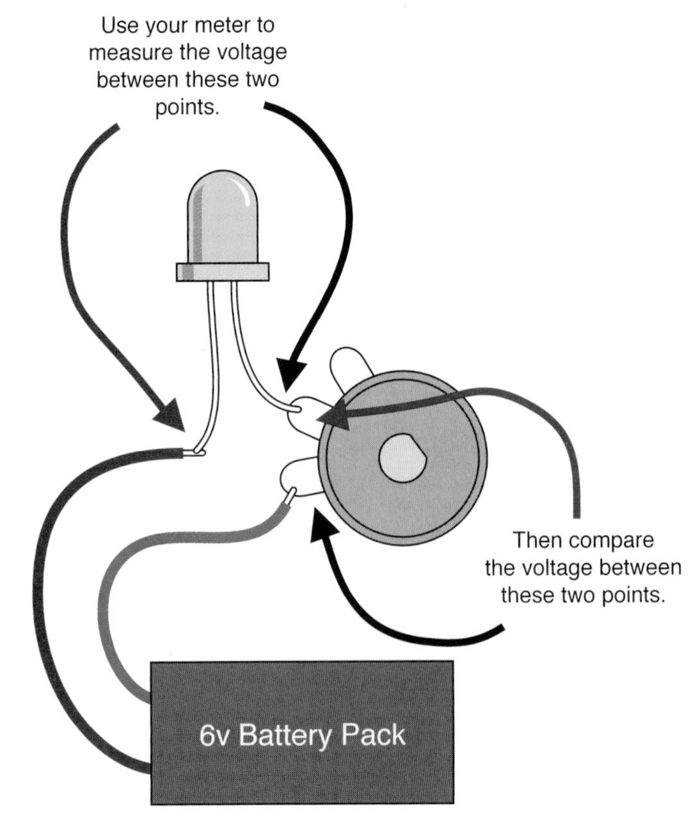

Figure 1-55. *How to measure voltage in a simple circuit.*

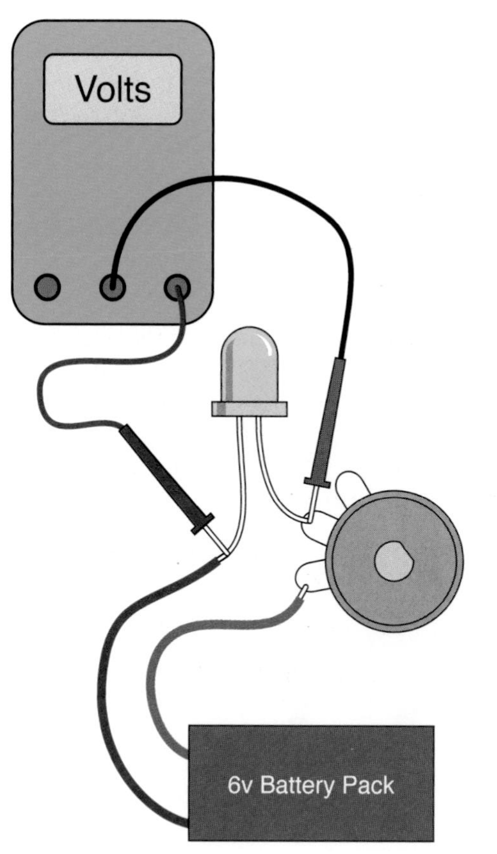

Figure 1-56. *The meter shows how much voltage the LED takes.*

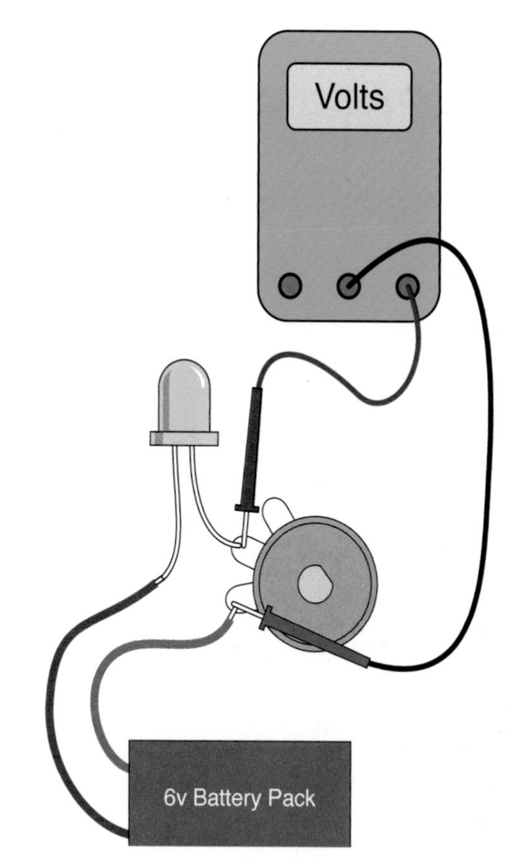

Figure 1-57. *The meter shows how much voltage the potentiometer takes.*

Checking the Flow

Now I want you to make a different measurement. I want you to measure the flow, or current, in the circuit, using your meter set to mA (milliamps). Remember, to measure current:

- You can only measure current when it passes *through* the meter.
- You have to insert your meter into the circuit.
- Too much current will blow the fuse inside your meter.

Make sure you set your meter to measure mA, not volts, before you try this. Some meters require you to move one of your leads to a different socket on the meter, to measure mA. See Figures 1-58 through 1-61.

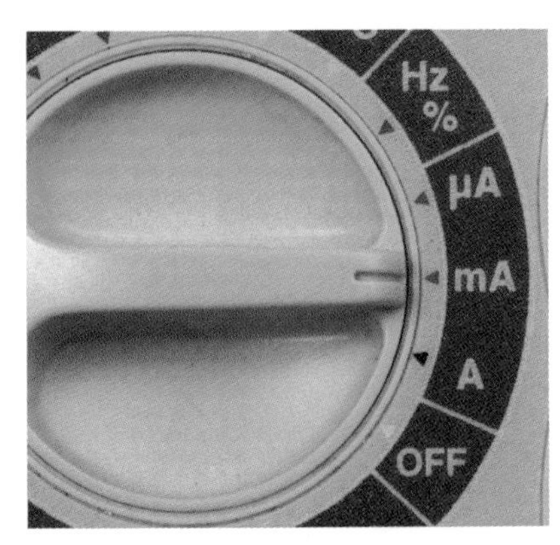

Figure 1-58. *Any meter will blow its internal fuse if you try to make it measure too high an amperage. In our circuit, this is not a risk as long as you keep the potentiometer in the middle of its range. Choose "mA" for milliamps and remember that the meter displays numbers that mean thousandths of an amp.*

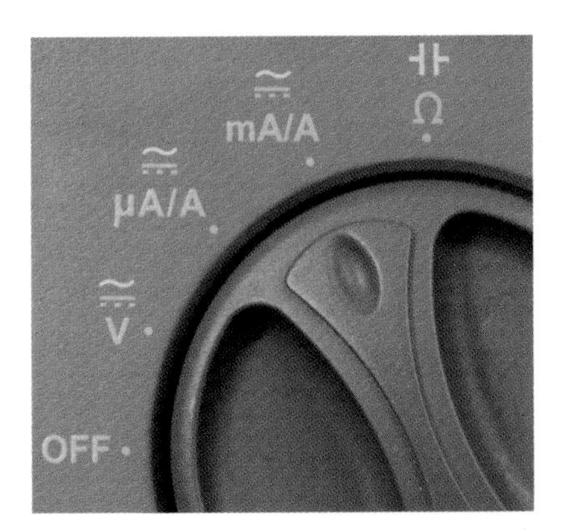

Figure 1-59

Figure 1-60

Figure 1-61. *A manual meter such as the one here may require you to shift the red lead to a different socket, to measure milliamps. Most modern meters don't require this until you are measuring higher currents.*

Insert your meter into the circuit, as shown in Figure 1-62. Don't turn the potentiometer more than halfway up. The resistance in the potentiometer will protect your meter, as well as the LED. If the meter gets too much current, you'll find yourself replacing its internal fuse.

As you adjust the potentiometer up and down a little, you should find that the varying resistance in the circuit changes the flow of current—the amperage. This is why the LED burned out in the previous experiment: too much current made it hot, and the heat melts it inside, just like the fuse in the previous experiment. *A higher resistance limits the flow of current, or amperage.*

Now insert the meter in another part of the circuit, as shown in Figure 1-63. As you turn the potentiometer up and down, you should get exactly the same results as with the configuration in Figure 1-62. This is because the current is the same at all points in a similar circuit. *It has to be, because the flow of electrons has no place else to go.*

It's time now to nail this down with some numbers. Here's one last thing to try. Set aside the LED and substitute a 1KΩ resistor, as shown in Figure 1-64. The total resistance in the circuit is now 1KΩ plus whatever the resistance the potentiometer provides, depending how you set it. (The meter also has some resistance, but it's so low, we can ignore it.)

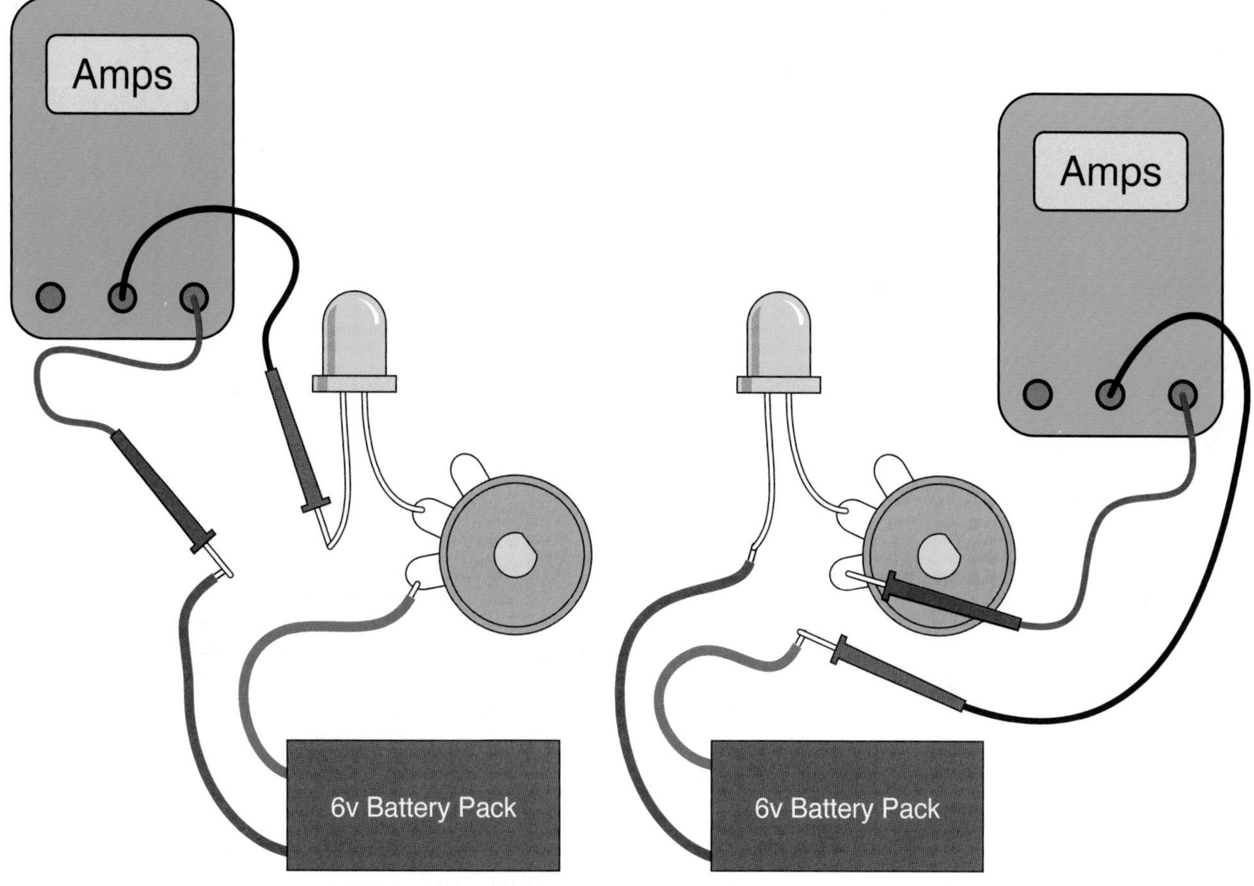

Figure 1-62. *To measure amps, as illustrated here and in Figure 1-63, the current has to pass through the meter. When you increase the resistance, you restrict the current flow, and the lower flow makes the LED glow less brightly.*

Figure 1-63

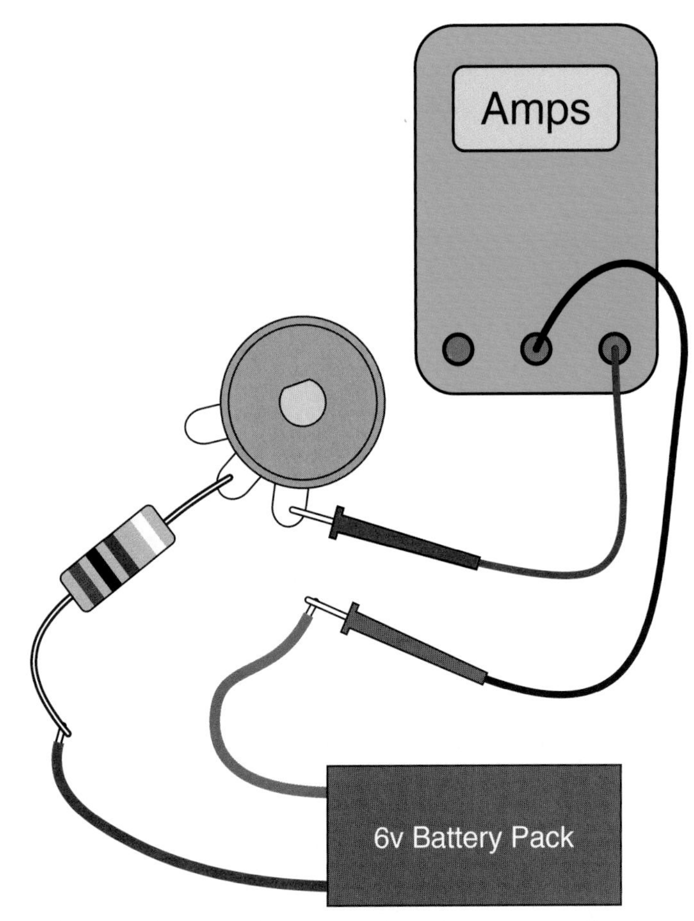

Figure 1-64. *If you substitute a resistor instead of the LED, you can confirm that the current flowing through the circuit varies with the total resistance in the circuit, if the voltage stays the same.*

Turn the potentiometer all the way counterclockwise, and you have a total of 3K resistance in the circuit. Your meter should show about 2 mA flowing. Now turn the potentiometer halfway, and you have about 2K total resistance. You should see about 3 mA flowing. Turn the potentiometer all the way clockwise, so there's a total of 1K, and you should see 6 mA flowing. You may notice that if we multiply the resistance by the amperage, we get 6 each time—which just happens to be the voltage being applied to the circuit. See the following table.

Total resistance	Current	Voltage
(KΩ)	(mA)	(Volts)
3	2	6
2	3	6
1	6	6

In fact, we could say:

voltage = kilohms × milliamps

But wait a minute: 1K is 1,000 ohms, and 1mA is 1/1,000 of an amp. Therefore, our formula should really look like this:

voltage = (ohms × 1,000) × (amps/1,000)

The two factors of 1,000 cancel out, so we get this:

volts = ohms × amps

This is known as Ohm's Law. See the section, "Fundamentals: Ohm's Law," on the following page.

FUNDAMENTALS

Series and parallel

Before we go any further, you should know how resistance in a circuit increases when you put resistors in series or in parallel. Figures 1-65 through 1-67 illustrate this. Remember:

- Resistors in series are oriented so that one follows the other.
- Resistors in parallel are oriented side by side.

When you put two equal-valued resistors in series, you double the total resistance, because electricity has to pass through two barriers in succession.

When you put two equal-valued resistors in parallel, you divide the total resistance by two, because you're giving the electricity two paths which it can take, instead of one.

In reality we don't normally need to put resistors in parallel, but we often put other types of components in parallel. Lightbulbs in your house, for instance, are all wired that way. So, it's useful to understand that resistance in a circuit goes down if you keep adding components in parallel.

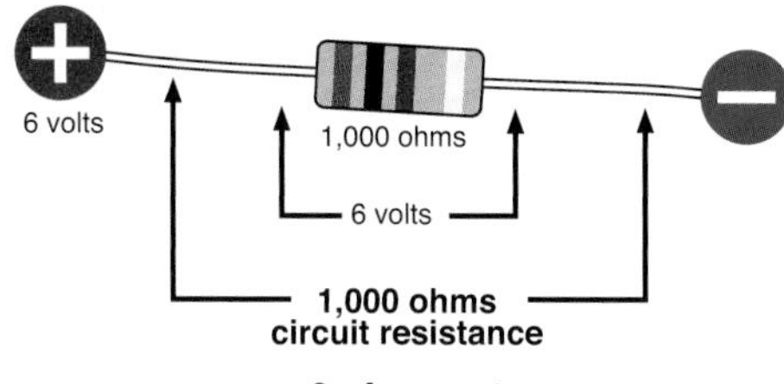

Figure 1-65. *One resistor takes the entire voltage, and according to Ohm's Law, it draws v/R = 6/1,000 = 0.006 amps = 6mA of current.*

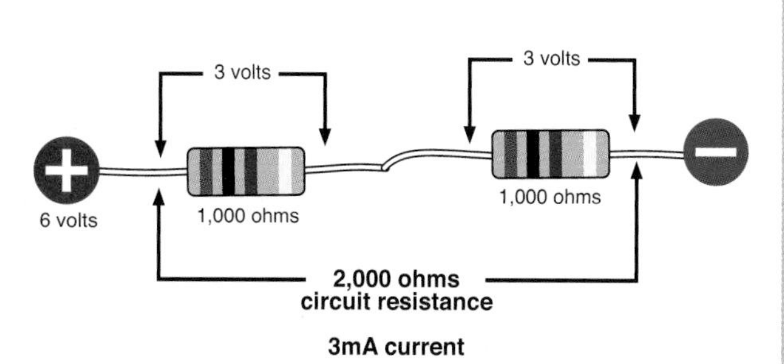

Figure 1-66. *When two resistors are in series, the electricity has to pass through one to reach the other, and therefore each of them takes half the voltage. Total resistance is now 2,000 ohms, and according to Ohm's Law, the circuit draws v/R = 6/2,000 = 0.003 amps = 3mA of current.*

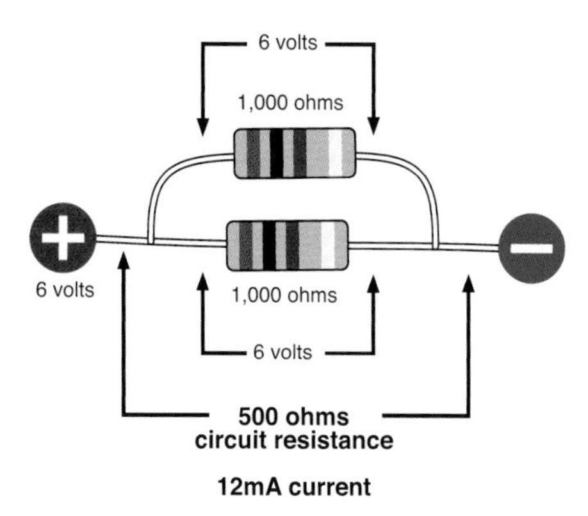

Figure 1-67. *When two resistors are in parallel, each is exposed to the full voltage, so each of them takes 6 volts. The electricity can now flow through both at once, so the total resistance of the circuit is half as much as before. According to Ohm's Law, the circuit draws v/R = 6/500 = 0.012 amps = 12mA of current.*

FUNDAMENTALS

Ohm's Law

For reasons I'll explain in a moment, amps are normally abbreviated with the letter I. V stands for volts and R stands for resistance in ohms (because the omega symbol, Ω, is not easily generated from most keyboards). Using these symbols, you can write Ohm's Law in three different ways:

$$V = I \times R$$
$$I = V/R$$
$$R = V/I$$

Remember, V is a *difference* in voltage between two points in a simple circuit, R is the resistance in ohms *between* the same two points, and I is the current in amps flowing *through* the circuit between the two points.

Letter I is used because originally current was measured by its *inductance*, meaning the ability to induce magnetic effects. It would be much less confusing to use A for amps, but unfortunately it's too late for that to happen.

Using Ohm's Law

Ohm's Law is extremely useful. For example, it helps us to figure out whether a component can be used safely in a circuit. Instead of stressing the component until we burn it out, we can predict whether it will work.

For instance, the first time you turned the potentiometer, you didn't really know how far you could go until the LED burned out. Wouldn't it be useful to know precisely what resistance to put in series with an LED, to protect it adequately while providing as much light as possible?

How to Read a Data Sheet

Like most information, the answer to this question is available online.

Here's how you find a manufacturer's data sheet (Figure 1-68). First, find the component that you're interested in from a mail-order source. Next, Google the part number and manufacturer's name. Usually the data sheet will pop up as the first hit. A source such as *Mouser.com* makes it even easier by giving you a direct link to manufacturers' data sheets for many products.

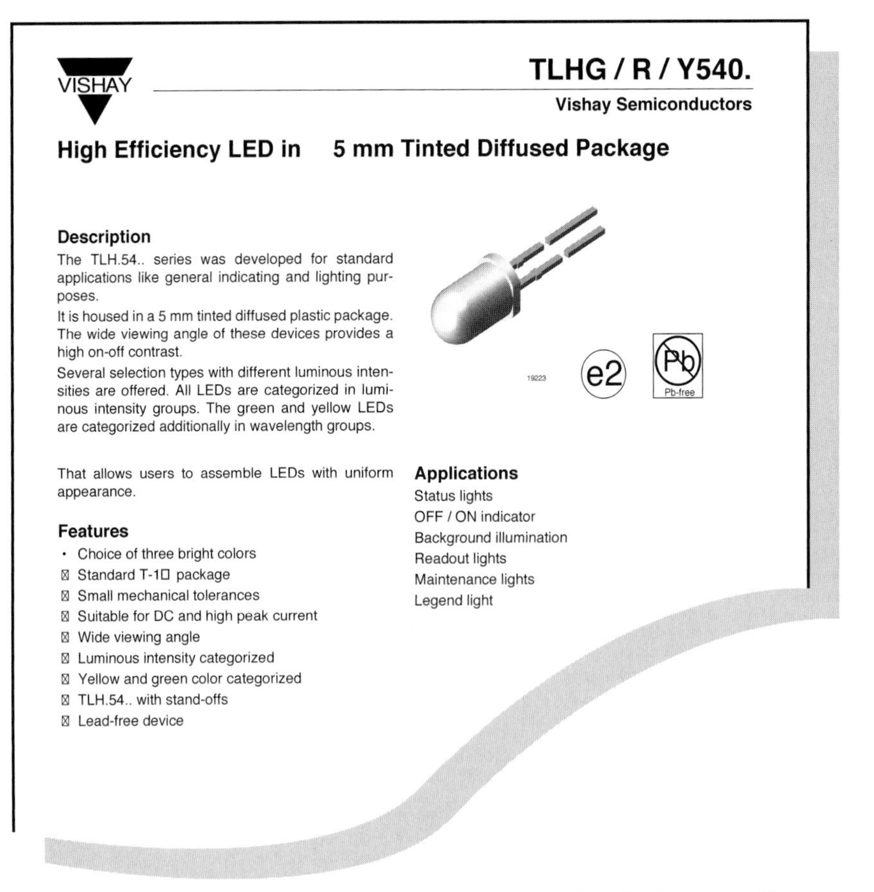

VISHAY

TLHG / R / Y540.

Vishay Semiconductors

High Efficiency LED in 5 mm Tinted Diffused Package

Description

The TLH.54.. series was developed for standard applications like general indicating and lighting purposes.

It is housed in a 5 mm tinted diffused plastic package. The wide viewing angle of these devices provides a high on-off contrast.

Several selection types with different luminous intensities are offered. All LEDs are categorized in luminous intensity groups. The green and yellow LEDs are categorized additionally in wavelength groups.

That allows users to assemble LEDs with uniform appearance.

Features

- Choice of three bright colors
- Standard T-1□ package
- Small mechanical tolerances
- Suitable for DC and high peak current
- Wide viewing angle
- Luminous intensity categorized
- Yellow and green color categorized
- TLH.54.. with stand-offs
- Lead-free device

Applications

Status lights
OFF / ON indicator
Background illumination
Readout lights
Maintenance lights
Legend light

Figure 1-68. *The beginning of a typical data sheet, which includes all relevant specifications for the product, freely available online.*

BACKGROUND

How much voltage does a wire consume?

Normally, we can ignore the resistance in electric wires, such as the little leads of wire that stick out of resistors, because it's trivial. However, if you try to force large amounts of current through long lengths of thin wire, the resistance of the wire can become important.

How important? Once again, we can use Ohm's Law to find out.

Suppose that a very long piece of wire has a resistance of 0.2Ω. And we want to run 15 amps through it. How much voltage will the wire steal from the circuit, because of its resistance?

Once again, you begin by writing down what you know:

$R = 0.2$

$I = 15$

We want to know V, the potential difference, for the wire, so we use the version of Ohm's Law that places V on the left side:

$V = I \times R$

Now plug in the values:

$V = 15 \times 0.2 = 3$ volts

Three volts is not a big deal if you have a high-voltage power supply, but if you are using a 12-volt car battery, this length of wire will take one-quarter of the available voltage.

Now you know why the wiring in automobiles is relatively thick—to reduce its resistance well below 0.2Ω. See Figure 1-69.

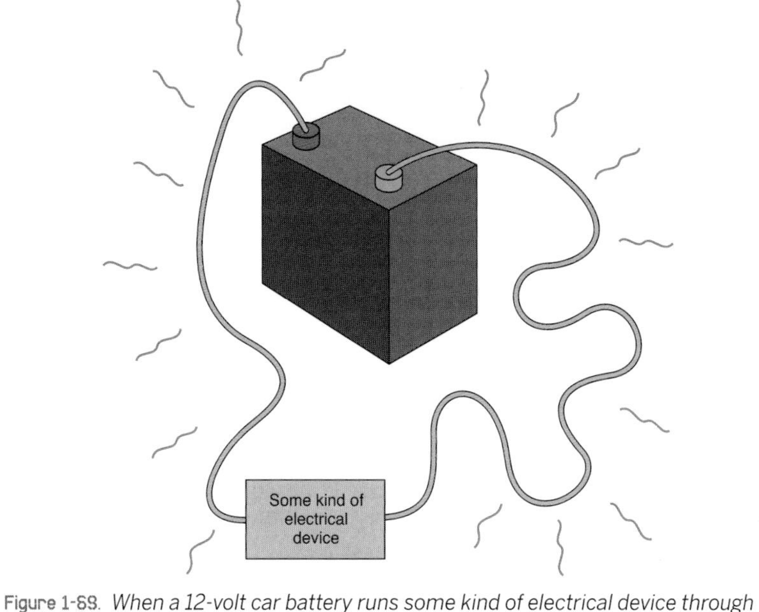

Figure 1-69. *When a 12-volt car battery runs some kind of electrical device through a long piece of thin wire, the resistance of the wire steals some of the voltage and dissipates it as heat.*

BACKGROUND

The origins of wattage

James Watt (Figure 1-70) is known as the inventor of the steam engine. Born in 1736 in Scotland, he set up a small workshop in the University of Glasgow, where he struggled to perfect an efficient design for using steam to move a piston in a cylinder. Financial problems and the primitive state of the art of metal working delayed practical applications until 1776.

Despite difficulties in obtaining patents (which could only be granted by an act of parliament in those times), Watt and his business partner eventually made a lot of money from his innovations. Although he predated the pioneers in electricity, in 1889 (70 years after his death), his name was assigned to the basic unit of electric power that can be defined by multiplying amperes by volts. See the Fundamentals section, "Watt Basics," on page 31.

Figure 1-70. *James Watt's development of steam power enabled the industrial revolution. After his death, he was honored by having his name applied to the basic unit of power in electricity.*

Here's an example. Suppose I want a red LED, such as the Vishay part TLHR5400, which has become such a common item that I can buy them individually for 9 cents apiece. I click the link to the data sheet maintained by the manufacturer, Vishay Semiconductor. Almost immediately I have a PDF page on my screen. This data sheet is for TLHR, TLHG, and TLHY types of LED, which are red, green, and yellow respectively, as suggested by the R, G, and Y in the product codes. I scroll down and look at the "Optical and Electrical Characteristics" section. It tells me that under conditions of drawing a current of 20 mA, the LED will enjoy a "Typ," meaning, typical, "forward voltage" of 2 volts. The "Max," meaning maximum, is 3 volts.

Let's look at one other data sheet, as not all of them are written the same way. I'll choose a different LED, the Kingbright part WP7113SGC. Click on the link to the manufacturer's site, and I find on the second page of the data sheet a typical forward voltage of 2.2, maximum 2.5, and a maximum forward current of 25 mA. I also find some additional information: a maximum reverse voltage of 5 and maximum reverse current of 10 uA (that's microamps, which are 1,000 times smaller than milliamps). This tells us that you should avoid applying excessive voltage to the LED the wrong way around. If you exceed the reverse voltage, you risk burning out the LED. Always observe polarity!

Kingbright also warns us how much heat the LED can stand: 260° C (500° F) for a few seconds. This is useful information, as we'll be putting aside our alligator clips and using hot molten solder to connect electrical parts in the near future. Because we have already destroyed a battery, a fuse, and an LED in just four experiments, maybe you won't be surprised when I tell you that we will destroy at least a couple more components as we test their limits with a soldering iron.

Anyway, now we know what an LED wants, we can figure out how to supply it. If you have any difficulties dealing with decimals, check the Fundamentals section "Decimals," on the next page, before continuing.

How Big a Resistor Does an LED Need?

Suppose that we use the Vishay LED. Remember its requirements from the data sheet? Maximum of 3 volts, and a safe current of 20mA.

I'm going to limit it to 2.5 volts, to be on the safe side. We have 6 volts of battery power. Subtract 2.5 from 6 and we get 3.5. So we need a resistor that will take 3.5 volts from the circuit, leaving 2.5 for the LED.

The current flow is the same at all places in a simple circuit. If we want a maximum of 20mA to flow through the LED, the same amount of current will be flowing through the resistor.

Now we can write down what we know about the resistor in the circuit. Note that we have to convert all units to volts, amps, and ohms, so that 20mA should be written as 0.02 amps:

$V = 3.5$ (the potential drop across the resistor)
$I = 0.02$ (the current flowing through the resistor)

We want to know R, the resistance. So, we use the version of Ohm's Law that puts R on the left side:

$$R = V/I$$

Now plug in the values:

$$R = 3.5/0.02$$

Run this through your pocket calculator if you find decimals confusing. The answer is:

$$R = 175\Omega$$

It so happens that 175Ω isn't a standard value. You may have to settle for 180 or 220Ω, but that's close enough.

Evidently the 470Ω resistor that you used in Experiment 3 was a very conservative choice. I suggested it because I said originally that you could use any LED at all. I figured that no matter which one you picked, it should be safe with 470Ω to protect it.

Cleanup and Recycling

The dead LED can be thrown away. Everything else is reusable.

FUNDAMENTALS

Decimals

Legendary British politician Sir Winston Churchill is famous for complaining about "those damned dots." He was referring to decimal points. Because Churchill was Chancellor of the Exchequer at the time, and thus in charge of all government expenditures, his difficulty with decimals was a bit of a problem. Still, he muddled through in time-honored British fashion, and so can you.

You can also use a pocket calculator—or follow two basic rules.

Doing multiplication: move the decimal points

Suppose you want to multiply 0.03 by 0.002:

1. Move the decimal points to the ends of both the numbers. In this case, you have to move the decimal points by a total of 5 places to get 3 and 2.
2. Do the multiplication of the whole numbers you have created and note the result. In this case, 3 x 2 = 6.
3. Move the decimal point back again by the same number of places you counted in step 1. In this case, you get 0.00006.

Doing division: cancel the zeros

Suppose you need to divide 0.006 by 0.0002:

1. Shift the decimal points to the right, in both the numbers, by the same number of steps, until both the numbers are greater than 1. In this case, shift the point four steps in each number, so you get 60 divided by 2.
2. Do the division. The result in this case is 30.

THEORY

Doing the math on your tongue

I'm going to go back to the question I asked in the previous experiment: why didn't your tongue get hot?

Now that you know Ohm's Law, you can figure out the answer in numbers. Let's suppose the battery delivered its rated 9 volts, and your tongue had a resistance of 50K, which is 50,000 ohms. Write down what you know:

V = 9

R = 50,000

We want to know the current, I, so we use the version of Ohm's Law that puts this on the left:

I = V/R

Plug in the numbers:

I = 9/50,000 = 0.00018 amps

Move the decimal point three places to convert to milliamps:

I = 0.18 mA

That's a tiny current that will not produce much heat at 9 volts.

What about when you shorted out the battery? How much current made the wires get hot? Well, suppose the wires had a resistance of 0.1 ohms (probably it's less, but I'll start with 0.1 as a guess). Write down what we know:

V = 1.5

R = 0.1

Once again we're trying to find I, the current, so we use:

I = V/R

Plug in the numbers:

I = 1.5/0.1 = 15 amps

That's 100,000 times the current that may have passed through your tongue, which would have generated much more heat, even though the voltage was lower.

Could that tiny little battery really pump out 15 amps? Remember that the battery got hot, as well as the wire. This tells us that the electrons may have met some resistance inside the battery, as well as in the wire. (Otherwise, where else did the heat come from?) Normally we can forget about the internal resistance of a battery, because it's so low. But at high currents, it becomes a factor.

I was reluctant to short-circuit the battery through a meter, to try to measure the current. My meter will fry if the current is greater than 10A. However I did try putting other fuses into the circuit, to see whether they would blow. When I tried a 10A fuse, it did not melt. Therefore, for the brand of battery I used, I'm fairly sure that the current in the short circuit was under 10A, but I know it was over 3A, because the 3A fuse blew right away.

The internal resistance of the 1.5-volt battery prevented the current in the short circuit from getting too high. This is why I cautioned against using a larger battery (especially a car battery). Larger batteries have a much lower internal resistance, allowing dangerously high currents which generate explosive amounts of heat. A car battery is designed to deliver literally hundreds of amps when it turns a starter motor. That's quite enough current to melt wires and cause nasty burns. In fact, you can weld metal using a car battery.

Lithium batteries also have low internal resistance, making them very dangerous when they're shorted out. High current can be just as dangerous as high voltage.

FUNDAMENTALS

Watt basics

So far I haven't mentioned a unit that everyone is familiar with: watts.

A watt is a unit of power, and when power is applied over a period of time, it performs work. Engineers have their own definition of work—they say that work is done when a person, an animal, or a machine pushes something to overcome mechanical resistance. Examples would be a steam engine pulling a train on a level track (overcoming friction and air resistance) or a person walking upstairs (overcoming the force of gravity).

When electrons push their way through a circuit, they are overcoming a kind of resistance, and so they are doing work, which can be measured in watts per second. The definition of a watt is easy:

watts = volts × amps

Or, using the symbols customarily assigned, these three formulas all mean the same thing:

$W = V \times I$

$V = W/I$

$I = W/V$

Watts can be preceded with an "m," for "milli," just like volts:

Number of watts	Usually expressed as	Abbreviated as
0.001 watts	1 milliwatt	1mW
0.01 watts	10 milliwatts	10 mW
0.1 watts	100 milliwatts	100 mW
1 watt	1,000 milliwatts	1W

Because power stations, solar installations, and wind farms deal with much larger numbers, you may also see references to kilowatts (using letter K) and megawatts (with a capital M, not to be confused with the lowercase m used to define milliwatts):

Number of watts	Usually expressed as	Abbreviated as
1,000 watts	1 kilowatt	1 KW
1,000,000 watts	1 megawatt	1 MW

Lightbulbs are calibrated in watts. So are stereo systems. The watt is named after James Watt, inventor of the steam engine. Incidentally, watts can be converted to horsepower, and vice versa.

THEORY

Power assessments

I mentioned earlier that resistors are commonly rated as being capable of dealing with 1/4 watt, 1/2 watt, 1 watt, and so on. I suggested that you should buy resistors of 1/4 watt or higher. How did I know this?

Go back to the LED circuit. Remember we wanted the resistor to drop the voltage by 3.5 volts, at a current of 20 mA. How many watts of power would this impose on the resistor?

Write down what you know:

$V = 3.5$ (the voltage drop imposed by the resistor)

$I = 20mA = 0.02$ amps (the current flowing through the resistor)

We want to know W, so we use this version of the formula:

$W = V \times I$

Plug in the values:

$W = 3.5 \times 0.02 = 0.07$ watts (the power being dissipated by the resistor)

Because 1/4 watt is 0.25 watts, obviously a 1/4 watt resistor will have about four times the necessary capacity. In fact you could have used a 1/8 watt resistor, but in future experiments we may need resistors that can handle 1/4 watt, and there's no penalty for using a resistor that is rated for more watts than will actually pass through it.

Experiment 5: Let's Make a Battery

Long ago, before web surfing, file sharing, or cell phones, kids were so horribly deprived that they tried to amuse themselves with kitchen-table experiments such as making a primitive battery by pushing a nail and a penny into a lemon. Hard to believe, perhaps, but true!

This is seriously old-school—but I want you to try it anyway, because anyone who wants to get a feel for electricity should see how easy it is to extract it from everyday objects around us. Plus, if you use enough lemons, you just *might* generate enough voltage to power an LED.

The basic components of a battery are two metal electrodes immersed in an electrolyte. I won't define these terms here (they're explained in the following section "Theory: The nature of electricity"). Right now all you need to know is that lemon juice will be your electrolyte, and copper and zinc will be your electrodes. A penny provides the necessary copper, as long as it is fairly new and shiny. Pennies aren't solid copper anymore, but they are still copper-plated, which is good enough.

To find some metallic zinc, you will have to make a trip to a hardware store, where you should ask for roofing nails. The nails are zinc-plated to prevent them from rusting. Small metal brackets or mending plates also are usually zinc-plated. They should have a slightly dull, silvery look. If they have a mirror-bright finish, they're more likely to be nickel-plated.

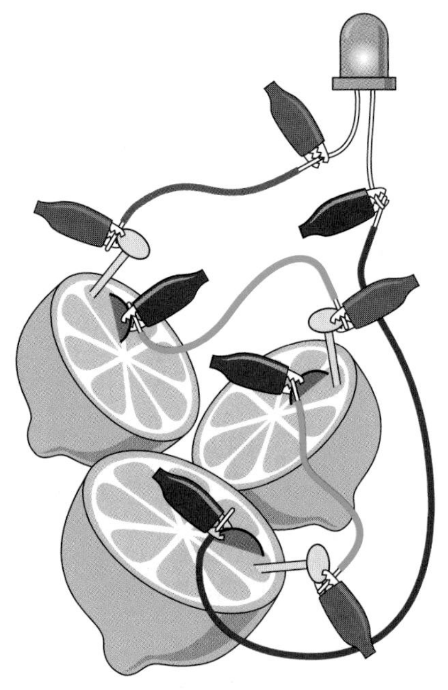

Figure 1-71. *A three-lemon battery. Don't be too disappointed if the LED fails to light up. The lemons have a high electrical resistance, so they can't deliver much current, especially through the relatively small surface area of the nails and the pennies. However, the lemon battery does generate voltage that you can measure with your meter.*

Cut a lemon in half, set your multimeter so that it can measure up to 2 volts DC, and hold one probe against a penny while you hold the other probe against a roofing nail (or other zinc-plated object). Now force the penny and the nail into the exposed juicy interior of the lemon, as close to each other as possible, but not actually touching. You should find that your meter detects between 0.8 volts and 1 volt.

You can experiment with different items and liquids to see which works best. Immersing your nail and penny in lemon juice that you have squeezed into a shot glass or egg cup may enhance the efficiency of your battery, although you'll have a harder time holding everything in place. Grapefruit juice and vinegar will work as substitutes for lemon juice.

Figure 1-72. *Bottled lemon juice seems to work just as well as fresh lemon juice. I cut the bottoms off three paper cups, inserted a galvanized bracket into each, and used heavyweight stranded copper wire to make the positive electrodes*

To drive a typical LED, you need more than 1 volt. How to generate the extra electrical pressure? By putting batteries in series, of course. In other words, more lemons! (Or more shot glasses or egg cups.) You'll also need lengths of wire to connect multiple electrodes, and this may entail skipping ahead to Chapter 2, where I describe how to strip insulation from hookup wire. Figures 1-71 and 1-72 show the configuration.

If you set things up carefully, making sure than none of the electrodes are touching, you may be able to illuminate your LED with two or three lemon-juice batteries in series. (Some LEDs are more sensitive to very low currents than others. Later in the book I'll be talking about very-low-current LEDs. If you want your lemon-juice battery to have the best chance of working, you can search online for low-current LEDs and buy a couple.)

THEORY

The nature of electricity

To understand electricity, you have to start with some basic information about atoms. Each atom consists of a nucleus at the center, containing protons, which have a positive charge. The nucleus is surrounded by electrons, which carry a negative charge.

Breaking up the nucleus of an atom requires a lot of energy, and can also liberate a lot of energy—as happens in a nuclear explosion. But persuading a couple of electrons to leave an atom (or join an atom) takes very little energy. For instance, when zinc reacts chemically with an acid, it can liberate electrons. This is what happens at the zinc electrode of the chemical battery in Experiment 5.

The reaction soon stops, as electrons accumulate on the zinc electrode. They feel a mutual force of repulsion, yet they have nowhere to go. You can imagine them like a crowd of hostile people, each one wanting the others to leave, and refusing to allow new ones to join them, as shown in Figure 1-73.

Figure 1-73. *Electrons on an electrode have a bad attitude known as mutual repulsion.*

Now consider what happens when a wire connects the zinc electrode, which has a surplus of electrons, to another electrode, made from a different material, that has a shortage of electrons. The electrons can pass through the wire very easily by jumping from one atom to the next, so they escape from the zinc electrode and run through the wire, propelled by their great desire to get away from each other. See Figure 1-74. This mutual force of propulsion is what creates an electrical current.

Now that the population of electrons on the zinc electrode has been reduced, the zinc-acid reaction can continue, replacing the missing electrons with new ones—which promptly imitate their predecessors and try to get away from each other by running away down the wire. The process continues until the zinc-acid reaction grinds to a halt, usually because it creates a layer of a compound such as zinc oxide, which won't react with acid and prevents the acid from reacting with the zinc underneath. (This is why your zinc electrode may have looked sooty when you pulled it out of the acidic electrolyte.)

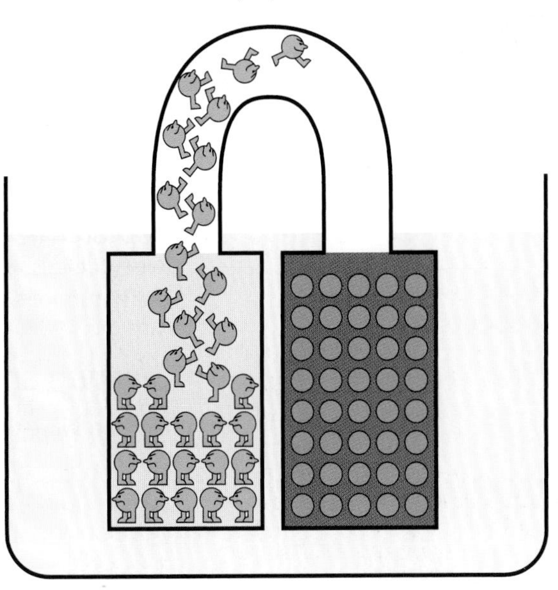

Figure 1-74. *As soon as we open up a pathway from a zinc electrode crowded with electrons to a copper electrode, which contains "holes" for the electrons, their mutual repulsion makes them try to escape from each other to their new home as quickly as possible.*

This description applies to a "primary battery," meaning one that is ready to generate electricity as soon as a connection between its terminals allows electrons to transfer from one electrode to the other. The amount of current that a primary battery can generate is determined by the speed at which chemical reactions inside the battery can liberate electrons. When the raw metal in the electrodes has all been used up in chemical reactions, the battery can't generate any more electricity and is dead. It cannot easily be recharged, because the chemical reactions are not easily reversible, and the electrodes may have oxidized.

In a rechargeable battery, also known as a secondary battery, a smarter choice of electrodes and electrolyte does allow the chemical reactions to be reversed.

How much current is being generated in your lemon battery? Set your meter to measure milliamps, and connect it between the nail and the penny. I measured about 2mA, but got 10mA when I used some #10 stranded copper wire instead of a penny and a large mending plate instead of a roofing nail, immersed in a cup of grapefruit juice. When a larger surface area of metal makes better contact with the electrolyte, you get a greater flow of current. (Don't ever connect your meter to measure amps directly between the terminals of a real battery. The current will be too high, and can blow the fuse inside your meter.)

What's the internal resistance of your lemon? Put aside the copper and zinc electrodes and insert your nickel-plated meter probes into the juice. I got a reading of around 30K when both probes were in the same segment of the lemon, but 40K or higher if the probes were in different segments. Is the resistance lower when you test liquid in a cup?

Here are a couple more questions that you may wish to investigate. For how long will your lemon battery generate electricity? And why do you think your zinc-plated electrode becomes discolored after it has been used for a while?

Electricity is generated in a battery by an exchange of ions, or free electrons, between metals. If you want to know more about this, check the section "Theory: The nature of electricity" on the previous page.

Cleanup and Recycling

The hardware that you immersed in lemons or lemon juice may be discolored, but it is reusable. Whether you eat the lemons is up to you.

BACKGROUND

Positive and negative

If electricity is a flow of electrons, which have a negative charge, why do people talk as if electricity flows from the positive terminal to the negative terminal of a battery?

The answer lies in a fundamental embarrassment in the history of research into electricity. For various reasons, when Benjamin Franklin was trying to understand the nature of electric current by studying phenomena such as lightning during thunderstorms, he believed he observed a flow of "electrical fluid" from positive to negative. He proposed this concept in 1747.

In fact, Franklin had made an unfortunate error that remained uncorrected until after physicist J. J. Thomson announced his discovery of the electron in 1897, 150 years later. Electricity actually flows from an area of greater negative charge, to some other location that is "less negative"—that is, "more positive." In other words, electricity is a flow of negatively charged particles. In a battery, they originate from the negative terminal and flow to the positive terminal.

You might think that when this fact was established, everyone should have discarded Franklin's idea of a flow from positive to negative. But when an electron moves through a wire, you can still think of an equal positive charge flowing in the opposite direction. When the electron leaves home, it takes a small negative charge with it; therefore, its home becomes a bit more positive. When the electron arrives at its destination, its negative charge makes the destination a bit less positive. This is pretty much what would happen if an imaginary positive particle traveled in the opposite direction. Moreover, all of the mathematics describing electrical behavior are still valid if you apply them to the imaginary flow of positive charges.

As a matter of tradition and convenience we still retain Ben Franklin's erroneous concept of flow from positive to negative, because it really makes no difference. In the symbols that represent components such as diodes and transistors, you will actually find arrows reminding you which way these components should be placed—and the arrows all point from positive to negative, even though that's not the way things really work at all! Ben Franklin would have been surprised to learn that although most lightning strikes occur when a negative charge in clouds discharges to neutralize a positive charge on the ground, some forms of lightning are actually a flow of electrons from the negatively charged surface of the earth, up to a positive charge in the clouds. That's right: someone who is "struck by lightning" may be hurt by *emitting* electrons rather than by receiving them, as shown in Figure 1-75.

Figure 1-75. *In some weather conditions, the flow of electrons during a lightning strike can be from the ground, through your feet, out of the top of your head, and up to the clouds. Benjamin Franklin would have been surprised.*

THEORY

Basic measurements

Electrical potential is measured by adding up the charges on individual electrons. The basic unit is the *coulomb*, equal to the total charge on about 6,250,000,000,000,000,000 electrons.

If you know how many electrons pass through a piece of wire each second, this establishes the flow of electricity, which can be expressed in amperes. In fact 1 ampere can be defined as 1 coulomb per second. Thus:

1 ampere = 1 coulomb/second

= about 6.25 quintillion electrons/second

There's no way to "see" the number of electrons running through a conductor (Figure 1-76), but there are indirect ways of getting at this information. For instance, when an electron goes running through a wire, it creates a wave of electromagnetic force around it. This force can be measured, and we can calculate the amperage from that. The electric meter installed at your home by the utility company functions on this principle.

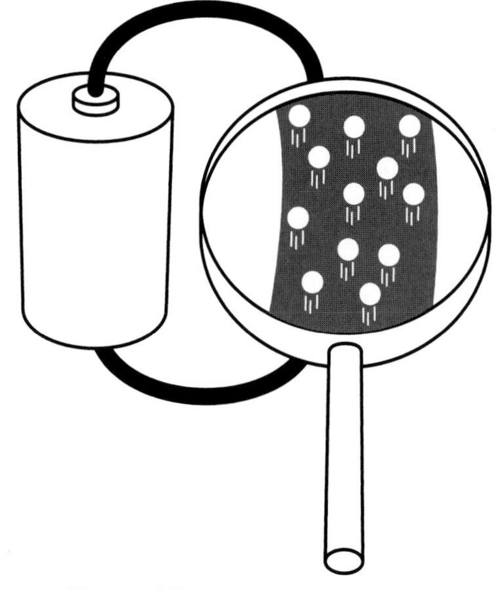

Figure 1-76. *If you could look inside an electric wire with a sufficiently powerful magnifying device, and the wire happened to be carrying 1 ampere of electron flow at the time, you might hope to see about 6.25 quintillion electrons speeding past each second.*

If electrons are just moving freely, they aren't doing any work. If you had a loop of wire of zero resistance, and you kick-started a flow of electrons somehow, they could just go buzzing around forever. (This is what happens inside a superconductor—almost.)

Under everyday conditions, even a copper wire has some resistance. The force that we need to push electrons through it is known as "voltage," and creates a flow that can create heat, as you saw when you shorted out a battery. (If the wire that you used had zero resistance, the electricity running through it would not have created any heat.) We can use the heat directly, as in an electric stove, or we can use the electrical energy in other ways—to run a motor, for instance. Either way, we are taking energy out of the electrons, to do some work.

One volt can be defined as the amount of pressure that you need to create a flow of 1 ampere, which does 1 watt of work. As previously defined, 1 watt = 1 volt × 1 ampere, but the definition actually originated the other way around:

1 volt = 1 watt/1 ampere

It's more meaningful this way, because a watt can be defined in nonelectrical terms. Just in case you're interested, we can work backward through the units of the metric system like this:

1 watt = 1 joule/second

1 joule = a force of 1 newton acting through 1 meter

1 newton = the force required to accelerate 1 kilogram by 1 meter per second, each second

On this basis, the electrical units can all be anchored with observations of mass, time, and the charge on electrons.

Practically Speaking

For practical purposes, an intuitive understanding of electricity can be more useful than the theory. Personally I like the water analogies that have been used for decades in guides to electricity. Figure 1-77 shows a tall tank half full of water, with a hole punched in it near the bottom. Think of the tank as being like a battery. The height of the water is comparable to voltage. The volume of flow through the hole, per second, is comparable to amperage. The smallness of the hole is comparable to resistance. See Figure 1-79 on the next page.

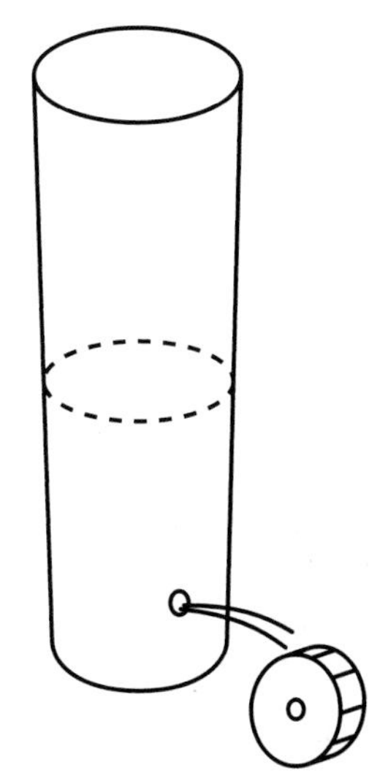

Figure 1-77. *If you want to get work out of a system...*

Where's the wattage in this picture? Suppose we place a little water wheel where it is hit by the flow from the hole. We can attach some machinery to the water wheel. Now the flow is doing some work. (Remember, wattage is a measurement of work.)

Maybe this looks as if we're getting something for nothing, extracting work from the water wheel without putting any energy back into the system. But remember, the water level in the tank is falling. As soon as I include some helpers hauling the waste water back up to the top of the tank (in Figure 1-78), you see that we have to put work in to get work out.

Figure 1-78. *. . . somehow or other you have to put work back into it.*

Similarly, a battery may seem to be giving power out without taking anything in, but the chemical reactions inside it are changing pure metals into metallic compounds, and the power we get out of a battery is enabled by this change of state. If it's a rechargeable battery, we have to push power back into it to reverse the chemical reactions.

Going back to the tank of water, suppose we can't get enough power out of it to turn the wheel. One answer could be to add more water. The height of the water will create more force. This would be the same as putting two batteries end to end, positive to negative, in series, to double the voltage. See Figure 1-80. As long as the resistance in the circuit remains the same, greater voltage will create more amperage, because amperage = voltage/resistance.

What if we want to run two wheels instead of one? We can punch a second hole in the tank, and the force (voltage) will be the same at each of them. However, the water level in the tank will drop twice as fast. Really, we'd do better to build a second tank, and here again the analogy with a battery is good. If you wire two batteries side by side, in parallel, you get the same voltage, but for twice as long. The two batteries may also be able to deliver more current than if you just used one. See Figure 1-81.

Summing up:

- Two batteries in series deliver twice the voltage.
- Two batteries in parallel can deliver twice the current.

All right, that's more than enough theory for now. In the next chapter, we'll continue with some experiments that will build on the foundations of knowledge about electricity, to take us gradually toward gadgets that can be fun and useful.

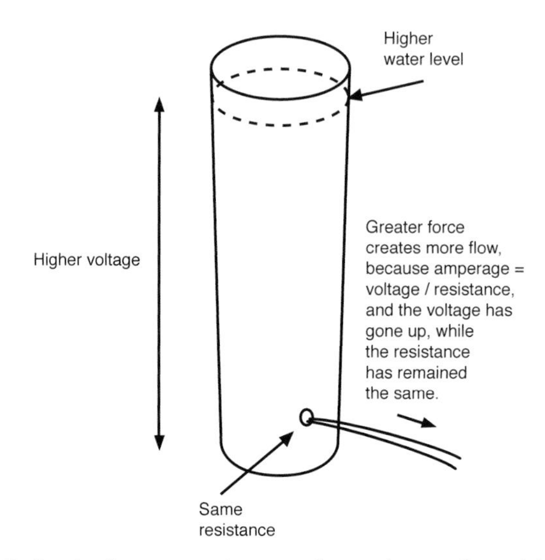

Figure 1-79. *Greater force generates more flow, as long as the resistance remains the same.*

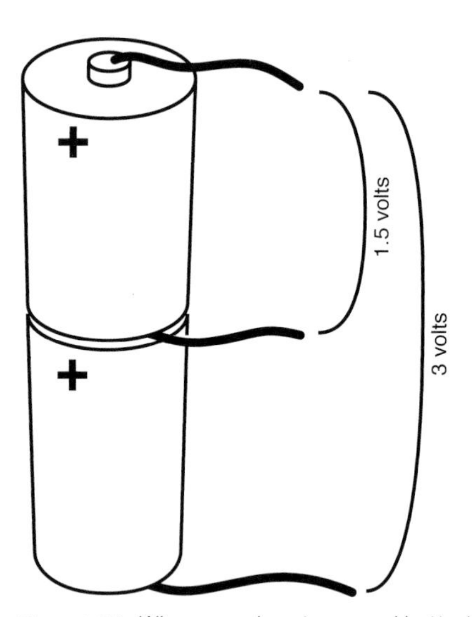

Figure 1-80. *When you place two equal batteries in series, you double the voltage.*

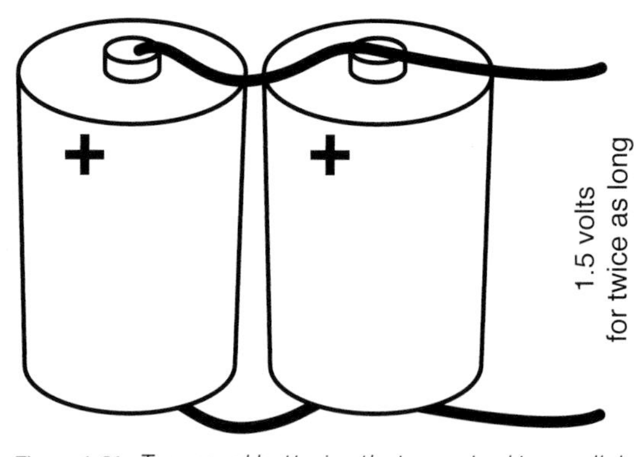

Figure 1-81. *Two equal batteries that are wired in parallel will deliver the same voltage for twice as long as one.*

2

Switching Basics and More

The concept of switching is fundamental in electronics, and I'm not just talking about power switches. By "switching," I mean using one flow of electricity to switch, or control, another. This is such an important principle that no digital device can exist without it.

Today, switching is mostly done with semiconductors. Before I deal with them, I'll back up and illustrate the concept by introducing you to relays, which are easier to understand, because you can see what's going on inside them. And before I get to relays, I'll deal with everyday on/off switches, which may seem very simple—but we have to nail down the basics.

Also in this chapter, I'll deal with capacitance, because capacitance and resistance are fundamental to electronic circuits. By the end of the chapter, you should have a basic grounding in electronics and be able to build the noise-making section of a simple intrusion alarm. This will be your first circuit that does something genuinely useful!

IN THIS CHAPTER

Shopping List: Experiments 6 Through 11

As in the previous shopping list, you should visit the various online suppliers for availability and pricing of components and devices. Manufacturers seldom sell small numbers of parts directly. Check the appendix for a complete list of URLs for all the companies mentioned here.

Devices

- Power supply/universal AC adapter, 3 to 12 volts at 1A (1,000 mA). See Figure 2-1. Part number 273-0316 from RadioShack, part PH-62092 by Philips, or similar.
- Breadboard suitable for integrated circuits. Quantity: 1. See Figures 2-2 and 2-3. Part 276-002 from RadioShack, model 383-X1000 made by PSP, part 923252-I by 3M, or similar. A breadboard that has screw terminals mounted beside it will be a little easier to use but more expensive than one that does not have terminals.

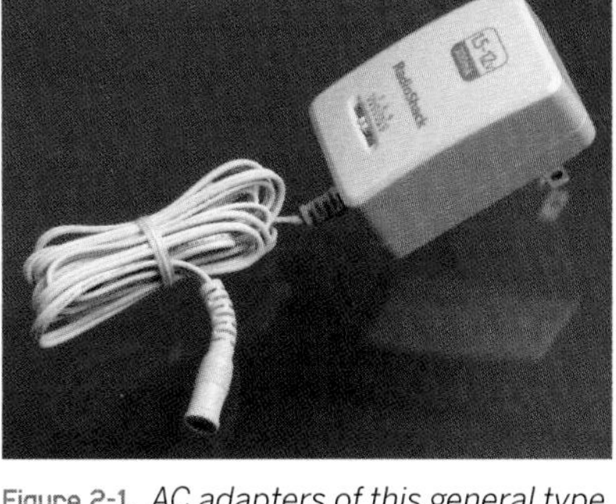

Figure 2-1. *AC adapters of this general type can supply a variety of voltages, with a range as wide as 3 to 12 volts.*

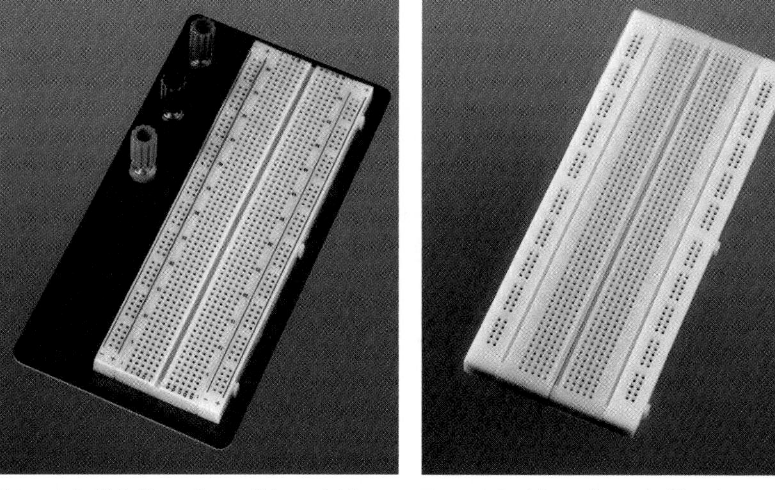

Figure 2-2. *This "breadboard" for quickly constructing electronic circuits has a metal base, and screw terminals for attaching wires from a power supply.*

Figure 2-3. *A breadboard without screw terminals is almost as convenient, and is cheaper.*

Tools

Wire strippers

Ideal model 45-121 wire strippers for 16- to 26-gauge wire, or similar. See Figure 2-4. (The "gauge" of the wire tells you how thick it is. A higher gauge means a thinner wire. In this book, we will mainly be using thin wire of 20- to 24-gauge.)

You may also consider the Kronus Automatic Wire Strippers, part 64-0083 from RadioShack, or GB Automatic Wire Strippers, part SE-92 from Amazon.com. See Figure 2-5.

The Kronus and GB wire strippers are functionally identical. The advantage of their design is that it enables you to strip insulation from a wire with one hand. But they do not work well on really thin wire.

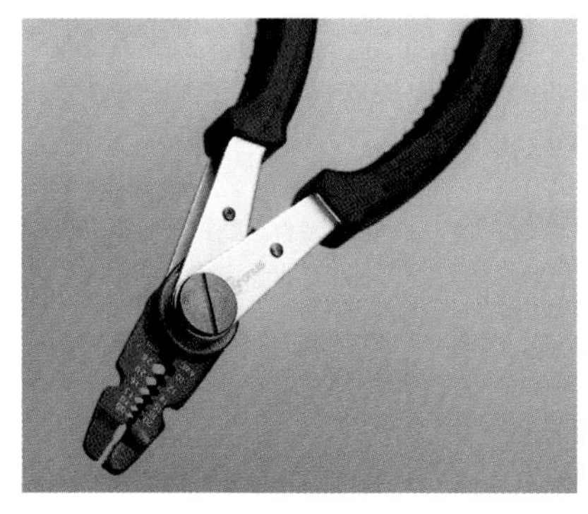

Figure 2-4. *To use these wire strippers, insert a piece of insulated wire in the appropriate-sized hole between the jaws, grip the handles, and pull a section of insulation away.* *See page 45.*

Figure 2-5. *These automatic wire strippers enable one-handed operation, but are not suitable for very small wire diameters.* *See page 44.*

Supplies

Hookup wire

Solid-conductor, 22-gauge, minimum 25 feet of each color. See Figure 2-6. Part 278-1221 from RadioShack, catalog item 9948T17 from McMaster-Carr, or check eBay for deals.

It's easy to buy the wrong kind of wire. You need solid-core wire, which has a single conductor inside the plastic insulation, not stranded, which has multiple, thinner conductors. See Figures 2-7 and 2-8. You're going to be pushing wires into little holes in a "breadboard," and stranded wire won't let you do this. You will also have problems if you buy wire thicker than 22-gauge. Remember: the lower the gauge number, the thicker the wire.

For a little extra money, you can buy an assortment of precut sections of wire, with ends stripped and ready for use. Try catalog item JW-140 (jumper wire assortment) from All Electronics or search eBay for "breadboard wire." See Figure 2-9.

Figure 2-6. *Using hookup wire with different colors of plastic insulation will help you to distinguish one wire from another in your circuits.*

Patch cords

Patch cords are not strictly necessary but very convenient. You don't want audio or video patch cords, which have a plug on each end; you want wires with alligator clips on each end, also sometimes referred to as "test leads." Try catalog item 461-1176-ND from Digi-Key or catalog item MTL-10 from All Electronics. See Figure 2-10.

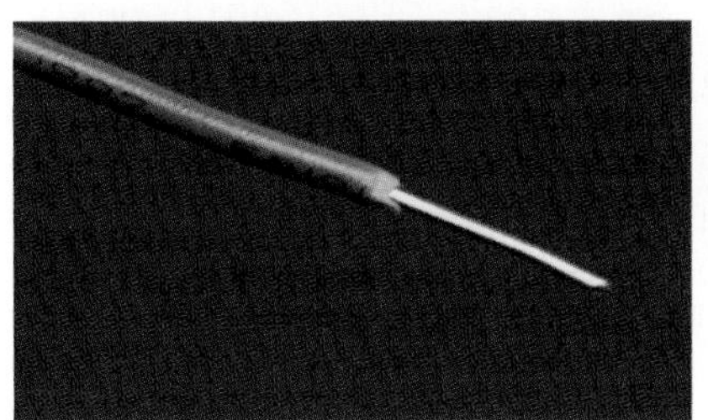

Figure 2-7. *Solid-conductor wire of 22 or 24 gauge is suitable for most of the experiments in this chapter.*

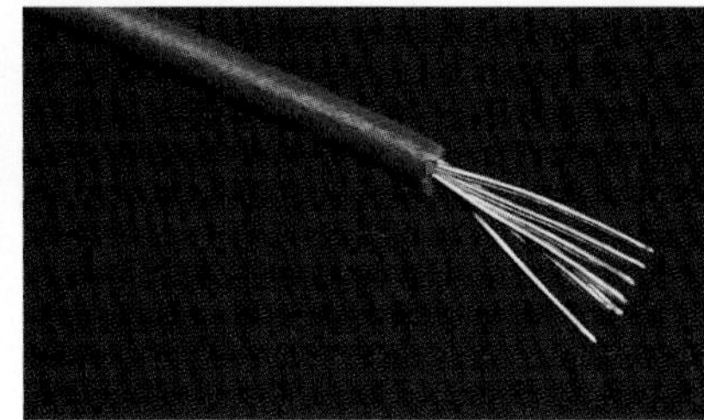

Figure 2-8. *Stranded is more flexible, but cannot be used easily with breadboards.*

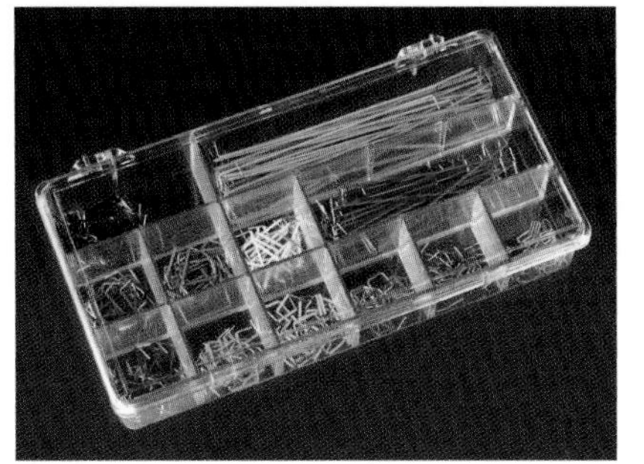

Figure 2-9. *Precut wires with stripped ends can save a lot of time and trouble—if you don't mind paying a little extra.*

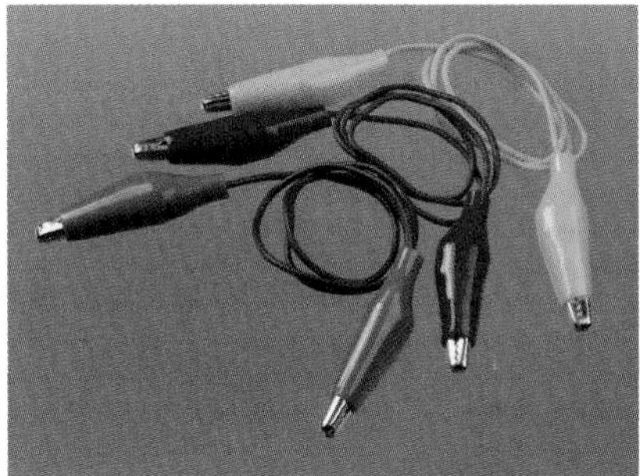

Figure 2-10. *Patch cords, sometimes known as test leads, consist of wires preattached to alligator clips. This is another of those little luxuries that reduces the hassle factor in hobby electronics.*

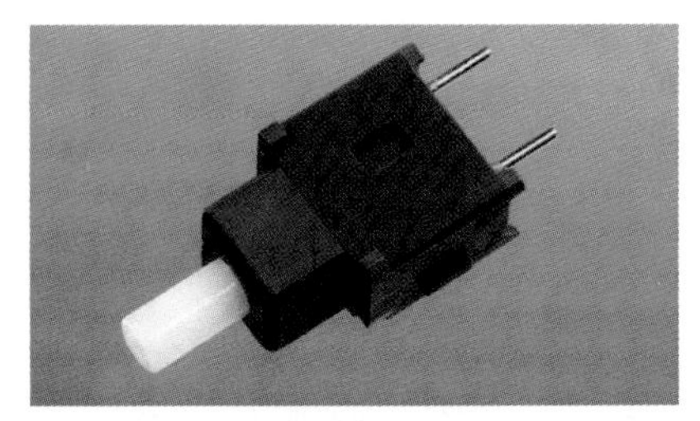

Figure 2-11. *The terminals protruding from this tiny pushbutton are spaced 0.2 inches apart, making it ideal for the "breadboard" that you'll be using.*

Figure 2-12. *This relatively large toggle switch made by NKK has screw terminals, which will reduce the inconvenience of attaching it to hookup wire.*

Figure 2-13. *Transistors are commonly sold either in little metal cans or sealed into little lumps of plastic. For our purposes, the packaging makes no difference.*

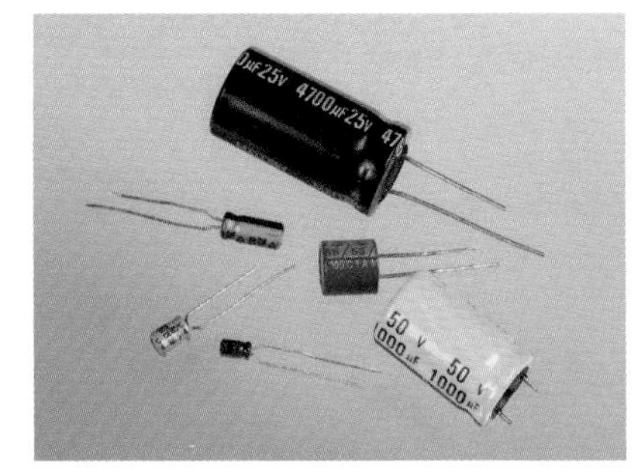

Figure 2-14. *An assortment of electrolytic capacitors.*

Components

Pushbutton

Momentary-on, SPST, sometimes referred to as OFF-(ON) or (ON)-OFF type. Must be PCB- or PC-mount, meaning is extremely small with thin spiky contacts on the bottom. Quantity: 1. See Figure 2-11.

Examples are part number AB11AP by NKK, part MPA103B04 by Alcoswitch, or part EP11SD1CBE by C&K. If you have a choice, buy the cheapest, as we're going to switch very low current.

Switches

Toggle switch, single-pole, double-throw (SPDT), sometimes referred to as ON-ON type. Quantity: 2. See Figure 2-12.

Model S302T-RO by NKK is ideal; it has screw terminals that will eliminate the need for alligator clips. Other options are catalog item MTS-4PC from All Electronics or part 275-603 from RadioShack.

We won't be switching large currents or high voltages, so the exact type of switch is unimportant. However, the terminals on larger-size switches are spaced wider apart, which makes them easier to deal with.

Relays

DPDT, nonlatching, 12v DC. Quantity: 2.

It's important to get the right kind of relay—one whose configuration matches the pictures I'll be using. Look for parts FTR-F1CA012V or FTR-F1CD012V by Fujitsu, G2RL-24-DC12 by Omron, or OMI-SH-212D by Tyco. Avoid substitutions.

Potentiometer

1 megohm linear potentiometer, Part number 271-211 from RadioShack, part number 24N-1M-15R-R from Jameco, or similar.

Transistors

NPN transistor, general-purpose, such as 2N2222 by STMicroelectronics, part PN2222 by Fairchild, or part 2N2222 from RadioShack. Quantity: 4. See Figure 2-13.

2N6027 programmable unijunction transistor manufactured by On Semiconductor or Motorola. Quantity: 4 (allowing for 2 spares in case of damage).

Capacitors

Electrolytic capacitors, assorted. Must be rated for a minimum of 25 volts and include at least one capacitor of 1,000 µF (microfarads) and two capacitors of 2.2 µF. If you search on eBay, make sure you find *electrolytic* capacitors. If they're rated for higher voltages, that's OK, although they will be physically larger than you need. See Figure 2-14.

Ceramic capacitors, assorted. Make sure you get at least one rated at 0.0047 µF (which can also be written as 4.7 nF). See Figure 2-15.

Resistors

If you bought only a minimal selection for experiments 1 through 5, now's the time to buy a larger assortment, so that you won't be stuck needing the one value that you don't have. 1/4-watt minimum.

Loudspeaker

Any 8Ω, 1-inch miniature loudspeaker such as part 273-092 from RadioShack. See Figure 2-16.

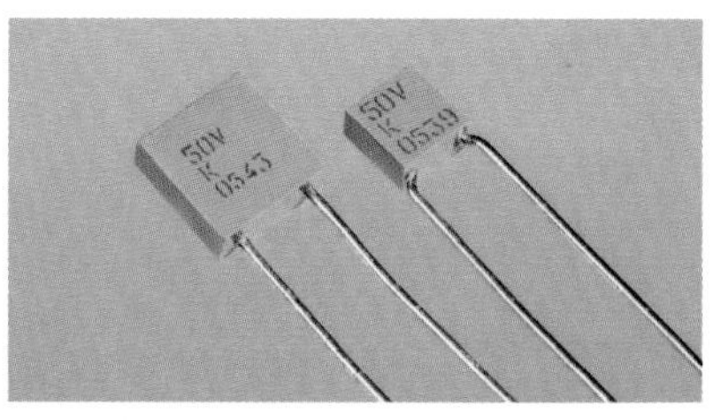

Figure 2-15. *Ceramic capacitors mostly look like this, although many of them are round or bead-shaped instead of square. The packaging shape is unimportant to us.*

Figure 2-16. *This miniature loudspeaker, just over 1 inch in diameter, is useful for verifying audio output direct from transistor circuits.*

Experiment 6: Very Simple Switching

You will need:

- AA batteries. Quantity: 2.
- Battery carrier for 2 AA batteries. Quantity: 1.
- LED. Quantity: 1.
- Toggle switches, SPDT. Quantity: 2. See Figure 2-12.
- 220Ω or similar value resistor, 1/4-watt minimum. Quantity: 1.
- Alligator clips. Quantity: 8.
- Wire or patch cords. See Figure 2-10, shown previously.
- Wire cutters and wire strippers if you don't use patch cords. See Figure 2-4, shown previously.

In Experiment 3, you illuminated an LED by attaching a battery, and switched it off by removing the battery. For greater convenience our circuits should have proper switches to control power, and while I'm dealing with the general topic of switches, I'm going to explore all the varieties, using a circuit to suggest some possibilities.

Assemble the parts as shown in Figures 2-17 and 2-18. The long lead on the LED must connect with the resistor, because that is the more positive side of the circuit.

You'll notice that you have to include a couple lengths of wire. I suggest green wire to remind you that these sections are not connected directly to positive or to negative power. But you can use any color you like. You can also substitute patch cords, if you have them. However, learning to strip insulation from pieces of wire is a necessary skill, so let's deal with that now.

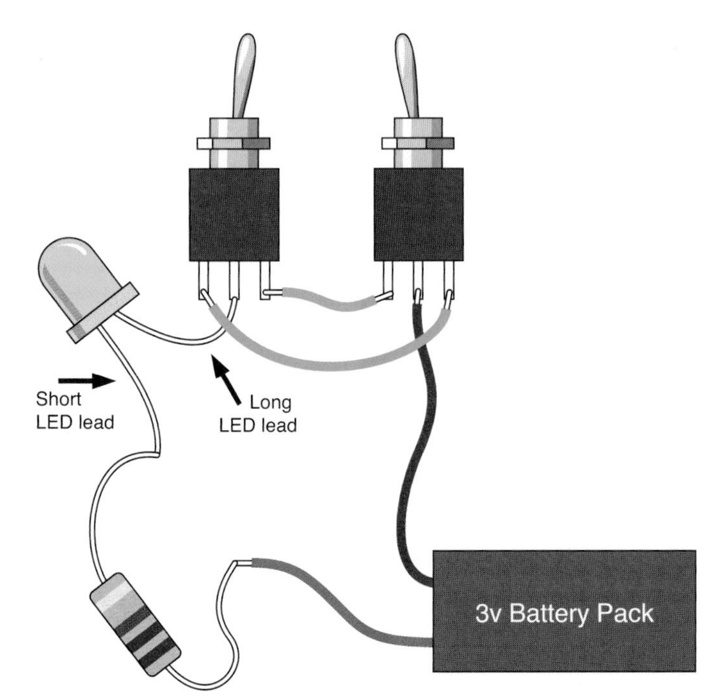

Figure 2-17. *If the LED is on, flipping either of the switches will turn it off. If the LED is off, either of the switches will turn it on. Use alligator clips to attach the wires to each other, and to the switches if your switches don't have screw terminals. Be careful that the clips don't touch each other.*

Figure 2-18. *Full-size toggle switches with screw terminals make it easy to hook up this simple circuit.*

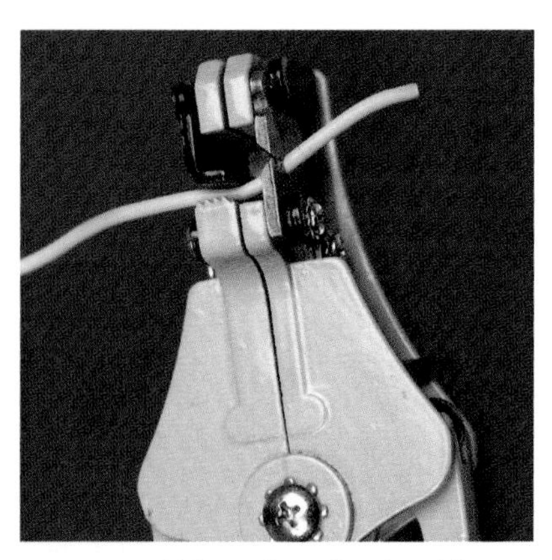

Figure 2-19. *Using automatic wire strippers, when you squeeze the handles the jaw on the left clamps the wire, the sharp grooves on the right bite into the insulation. Squeeze harder and the jaws pull away from each other, stripping the insulation from the wire.*

Tools

If automatic wire strippers (Figure 2-19) don't grip skinny 22-gauge wire very effectively, try the Ideal brand of wire strippers shown back in Figure 2-4, or use plain and simple wire cutters as shown in Figure 2-20. When using wire cutters, you hold the wire in one hand and apply the tool in your other hand, squeezing the handles with moderate pressure—just enough to bite into the insulation, but not so much that you chop the wire. Pull the wire down while you pull the cutters up, and with a little practice you can rip the insulation off to expose the end of the wire.

Macho hardware nerds may use their teeth to strip insulation from wires. When I was younger, I used to do this. I have two slightly chipped teeth to prove it. Really, it's better to use the right tool for the job.

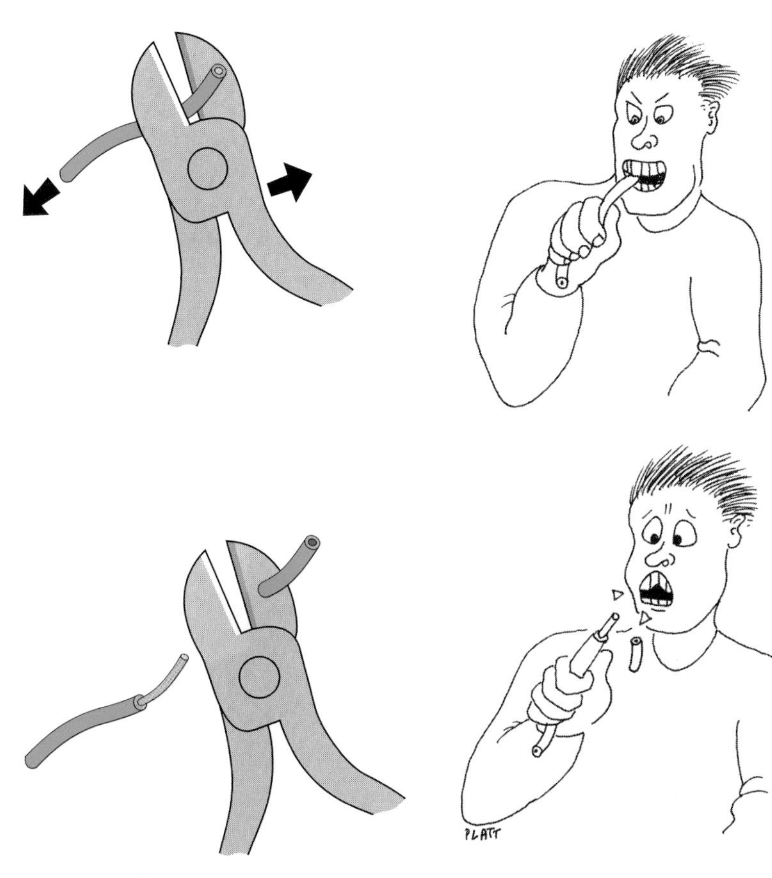

Figure 2-20. *To remove insulation from the end of a thin piece of wire, you can also use wire cutters. This takes a little practice.*

Figure 2-21. *Those who tend to misplace tools, and feel too impatient to search for them, may feel tempted to use their teeth to strip insulation from wire. This may not be such a good idea.*

Connection Problems

Depending on the size of toggle switches that you are using, you may have trouble fitting in all the alligator clips to hold the wires together. Miniature toggle switches, which are more common than the full-sized ones these days, can be especially troublesome (see Figure 2-22). Be patient: fairly soon we'll be using a breadboard, which will eliminate alligator clips almost completely.

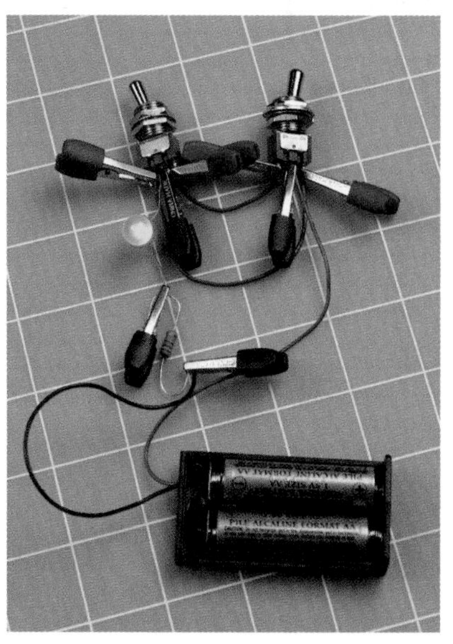

Figure 2-22. *Miniature toggle switches can be used—ideally, with miniature alligator clips—but watch out for short circuits.*

Testing

Make sure that you connect the LED with its long wire toward the positive source of power (the resistor, in this case). Now flip either of the toggle switches. If the LED was on, it will go off, and if it was off, it will go on. Flip the other toggle switch, and it will have the same effect. If the LED does not go on at all, you've probably connected it the wrong way around. Another possibility is that two of your alligator clips may have shorted out the battery.

Assuming your two switches do work as I described them, what's going on here? It's time to nail down some basic facts.

FUNDAMENTALS

All about switches

When you flip the type of toggle switch that you used in Experiment 6, it connects the center terminal with one of the outer terminals. Flip the switch back, and it connects the center terminal with the other outer terminal, as shown in Figure 2-23.

The center terminal is called the pole of the switch. Because you can flip, or throw, this switch to make two possible connections, it is called a *double-throw switch*. As mentioned earlier, a single-pole, double-throw switch is abbreviated *SPDT*.

Some switches are on/off, meaning that if you throw them in one direction they make a contact, but in the other direction, they make no contact at all. Most of the light switches in your house are like this. They are known as *single-throw switches*. A single-pole, single-throw switch is abbreviated *SPST*.

Some switches have two entirely separate poles, so you can make two separate connections simultaneously when you flip the switch. These are called *double-pole switches*. Check the photographs in Figures 2-24 through 2-26 of old-fashioned "knife" switches (which are still used to teach electronics to kids in school) and you'll see the simplest representation of single and double poles, and single and double throws. Various toggle switches that have contacts sealed inside them are shown in Figure 2-27.

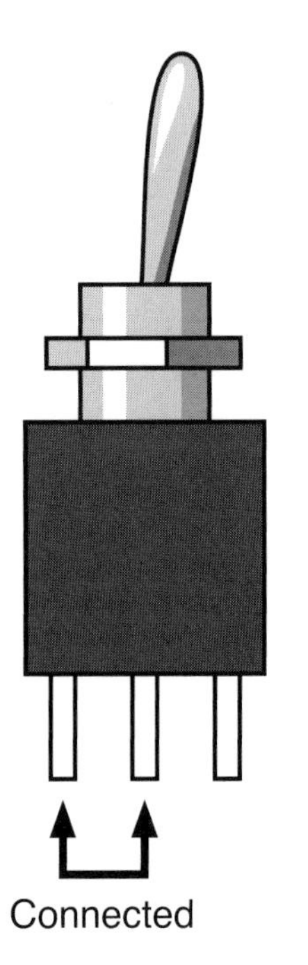

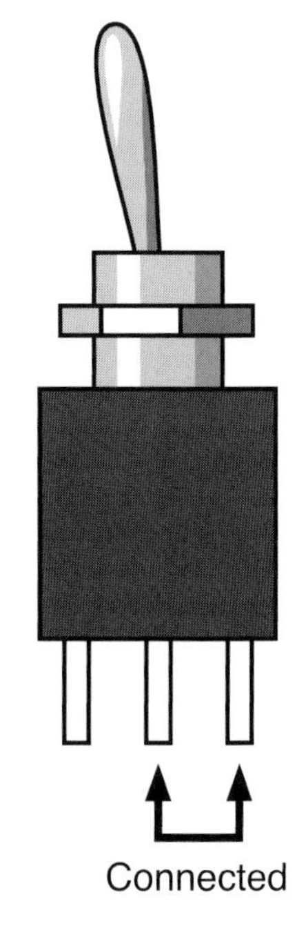

Figure 2-23. *The center terminal is the pole of the switch. When you flip the toggle, the pole changes its connection.*

Figure 2-24. *This primitive-looking single-pole, double-throw switch does exactly the same thing as the toggle switches in Figures 2-23 and 2-27.*

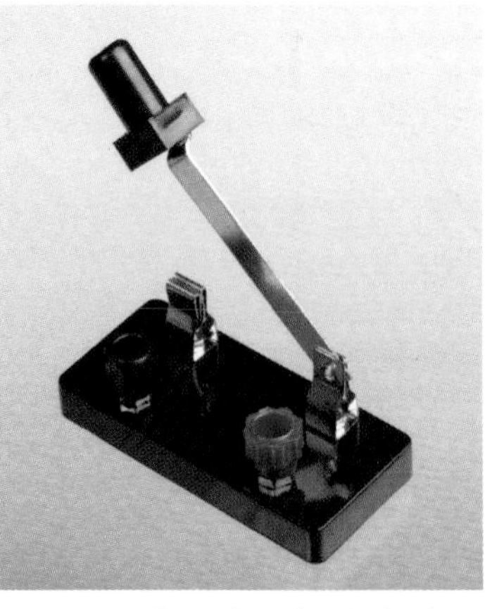

Figure 2-25. *A single-pole, single-throw switch makes only one connection with one pole. Its two states are simply open and closed, on and off.*

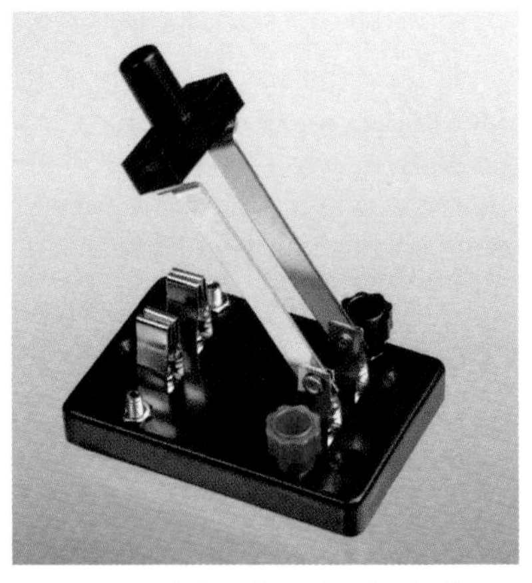

Figure 2-26. *A double-pole, single-throw switch makes two separate on/off connections.*

FUNDAMENTALS

All about switches (continued)

Figure 2-27. *These are all toggle switches. Generally, the larger the switch, the more current it can handle.*

To make things more interesting, you can also buy switches that have three or four poles. (Some rotary switches have even more, but we won't be using them.) Also, some double-throw switches have an additional "center off" position.

Putting all this together, I made a table of possible types of switches (Figure 2-28). When you're reading a parts catalog, you can check this table to remind yourself what the abbreviations mean.

	Single Pole	Double Pole	3-Pole	4-Pole
Single Throw	SPST ON-OFF	DPST ON-OFF	3PST ON-OFF	4PST ON-OFF
Double Throw	SPDT ON-ON	DPDT ON-ON	3PDT ON-ON	4PDT ON-ON
Double Throw with Center Off	SPDT ON-OFF-ON	DPDT ON-OFF-ON	3PDT ON-OFF-ON	4PDT ON-OFF-ON

Figure 2-28. *This table summarizes all the various options for toggle switches and pushbuttons.*

Now, what about pushbuttons? When you press a door bell, you're making an electrical contact, so this is a type of switch—and indeed the correct term for it is a momentary switch, because it makes only a momentary contact. Any spring-loaded switch or button that wants to jump back to its original position is known as a momentary switch. We indicate this by putting its momentary state in parentheses. Here are some examples:

- OFF-(ON): Because the ON state is in parentheses, it's the momentary state. Therefore, this is a single-pole switch that makes contact only when you push it, and flips back to make no contact when you let it go. It is also known as a "normally open" momentary switch, abbreviated "NO."
- ON-(OFF): The opposite kind of momentary single-pole switch. It's normally ON, but when you push it, you break the connection. So, the OFF state is momentary. It is known as a "normally closed" momentary switch, abbreviated "NC."
- (ON)-OFF-(ON): This switch has a center-off position. When you push it either way, it makes a momentary contact, and returns to the center when you let it go.

Other variations are possible, such as ON-OFF-(ON) or ON-(ON). As long as you remember that parentheses indicate the momentary state, you should be able to figure out what these switches are.

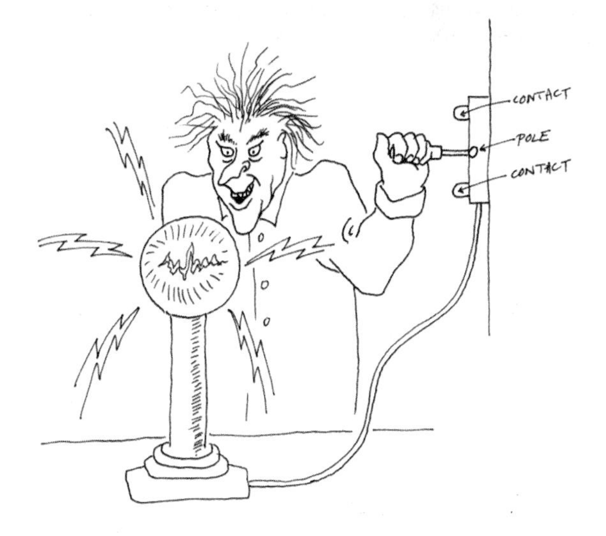

Figure 2-29. *This evil mad scientist is ready to apply power to his experiment. For this purpose, he is using a single-pole, double-throw knife switch, conveniently mounted on the wall of his basement laboratory.*

FUNDAMENTALS

All about switches (continued)

Sparking

When you make and break an electrical connection, it tends to create a spark. Sparking is bad for switch contacts. It eats them until the switch doesn't make a reliable connection anymore. For this reason, you must use a switch that is appropriate for the voltage and amperage that you are dealing with. Electronic circuits generally are low-current, and low-voltage, so you can use almost any switch, but if you are switching a motor, it will tend to suck an initial surge of current that is at least double the rating of the motor when it is running constantly. You should probably use a 4-amp switch to turn a 2-amp motor on and off.

Checking a switch

You can use your meter to check a switch. Doing this helps you find out which contacts are connected when you turn a switch one way or the other. It's also useful if you have a pushbutton and you can't remember whether it's the type that is normally open (you press it to make a connection) or normally closed (you press it to break the connection). Set your meter to measure ohms, and touch the probes to the switch terminals while you work the switch.

This is a hassle, though, because you have to wait while the meter makes an accurate measurement. When you just want to know whether there is a connection, your meter has a "continuity tester" setting. It beeps if it finds a connection, and stays silent if it doesn't. See Figures 2-30 through 2-32 for examples of meters set to test continuity. Figure 2-33 offers an example of a toggle switch being tested for continuity.

Use the continuity-testing feature on your meter only on circuits or components that have no power in them at the time.

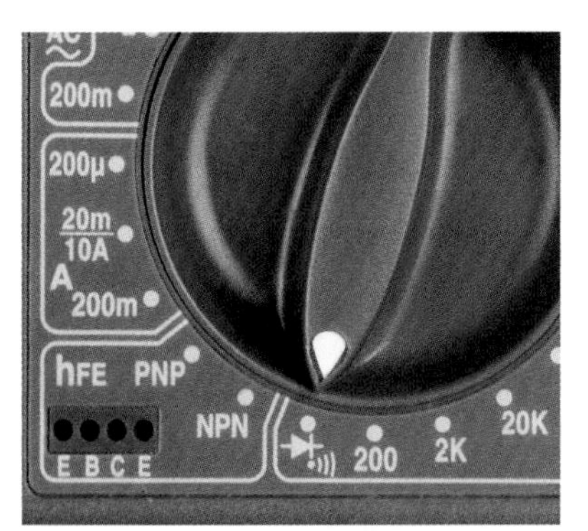

Figure 2-30

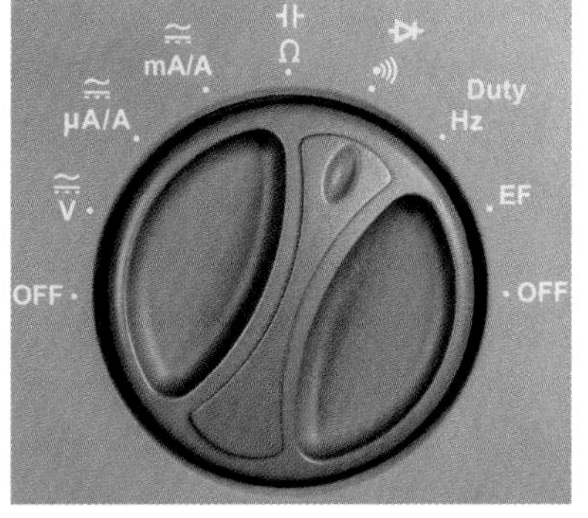

Figure 2-31

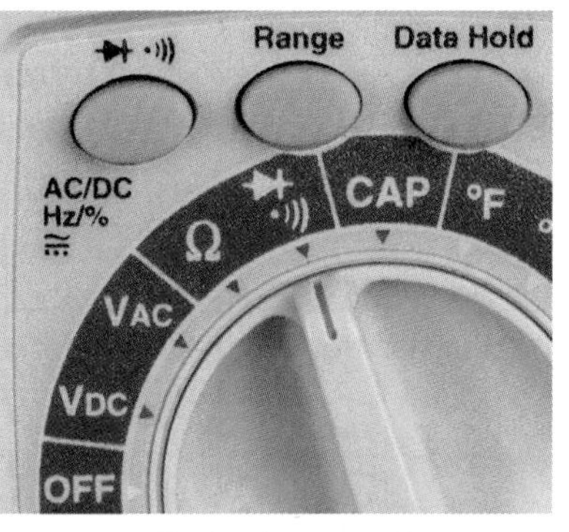

Figure 2-32. *To check a circuit for continuity, turn the dial of your meter to the symbol shown. Only use this feature when there is no power in the component or the circuit that you are testing.*

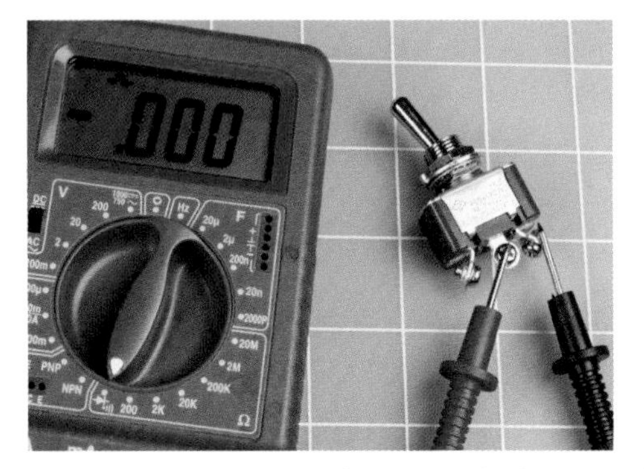

Figure 2-33. *When the switch connects two of its terminals, the meter shows zero resistance between them and will beep if you have set it to verify continuity.*

BACKGROUND

Early switching systems

Switches seem to be such a fundamental feature of our world, and their concept is so simple that it's easy to forget that they went through a gradual process of development and refinement. Primitive knife switches were quite adequate for pioneers of electricity who simply wanted to connect and disconnect electricity to some apparatus in a laboratory, but a more sophisticated approach was needed when telephone systems began to proliferate. Typically, an operator at a "switchboard" needed a way to connect any pair of 10,000 lines on the board. How could it be done?

In 1878, Charles E. Scribner (Figure 2-34) developed the "jack-knife switch," so called because the part of it that the operator held looked like the handle of a jackknife. Protruding from it was a plug, and when the plug was pushed into a socket, it made contact inside the socket. The socket, in fact, was the switch.

Figure 2-34. *Charles E. Scribner invented the "jack-knife switch" to satisfy the switching needs of telephone systems in the late 1800s. Today's audio jacks still work on the same basis.**

Audio connectors on guitars and amplifiers still work on the same principle, and when we speak of them as being "jacks," the term dates back to Scribner's invention. Switch contacts still exist inside a jack socket.

Today, of course, telephone switchboards have become as rare as telephone operators. First they were replaced with relays—electrically operated switches, which I'll talk about later in this chapter. And then the relays were superceded by transistors, which made everything happen without any moving parts. Before the end of this chapter, you'll be switching current using transistors.

* The photo on which this drawing is based first appeared in *The History of the Telephone* by Herbert Newton Casson in 1910 (Chicago: A. C. McClurg & Co.).

Introducing Schematics

In Figure 2-35, I've redrawn the circuit from Experiment 6 in a simplified style known as a "schematic." From this point onward, I will be illustrating circuits with schematics, because they make circuits easier to understand. You just need to know a few symbols to interpret them.

Larger versions of all schematics and breadboard photos are available online at this book's website: http://oreilly.com/catalog/9780596153748.

Compare the schematic here with the drawing of the circuit in Figure 2-17. They both show exactly the same thing: Components, and connections between them. The gray rectangles are the switches, the zigzag thing is the resistor, and the symbol with two diagonal arrows is the LED.

The schematic LED symbol includes two arrows indicating that it emits light, because there are some kinds of diodes, which we'll get to later, that don't. The triangle inside the diode symbol always points from positive to negative.

Trace the path that electricity can take through the circuit and imagine the switches turning one way or the other. You should see clearly now why either switch will reverse the state of the LED from on to off or off to on.

This same circuit is used in houses where you have a switch at the bottom of a flight of stairs, and another one at the top, both controlling the same lightbulb. The wires in a house are much longer, and they snake around behind the walls, but because their connections are still the same, they could be represented with the same basic schematic. See Figure 2-36.

A schematic doesn't tell you exactly where to put the components. It just tells you how to join them together. One problem: Different people use slightly different schematic symbols to mean the same thing. Check the upcoming section, "Fundamentals: Basic schematic symbols," for the details.

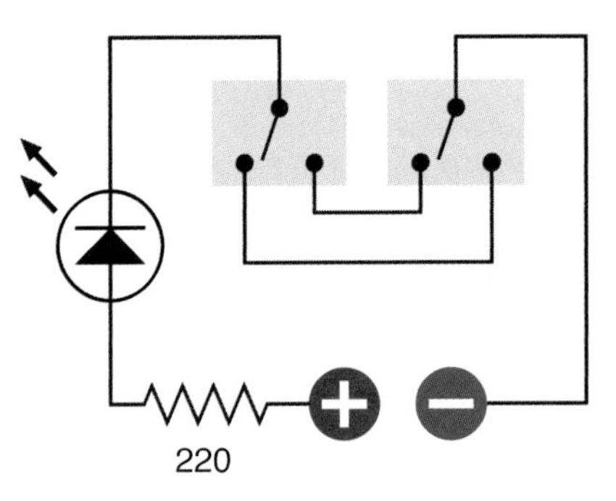

Figure 2-35. *This schematic shows the same circuit as in Figure 2-17 and makes it easier to see how the switches function.*

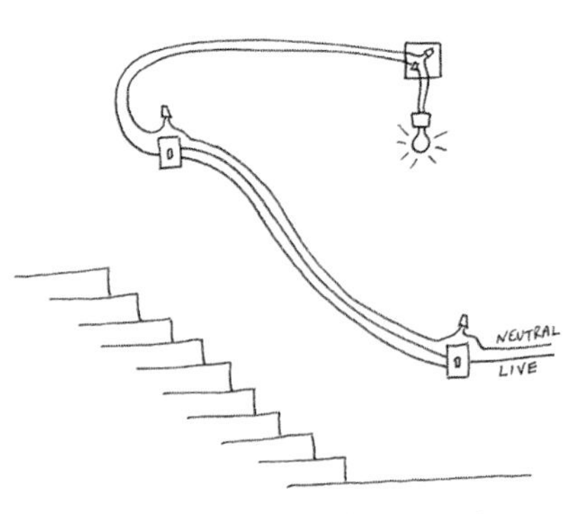

Figure 2-36. *The two-switch circuit shown in Figures 2-17 and 2-35 is often found in house wiring, especially where switches are located at the top and bottom of a flight of stairs. This sketch shows what you might find inside the walls. Wires are joined with "wire nuts" inside boxes that are hidden from everyday view.*

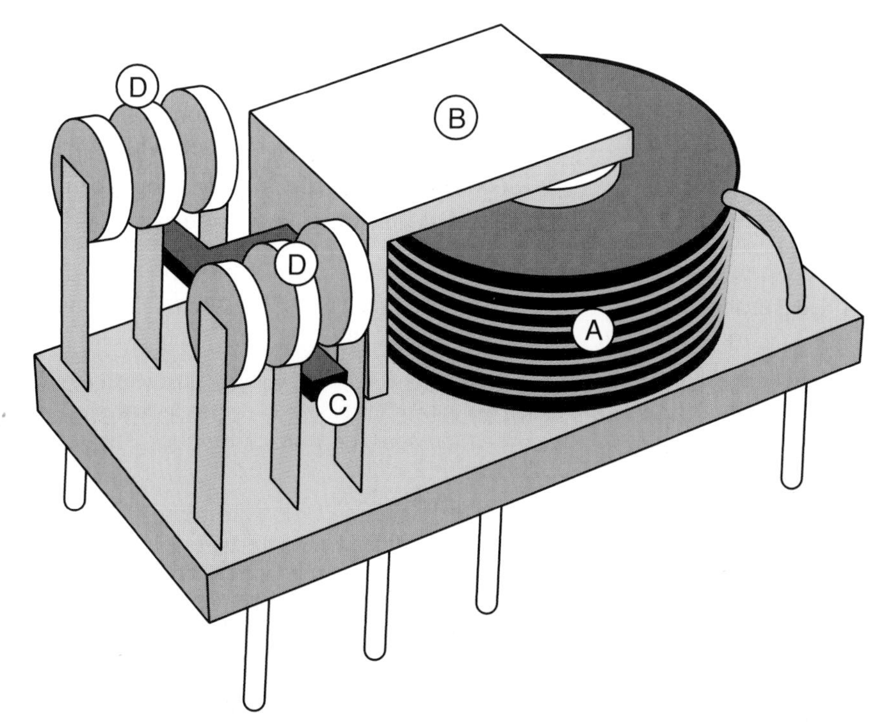

Figure 2-51. *This is one way that the parts inside a relay can be arranged. The coil, A, generates a magnetic attraction pulling lever B downward. A plastic extension, C, pushes outward against flexible metal strips and moves the poles of the relay, D, between the contacts.*

Figure 2-52. *To look inside a sealed relay, shave the top edges of the plastic package with a utility knife til you open a thin crack.*

Figure 2-53. *Insert the blade of your knife to pry open the top, then repeat the procedure for the sides.*

Figure 2-54. *If you are really, really careful, the relay should still work after you open it.*

Figure 2-55

Figure 2-56. *Patience is essential when carving the edges of a relay package in order to open it. Faster methods such as a tomahawk or a flamethrower will satisfy the emotional needs of those with a short attention span, but results may be unpredictable.*

Figure 2-57. *Four assorted 12-volt relays, shown with and without their packages. The automotive relay (far left) is the simplest and easiest to understand, because it is designed without much concern for the size of the package. Smaller relays are more ingeniously designed, more complex, and more difficult to figure out. Usually, but not always, a smaller relay is designed to switch less current than a larger one.*

FUNDAMENTALS

Inside a relay

A relay contains a coil of wire wrapped around an iron core. When electricity runs through the coil, the iron core exerts a magnetic force, which pulls a lever, which pushes or pulls a springy strip of metal, closing two contacts. So as long as electricity runs through the coil, the relay is "energized" and its contacts remain closed.

When the power stops passing through the coil, the relay lets go and the springy strip of metal snaps back into its original position, opening the contacts. (The exception to this rule is a latching relay, which requires a second pulse through a separate coil to flip it back to its original position; but we won't be using latching relays until later in the book.)

Relays are categorized like switches. Thus, you have SPST relays, DPST, SPDT, and so on.

Compare the schematics in Figure 2-58 with the schematics of switches in Figure 2-38. The main difference is that the relay has a coil that activates the switch. The switch is shown in its "relaxed" mode, when no power flows through the coil.

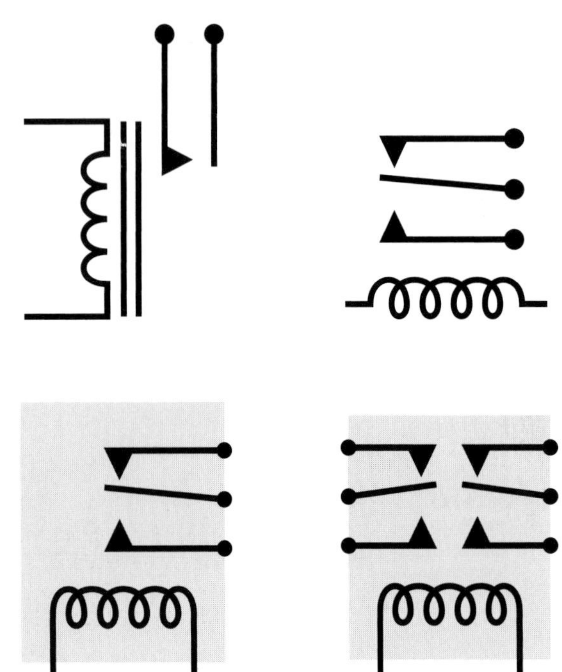

Figure 2-58. *Various ways to show a relay in a schematic. Top left: SPST. Top right and bottom left: SPDT. Bottom right: DPDT. The styles at bottom-left and bottom-right will be used in this book.*

The contacts are shown as little triangles. When there are two poles instead of one, the coil activates both switches simultaneously.

Most relays are nonpolarized, meaning that you can run electricity through the coil in either direction, and the relay doesn't care. You should check the data sheet to make sure, though. Some relay coils work on AC voltage, but almost all low-voltage relays use direct current—a steady flow of electricity, such as you would get from a battery. We'll be using DC relays in this book.

Relays suffer from the same limitations as switches: their contacts will be eroded by sparking if you try to switch too much voltage. It's not worth saving a few dollars by using a relay that is rated for less current or voltage than your application requires. The relay will fail you when you need it most, and may be inconvenient to replace.

Because there are so many different types of relays, read the specifications carefully before you buy one. Look for these basics:

Coil voltage
: The voltage that the relay is supposed to receive when you energize it.

Set voltage
: The minimum voltage that the relay needs to close its switch. This will be a bit less than the ideal coil voltage.

Operating current
: The power consumption of the coil, usually in milliamps, when the relay is energized. Sometimes the power is expressed in milliwatts.

Switching capacity
: The maximum amount of current that you can switch with contacts inside the relay. Usually this is for a "resistive load," meaning a passive device such as light bulb. When you use a relay to switch on a motor, the motor takes a big initial surge of current before it gets up to speed. In this case, you should choose a relay rated for double the current that the motor draws when it is running.

Procedure

Turn the relay with its legs in the air and attach wires and LEDs as shown in Figure 2-59, with a 680Ω resistor (a 1K resistor will be OK if you don't have the correct value). Also attach a pushbutton switch. (Your pushbutton switch may look different from the one shown, but as long as it is a SPST pushbutton with two contacts at the bottom, it will work the same way.) When you press the pushbutton, the relay will make the first LED go out and the second LED light up. When you release the pushbutton, the first LED lights up and the second one goes out.

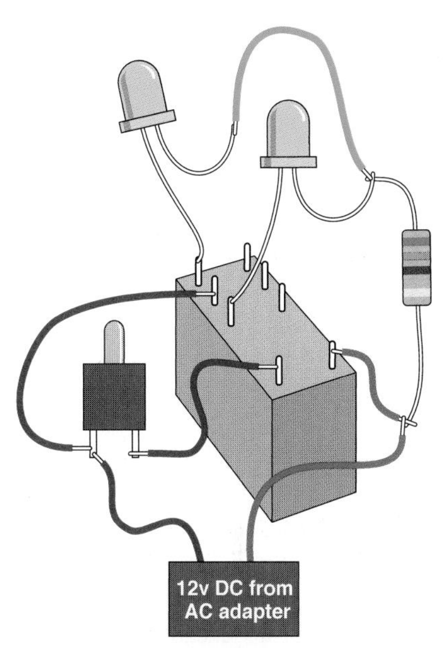

Figure 2-59. *As before, you can use patch cords, if you have them, instead of some of the wired connections shown here.*

How It Works

Check the schematic in Figure 2-60 and compare it with Figure 2-59. Also see Figure 2-62, which shows how the pins outside the relay make connections inside the relay when its coil is energized, and when it is not energized.

This is a DPDT relay, but we are only using one pole and ignoring the other. Why not buy a SPDT relay? Because I want the pins to be spaced the way they are when you will upgrade this circuit by transferring it onto a breadboard, which will happen very shortly.

On the schematic, I have shown the switch inside the relay in its relaxed state. When the coil is energized, the switch flips upward, which seems counterintuitive, but just happens to be the way that this particular relay is made.

When you're sure you understand how the circuit works, it's time to move on to the next step: making a small modification to get the relay to switch itself on and off, as we'll do in Experiment 8.

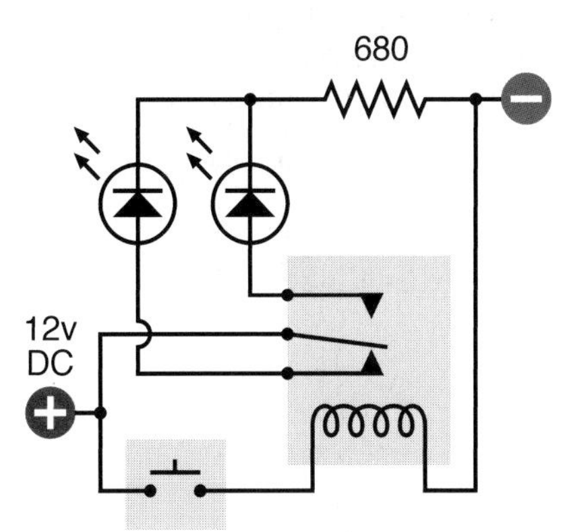

Figure 2-60. *Same circuit, shown in schematic form.*

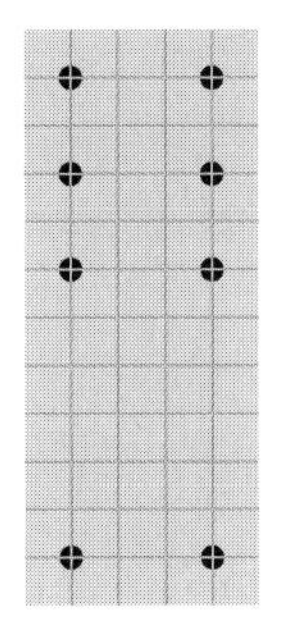

Figure 2-61. *The layout of the pins of the relay, superimposed on a grid of 1/10-inch squares. This is the type of relay that you will need in Experiment 8.*

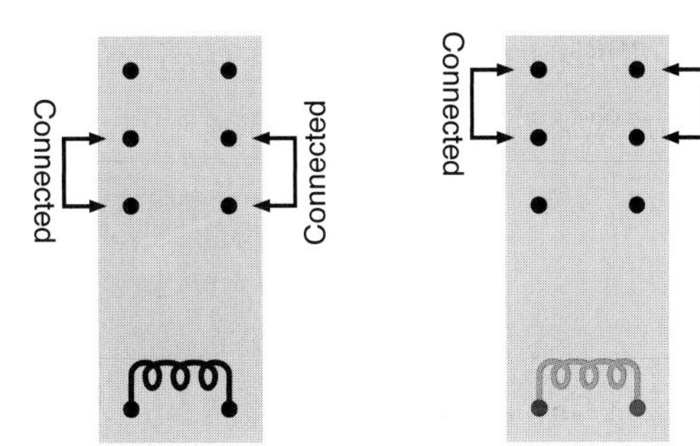

Figure 2-62. *How the relay connects the pins, when it is not energized (left) and when it is energized (right).*

Larger versions of all schematics and breadboard photos are available online at this book's website: http://oreilly.com/catalog/9780596153748.

Experiment 8: A Relay Oscillator

You will need:

- AC adapter, breadboard, wire, wire cutters and strippers.
- DPDT relay. Quantity: 1.
- LEDs. Quantity: 2.
- Pushbutton, SPST. Quantity: 1.
- Alligator clips. Quantity: 8.
- Resistor, approximately 680Ω. Quantity: 1.
- Capacitor, electrolytic, 1,000 μF. Quantity: 1.

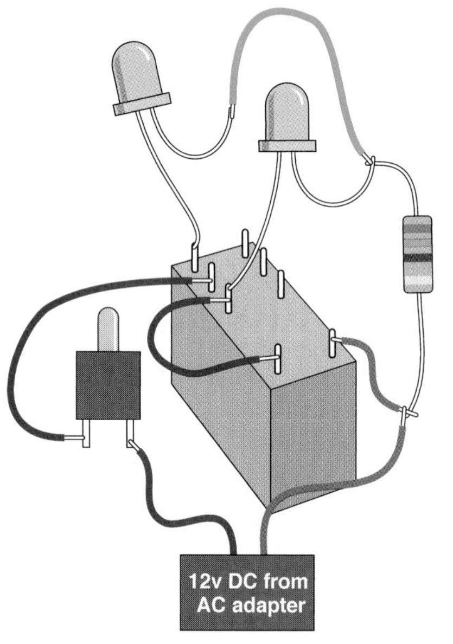

Figure 2-63. *A small revision to the previous circuit causes the relay to start oscillating when power is applied.*

Look at the revised drawing in Figure 2-63 and the revised schematic in Figure 2-64 and compare them with the previous ones. Originally, there was a direct connection from the pushbutton to the coil. In the new version, the power gets to the coil by going, first, through the contacts of the relay.

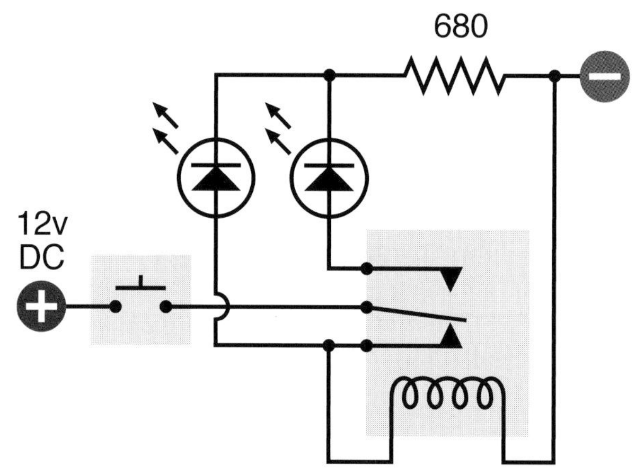

Figure 2-64. *The oscillator circuit shown in schematic form.*

Now, when you press the button, the contacts in their relaxed state feed power to the coil as well as to the lefthand LED. But as soon as the coil is energized, it opens the contacts. This interrupts the power to the coil—so the relay relaxes, and the contacts close again. They feed another pulse of power to the coil, which opens the contacts again, and the cycle repeats endlessly.

Because we're using a very small relay, it switches on and off extremely fast. In fact, it oscillates perhaps 50 times per second (too fast for the LEDs to show what's really happening). Make sure your circuit looks like the one in the diagram, and then press the pushbutton very briefly. You should hear the relay make a buzzing sound. If you have impaired hearing, touch the relay lightly with your finger, and you should feel the relay vibrating.

When you force a relay to oscillate like this, it's liable to burn itself out or destroy its contacts. That's why I asked you to press the pushbutton briefly. To make the circuit more practical, we need something to slow the relay down and prevent it from self-destructing. That necessary item is a capacitor.

Adding Capacitance

Add a 1,000 μF electrolytic capacitor in parallel with the coil of the relay as shown in the diagram in Figure 2-65 and the schematic in Figure 2-66. Check Figure 2-14 if you're not sure what a capacitor looks like. The 1,000 μF value will be printed on the side of it, and I'll explain what this means a little later.

Make sure the capacitor's *short* wire is connected to the *negative* side of the circuit; otherwise, it won't work. In addition to the short wire, you should find a minus sign on the body of the capacitor, which is there to remind you which side is negative. Electrolytic capacitors are fussy about this.

When you press the button now, the relay should click slowly instead of buzzing. What's happening here?

A capacitor is like a tiny rechargeable battery. It's so small that it charges in a fraction of a second, before the relay has time to open its lower pair of contacts. Then, when the contacts are open, the capacitor acts like a battery, providing power to the relay. It keeps the coil of the relay energized for about one second. After the capacitor exhausts its power reserve, the relay relaxes and the process repeats.

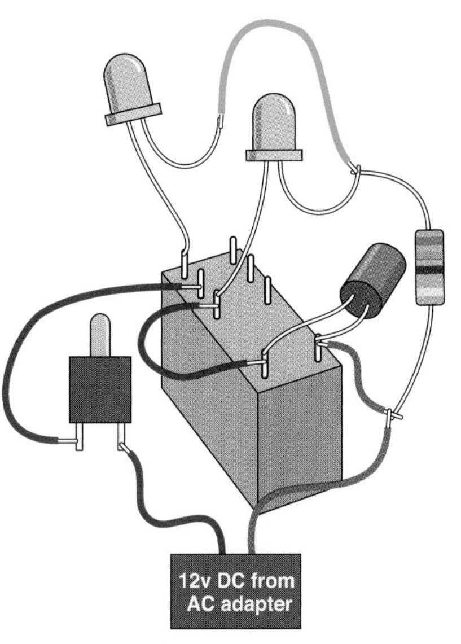

Figure 2-65. *Adding a capacitor makes the relay oscillate more slowly.*

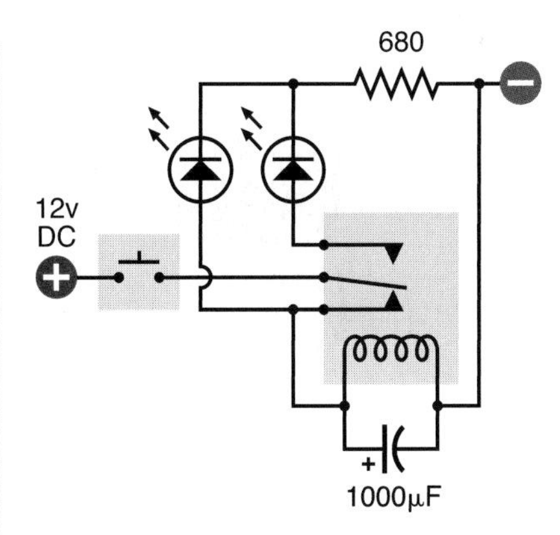

Figure 2-66. *The capacitor appears at the bottom of this schematic diagram.*

FUNDAMENTALS

Farad basics

The Farad is an international unit to measure capacitance. Modern circuits usually require small capacitors. Consequently it is common to find capacitors measured in microfarads (one-millionth of a farad) and even picofarads (one-trillionth of a farad). Nanofarads are also used, more often in Europe than in the United States. See the following conversion table.

0.001 nanofarad	1 picofarad	1 pF
0.01 nanofarad	10 picofarads	10 pF
0.1 nanofarad	100 picofarads	100 pF
1 nanofarad	1,000 picofarads	1,000 pF
0.001 microfarad	1 nanofarad	1 nF
0.01 microfarad	10 nanofarads	10 nF
0.1 microfarad	100 nanofarads	100 nF
1 microfarad	1,000 nanofarads	1,000 nF
0.000001 Farad	1 microfarad	1 μF
0.00001 Farad	10 microfarads	10 μF
0.0001 Farad	100 microfarads	100 μF
0.001 Farad	1,000 microfarads	1,000 μF

(You may encounter capacitances greater than 1,000 microfarads, but they are uncommon.)

Getting Zapped by Capacitors

If a large capacitor is charged with a high voltage, it can retain that voltage for a long time. Because the circuits in this book use low voltages, you don't have to be concerned about that danger here, but if you are reckless enough to open an old TV set and start digging around inside (which I do not recommend), you may have a nasty surprise. An undischarged capacitor can kill you as easily as if you stick your finger into an electrical outlet. Never touch a large capacitor unless you really know what you're doing.

FUNDAMENTALS

Capacitor basics

DC current does not flow through a capacitor, but voltage can accumulate very quickly inside it, and remains after the power supply is disconnected. Figures 2-67 and 2-68 may help to give you an idea of what happens inside a capacitor when it is fully charged.

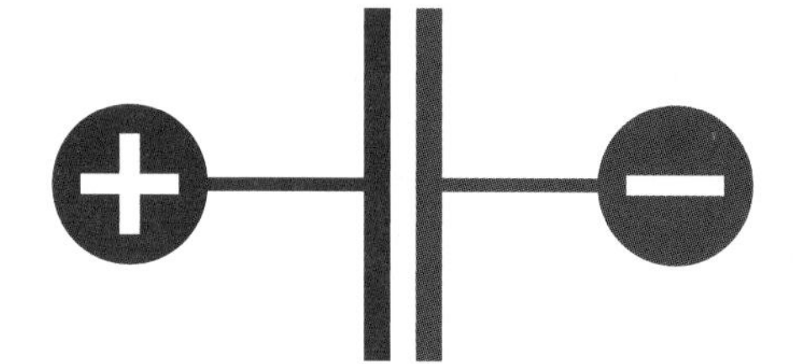

Figure 2-67. *When DC voltage reaches a capacitor, no current flows, but the capacitor charges itself like a little battery. The positive and negative charges are equal and opposite.*

Figure 2-68. *You can imagine positive "charge particles" accumulating on one side of the capacitor and attracting negative "charge particles" to the opposite side.*

In most modern electrolytic capacitors, the plates have been reduced to two strips of very thin, flexible, metallic film, often wrapped around each other, separated by an equally thin insulator. Disc ceramic capacitors typically consist of just a single disc of nonconductive material with metal painted on both sides and leads soldered on.

The two most common varieties of capacitors are ceramic (capable of storing a relatively small charge) and electrolytic (which can be much larger). Ceramics are often disc-shaped and yellow in color; electrolytics are often shaped like miniature tin cans and may be just about any color. Refer back to Figures 2-14 and 2-15 for some examples.

FUNDAMENTALS

Capacitor basics (continued)

Ceramic capacitors have no polarity, meaning that you can apply negative voltage to either side of them. Electrolytics do have polarity, and won't work unless you connect them the right way around.

The schematic symbol for a capacitor has two significant variants: with two straight lines (symbolizing the plates inside a capacitor), or with one straight line and one curved line, as shown in Figure 2-69. When you see a curved line, that side of the capacitor should be more negative than the other. The schematic symbol may also include a + sign. Unfortunately, some people don't bother to draw a curved plate on a polarized capacitor, yet others draw a curved plate even on a nonpolarized capacitor.

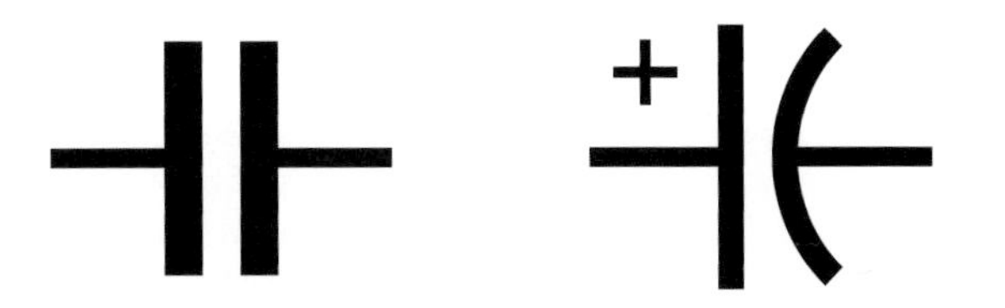

Figure 2-69. *The generic schematic for a capacitor is on the left. The version on the right indicates a polarized capacitor which requires its left plate to be "more positive" than its right plate. The plus sign is often omitted.*

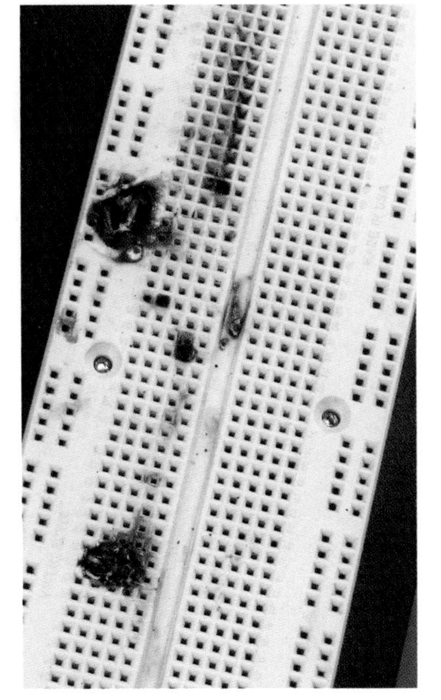

Figure 2-70. *A tantalum capacitor was plugged into this breadboard, accidentally connected the wrong way around to a power source capable of delivering a lot of current. After a minute or so of this abuse, the capacitor rebelled by popping open and scattering small flaming pieces, which burned their way into the plastic of the breadboard. Lesson learned: observe polarity!*

Capacitor Polarity

You must connect an electrolytic capacitor so that its longer wire is more positive than its shorter wire. The shell of the capacitor is usually marked with a negative sign near the shorter wire.

Some capacitors may behave badly if you don't observe their polarity. One time I connected a tantalum capacitor to a circuit, using a power supply able to deliver a lot of current, and was staring at the circuit and wondering why it wasn't working when the capacitor burst open and scattered little flaming fragments of itself in a 3-inch radius. I had forgotten that tantalum capacitors can be fussy about positive and negative connections. Figure 2-70 shows the aftermath.

BACKGROUND

Michael Faraday and capacitors

The earliest capacitors consisted of two metal plates with a very small gap between them. The principle of the thing was simple:

- If one plate was connected to a positive source, the positive charges attracted negative charges onto the other plate.
- If one plate was connected to a negative source, the negative charges attracted positive charges onto the other plate.

Figures 2-67 and 2-68, shown previously, convey the basic idea.

The electrical storage capacity of a capacitor is known as its *capacitance*, and is measured in farads, named after Michael Faraday (Figure 2-71), another of the pantheon of electrical pioneers. He was an English chemist and physicist who lived from 1791 to 1867.

Although Faraday was relatively uneducated and had little knowledge of mathematics, he had an opportunity to read a wide variety of books while working for seven years as a bookbinder's apprentice, and thus was able to educate himself. Also, he lived at a time when relatively simple experiments could reveal fundamental properties of electricity. Thus he made major discoveries including electromagnetic induction, which led to the development of electric motors. He also discovered that magnetism could affect rays of light.

His work earned him numerous honors, and his picture was printed on English 20-pound bank notes from 1991 through 2001.

Figure 2-71. *Michael Faraday*

Breadboarding the Circuit

I promised to free you in time from the frustrations of alligator clips, and that time has come. Please turn your attention to the block of plastic with lots of little holes in it that I asked you to buy. For reasons that I do not know, this is called a *breadboard*. When you plug components into the holes, hidden metal strips inside the breadboard connect the components for you, allowing you to set up a circuit, test it, and modify it very easily. Afterward you can pull the components off the breadboard and put them away for future experiments.

Without a doubt, breadboarding is the most convenient way to test something before you decide whether you want to keep it.

Almost all breadboards are designed to be compatible with integrated circuit chips (which we will be using in Chapter 4 of this book). The chip straddles an empty channel in the center of the breadboard with rows of little holes either side—usually five holes per row. You insert other components into these holes.

In addition, the breadboard should have columns of holes running down each side. These are used to distribute positive and negative power.

Take a look at Figures 2-72 and 2-73, which show the upper part of a typical breadboard seen from above, and the same breadboard seen as if with X-ray vision, showing the metal strips that are embedded behind the holes.

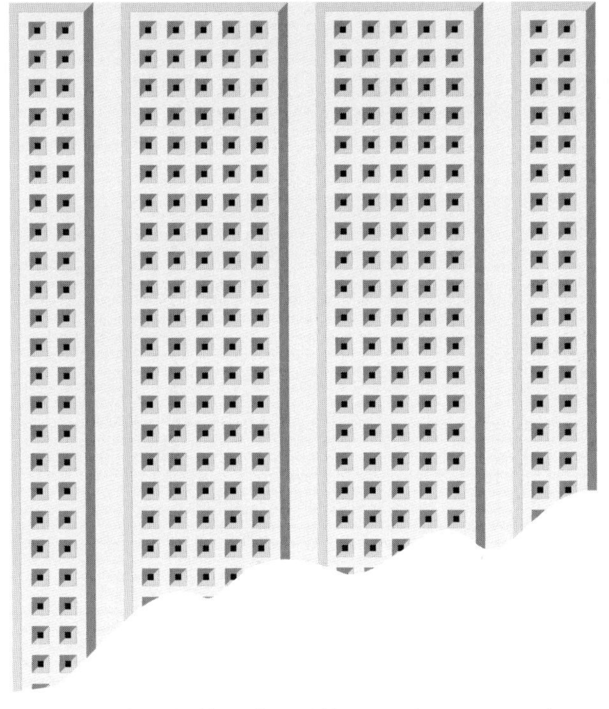

Figure 2-72. *A typical breadboard. You can plug components into the holes to test a circuit very quickly.*

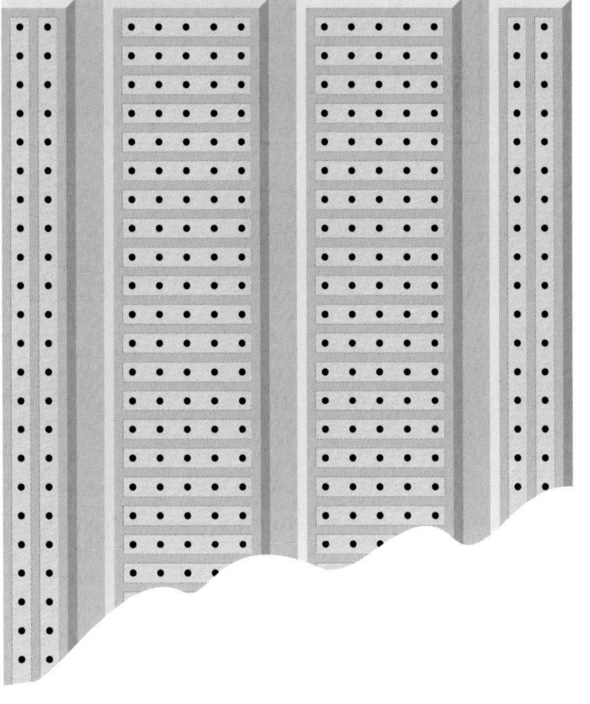

Figure 2-73. *This X-ray-vision view of the breadboard reveals the copper strips that are embedded in it. The strips conduct electricity from one component to another.*

Important note: some breadboards divide each vertical column of holes, on the left and the right, into two separate upper and lower sections. Use your meter's continuity testing feature to find out if your breadboard conducts power along its full length, and add jumper wires to link the upper and lower half of the breadboard if necessary.

Figure 2-74 shows how you can use the breadboard to replicate your oscillating relay circuit. To make this work, you need to apply the positive and negative power from your AC adapter. Because the wire from your AC adapter is almost certainly stranded, you'll have difficulty pushing it into the little holes. A way around this is to set up a couple of pieces of bare 22-gauge wire, and use them as terminals to which you clip the wire from the adapter, as in Figure 2-75. (Yes, you still need just a couple of alligator clips for this purpose.) Alternatively, you can use a breadboard with power terminals built into it, which is more convenient.

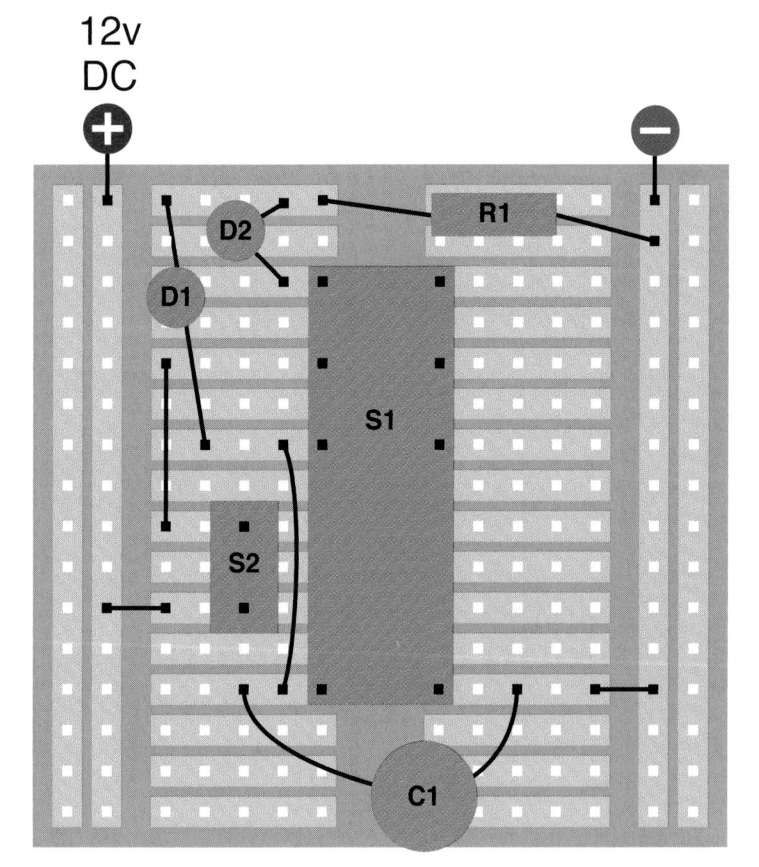

Figure 2-74. *If you place the components on your breadboard in the positions shown, they will create the same circuit that you built from wire and alligator clips in Experiment 8. Component values:*

D1, D2: Light-emitting diodes
S1: DPDT relay
S2: SPST momentary switch
C1: Electrolytic capacitor, 1,000 μF
R1: Resistor, 680Ω minimum

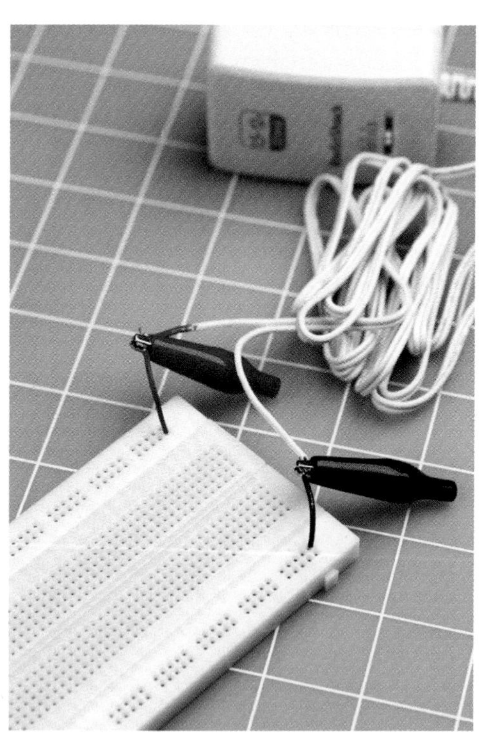

Figure 2-75. *If your breadboard doesn't have screw terminals, insert two short pieces of solid-core wire with stripped ends and then attach the stranded wires from the adapter using alligator clips.*

You'll need some more 22-gauge wire, or some precut hookup wire, to supply the power to your components, which are plugged into the breadboard as shown in Figures 2-76 and 2-77. If you get all the connections right, the circuit should function the same way as before.

The geometry of the metal connecting strips in the breadboard often forces you to connect components in a roundabout way. The pushbutton, for instance, supplies power to the pole of the relay but cannot be connected directly opposite, because there isn't room for it.

Remember that the strips inside the breadboard that don't have any wires or components plugged into them are irrelevant; they don't do anything.

I'll include some suggested breadboard layouts for circuits as you continue through this book, but eventually you'll have to start figuring out breadboard layouts for yourself, as this is an essential part of hobby electronics.

Larger versions of all schematics and breadboard photos are available online at this book's website: http://oreilly.com/catalog/9780596153748.

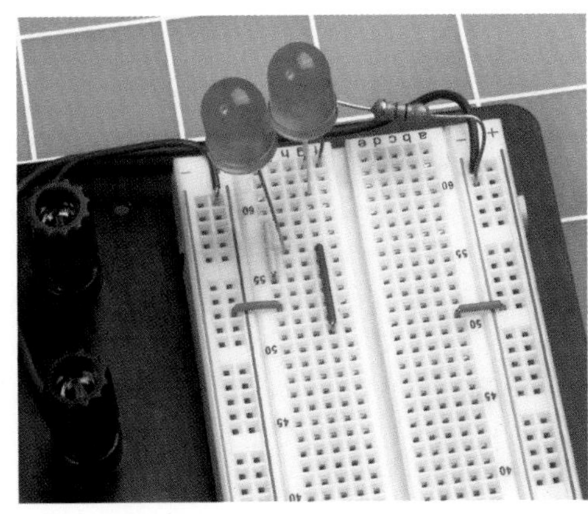

Figure 2-76. *Two oversized LEDs, one resistor, and the necessary jumper wires have been added to the breadboard.*

Figure 2-77. *Now the pushbutton, relay, and capacitor have been added to complete the circuit shown in the diagram and the schematic. When the pushbutton is pressed, the relay oscillates and the LEDs flash.*

Experiment 9: Time and Capacitors

You will need:

- AC adapter, breadboard, wire, wire cutters, and strippers.
- Multimeter.
- Pushbutton, SPST. Quantity: 1.
- Resistors and electrolytic capacitors, assorted.

In Experiment 8, when you put a capacitor in parallel with the coil of the relay, the capacitor charged almost instantly before discharging itself through the relay coil. If you add a resistor in series with a capacitor, the capacitor will take longer to charge. By making a capacitor take longer to charge, you can measure time, which is a very important concept.

Clean the components off your breadboard and use it to set up the very simple circuit shown in Figure 2-78, where C1 is a 1,000 μF capacitor, R1 is a 100K resistor, R2 is a 100Ω resistor, and S1 is the pushbutton that you used previously. Set your meter to measure volts DC, place the probes around the capacitor, and hold down the pushbutton. You should see the meter counting upward as the voltage accumulates on the capacitor. (This is easier with a meter that doesn't have autoranging, because you won't have to wait while the meter figures out which range to apply.) Resistor R1 slows the charging time for the capacitor.

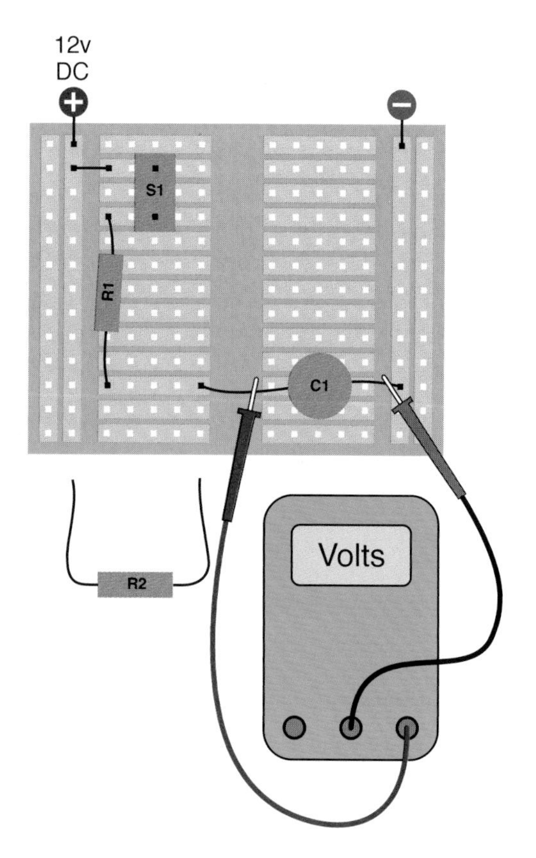

Figure 2-78. *Watch the voltage building up on the capacitor while you hold down the pushbutton. Substitute other values for R1, discharge the capacitor by touching R2 across it, and then repeat your measurement process.*

S1: Momentary pushbutton, OFF (ON)
R1: 100K initially
R2: 100Ω
C1: 1,000 μF

Release the pushbutton, set aside your meter, and discharge the capacitor by touching R2 across it for a second or two. Now substitute a 50K resistor for R1, and repeat the measurement. The meter should count upward almost twice as fast as before.

Voltage, Resistance, and Capacitance

Think of the resistor as a faucet, and the capacitor as a balloon that you are trying to fill with water. When you screw down the faucet until only a trickle comes through, the balloon will take longer to fill. But a slow flow of water will still fill the balloon completely if you wait long enough, and (assuming the balloon doesn't burst) the process ends when the pressure inside the balloon is equal to the water pressure in the pipe supplying the faucet. See Figure 2-79.

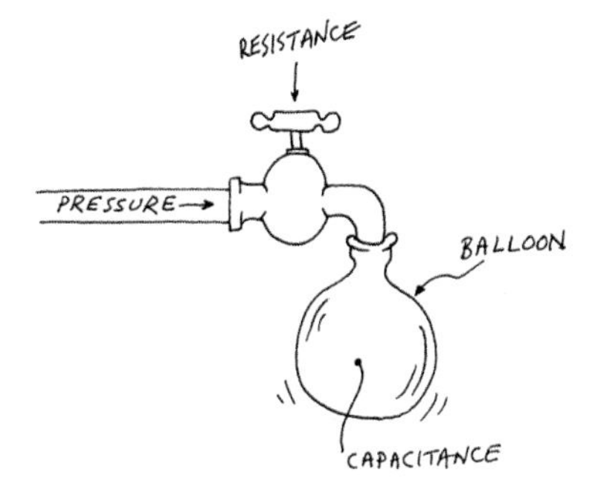

Figure 2-79. *When the faucet is closed half-way, the balloon will take longer to fill, but will still contain as much water and as much pressure in the end.*

Similarly, in your circuit, if you wait long enough, eventually the voltage across the capacitor should reach the same value as the voltage of the power supply. In a 12-volt circuit, the capacitor should eventually acquire 12 volts (although "eventually" may take longer than you think).

This may seem confusing, because earlier you learned that when you apply voltage at one end of a resistor, you get less voltage coming out than you have going in. Why should a resistor deliver the full voltage when it is paired with a capacitor?

Forget the capacitor for a moment, and remember how you tested just two 1K resistors. In that situation, each resistor contained half the total resistance of the circuit, so each resistor dropped half the voltage. If you held the negative probe of your meter against the negative side of your power supply and touched the positive probe to the center point between the two resistors, you would measure 6 volts. Figure 2-80 illustrates this.

Now, suppose you remove one of the 1K resistors and substitute a 9K resistor. The total resistance in the circuit is now 10K, and therefore the 9K resistor drops 90% of the 12 volts. That's 10.8 volts. You should try this and check it with your meter. (You are unlikely to find a 9K resistor, because this is not a standard value. Substitute the nearest value you can find.)

Now suppose you remove the 9K resistor and substitute a 99K resistor. Its voltage drop will be 99% of the available voltage, or 11.88 volts. You can see where this is heading: the larger the resistor, the larger its contribution to voltage drop.

However, I noted previously that a capacitor blocks DC voltage completely. It can accumulate an electrical charge, but no current passes through it. Therefore, a capacitor behaves like a resistor that has infinite resistance to DC current.

(Actually the insulation inside the capacitor allows a little bit of "leakage," but a perfect capacitor would have infinite resistance.)

The value of any resistor that you put in series with the capacitor is trivial by comparison. No matter how high the value of the resistor is, the capacitor still provides much more resistance in the circuit. This means that the capacitor steals almost the complete voltage drop in the circuit, and the voltage difference between one end of the resistor and the other will be zero (assuming that we ignore little imperfections in the components). Figure 2-80 may help to clarify this concept.

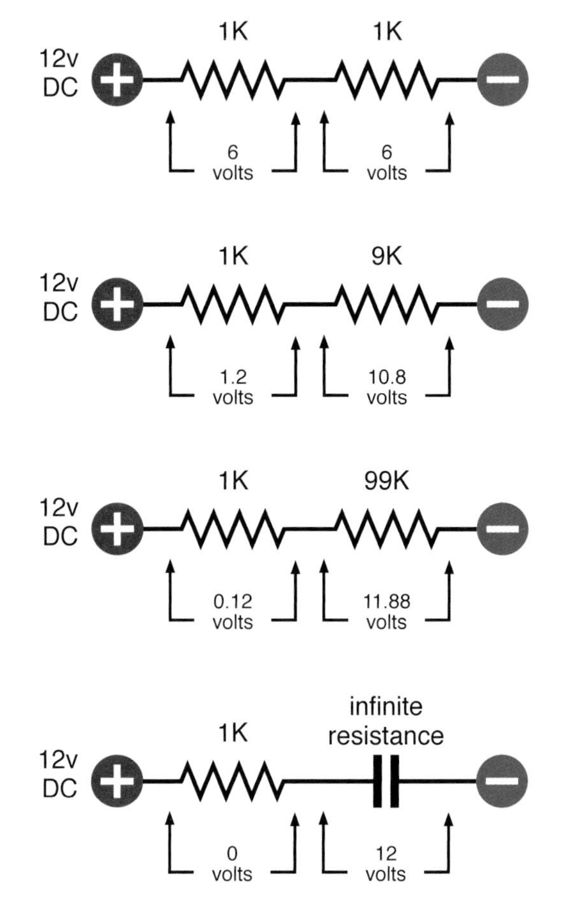

Figure 2-80. *When two resistances are in series, the larger one drops the voltage more than the smaller one. If the larger resistance becomes infinite (as in the case of a capacitor), the smaller one no longer has any measurable contribution to the voltage drop, and the voltage is almost exactly the same at both ends.*

You should try this using real resistors and capacitors—although if you do, you will run into a little problem. When you use your meter in its "DC volts" mode, it diverts a little of the current in the circuit—just a tiny taste—in the process of measuring it. The meter steals such a small amount, it doesn't affect the reading significantly when you are checking voltage across a resistor. The internal resistance of the meter is higher than the values of most resistors. However, remember that the internal resistance of a capacitor is almost infinite. Now the internal resistance of the meter becomes significant. Because you can never have an ideal meter, any more than you can have an ideal capacitor or resistor, your meter will always interfere with the circuit slightly, and you will get only an approximate indication.

If you try to measure the voltage on a capacitor that has been charged but is now not connected to anything else, you'll see the number slowly falling, as the capacitor discharges itself through the meter.

THEORY

The time constant

You may be wondering if there's a way to predict exactly how much time it takes for various capacitors to charge, when they are paired with various resistors. Is there a formula to calculate this?

Of course, the answer is yes, but the way we measure it is a bit tricky, because a capacitor doesn't charge at a constant rate. It accumulates the first volt very quickly, the second volt not quite as quickly, the third volt even less quickly—and so on. You can imagine the electrons accumulating on the plate of a capacitor like people walking into an auditorium and looking for a place to sit. The fewer seats that are left, the longer people take to find them.

The way we describe this is with something called a "time constant." The definition is very simple:

$$TC = R \times C$$

where TC is the time constant, in seconds, and a capacitor of C farads is being charged through a resistor of R ohms.

Going back to the circuit you just tested, try using it again, this time with a 1K resistor and the 1,000 μF capacitor. We have to change those numbers to farads and ohms before we can put them in the formula. Well, 1,000 μF is 0.001 farads, and 1K is 1,000 ohms, so the formula looks like this:

$$TC = 1{,}000 \times 0.001$$

In other words, TC = 1—a lesson that could not be much easier to remember:

> ***A 1K resistor in series with a 1,000 μF capacitor has a time constant of 1.***

Does this mean that the capacitor will be fully charged in 1 second? No, it's not that simple. TC, the time constant, is the time it takes for a capacitor to acquire *63%* of the voltage being supplied to it, if it starts with zero volts.

(Why 63%? The answer to that question is too complicated for this book, and you'll have to read about time constants elsewhere if you want to know more. Be prepared for differential equations.) Here's a formal definition for future reference:

> TC, the time constant, is the time it takes for a capacitor to acquire 63% of the difference between its current charge and the voltage being applied to it. When TC=1, the capacitor acquires 63% of its full charge in 1 second. When TC=2, the capacitor acquires 63% of its full charge in 2 seconds. And so on.

What happens if you continue to apply the voltage? History repeats itself. The capacitor accumulates *another* 63% of the *remaining* difference between its current charge, and the voltage being applied to it.

Imagine someone eating a cake. In his first bite he's ravenously hungry, and eats 63% of the cake in one second. In his second bite, not wanting to seem too greedy, he takes just another 63% of the cake that is left—and because he's not feeling so hungry anymore, he requires the same time to eat it as he took to eat the first bite. In his third bite, he takes 63% of what still remains, and still takes the same amount of time. And so on. He is behaving like a capacitor eating electricity (Figure 2-81).

Figure 2-81. *If our gourmet always eats just 63% of the cake still on the plate, he "charges up" his stomach in the same way that a capacitor charges itself. No matter how long he keeps at it, his stomach is never completely filled.*

THEORY

The time constant (continued)

The cake eater will always have a few crumbs to eat, because he never takes 100% of the remainder. Likewise, the capacitor will never acquire a full charge. In a perfect world of perfect components, this process would continue for an infinite time.

In the real world, we say rather arbitrarily:

> After 5 × TC the capacitor will be so nearly fully charged, we won't care about the difference.

In the table is a calculation (rounded to two decimal places) showing the charge accumulating on a capacitor in a 12-volt circuit where the time constant is 1 second.

Here's how to understand the table. V1 is the current charge on the capacitor. Subtract this from the supply voltage (12 volts) to find the difference. Call the result V2. Now take 63% of V2, and add this to the current charge (V1) and call the result V4. This is the new charge that the capacitor will have after 1 second, so we copy it down to the next line in the table, and it becomes the new value for V1.

Now we repeat the same process all over again. Figure 2-82 shows this in graphical form. Note that after 5 seconds, the capacitor has acquired 11.92 volts, which is 99% of the power supply voltage. This should be close enough to satisfy anyone's real-world requirements.

If you try to verify these numbers by measuring the voltage across the capacitor as it charges, remember that because your meter steals a little current, there will be a small discrepancy that will increase as time passes. For practical purposes, the system works well enough.

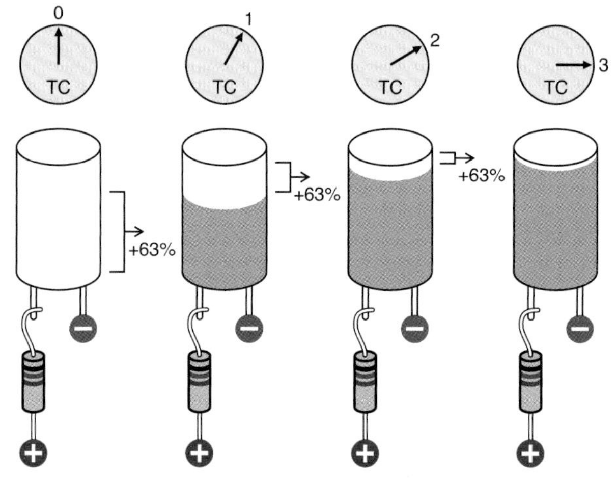

Figure 2-82. *A capacitor starts with 0 volts. After 1 time constant it adds 63% of the available voltage. After another time constant, it adds another 63% of the remaining voltage difference, and so on.*

Time in secs	V1: Charge on capacitor	V2: 12 – V1	V3: 63% of V2	V4: V1 + V3
0	0.00	12.00	7.56	7.56
1	7.56	4.44	2.80	10.36
2	10.36	1.64	1.03	11.39
3	11.39	0.61	0.38	11.77
4	11.77	0.23	0.15	11.92
5	11.92			

Experiment 10: Transistor Switching

You will need:

- AC adapter, breadboard, wire, and meter.
- LED. Quantity: 1.
- Resistors, various.
- Pushbutton, SPST. Quantity: 1.
- Transistor, 2N2222 or similar. Quantity: 1.
- Potentiometer, 1 megohm, linear.

A transistor can switch a flow of electricity, just like a relay. But it's much more sensitive and versatile, as this first ultra-simple experiment will show.

We'll start with the 2N2222 transistor, which is the most widely used semiconductor of all time (it was introduced by Motorola in 1962 and has been in production ever since).

First, you should get acquainted with the transistor. Because Motorola's patents on the 2N2222 ran out long ago, any company can manufacture their own version of it. Some versions are packaged in a little piece of black plastic; others are enclosed in a little metal "can." (See Figure 2-83.) Either way, it contains a piece of silicon divided into three sections known as the collector, the base, and the emitter. I'll describe their function in more detail in a moment, but initially you just need to know that in this type of transistor, the collector receives current, the base controls it, and the emitter sends it out.

Use your breadboard to set up the circuit shown in Figure 2-85. Be careful to get the transistor the right way around! (See Figure 2-84.) For the three brands I have mentioned in the shopping list, the flat side should face right, if the transistor is packaged in black plastic, or the little tab should face toward the lower left, if the transistor is packaged in metal.

Figure 2-83. *A typical transistor is packaged either in a little metal can or a molded piece of black plastic. The manufacturer's data sheet tells you the identities of the three wire leads, relative to the flat side of a black plastic transistor or the tab that sticks out of a metal-can transistor.*

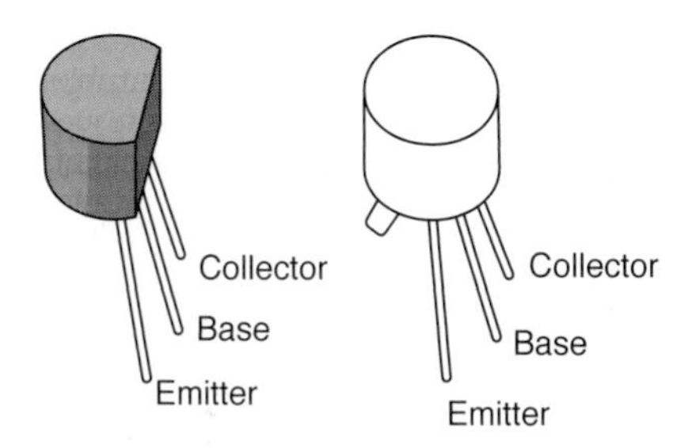

Figure 2-84. *The 2N2222 transistor may be packaged in either of these formats. Left: RadioShack or Fairchild. Right: STMicro (note the tab sticking out at lower-left). If you use a different brand, check the manufacturer's data sheet. Looking from above, insert the transistor in your breadboard with the flat side facing right, or the tab pointing down and to the left.*

Beware of confusion using part numbers P2N2222 and P2N2222A. Some were made by On Semiconductor and may be supplied by Radio Shack. Their pins are reversed, so that the part should be used with its flat side on the left, not on the right. The diagram is correct for parts such as 2N2222, PN2222, or PN2222A, but not for parts beginning with P2N.

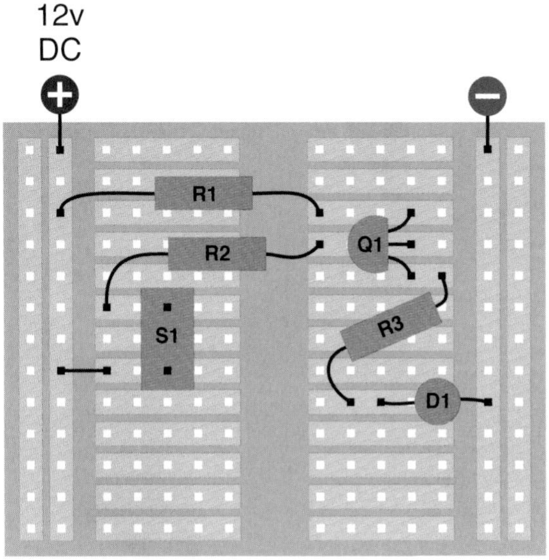

12v DC

Figure 2-85. *The transistor blocks voltage that reaches it through R1. But when pushbutton S1 is pressed, this tells the transistor to allow current to pass through it. Note that transistors are always identified with letter Q in wiring diagrams and schematics.*

S1: Pushbutton, momentary, OFF (ON)
R1: 180Ω
R2: 10K
R3: 680Ω
Q1: 2N2222 or similar
D1: LED

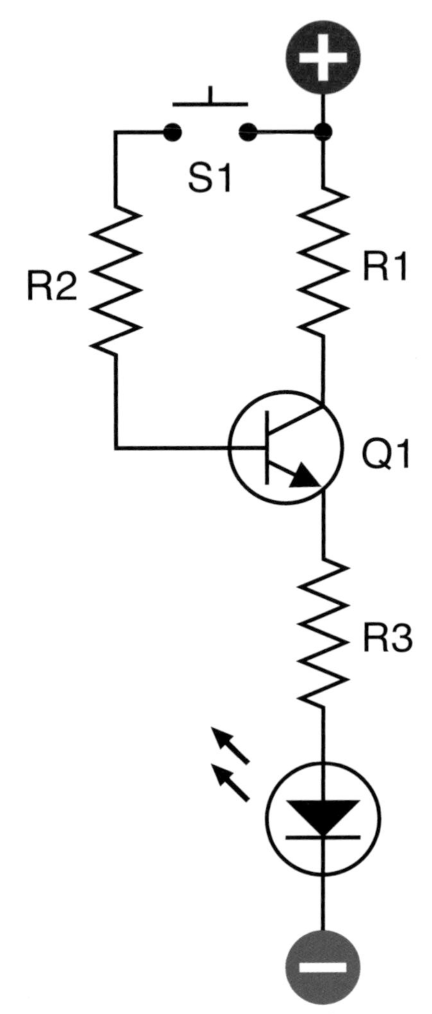

Figure 2-86. *This shows the same circuit as the breadboard diagram in Figure 2-85.*

Initially, the LED should be dark. Now press the pushbutton and the LED should glow brightly. Electricity is following two paths here. Look at the schematic in Figure 2-86, which shows the same circuit more clearly. I've shown positive at the top and negative at the bottom (the way most schematics do it) because it helps to clarify the function of this particular circuit. If you view the schematic from the side, the similarity with the breadboard layout is easier to see.

Through R1, voltage reaches the top pin (the collector) of the transistor. The transistor only lets a tiny trickle of it pass through, so the LED stays dark. When you press the button, voltage is also applied along a separate path, through R2 to the middle pin (the base) of the transistor. This tells the transistor to close its solid-state switch and allow current to flow out through its third pin (the emitter), and through R3, to the LED.

You can use your meter in volts DC mode to check the voltage at points in the circuit. Keep the negative probe from the meter touching the negative voltage source while you touch the positive probe to the top pin of the transistor, the middle pin, and the bottom pin. When you press the button, you should see the voltage change.

Fingertip Switching

Now here's something more remarkable. Remove R2 and the pushbutton, and insert two short pieces of of wire as shown in Figure 2-87. The upper piece of wire connects with the positive voltage supply; the lower piece connects with the middle pin of the transistor (its base). Now touch the tip of your finger to the two wires. Once again, the LED should glow, although not as brightly as before. Lick the tip of your finger, try again, and the LED should glow more brightly.

Never Use Two Hands

The fingertip switching demo is safe if the electricity passes just through your finger. You won't even feel it, because it's 12 volts DC from a power supply of 1 amp or less. But it's not a good idea to put the finger of one hand on one wire, and the finger of your other hand on the other wire. This would allow the electricity to pass through your body. Although the chance of hurting yourself this way is extremely small, you should never allow electricity to run through you from one hand to the other. Also, when touching the wires, don't allow them to penetrate your skin.

Your finger is conducting positive voltage to the base of the transistor. Even though your skin has a high resistance, the transistor still responds. It isn't just switching the LED on and off; it is *amplifying the current* applied to its base. This is an essential concept: *a transistor amplifies any changes in current that you apply to its base.*

Check Figure 2-88 to see more clearly what's happening.

If you studied the section "Background: Positive and negative" in Chapter 1, you learned that there is really no such thing as positive voltage. All we really have is negative voltage (created by the pressure of free electrons) and an absence of negative voltage (where there are fewer free electrons). But because the idea of a flow of electricity from positive to negative was so widely believed before the electron was discovered, and because the inner workings of a transistor involve "holes" which are an absence of electrons and can be thought of as positive, we can still pretend that electricity flows from positive to negative. See the following section, "Essentials: All about NPN and PNP transistors," for more details.

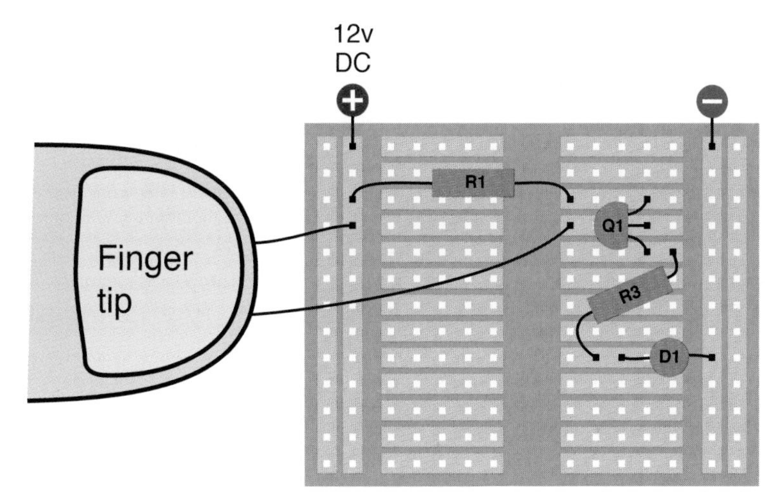

Figure 2-87

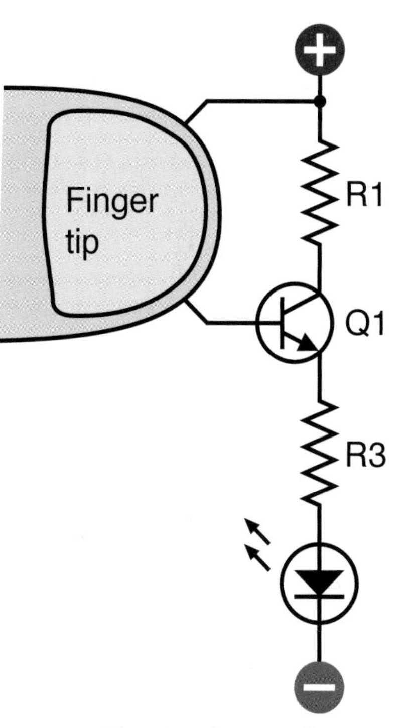

Figure 2-88. *These two diagrams show the same components as before, with a fingertip substituted for R2. Although only a trickle of voltage now reaches the base of the transistor, it's enough to make the transistor respond.*

ESSENTIALS

All about NPN and PNP transistors

A transistor is a semiconductor, meaning that sometimes it conducts electricity, and sometimes it doesn't. Its internal resistance varies, depending on the power that you apply to its base.

NPN and PNP transistors are bipolar semiconductors. They contain two slightly different variants of silicon, and conduct using both polarities of carriers—holes and electrons.

The NPN type is a sandwich with P-type silicon in the middle, and the PNP type is a sandwich with N-type silicon in the middle. If you want to know more about this terminology, and the behavior of electrons when they try to cross an NP junction or a PN junction, you'll have to read a separate source on this subject. It's too technical for this book. All you need to remember is:

- All bipolar transistors have three connections: Collector, Base, and Emitter, abbreviated as C, B, and E on the manufacturer's data sheet, which will identify the pins for you.
- NPN transistors are activated by *positive* voltage on the base relative to the emitter.
- PNP transistors are activated by *negative* voltage on the base relative to the emitter.

In their passive state, both types block the flow of electricity between the collector and emitter, just like an SPST relay in which the contacts are normally open. (Actually a transistor allows a tiny bit of current known as "leakage.")

You can think of a bipolar transistor as if it contains a little button inside, as shown in Figures 2-89 and 2-90. When the button is pressed, it allows a large current to flow. To press the button, you inject a much smaller current into the base by applying a small voltage to the base. In an NPN transistor, the control voltage is positive. In a PNP transistor, the control voltage is negative.

NPN transistor basics

- To start the flow of current from collector to emitter, apply a relatively positive voltage to the base.
- In the schematic symbol, the arrow points from base to emitter and shows the direction of positive current.
- The base must be at least 0.6 volts "more positive" than the emitter, to start the flow.
- The collector must be "more positive" than the emitter.

PNP transistor basics

- To start the flow of current from emitter to collector, apply a relatively negative voltage to the base.
- In the schematic symbol, the arrow points from emitter to base and shows the direction of positive current.
- The base must be at least 0.6 volts "more negative" than the emitter, to start the flow.
- The emitter must be "more positive" than the collector.

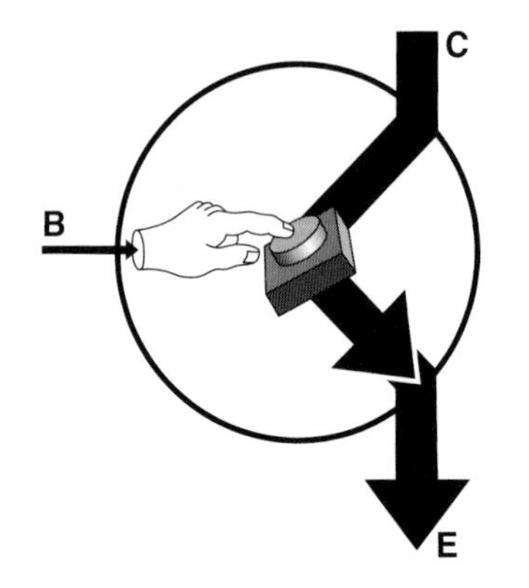

Figure 2-89. *You can think of a bipolar transistor as if it contains a button that can connect the collector and the emitter. In an NPN transistor, a small positive potential presses the button.*

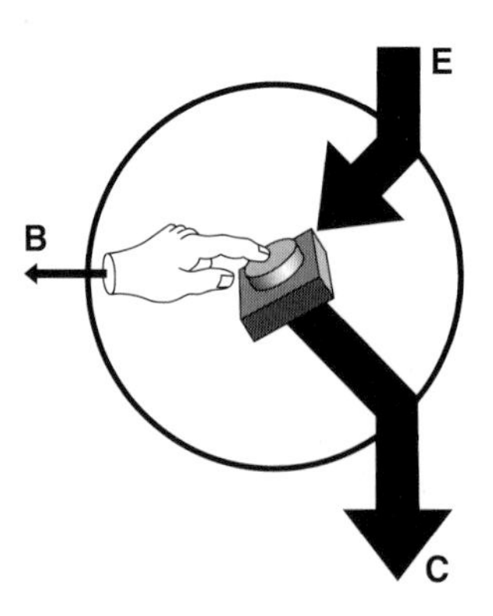

Figure 2-90. *In a PNP transistor, a small negative potential has the same effect. The arrows point in the direction of "positive current flow."*

ESSENTIALS

All about NPN and PNP transistors (continued)

All-transistor basics

- Never apply a power supply directly across a transistor. You can burn it out with too much current.
- Protect a transistor with a resistor, in the same way you would protect an LED.
- Avoid reversing the connection of a transistor between positive and negative voltages.
- Sometimes an NPN transistor is more convenient in a circuit; sometimes a PNP happens to fit more easily. They both function as switches and amplifiers, the only difference being that you apply a relatively positive voltage to the base of an NPN transistor, and a relatively negative voltage to the base of a PNP transistor.
- PNP transistors are used relatively seldom, mainly because they were more difficult to manufacture in the early days of semiconductors. People got into the habit of designing circuits around NPN transistors.
- Remember that bipolar transistors amplify current, not voltage. A small fluctuation of current through the base enables a large change in current between emitter and collector.
- Schematics sometimes show transistors with circles around them, and sometimes don't. In this book, I'll use circles to draw attention to them. See Figures 2-91 and 2-92.
- Schematics may show the emitter at the top and the collector at the bottom, or vice versa. The base may be on the left, or on the right, depending on what was most convenient for the person drawing the schematic. Be careful to look carefully at the arrow in the transistor to see which way up it is, and whether it is NPN or PNP. You can damage a transistor by connecting it incorrectly.
- Transistors come in various different sizes and configurations. In many of them, there is no way to tell which wires connect to the emitter, the collector, or the base, and some transistors have no part numbers on them. Before you throw away the packaging that came with a transistor, check to see whether it identifies the terminals.
- If you forget which wire is which, some multimeters have a function that will identify emitter, collector, and base for you. Check your multimeter instruction booklet for more details.

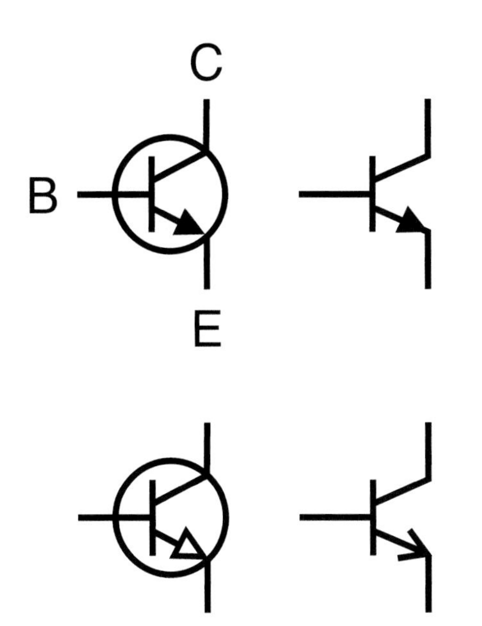

Figure 2-91. *The symbol for an NPN transistor always has an arrow pointing from its base to its emitter. Some people include a circle around the transistor; others don't bother. The style of the arrow may vary. But the meaning is always the same. The top-left version is the one I use in this book.*

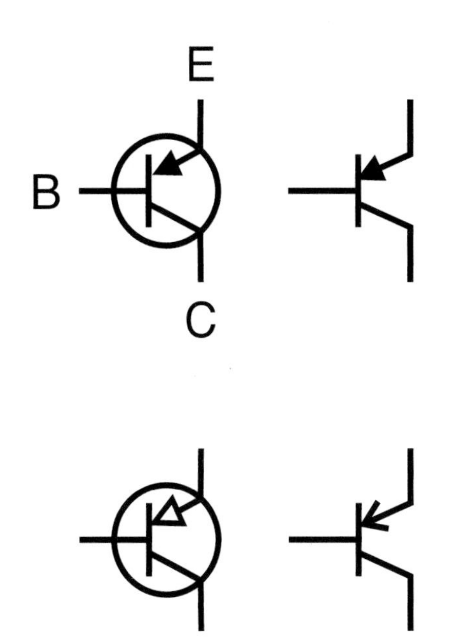

Figure 2-92. *The symbol for a PNP transistor always has an arrow pointing from its emitter to its base. Some people include a circle around the transistor; others don't bother. The style of the arrow may vary. But the meaning is always the same. The top-left version is the one I use in this book.*

BACKGROUND

Transistor origins

Though some historians trace the origins of the transistor back to the invention of diodes (which allow electricity to flow in one direction while preventing reversal of the flow), there's no dispute that the first working transistor was developed at Bell Laboratories in 1948 by John Bardeen, William Shockley, and Walter Brattain (Figure 2-93).

Shockley was the leader of the team, who had the foresight to see how potentially important a solid-state switch could be. Bardeen was the theorist, and Brattain actually made it work. This was a hugely productive collaboration—until it succeeded. At that point, Shockley started maneuvering to have the transistor patented exclusively under his own name. When he notified his collaborators, they were—naturally—unhappy about this idea.

A widely circulated publicity photograph didn't help, in that it showed Shockley sitting at the center in front of a microscope, as if he had done the hands-on work, while the other two stood behind him, implying that they had played a lesser role. In fact Shockley, as the supervisor, was seldom present in the laboratory where the real work was done.

The productive collaboration quickly disintegrated. Brattain asked to be transferred to a different lab at AT&T. Bardeen moved to the University of Illinois to pursue theoretical physics. Shockley eventually left Bell Labs and founded Shockley Semiconductor in what was later to become Silicon Valley, but his ambitions outstripped the capabilities of the technology of his time. His company never manufactured a profitable product.

Eight of Shockley's coworkers in his company eventually betrayed him by quitting and establishing their own business, Fairchild Semiconductor, which became hugely successful as a manufacturer of transistors and, later, integrated circuit chips.

Figure 2-93. *Photographs provided by the Nobel Foundation show, left to right, John Bardeen, William Shockley, and Walter Brattain. For their collaboration in development of the world's first working transistor in 1948, they shared a Nobel prize in 1956.*

ESSENTIALS

Transistors and relays

One limitation of NPN and PNP transistors is that they are naturally "off" until you turn them "on." They behave like a normally open pushbutton, which conducts electricity only for as long as you hold it down. They don't normally behave like a normal on switch, which stays on until you apply a signal to turn it off.

A relay offers more switching options. It can be normally open, normally closed, or it can contain a double-throw switch, which gives you a choice of two "on" positions. It can also contain a double-pole switch, which makes (or breaks) two entirely separate connections when you energize it. Single-transistor devices cannot provide the double-throw or double-pole features, although you can design more complex circuits that emulate this behavior.

Here's a list of transistor and relay characteristics.

	Transistor	Relay
Long-term reliability	Excellent	Limited
Configurable for DP and DT switching	No	Yes
Ability to switch large currents	Limited	Good
Able to switch alternating current (AC)	Usually not	Yes
Can be triggered by alternating current (AC)	Usually not	Optional
Suitability for miniaturization	Excellent	Very limited
Sensitive to heat	High	Moderate
Ability to switch at high speed	Excellent	Limited
Price advantage for low-voltage low-current	Yes	No
Price advantage for high-voltage high-current	No	Yes
Current leakage when "off"	Yes	No

The choice between relays or transistors will depend on each particular application.

THEORY

See the current

If you want to get a more precise understanding of how a transistor works, you should try this little test. It shows the precise behavior and limits of the 2N2222 transistor that you used in the previous experiment.

I've said that in an NPN transistor, the collector should always be more positive than the emitter and that the base should have a potential somewhere between those two voltages. Figure 2-94 shows this rather vague relationship. Now I want to substitute some numbers for these general statements.

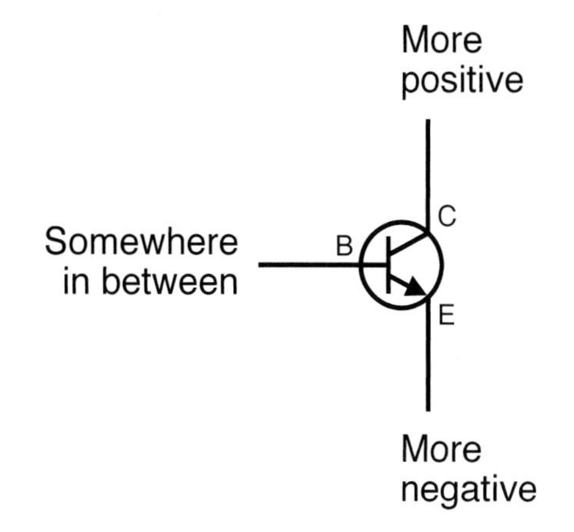

Figure 2-94. *The proper functioning of an NPN transistor requires you to maintain these voltage relationships.*

Take a look at the schematic in Figure 2-95, and check the component values. Notice that the total resistance above the transistor, from R1 + R2, is the same as the total resistance below it, from R3 + R4. Therefore the potential on the base of the transistor should be halfway between the two extremes—until you use potentiometer P1 to adjust the voltage of the base of the transistor up and down.

The two 180Ω resistors, R1 and R3, protect the transistor from passing excessive current. The two 10K resistors, R2 and R4, protect the base when the potentiometer is turned all the way up or all the way down.

I would like you to see what the transistor is doing by measuring the amperage flowing into the base at the position marked A1, and the total amperage flowing out through the emitter at the position marked A2. To do this, it would be really helpful if you had two meters. As that may be impractical, the breadboard diagrams in Figures 2-96 and 2-97 show how you can swap one meter between the two locations.

Remember that to measure milliamps, you have to pass electricity through the meter. This means that the meter must be inserted into the circuit, and whenever you remove the meter, you have to remake the connection where the meter was. The breadboard diagram shows how you can do this. Fortunately, it's very easy to remove and replace wires in a breadboard. Where wires are connected to the potentiometer, you may need to revert to using alligator clips.

Begin with the potentiometer turned about halfway through its range. Measure at A1 and A2. Turn the potentiometer up a bit, and measure current at the two locations again. Following is a table showing some actual readings I obtained at those two locations, using two digital meters simultaneously.

Milliamps passing through location A1	Milliamps passing through location A2
0.01	1.9
0.02	4.9
0.03	7.1
0.04	9.9
0.05	12.9
0.06	15.5
0.07	17.9
0.08	19.8
0.09	22.1
0.10	24.9
0.11	26.0
0.12	28.3

There's a very obvious relationship. The current emerging from the emitter of the transistor, through location A2, is about 240 times the current passing through location A1, into the base. The ratio of current coming out from the emitter of an NPN transistor to current going into the base is known as the *beta value* for a transistor. The beta value expresses the transistor's amplifying power.

It's a very constant ratio, until you push it a little too far. Above 0.12 mA, this particular transistor becomes "saturated," meaning that its internal resistance cannot go any lower.

THEORY

See the current (continued)

In my little experiment, I found that the maximum current at A2 was 33mA. A simple calculation using Ohm's Law showed me that this meant the transistor's internal resistance was near zero. This is why you should protect a transistor with some additional resistance in the circuit. If you don't, its low internal resistance would allow a huge current flow that would immediately burn it out.

What about the other end of its range? When it passes only 1.9 mA, the transistor has an internal resistance of around 6,000Ω. The conclusion is that depending how much current you apply to this transistor, its internal resistance varies between zero and 6,000Ω, approximately.

So much for the theory. Now what can we do with a transistor that's fun, or useful, or both? We can do Experiment 11!

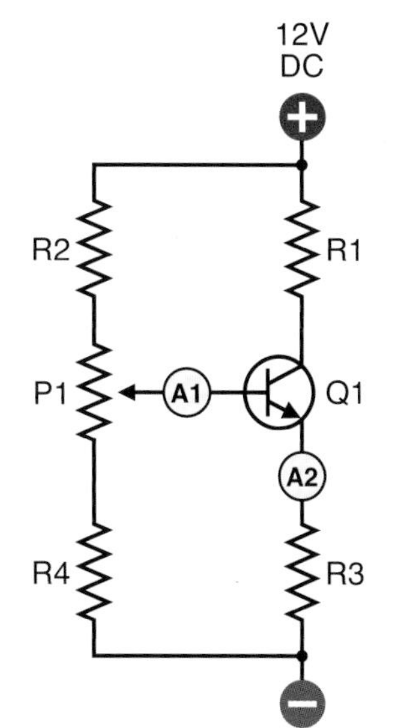

Figure 2-95. *This is basically the same as the previous circuit, with a potentiometer added and the LED removed. Component values:*

R1: 180Ω
R2: 10K
R3: 180Ω
R4: 10K
P1: 1M linear potentiometer
Q1: 2N2222 transistor

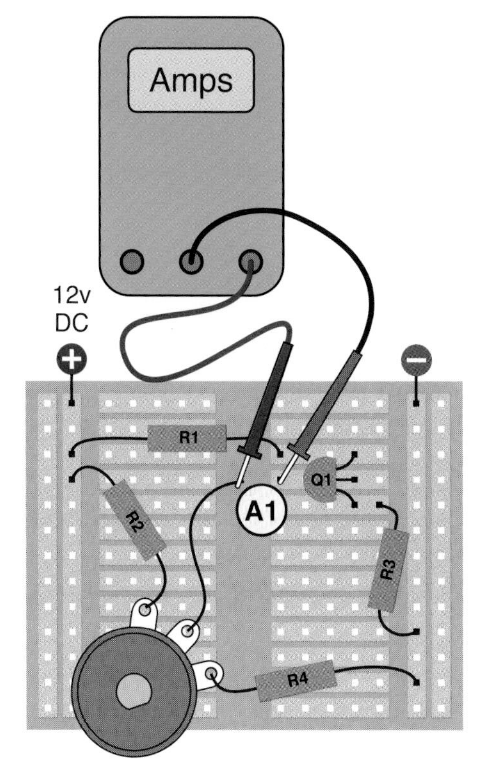

Figure 2-96. *The meter is measuring current flowing from the potentiometer into the base of the transistor at position A1 (see Figure 2-95).*

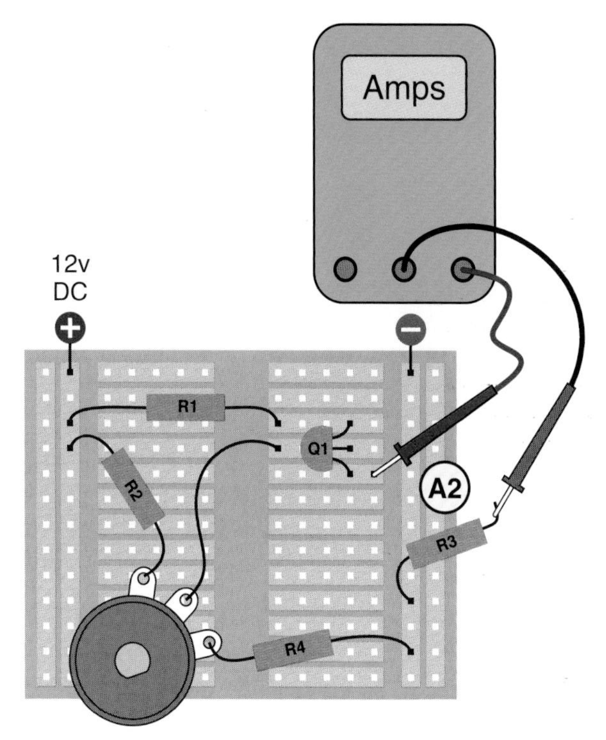

Figure 2-97. *One end of resistor R3 has been unplugged from the breadboard so that the meter now measures current flowing out through the emitter of the transistor, into R3, at position A2.*

Experiment 11: A Modular Project

You will need:

- AC adapter, breadboard, wire, and meter.
- LED. Quantity: 1.
- Resistors, various.
- Capacitors, various.
- Transistor, 2N2222 or similar. Quantity: 2.
- 2N6027 programmable unijunction transistor (PUT). Quantity: 2.
- Miniature 8Ω loudspeaker. Quantity: 1.

So far, I've described small circuits that perform very simple functions. Now it's time to show how modules can be combined to create a device that does a bit more.

The end product of this experiment will be a circuit that makes a noise like a small siren, which could be used in an intrusion alarm. You may or may not be interested in owning an alarm, but the four-step process of developing it is important, because it shows how individual clusters of components can be persuaded to communicate with each other.

I'll begin by showing how to use a transistor to make a solid-state version of the oscillating circuit that you built with a relay in Experiment 8. The relay, you may remember, was wired in such a way that the coil received power through the contacts of the relay. As soon as the coil was energized, it opened the contacts, thus cutting off its own power. As soon as the contacts relaxed they restored the power, and the process repeated itself.

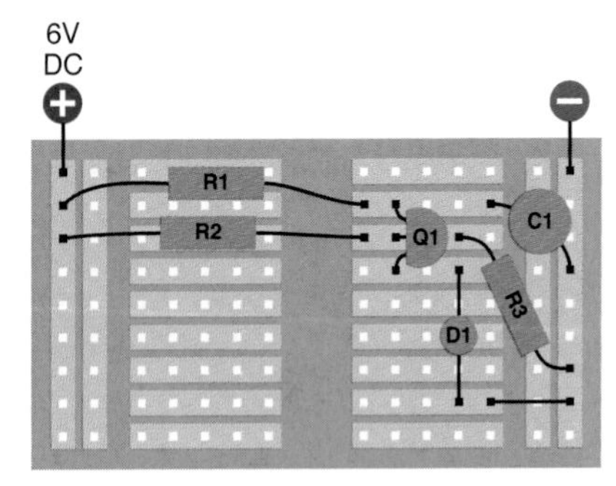

Figure 2-98. *Assemble these components, apply power, and the LED should start flashing.*

R1: 470K
R2: 15K
R3: 27K
C1: 2.2 μF electrolytic capacitor
D1: LED
Q1: 2N6027 programmable unijunction transistor

There's no way to do this with a single bipolar transistor. You actually need two of them, switching each other on and off, and the way that this works is quite hard to understand. An easier option is to use a different thing known as a programmable unijunction transistor, or PUT.

Unijunction transistors were developed during the 1950s, but fell into disuse when simple silicon chips acquired the ability to perform the same kinds of functions, more accurately and more cheaply. However, the so-called programmable unijunction transistor is still widely available, often used in applications such as lamp dimmers and motor controllers. Because its primary use is in generating a stream of pulses, it's ideal for our purposes.

If you put together the components shown in Figure 2-98, the LED should start flashing as soon as you apply power.

Note that this circuit will work on 6 volts. You won't damage anything if you run it with 12 volts, but as we continue adding pieces to it, you'll find that it actually performs better at 6 volts than at 12. If you read the next section, "Essentials: All about programmable unijunction transistors," you'll find out how the circuit works.

ESSENTIALS

All about programmable unijunction transistors

The schematic symbol for a programmable unijunction transistor, or PUT, looks very different from the symbol for a bipolar transistor, and its parts are named differently, too. Nevertheless, it does have a similar function as a solid-state switch. The symbol and the names of the three connections are shown in Figure 2-99.

Note that this is a rare case (maybe the only one in the whole of electronics!) in which you won't run into confusing variations of the basic schematic symbol. A PUT always seems to look the way I've drawn it here. Personally I think it would be clearer if we added a circle around it, but no one seems to do that, so I won't, either.

The 2N6027 is probably the most common PUT, and seems to be standardized in its packaging and pin-outs. I've only seen it in a plastic module rather than a little tin can. Figure 2-100 shows the functions of the leads if your 2N6027 is manufactured by Motorola or On Semiconductor. If you have one from another source, you should check the data sheet.

Note that the flat side of the plastic module faces the opposite way around compared with the 2N2222 bipolar transistor, when the two devices are functioning similarly.

The PUT blocks current until its internal resistance drops to allow flow from the "anode" to the "cathode." In this way, it seems very similar to an NPN transistor, but there's a big difference. When the voltage at the anode increases beyond a threshold, it forces the PUT to allow current to flow. (This threshold is programmed by voltage on the gate.)

Suppose you start with, say, 1 volt at the anode. Slowly, you increase this voltage. The transistor blocks it until the anode is above 4 volts. Suddenly this pressure breaks down the resistance and current surges from the anode to the cathode. If the voltage goes back down again, the transistor reverts to its original state and blocks the flow.

I've included another version of the "finger on the button" drawing to convey this concept. The voltage on the anode is itself responsible for pushing the button that opens the pathway to the cathode. See Figure 2-101.

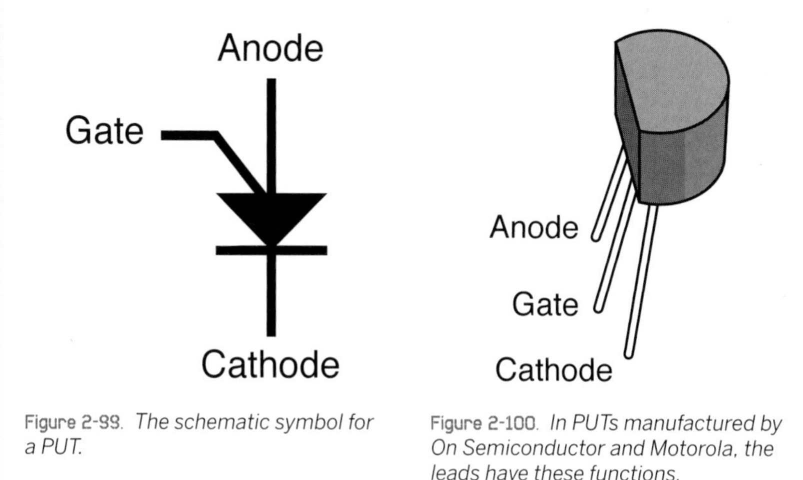

Figure 2-99. *The schematic symbol for a PUT.*

Figure 2-100. *In PUTs manufactured by On Semiconductor and Motorola, the leads have these functions.*

ESSENTIALS

All about programmable unijunction transistors (continued)

This may cause you to wonder what the function of the gate is. You can think of it as "assisting" the finger on the button. In fact, the gate is the "programmable" part of a PUT. By choosing a voltage for the gate, you establish the threshold point when current starts to flow.

Here's a simple take-home summary:

- The anode has to be more positive than the cathode, and the gate should be between those two extremes.
- If anode voltage increases above a threshold point, current bursts through and flows from the anode to the cathode.
- If anode voltage drops back down below the threshold, the transistor stops the flow.
- The voltage you apply to the gate determines how high the threshold is.
- The gate voltage is adjusted with two resistors, shown as R1 and R2 in the simple schematic in Figure 2-102. Typically, each resistor is around 20K. The PUT is protected from full positive voltage by R3, which can have a high value, 100K or greater, because very little current is needed to bias the transistor.
- You add your input signal in the form of positive voltage at the anode. When it exceeds the threshold, it flows out of the cathode and can work some kind of output device.

The only remaining question is how we make a PUT oscillate, to create a stream of on/off pulses. The answer is the capacitor that you included in the circuit that you breadboarded at the beginning of Experiment 11.

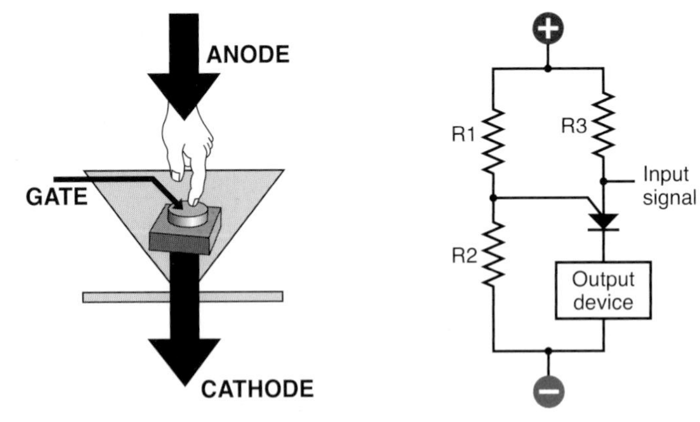

Figure 2-101. *When voltage at the anode of a PUT crosses a threshold (determined by a preset voltage at the gate), current breaks through and surges from the anode to the cathode. In this sense, the anode voltage acts as if it presses a button itself to open a connection inside the PUT, with some assistance from control voltage at the gate.*

Figure 2-102. *This simple schematic shows how a PUT is used. R1 and R2 determine the voltage at the gate, which sets the threshold point for the input at the anode. Above the threshold, current flows from anode to cathode.*

Step 1: Slow-Speed Oscillation

Figure 2-103 is a schematic version of the previous PUT breadboard circuit shown in Figure 2-98, drawn so that the layout looks as much like the breadboard as possible.

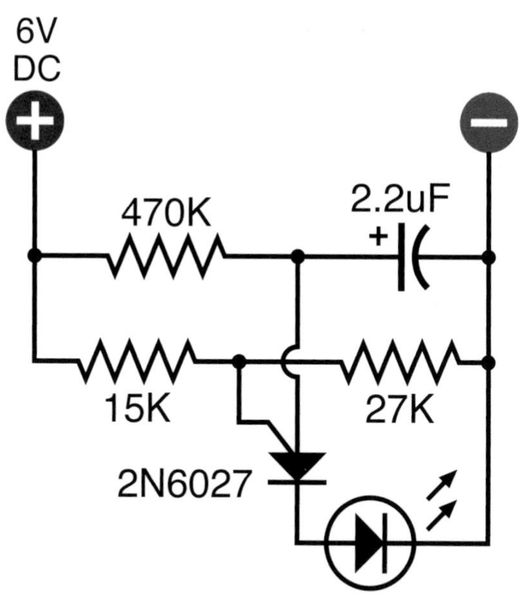

Figure 2-103. *This makes it easier to see what's happening in the breadboard version.*

The 15K resistor and 27K resistor establish the voltage at the gate. The 470K resistor supplies the anode of the PUT, but the PUT begins in its "off" condition, blocking the voltage. So the voltage starts to charge the 2.2 μF capacitor.

You may remember that a resistor slows the rate at which a capacitor accumulates voltage. The bigger the resistor and/or the larger the capacitor, the longer the capacitor takes to reach a full charge.

But notice that the PUT is connected directly with the capacitor. Therefore, whatever voltage accumulates on the capacitor is also experienced by the PUT. As the voltage gradually increases, finally it reaches the threshold, which flips the PUT into its "on" state. The capacitor immediately discharges itself through the PUT, through the LED (which flashes), and from there to the negative side of the power supply.

The surge depletes the capacitor. The voltage drops back down (although not to zero, so the ongoing cycle length cannot be predicted from the formula to establish the time constant). Now the capacitor has to recharge itself all over again, until the whole process repeats itself.

If you substitute a 22 μF capacitor, the charge/discharge cycle should take about 10 times as long, which will give you time to measure it. Set your meter to measure volts DC and place its probes on either side of the capacitor. You can actually watch the charge increasing until it reaches the threshold, at which point the capacitor discharges and the voltage drops back down again.

So now we have an oscillator. What's next?

THEORY

Capacitor Charge Time

The amount of time it takes for a capacitor to reach its threshold is calculated with 5RC, where R is the resistance (in ohms) of the resistor, and C is the capacitance (in farads) of the capacitor. So in this case, you'd multiply 5 by 470,000 by 0.0000022, which gives us 5.17 seconds.

Step 2: Beyond the Persistence of Vision

If you substitute a much smaller capacitor, it will charge much more quickly, and the LED will flash faster. Suppose you use a capacitor of 0.0047 μF (which can also be expressed as 4.7 nanofarads, or 4.7 nF). This seems like an odd number, but it's a standard value for a capacitor. This will reduce the capacitance by a factor of more than 500, and therefore the LED should flash about 500 times as fast, which should be about 1,000 times per second. The human eye cannot detect such rapid pulses. The human ear, however, can hear frequencies up to 10,000 per second and beyond. If we substitute a miniature loudspeaker for the LED, we should be able to hear the oscillations.

Figure 2-104 shows how I'd like you to make this happen. Please leave your original, slow-flashing circuit untouched, and make a duplicate of it farther down the breadboard, changing a couple of component values as indicated. In the schematic in Figure 2-105, the new part of the circuit is in solid black, while the previous section is in gray.

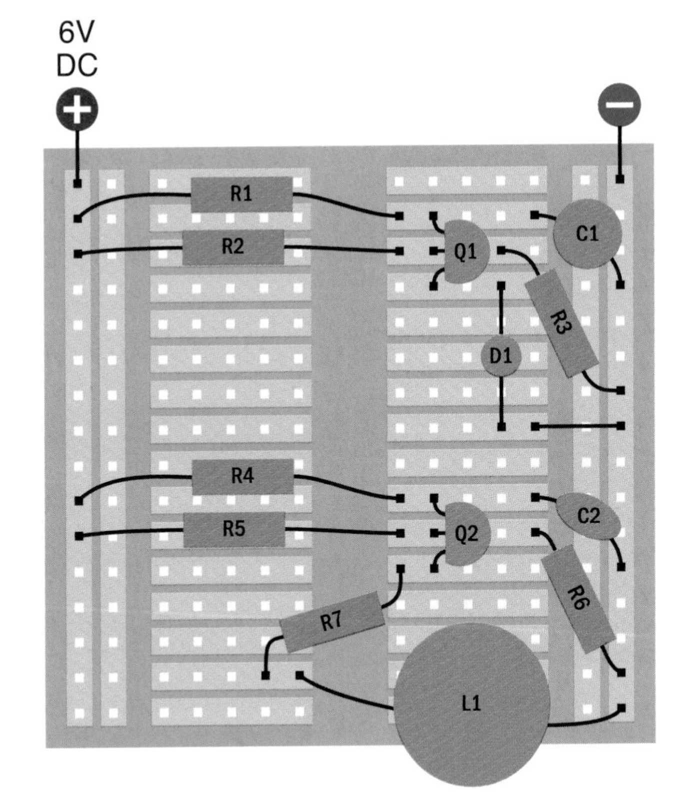

Figure 2-104. *The extra components which have been added at the lower half of the breadboard have the same functions as the components at the top, but some values are slightly different:*

R4: 470K
R5: 33K
R6: 27K
R7: 100Ω
C2: 0.0047 μF
Q2: 2N6027
L1: 8Ω 1-inch loudspeaker

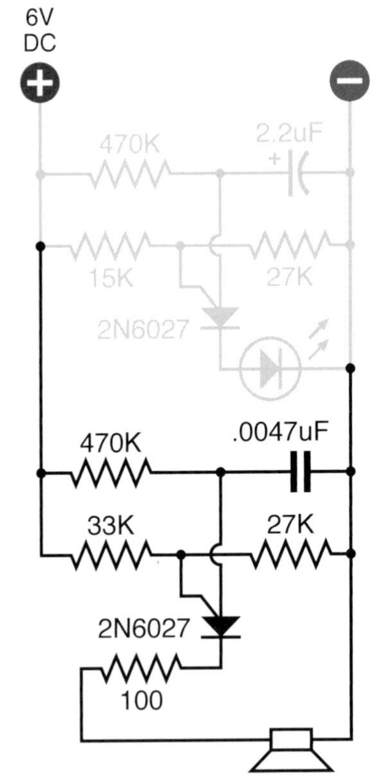

Figure 2-105. *The previous section that you built is shown in gray. Just add the new section in black.*

I want you to keep the slow-flashing circuit separately, untouched, because I have an idea to make use of it a little later. You can leave the LED blinking.

The loudspeaker should be wired in series with a 100Ω resistor to limit the current that flows out of the PUT. The loudspeaker doesn't have any polarity, even though it is fitted with a red wire and a black wire. You can connect it either way around.

Initially, you may be disappointed, because the circuit will not seem to be doing anything. However, if you place your ear very, very close to the loudspeaker, and if you wired the circuit correctly, you should hear a faint buzz, like a mosquito. Obviously, this isn't loud enough to serve any practical purpose. We need to make it louder. In other words, we need to amplify it.

Maybe you remember that the 2N2222, which you played with previously, can function as an amplifier. So let's try using that.

Step 3: Amplification

Disconnect the loudspeaker and its 100Ω series resistor. Then add the 2N2222, which is linked with the output from the PUT via a 1K resistor to protect it from excessive current. See Figure 2-107.

The emitter of the 2N2222 is connected to ground, and the collector is supplied through the loudspeaker and its 100Ω series resistor. This way, small fluctuations in the output from the PUT are sensed by the base of the 2N2222 which converts them into bigger fluctuations between the collector and the emitter, which draw current through the loudspeaker. Check the schematic in Figure 2-108.

Now the sound should be louder than an insect buzz, but still not really loud enough to be useful. What to do?

Well—how about if we add another 2N2222? Bipolar transistors can be placed in series, so that the output from the first one goes to the base of the second one. The 240:1 amplification of the first one is multiplied by another 240:1, giving a total amplification of more than 50,000:1.

There are limits to this technique. The 2N2222 can only conduct so much current before getting overloaded, and excess amplification can cause distortion. But when I built this circuit, I used a meter to verify that we're still within the design limits of a 2N2222, and for this project, I don't care whether the sound is slightly distorted.

BACKGROUND

Mounting a loudspeaker

The diaphragm or cone of a loudspeaker is designed to radiate sound, but as it oscillates to and fro, it emits sound from its back side as well as its front side. Because the sounds are opposite in phase, they tend to cancel each other out.

The perceived output from a loudspeaker can increase dramatically if you add a horn around it in the form of a tube to separate the output from the front and back of the speaker. For a miniature 1-inch loudspeaker, you can bend and tape a file card around it. See Figure 2-106.

Better still, mount it in a box so that the box absorbs the sound from the rear of the loudspeaker. For purposes of these simple experiments, I won't bother to go into the details of vented enclosures and bass-reflex designs.

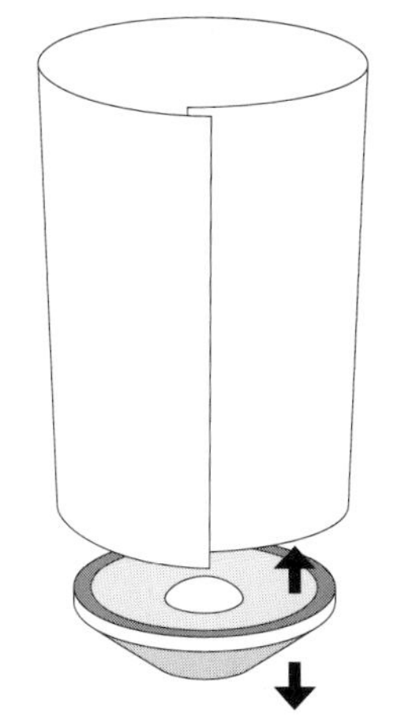

Figure 2-106. *A loudspeaker emits sound from its bottom surface as well as its top surface. To increase the perceived audio volume, use a cardboard tube to separate the two sound sources, or mount the speaker in a small box.*

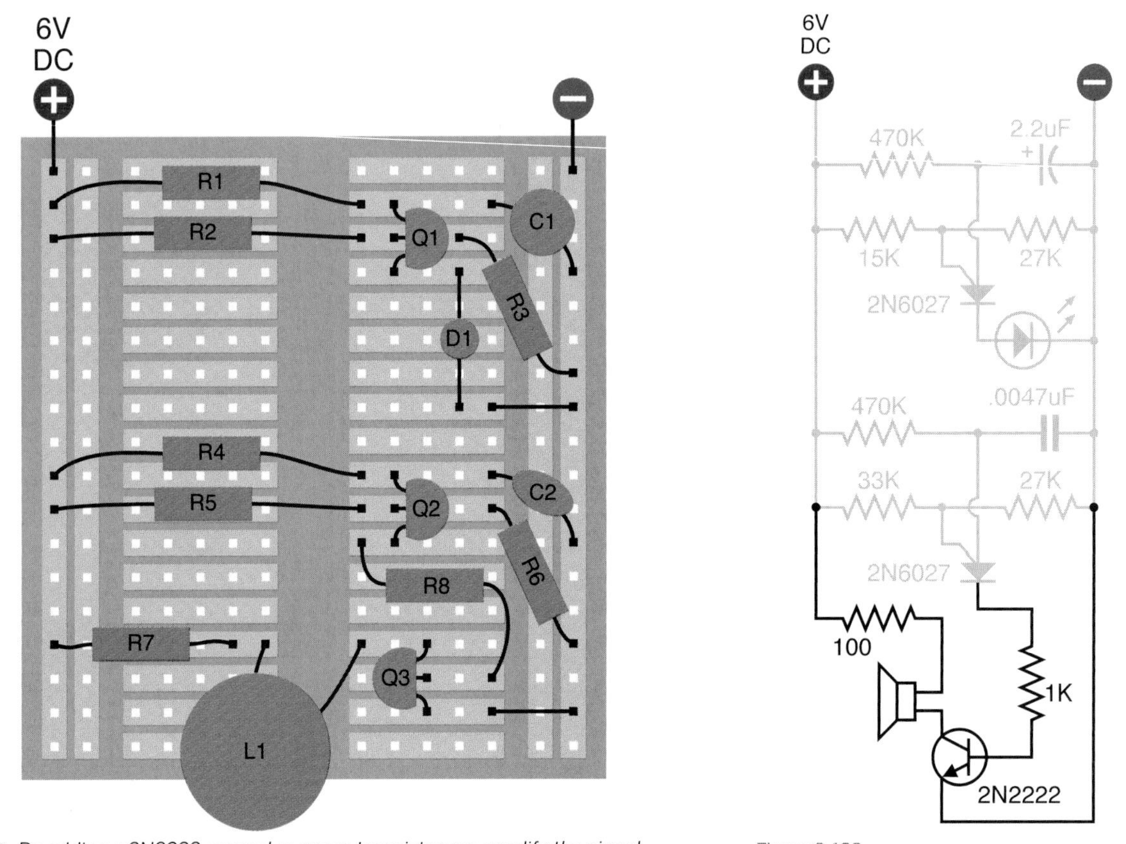

Figure 2-107. *By adding a 2N2222 general-purpose transistor, we amplify the signal from Q2:*

R8: 1K
Q3: 2N2222

Other components are the same as in the previous step in constructing this circuit.

Figure 2-108

Add the second 2N2222 as shown in Figure 2-109. In Figure 2-110, once again the previously wired section is in gray.

If the accumulation of electrical components is beginning to seem confusing, remember that each cluster of parts has a separate defined function. We can draw a block diagram to illustrate this, as in Figure 2-112.

Using the second 2N2222, you should find that the output is more clearly audible, at least within the limits of your tiny 1-inch loudspeaker. Cup your hands around it to direct the sound, and you'll find that the volume seems to increase. You can also try using a 3-inch loudspeaker, which will create a generally better audio output while still remaining within the limits of the little 2N2222 transistor. See Figure 2-106, shown previously, and Figure 2-111.

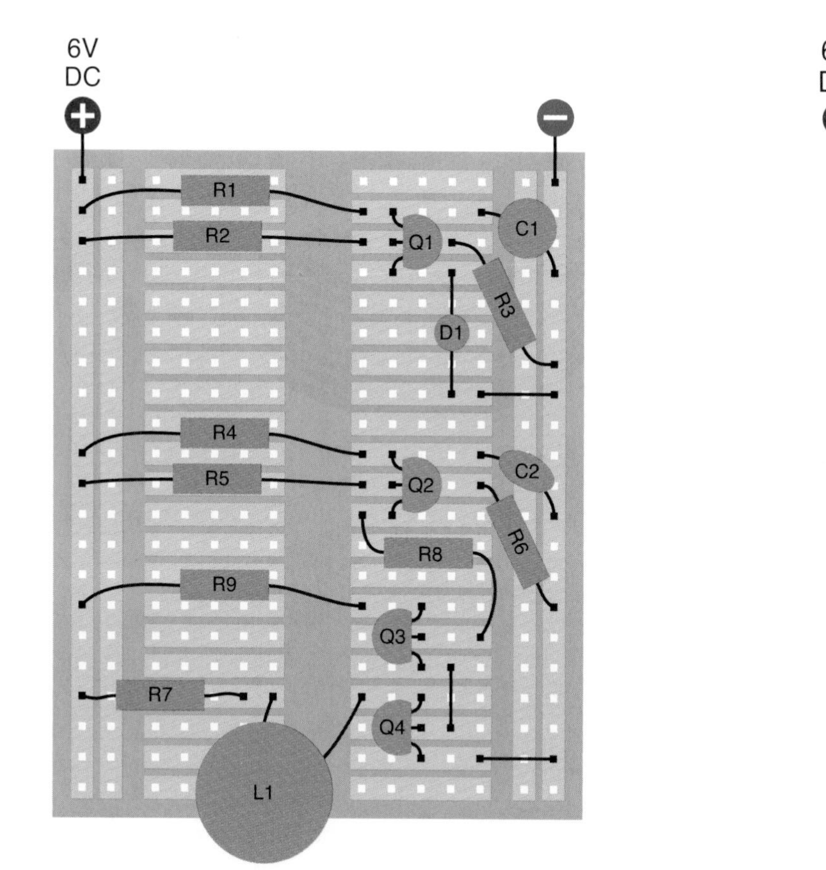

Figure 2-109. *Q4 is another 2N2222 transistor that further amplifies the signal. It receives power through R9: 2.2K.*

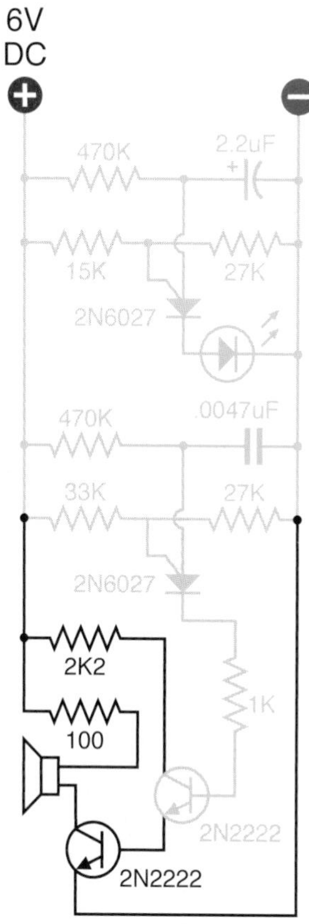

Figure 2-110. *This schematic is comparable with the component layout in Figure 2-109.*

Figure 2-111. *The 2N2222 transistor is quite capable of driving a 3-inch loudspeaker, which will create much better sound than a 1-inch speaker.*

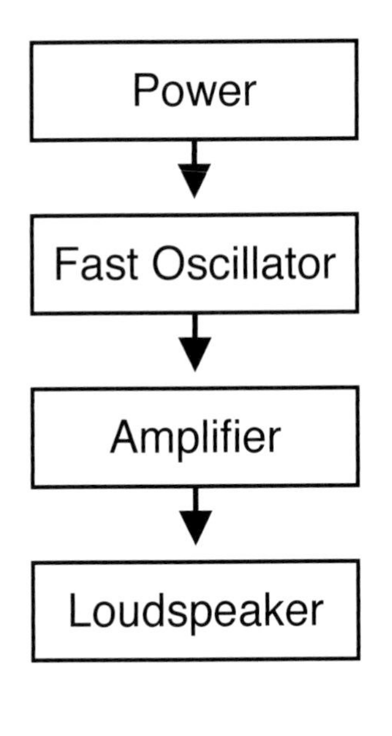

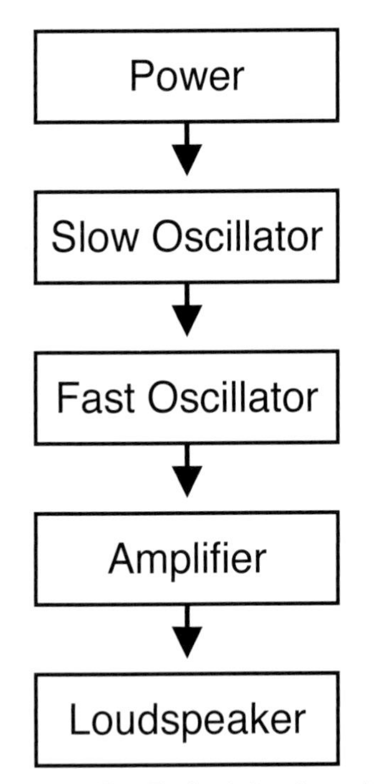

Figure 2-112. *Top: The basic functions of the noisemaking oscillator circuit shown as a block diagram. Bottom: The same functions with a slow oscillator added to control the fast oscillator.*

Step 4: Pulsed Output

If you wanted to use this audio signal as some kind of an alarm, a steady droning noise is not very satisfactory. A pulsing output would be a much better attention-getter.

Well, the first section of the circuit that you assembled created a pulsing signal about twice per second. You used it to flash an LED. Maybe we can get rid of the LED and feed the output from the first section to the second section. The lower block diagram in Figure 2-112 explains this concept.

Can it really be that simple? Well, yes and no. The trick is to make the output from the first section compatible with the input to the second section. If you simply connect a wire from the cathode of the first PUT to the anode of the second PUT, that's not going to work, because the second PUT is already oscillating nicely between low and high voltage, about 1,000 times each second. Add more voltage, and you will disrupt the balance that enables oscillation.

However, remember that the voltage on the gate of a PUT affects its threshold for conducting electricity. Maybe if we connect the output from Q1 to the gate of Q2, we'll be able to adjust that threshold automatically. The voltage still has to be in a range that the PUT finds acceptable, though. We can try various resistors to see which one works well.

This sounds like trial and error—and that's exactly what it is. Doing the math to predict the behavior of a circuit like this is far too complicated—for me, anyway. I just looked at the manufacturer's data sheet, saw the range of resistor values that the PUT would tolerate, and chose one that seemed as if it should work.

If you remove the LED and substitute R10 as shown in the breadboard diagram in Figure 2-113, you'll find that the fluctuating output from Q1 makes Q2 emit a two-tone signal. This is more interesting, but still not what I want. I'm thinking that if I make the pulses out of Q1 less abrupt, the result could be better, and the way to smooth a pulsing output is to hook up another capacitor that will charge at the beginning of each pulse and then release its charge at the end of each pulse. This is the function of C3 in Figure 2-114, and it completes the circuit so that it makes a whooping sound almost like a "real" alarm.

If you don't get any audio output, check your wiring very carefully. It's easy to make a wrong connection on the breadboard, especially between the three legs of each transistor. Use your meter, set to DC volts, to check that each section of the circuit has a positive voltage relative to the negative side of the power supply.

Figure 2-115 shows how your circuit should actually look on the breadboard.

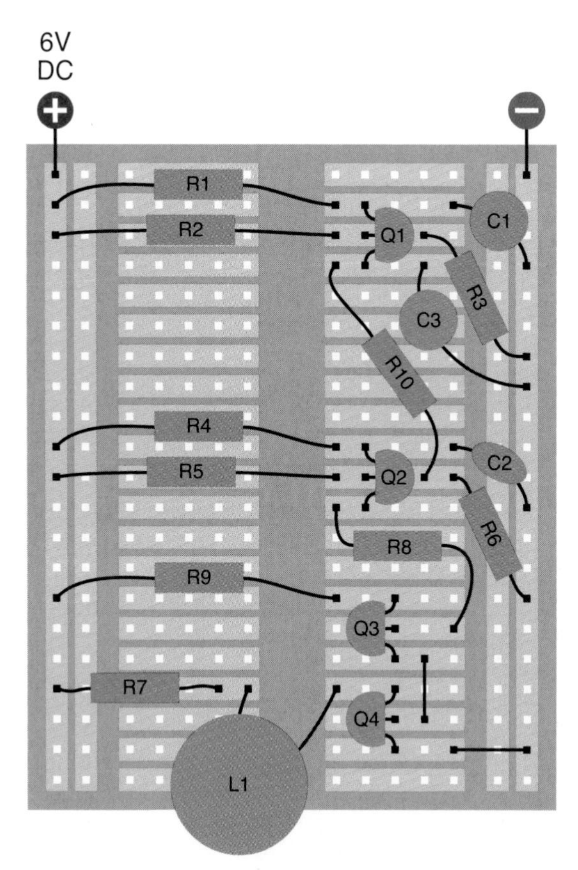

Figure 2-113. *R10 connects the slow-running oscillator at the top of the breadboard to the gate of Q2, the PUT in the middle of the breadboard. This modulates the audio oscillator, with addition of a smoothing capacitor.*

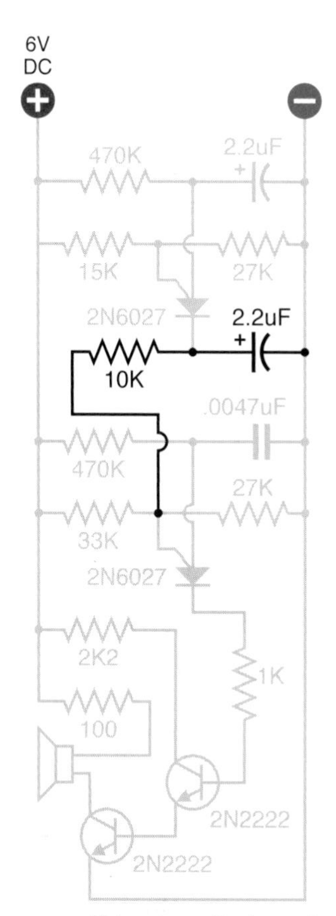

Figure 2-114. *This schematic shows the same circuit as in Figure 2-113:*

R10: 10K
C3: 2.2 μF

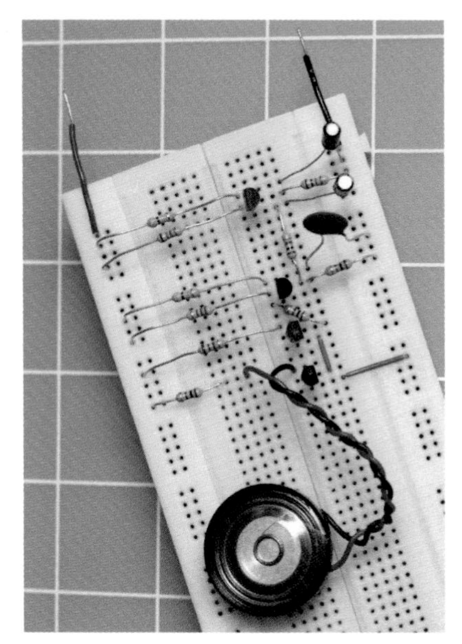

Figure 2-115. *This photograph shows the complete alarm-audio circuit on a breadboard.*

Tweaking it

There's still a lot of room for creativity here:

- Adjust the frequency of the sound: Use a smaller or larger capacitor instead of C2 (half or twice the current value). Use a smaller or larger value for R5.
- Adjust the pulsing feature: Use a smaller or larger capacitor instead of C1 (half or twice the current value). Use a smaller or larger value for R2.
- General performance adjustments: try a larger value for R1. Try smaller or larger values for C3.
- Try running the circuit at 7.5 volts, 10 volts, and 12 volts.

The circuits in this book are suggested as only a starting point. You should always try to tweak them to make them your own. As long as you follow the general rule of protecting transistors and LEDs with resistors, and respecting their requirements for positive and negative voltage, you're unlikely to burn them out. Of course, accidents will happen—I myself tend to be careless, and fried a couple of LEDs while working on this circuit, just because I connected them the wrong way around.

Step 5: Enhancements

A noisemaking circuit is just the output of an alarm. You would need several enhancements to make it useful:

1. Some kind of an intrusion sensor. Maybe magnetic switches for windows and doors?
2. A way to start the sound if any one of the sensors is triggered. The way this is usually done is to run a very small but constant current through all of the switches in series. If any one switch opens, or if the wire itself is broken, this interrupts the current, which starts the alarm. You could make this happen with a double-throw relay, keeping the relay energized all the time until the circuit is broken, at which point, the relay relaxes, opening one pair of contacts and closing the other pair, which can send power to the noisemaker.

   The trouble is that a relay draws significant power while it's energized, and it also tends to get hot. I want my alarm system to draw very little current while it's in "ready" mode, so that it can be powered by a battery. Alarm systems should never depend entirely on AC house current.

   If we don't use a relay, can we use a transistor to switch on the rest of the circuit when the power is interrupted? Absolutely; in fact, one transistor will do it.
3. But how do we arm the alarm in the first place? Really, we need a three-step procedure. First, check a little light that comes on when all the doors and windows are closed. Second, press a button that starts a 30-second countdown, giving you time to leave, if that's what you want to do. And third, after 30 seconds, the alarm arms itself.

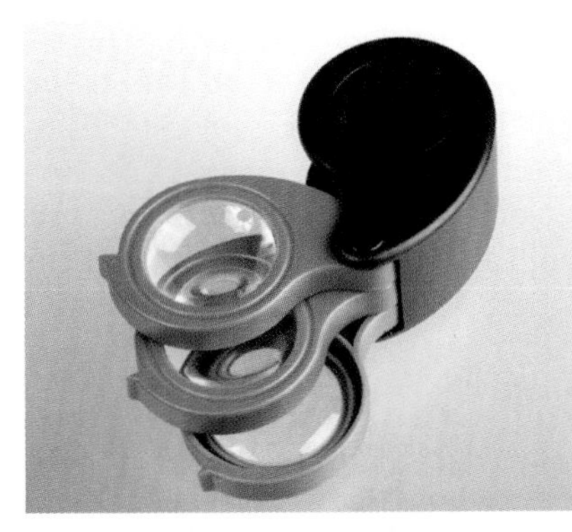

Figure 3-4. *As long as you treat it carefully, a cheap set of plastic magnifying lenses is perfectly acceptable. Handheld magnification is essential for inspecting the solder joints that you make on perforated board.*

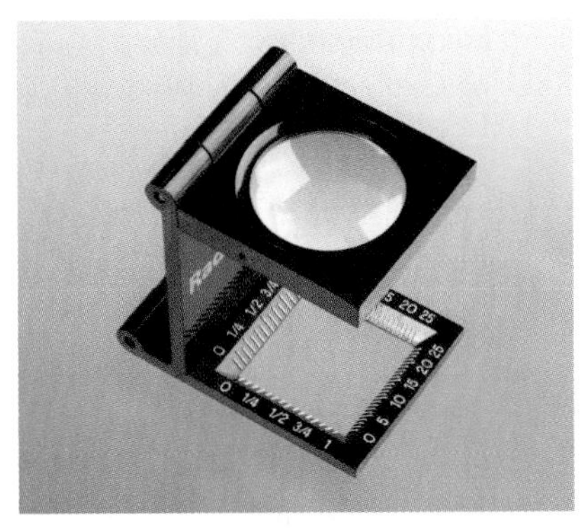

Figure 3-5. *This kind of folding magnifier can stand on your desktop and is useful for checking part numbers on tiny components.*

Essential: Clip-on meter test leads

The probes that came with your multimeter require you to hold them in contact while you make a reading. This requires both hands, preventing you from doing anything else at the same time.

When you use a pair of "minigrabber" probes with little spring-loaded clips at the end, you can attach the Common (negative) lead from your meter to the negative side of your circuit and leave it there, while you touch or attach the positive probe elsewhere.

The Pomona model 6244-48-0 (shown in Figure 3-6) from Meter Superstore and some other suppliers is what you need. If you have trouble finding it or you object to the cost, you may consider making your own by buying a couple of "banana plugs" (such as RadioShack part 274-721) that will fit the sockets on your meter, and then use 16-gauge or thicker stranded wire to connect the plugs with IC test clips, such as Kobiconn 13IC331 or RadioShack "mini hook clips," part number 270-372. See Figures 3-7 and 3-8.

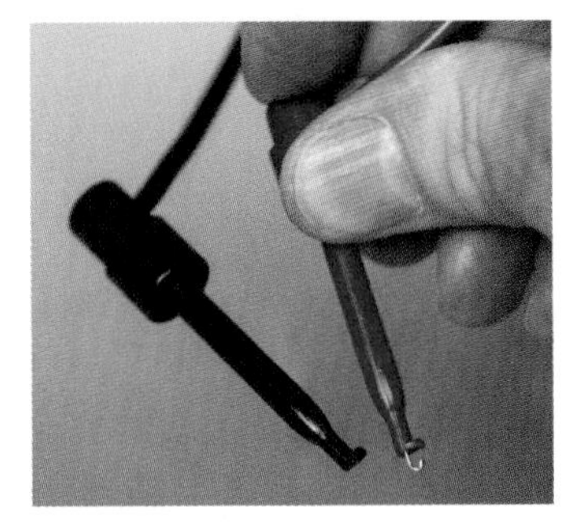

Figure 3-6. *These "minigrabber" add-ons for meter leads make it much easier to measure voltage or current. Push the spring-loaded button, and a little copper hook slides out. Attach it to a wire, release the button, and you have your hands free for other tasks. It's a mystery that meters are not supplied with these grabbers as standard equipment.*

Figure 3-7. *To make your own minigrabber meter leads, first attach a banana plug to a wire by sliding the wire through the cap, into the plug, and out through a hole in the side.*

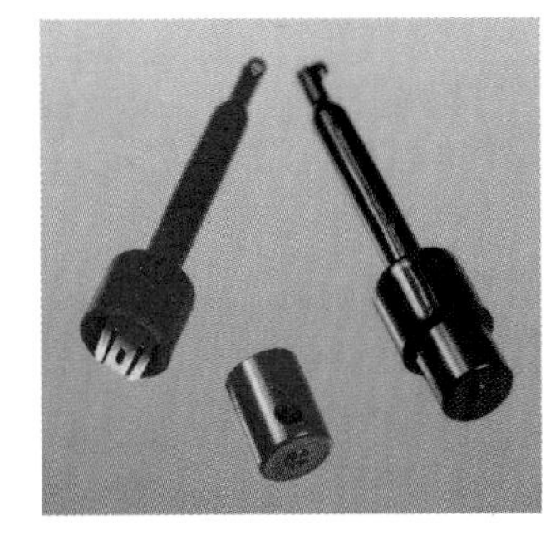

Figure 3-8. *Then screw a collar over the protruding piece of wire, and screw on the cap. The other end of the wire is soldered to a probe.*

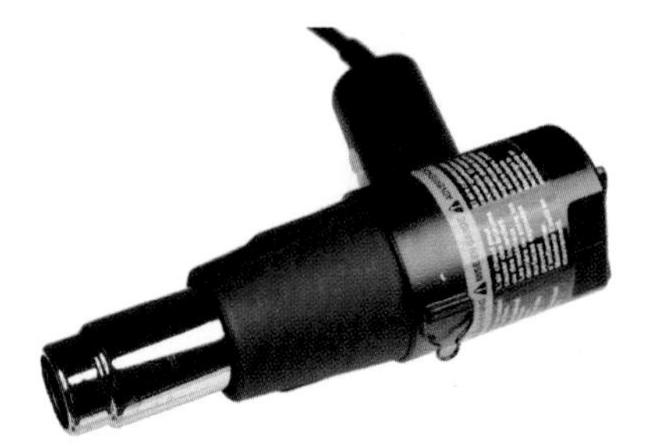

Figure 3-9. *Like an overpowered hair dryer, the heat gun is used with heat-shrink tubing to create a snug, insulated sheath around bare wire.*

Essential: Heat gun

After you join two wires with solder, you often need to insulate them. Electrical tape, sometimes called insulating tape, is messy and tends to come unstuck. You'll be using heat-shrink tube, which forms a safe, permanent sheath around a bare-metal joint. To make the tube shrink, use a heat gun, which is like a very powerful hair dryer. They're available from any hardware supply source, and I suggest you buy the cheapest one you can find. See Figure 3-9.

Essential: Solder pump

This little gadget sucks up hot, melted solder when you are trying to remove a solder joint that you made in the wrong place. Available from All Electronics (catalog item SSR-1) or RadioShack 64-2086. See Figure 3-10.

Essential: Desoldering wick

Also known as desoldering braid. See Figure 3-11. You use this to soak up solder, in conjunction with the Solder Pump. Available from All Electronics (catalog item SWK) or RadioShack (part 64-2090).

Essential: Miniature screwdriver set

Dinky little electronic parts often have dinky little screws in them, and if you try to use the wrong size of screwdriver, you'll tend to mash the heads of the screws. I like the Stanley precision set, part number 66-052, shown in Figure 3-12. But any set will do as long as it has both small Phillips and straight-blade screwdrivers.

Recommended: Soldering stand

Like a holster for a gun, you rest your soldering iron in this stand when the iron is hot but not on use. Examples are catalog item 50B-205 from All Electronics, RadioShack model 64-2078, or check eBay. See Figure 3-13. This item may be built into the helping hand, but you need an extra one for your second soldering iron.

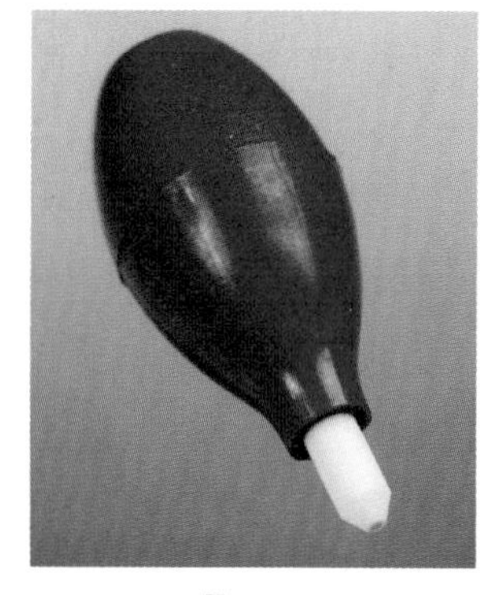

Figure 3-10. *To remove a solder joint, you can heat the solder until it's liquid, then suck it up into this squeezable rubber bulb.*

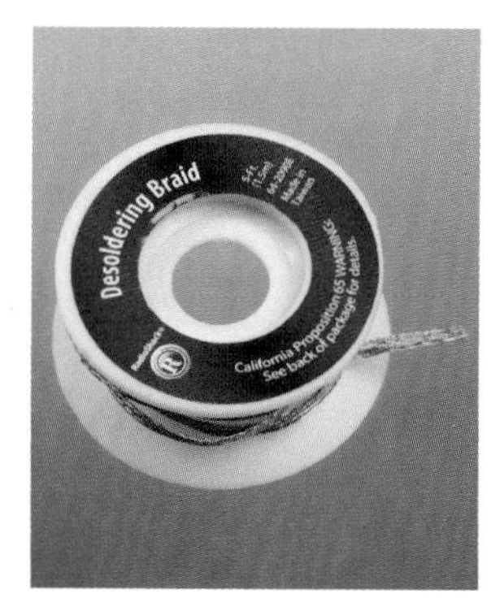

Figure 3-11. *An additional option for removing liquid solder is to soak it up in this copper braid.*

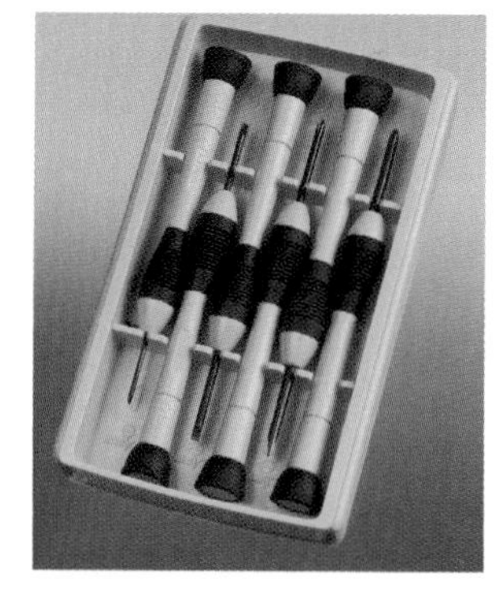

Figure 3-12. *A set of small screwdrivers is essential.*

Figure 3-13. *A safe and simple additional stand for a hot soldering iron.*

Recommended: Miniature hand saw

I assume that you will want to mount a finished electronics project in a decent-looking enclosure. Consequently, you are likely to need tools to cut, shape, and trim thin plastic. For example, you may want to cut a square hole so that you can mount a square power switch in it.

Power tools are not suitable for this kind of delicate work. A miniature handsaw (a.k.a. a "hobby saw") is ideal for trimming things to fit. X-Acto makes a range of tiny saw blades. I suggest the #15 blade, plus the handle that it fits in, shown in Figure 3-14. Available online from Tower Hobbies, Hobbylinc, ArtCity, and many other arts/crafts sources. Also look for the larger X-Acto saw blade, #234 or #239, which you can use for cutting perforated board.

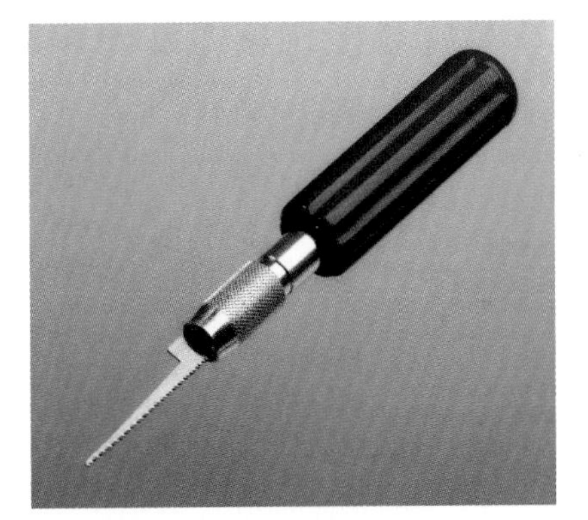

Figure 3-14. *X-Acto makes a range of small saw blades that are ideal for cutting square holes to mount components in plastic panels.*

Recommended: Miniature vise

A miniature vise can do things that the helping hand cannot. I use mine when I'm sawing small pieces of plastic and as a dead weight to anchor a piece of perforated board while I'm working on it. See Figure 3-15.

Look for a cast-iron vise that is listed as being 1 inch or slightly larger, available from Megahobby, eBay, and other arts/crafts sources. Also consider the PanaVise, which has a tilting head to allow you to turn your work to any angle.

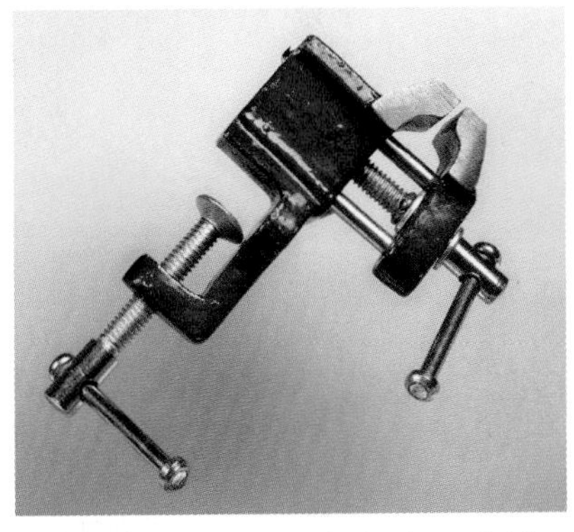

Figure 3-15. *This one-inch vise is available from the McMaster-Carr catalog.*

Recommended: Deburring tool

A deburring tool instantly smoothes and bevels any rough edge (when you have sawn or drilled a piece of plastic, for instance) and also can enlarge holes slightly. This may be necessary because some components are manufactured to metric sizes, which don't fit in the holes that you drill with American bits. Your small local hardware store may not stock deburring tools, but they are very inexpensively available from Sears, McMaster-Carr, KVM Tools, or Amazon. See Figure 3-16.

Figure 3-16. *This cunning little blade, safety-tipped with a round bump on the end, removes rough edges from saw cuts with a single stroke, and can enlarge holes that are almost big enough—but not quite.*

Optional: Hand-cranked countersink

You need a countersink to bevel the edges of screw holes to accept flat-headed screws. If you use a countersink bit in an electric drill, it won't give you precise control when you're working with thin, soft plastic.

Handheld countersinks that you grasp and turn like a screwdriver are easy to find, but McMaster-Carr (catalog item 28775A61) is the only source I've found for a hand-cranked tool that is much quicker to use. It comes with a set of bits, as shown in Figure 3-17.

Optional: Pick and hook set

Made by Stanley, part number 82-115, available from Amazon and hardware stores. You can find imported imitations for a few dollars less. See Figure 3-18.

Optional: Calipers

These may seem like a luxury, but are useful for measuring the external diameter of a round object (such as the screw thread on a switch or a potentiometer) or the internal diameter of a hole (into which you may want a switch or potentiometer to fit).

I like Mitutoyo calipers, and the low-end model 505-611 (shown in Figure 3-19) does everything I need. You can find cheaper brands, but economizing on precision measuring tools may not be a wise policy in the long term. The manufacturer's site will show you all their available models, after which you can Google "Mitutoyo" to find retail sources.

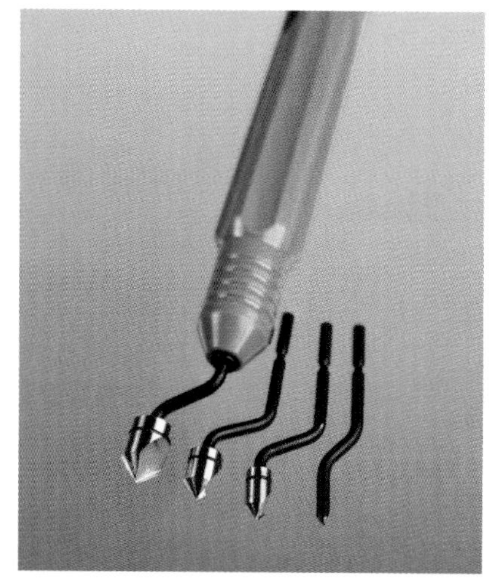

Figure 3-17. *You spin this countersink tool like a hand crank to add just the right amount of bevel to a hole, so that it will accommodate a flat-head screw.*

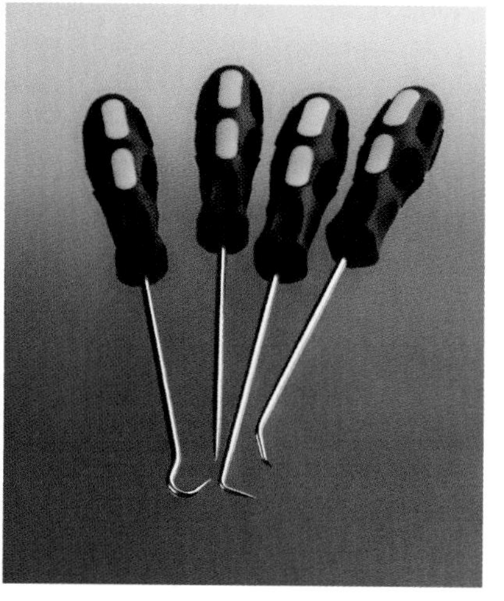

Figure 3-18. *This pick-and-hook set is useful in many unexpected ways.*

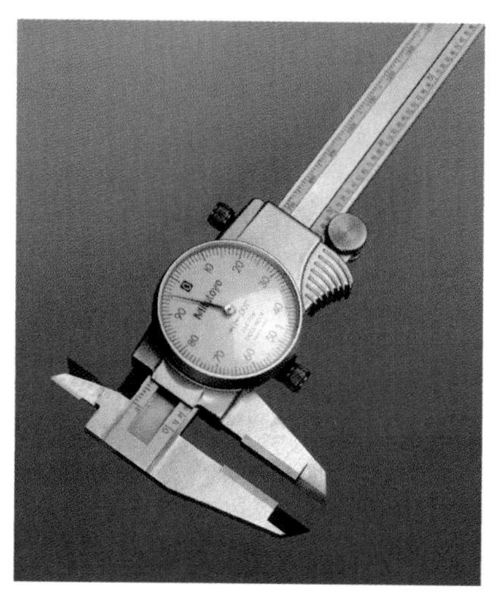

Figure 3-19. *Calipers can be digital (which automatically convert from millimeters to 1/64 inch to 1/1,000 inch), or analog like these (so you never need to worry about a dead battery).*

Supplies

Solder

This is the stuff that you will melt to join components together on a permanent (we hope) basis. You need some very thin solder, size 0.022 inches, for very small components, and thicker solder, 0.05 inches, for heavier items. Avoid buying solder that is intended for plumbers, or for craft purposes such as creating jewelry. A range of solder thicknesses is shown in Figure 3-20. You want to make sure to get lead-free solder.

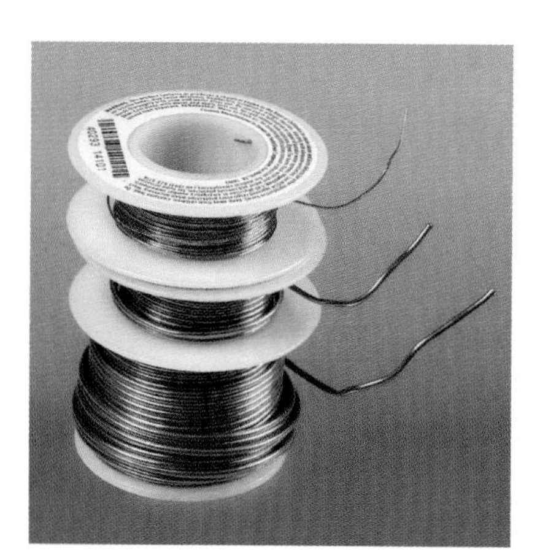

Figure 3-20. *Spools of solder in various thicknesses.*

Electronics solder has a nonacidic rosin core that is appropriate for electronic components. Rolls of solder are available from all hobby-electronics sources including All Electronics, RadioShack, and Jameco, or search for "electronic solder" on Amazon.

Wire

You'll need some stranded wire to make flexible external connections with the circuit that you'll be building. Look for 22-gauge stranded hookup wire, in red, black, and green, 10 feet (minimum) of each.

If you want to install the intrusion alarm after completing that project in Experiment 15, you'll need white-insulated two-conductor wire of the type sold for doorbells or furnace controls. This is available by the foot from Lowe's, Home Depot, Ace Hardware, and similar stores. You'll decide how much to buy after you measure the distances between the magnetic sensor switches that you decide to install.

Heat-shrink tube

For use in conjunction with your heat gun, described previously. You'll need a range of sizes in any colors of your choice. See Figure 3-21. Check RadioShack part 278-1627, other electronics suppliers, or your local hardware store. Prices will vary widely. You can buy the cheapest.

Copper alligator clips

These absorb heat when you are soldering delicate components. The Mueller BU-30C is a full-size solid copper alligator clip for maximum heat absorption. RadioShack sells smaller clips (part number 270-373, shown in Figure 3-22) that are suitable for tiny components.

Figure 3-21. *Slide heat-shrink tubing over a bare joint and apply heat from a heat gun to make a tight insulating seal around the joint.*

Figure 3-22. *These small clips absorb heat to protect components when you're soldering them.*

Perforated board

When you're ready to move your circuit from a breadboard to a more permanent location, you'll want to solder it to a piece of perforated board, often known as "prototyping board" but also called "perfboard."

You need the type that has copper strips etched onto the back, in exactly the same "breadboard layout" as the conductors hidden inside a breadboard, so that you can retain the same layout of your components when you solder them into place. Examples are RadioShack part 276-150 (shown in Figure 3-23) for small projects and part 276-170 (in Figure 3-24) for larger projects, such as Experiment 15.

For very small projects in which you will connect components using their wires alone, you need perfboard that isn't etched with copper strips connecting the holes. I like the Twin Industries 7100 range (available from Mouser.com) or Vectorboard from Newark Electronics, shown in Figure 3-25. You use a saw to cut out as small a piece as you need. Cheaper options are RadioShack part 276-149 (shown in Figure 3-26), or PC-1 from All Electronics. These have little copper circles around each hole that are not necessary for our purposes, but not a problem, either.

Figure 3-23. *This perforated board has a pattern of copper traces similar to the pattern inside a breadboard, so that you can lay out the components with minimal risk of wiring errors, when you're ready to create a permanently soldered version of your project.*

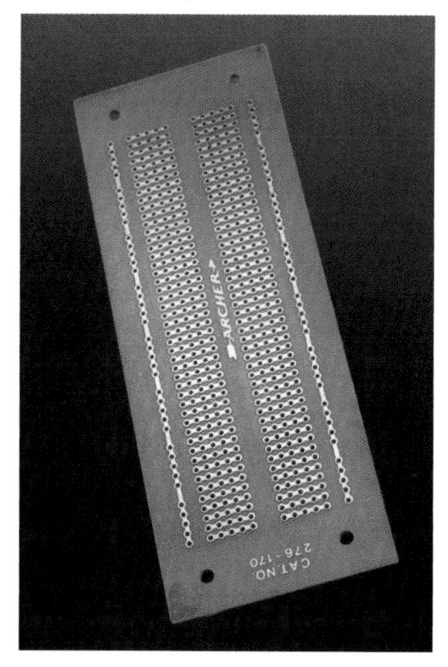

Figure 3-24. *A larger example of perforated board with breadboard geometry.*

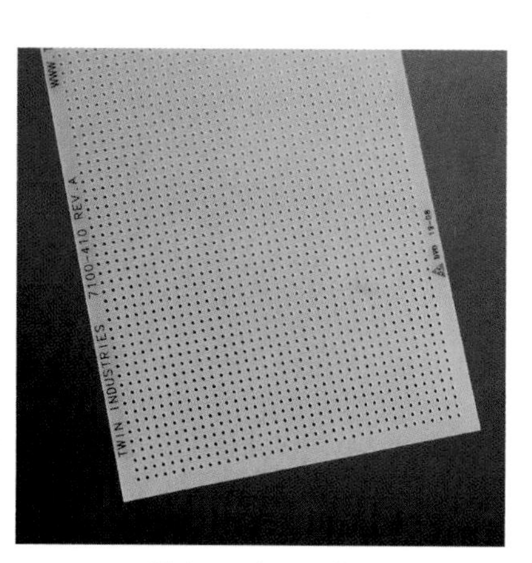

Figure 3-25. *Plain perforated board (with no copper traces) can be used for mounting components when you want to do point-to-point wiring.*

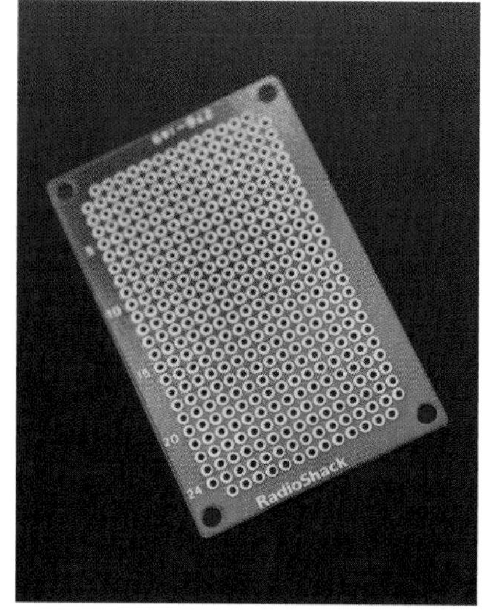

Figure 3-26. *A small piece of perforated board with individual copper solder pads to assist you in mounting components.*

Plywood

When you use a soldering iron, hot drops of solder tend to fall onto your table or workbench. The solder solidifies almost instantly, can be difficult to remove, and will leave a scar. Consider using a 2-foot square of half-inch plywood to provide disposable protection. You can buy it precut at Home Depot or Lowe's.

Machine screws

To mount components behind a panel, you need small machine screws (or "bolts"). They look nice if they have flat heads that fit flush against the panel. I suggest stainless-steel machine screws, #4 size, in 1/2-, 5/8-, 3/4-, and 1-inch lengths, 100 of each, plus 400 washers and 400 #4 locknuts of the type that have nylon inserts, so that they won't work loose. Check McMaster-Carr for a large and reasonably priced selection.

Project boxes

A project box is just a small box (usually plastic) with a removable lid. You mount your switches, potentiometers, and LEDs in holes that you drill through the box, and you attach your circuit on a perforated board that goes inside the box. Search All Electronics for "project box" or RadioShack for "project enclosure."

You need a box measuring approximately 6 inches long, 3 inches wide, and 2 inches high, such as RadioShack part 270-1805. Anything similar will do. I suggest you buy a couple other sizes as well, as they will be useful in the future.

Components

Power plugs, sockets, and binding posts

After you finish a project and put it in a box, you'll need a convenient way to supply it with power. Buy yourself a pair of insulated binding posts, such as RadioShack part 274-661, shown in Figure 3-27. Also obtain a panel-mounted power jack, size N, such as RadioShack part 274-1583, and DC power plug, size N, such as RadioShack 274-1573. The plug-and-socket pair is pictured in Figure 3-28.

Finally, you will need interconnects that are sized to fit a perforated board that is drilled at intervals of 1/10 inch. Sometimes known as "single inline sockets and headers," but also known as "boardmount sockets and pin-strip headers," they come in strips of 36 or more, and you can snip off as many as you need. Examples are Mill-Max part numbers 800-10-064-10-001000 and 801-93-050-10-001000, or 3M part numbers 929974-01-36-RK and 929834-01-36-RK. You can buy them from the usual electronics suppliers. Figure 3-29 shows headers before and after being snapped into small sections. Make sure that the interconnects have a terminal spacing of 0.1 inch.

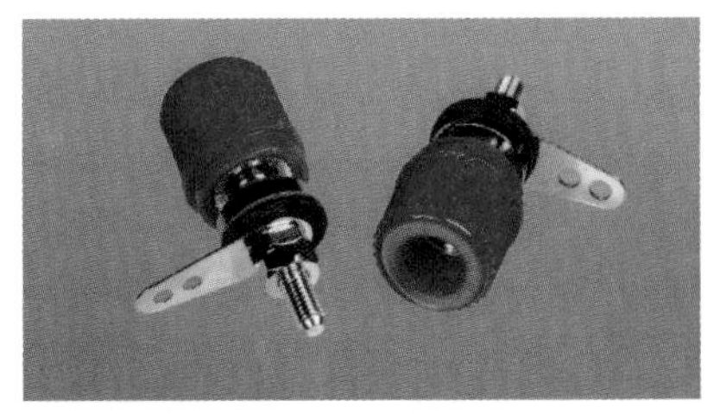

Figure 3-27. *These terminals, also known as binding posts, enable a solderless connection with wires that have stripped ends. Also available in black.*

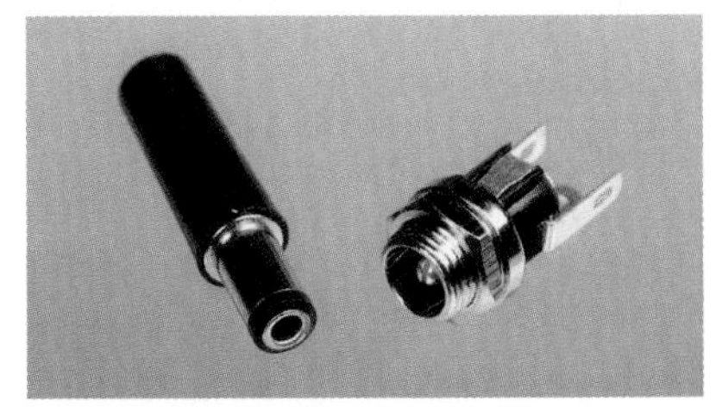

Figure 3-28. *The socket on the right can be mounted in a project box to receive power from the plug on the left.*

Battery

After you complete Experiment 15 at the end of this section of the book, if you want to use the project on a practical basis, you'll need a 12-volt battery. Search online for "12v battery" and you'll find many sealed, re-chargeable lead-acid batteries that are designed for alarm systems, some measuring as small as 1×2×3-inch and costing under $10. You need a charger with it, which will probably cost you about $10.

Switches and relays

You will need the same DPDT relay and the same SPDT toggle switch that were mentioned in Chapter 2 shopping list.

For Experiment 15, you'll need magnetic switches that you can apply to doors or windows, such as the Directed model 8601, available from dozens of sources online.

Also you will need a DPDT pushbutton switch, ON-(ON) type, with solder terminals. Examples are model MPG206R04 by Tyco or model MB2061SS1W01-RO by NKK (with optional cap). Or search eBay for "DPDT pushbutton."

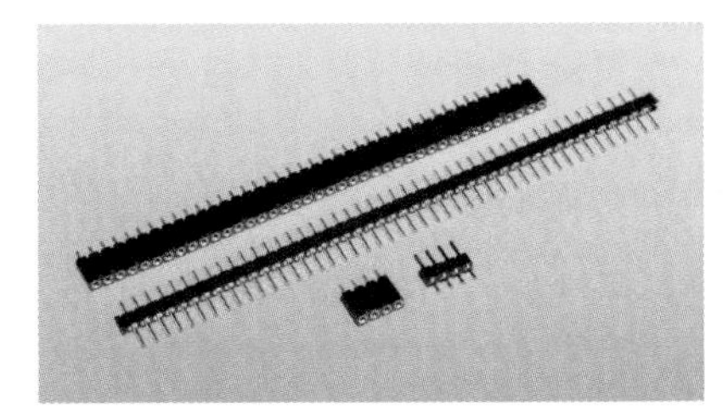

Figure 3-29. *Single inline sockets (top) and headers (middle) allow you to make very compact plug-and-socket connections to a PC board. They can be sawn, cut, or snapped into smaller sections (bottom). The terminals are 0.1 inch apart.*

Diodes

Buy at least half-a-dozen red 5 mm LEDs rated for approximately 2 volts, such as the Optek part number OVLFR3C7, Lumex part number SSL-LX-5093IT, or Avago part HLMP-D155. Buy half-a-dozen similar green LEDs at the same time.

In addition, you'll need a signal diode, type 1N4001 (any brand will do). Figure 3-30 shows an example, highly magnified. They're cheap, and likely to be useful in the future, so buy 10 of them.

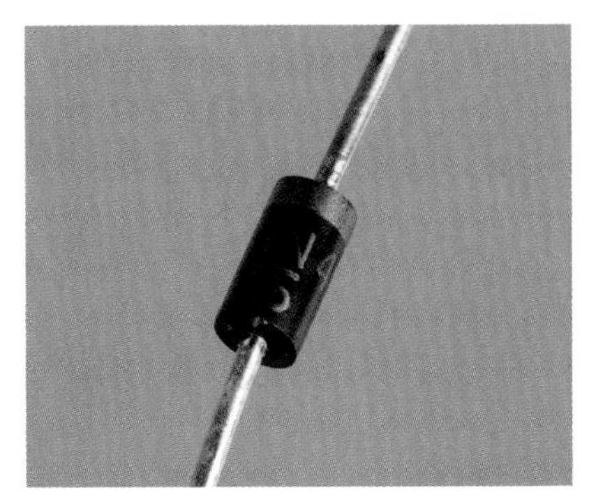

Figure 3-30. *This 1N4001 diode is about 1/4 inch long and can handle up to 50 volts.*

Soldering Irons Get Hot!

Please take these basic precautions:

Use a proper stand (such as the one incorporated in your helping hands) to hold your soldering iron. Don't leave it lying on a workbench.

If you have infants or pets, remember that they may play with, grab, or snag the wire to your soldering iron. They could injure themselves (or you).

Be careful never to rest the hot tip of the iron on the power cord that supplies electricity to the iron. It can melt the plastic in seconds and cause a dramatic short circuit.

If you drop a soldering iron, don't be a hero and try to catch it. Most likely you will grab the hot part, which hurts. (I speak from experience.) When you burn your hand, you will instinctively let go of the iron, so you may as well let it drop freely without the intermediate step of burning yourself while it's on its way to the floor. Naturally, you should pick it up quickly after it does hit the floor, but by then you will have gained the necessary time in which to make a sensible decision to grab it by the cool end.

Always bear in mind that others in your home are more at risk of hurting themselves on a soldering iron than you are, because they won't know that it's hot. Most soldering irons have no warning lights to tell you that they're plugged in. As a general rule, always assume that a soldering iron is hot, even if it's unplugged. It may retain sufficient heat to burn you for longer than you expect.

Loudspeaker

To complete the project in Experiment 15, you'll need a loudspeaker small enough to fit inside your project box but louder than the 1-inch speaker that you used previously. It should be 2 inches or 2.5 inches (50 to 60 mm) in diameter. If you can find a 100Ω speaker, it will give you more output, but an 8Ω speaker will be acceptable.

Experiment 12: Joining Two Wires Together

Your adventure into soldering begins with the prosaic task of joining one wire to another, but will lead quickly to creating a full electronic circuit on perforated board. So let's get started!

You will need:

- 30-watt or 40-watt soldering iron
- 15-watt pencil-type soldering iron
- Thin solder (0.022 inches or similar)
- Medium solder (0.05 inches or similar)
- Wire strippers and cutters
- "Helping hand" gadget to hold your work
- Shrink-wrap tubing, assorted
- Heat gun
- Something to protect your work area from drops of solder

Your First Solder Joint

We'll start with your general-duty soldering iron—the one rated for 30 or 40 watts. Plug it in, leave it safely in its holder, and find something else to do for five minutes. If you try to use a soldering iron without giving it time to get fully hot, you will *not* make good joints.

Strip the insulation from the ends of two pieces of 22-gauge solid wire and clamp them in your helping hand so that they cross each other and touch each other, as shown in Figure 3-31.

To make sure that the iron is ready, try to melt the end of a thin piece of solder on the tip of the iron. The solder should melt instantly. If it melts slowly, the iron isn't hot enough yet.

Now follow these steps (shown in Figures 3-32 through 3-36):

1. Make sure the tip of the soldering iron is clean (wipe it on the moistened sponge in the base of your helping hand if necessary), then touch it against the intersection of the wires steadily for three seconds to heat them. If you have hard tap water, use distilled water to wet the sponge to avoid a buildup of mineral deposits on the tip of your soldering iron.
2. While maintaining the iron in this position, feed a little solder onto the intersection of the wires, also touching the tip of the soldering iron. Thus, the two wires, the solder, and the tip of the iron should all come together at one point. The solder should spread over the wires within another two seconds.
3. Remove the iron and the solder. Blow on the joint to cool it. Within 10 seconds, it should be cool enough to touch.
4. Unclamp the wires and try to tug them apart. Tug hard! If they defeat your best attempts to separate them, the wires are electrically joined and should stay joined. If you didn't make a good joint, you will be able to separate the wires relatively easily, probably because you didn't apply enough heat or enough solder to connect them.

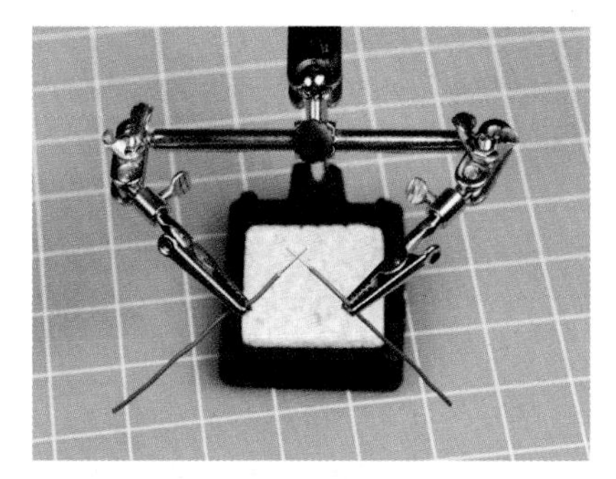

Figure 3-31. *A helping hand work aid is shown here holding two wires with their stripped ends touching. The magnifying glass has been hinged out of the way.*

The reason I asked you to begin by using the higher-powered soldering iron is that it delivers more heat, which makes it easier to use.

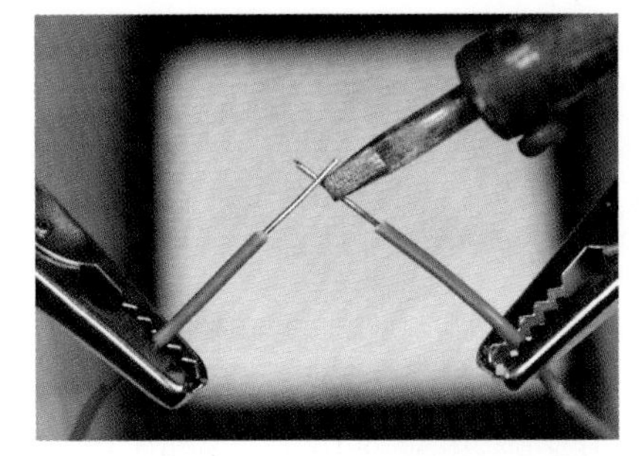

Figure 3-32

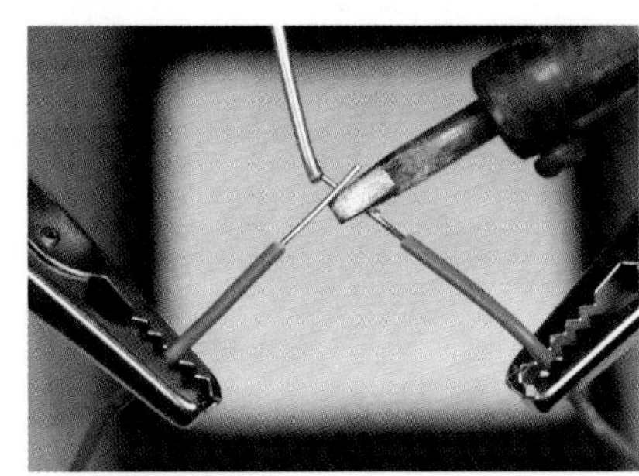

Figure 3-33

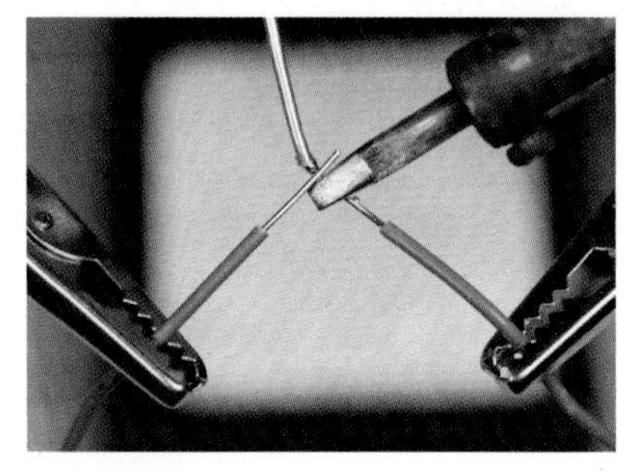

Figure 3-34

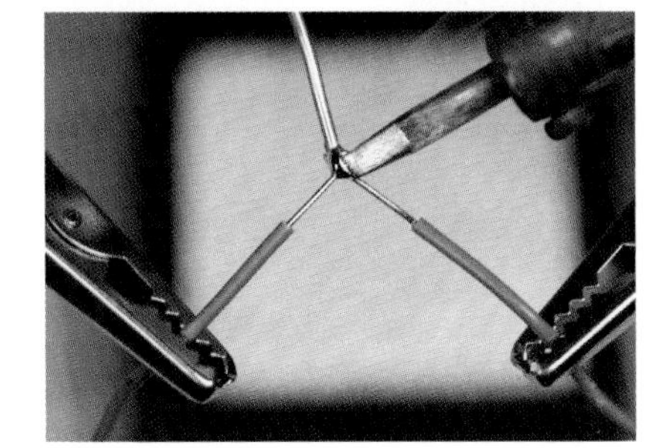

Figure 3-35. *This and the preceding three figures illustrate four steps to making a solder joint: apply heat to the wires, bring in the solder while maintaining the heat, wait for the solder to start to melt, and wait a moment longer for it to form a completely molten bead. The whole process should take between 4 and 6 seconds.*

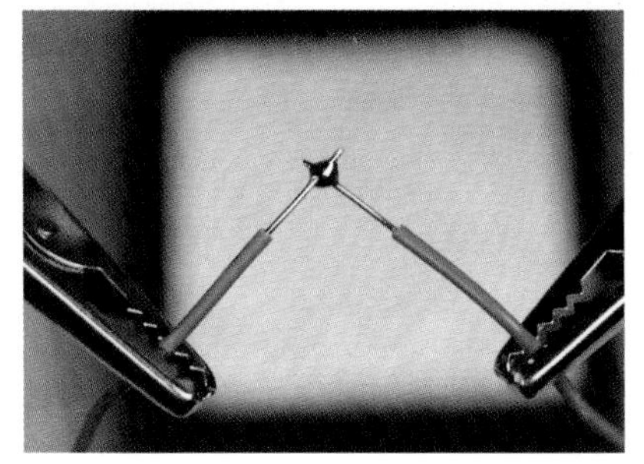

Figure 3-36. *The completed joint should be shiny, uniform, and rounded in shape.*

BACKGROUND

Soldering myths

Myth #1: Soldering is very difficult.
Millions of people have learned how to do it, and statistically, you are unlikely to be less coordinated than all of them. I have a lifelong problem with a tremor in my hands that makes it difficult for me to hold small things steadily. I also get impatient with repetitive detail work. If I can solder components, almost anyone should be able to.

Myth #2: Soldering involves poisonous chemicals.
Modern solder contains no lead. You should avoid inhaling the fumes for prolonged periods, but that also applies to everyday products such as bleach and paint. If soldering was a significant health hazard, we should have seen a high death rate among electronics hobbyists decades ago.

Myth #3: Soldering is hazardous.
A soldering iron is less hazardous than the kind of iron that you might use to iron a shirt, because it delivers less heat. In fact, in my experience, soldering is safer than most activities in a typical home or basement workshop. That doesn't mean you can be careless!

Soldering alternatives

As recently as the 1950s, connections inside electronic appliances such as radio sets were still being hand-soldered by workers on production lines. But the growth of telephone exchanges created a need for a faster way to make large numbers of rapid, reliable point-to-point wiring connections, and "wire wrap" became a viable alternative.

In a wire-wrapped electronics project, components are mounted on a circuit board that has long, gold-plated, sharp-cornered square pins sticking out of the rear. Special silver-plated wire is used, with an inch of insulation stripped from its ends. A manual or power-driven wire-wrap tool twirls the end of a wire around one of the pins, applying sufficient tension to "cold-weld" the soft silver plating of the wire to the pin. The wrapping process exerts sufficient pressure to make a very reliable joint, especially as 7 to 9 turns of wire are applied, each turn touching all four corners of the pin.

During the 1970s and 1980s, this system was adopted by hobbyists who built their own home computers. A wire-wrapped circuit board from a hand-built computer is shown in Figure 3-37. The technique was used by NASA to wire the computer in the Apollo spacecraft that went to the moon, but today, wire-wrapping has few commercial applications.

The widespread industrial use of "through-hole" components, such as the chips on early desktop computers, encouraged development of wave soldering, in which a wave or waterfall of molten solder is applied to the underside of a preheated circuit board where chips have been inserted. A masking technique prevents the solder from sticking where it isn't wanted.

Today, surface-mount components (which are significantly smaller than their through-hole counterparts) are glued to a circuit board with a solder paste, and the entire assembly is then heated, melting the paste to create a permanent connection.

Figure 3-37. *This picture shows some of the wire-wrapping in Steve Chamberlin's custom-built, retro 8-bit CPU and computer. "Back in the day," connecting such a network of wires with solder joints would have been unduly time-consuming and prone to faults. Photo credit: Steve Chamberlin.*

TOOLS

Eight most common soldering errors

1. Not enough heat.

   The joint looks OK, but because you didn't apply quite enough heat, the solder didn't melt sufficiently to realign its internal molecular structure. It remained granular instead of becoming a solid, uniform blob, and you end up with a "dry joint," also known as a "cold joint," which will come apart when you pull the wires away from each other. Reheat the joint thoroughly and apply new solder.

   A leading cause of underheated solder is the temptation to use the soldering iron to carry solder to the joint. This results in the cold wires reducing the temperature of the solder. What you should do is touch the soldering iron to heat the wires first, and then apply the solder. This way, the wires are hot and help to melt the solder, which wants to stick to them.

   Because this is such a universal problem, I'll repeat myself: *Never melt solder on the tip of the iron and then use it to carry the solder to the joint.*

   You don't want to put hot solder on cold wires. You want to put cold solder on hot wires.

2. Too much heat.

   This may not hurt the joint, but can damage everything around it. Vinyl insulation will melt, exposing the wire and raising the risk of short circuits. You can easily damage semiconductors, and may even melt the internal plastic components of switches and connectors.

   Damaged components must be desoldered and replaced, which will take time and tends to be a big hassle (see "Tools: Desoldering" on page 109 for advice).

3. Not enough solder.

   A thin connection between two conductors may not be strong enough. When joining two wires, always check the underside of the joint to see whether the solder penetrated completely.

4. Moving the joint before the solder solidifies.

   You may create a fracture that you won't necessarily see. It may not stop your circuit from working, but at some point in the future, as a result of vibration or thermal stresses, the fracture can separate just enough to break electrical contact. Tracking it down will then be a chore. If you clamp components before you join them, or use perforated board to hold the components steady, you can avoid this problem.

5. Dirt or grease.

   Electrical solder contains rosin that cleans the metal that you're working with, but contaminants can still prevent solder from sticking. If any component looks dirty, clean it with fine sandpaper before joining it.

6. Carbon on the tip of your soldering iron.

   The iron gradually accumulates flecks of black carbon during use, and they can act as a barrier to heat transfer. Wipe the tip of the iron on the little sponge mounted in the base of your soldering iron stand or your helping hand.

7. Inappropriate materials.

   Electronic solder is designed for electronic components. It will not work with aluminum, stainless steel, or various other metals. You may be able to make it stick to chrome-plated items, but only with difficulty.

8. Failure to test the joint.

   Don't just assume that it's OK. Always test it, by applying manual force if you can (see Figures 3-38 and 3-39 for the ideal protocol) or, if you can't get a grip on the joint, slip a screwdriver blade under it and flex it just a little, or use small pliers to try to pull it apart. Don't be concerned about ruining your work. If your joint doesn't survive rough treatment, it wasn't a good joint.

Of the eight errors, dry/cold joints are by far the worst, because they are easy to make and can look OK.

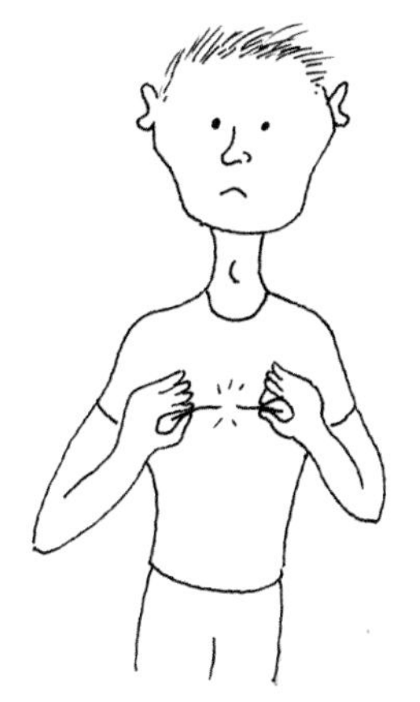

Figure 3-38. *Test result of a bad solder joint.*

Figure 3-39. *Test result of a good solder joint.*

Your Second Solder Joint

Time now to try your pencil-style soldering iron. Once again, you must leave it plugged in for a good five minutes to make sure it's hot enough. In the meantime, don't forget to unplug your other soldering iron, and put it somewhere safe while it cools.

This time I'd like you to align the wires parallel with each other. Joining them this way is a little more difficult than joining them when they cross each other, but it's a necessary skill. Otherwise, you won't be able to slide heat-shrink tubing over the finished joint to insulate it.

Figures 3-40 through 3-44 show a successful joint of this type. The two wires do not have to make perfect contact with each other; the solder will fill any small gaps. But the wires must be hot enough for the solder to flow, and this can take an extra few seconds when you use the low-wattage pencil-style iron.

Be sure to feed the solder in as shown in the pictures. Remember: don't try to carry the solder to the joint on the tip of the iron. Heat the wires first, and then touch the solder to the wires and the tip of the iron, while keeping it in contact with the wires. Wait until the solder liquifies, and you will see it running eagerly into the joint. If this doesn't happen, be more patient and apply the heat for a little longer.

Figure 3-40

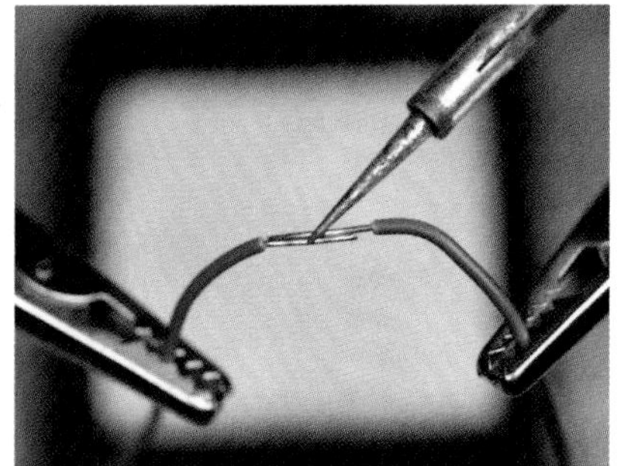

Figure 3-41

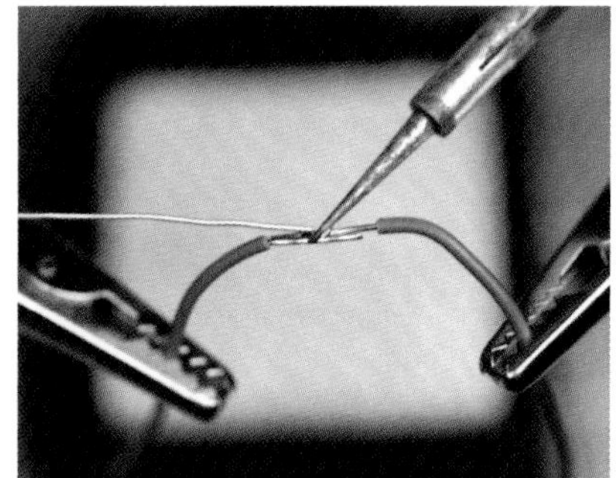

Figure 3-42

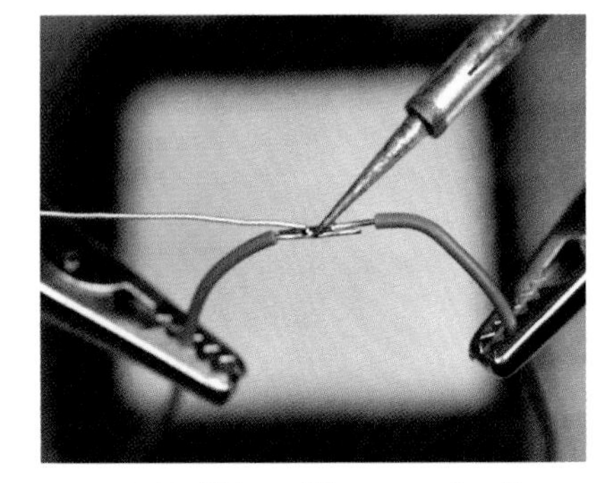

Figure 3-43. *This and the preceding three figures show how joining two wires that are parallel is more difficult, and the low-wattage, pencil-type soldering iron will require longer to heat them sufficiently for a good joint. Thinner solder can be used.*

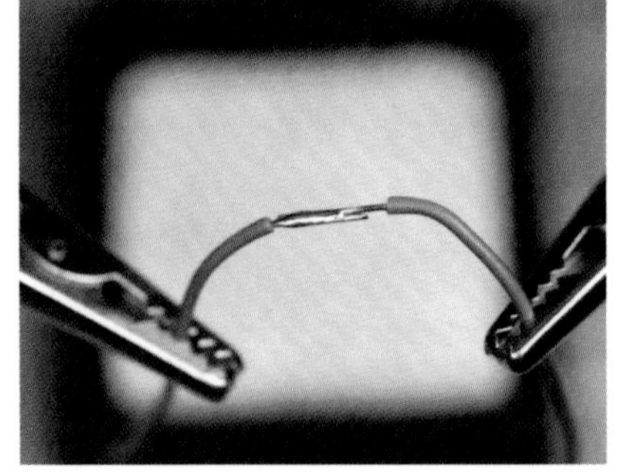

Figure 3-44. *The finished joint has enough solder for strength, but not so much solder that it will prevent heat-shrink tubing from sliding over it.*

THEORY

Soldering theory

The better you understand the process of soldering, the easier it should be for you to make good solder joints.

The tip of the soldering iron is hot, and you want to transfer that heat into the joint that you are trying to make. In this situation, you can think of the heat as being like a fluid. The larger the connection is between the soldering iron and the joint, the greater the quantity of heat, per second, that can flow through it.

For this reason, you should adjust the angle of the soldering iron so that it makes the widest possible contact. If it touches the wires only at a tiny point, you'll limit the amount of heat flow. Figures 3-45 and 3-46 illustrate this concept. Once the solder starts to melt, it broadens the area of contact, which helps to transfer more heat, so the process accelerates naturally. Initiating it is the tricky part.

The other aspect of heat flow that you should consider is that it can suck heat away from the places where you want it, and deliver it to places where you don't want it. If you're trying to solder a very heavy piece of copper wire, the joint may never get hot enough to melt the solder, because the heavy wire conducts heat away from the joint. You may find that even a 40-watt iron isn't powerful enough to overcome this problem, and if you are doing heavy work, you may need a more powerful iron.

As a general rule, if you can't complete a solder joint in 10 seconds, you aren't applying enough heat.

Figure 3-45. *With only a small surface area of contact between the iron and the working surface, an insufficient amount of heat is transferred.*

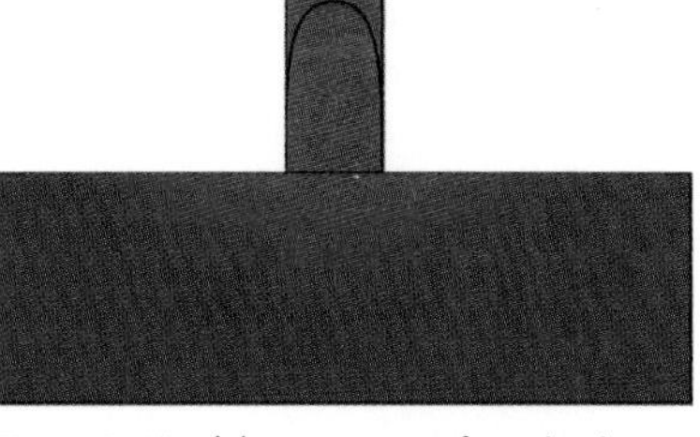

Figure 3-46. *A larger area of contact between the soldering iron and its target will greatly increase the heat transfer.*

TOOLS

Desoldering

Desoldering is much, much harder than soldering. Two simple tools are available:

- *Suction pump*. First, you apply the soldering iron to make the solder liquid. Then you use this simple gadget to try to suck up as much of the liquid as possible. Usually it won't remove enough metal to allow you to pull the joint apart, and you will have to try the next tool. Refer back to Figure 3-10.
- *Desoldering wick or braid*. Desoldering wick, also known as braid, is designed to soak up the solder from a joint, but again, it won't clean the joint entirely, and you will be in the awkward position of trying to use both hands to pull components apart while simultaneously applying heat to stop the solder from solidifying. Refer back to Figure 3-11.

I don't have much advice about desoldering. It's a frustrating experience (at least, I think so) and can damage components irrevocably.

Heat Guns Get Hot, Too!

Notice the chromed steel tube at the business end of your heat gun. Steel costs more than plastic, so the manufacturer must have put it there for a good reason—and the reason is that the air flowing through it becomes so hot that it would melt a plastic tube.

The metal tube stays hot enough to burn you for several minutes after you've used it. And, as in the case of soldering irons, other people (and pets) are vulnerable, because they won't necessarily know that the heat gun is hot. Most of all, make sure that no one in your home ever makes the mistake of using a heat gun as a hair dryer (Figure 3-47).

This tool is just a little more hazardous than it appears.

Adding Insulation

After you've succeeded in making a good inline solder connection between two wires, it's time for the easy part. Choose some heat-shrink tubing that is just big enough to slide over the joint with a little bit of room to spare.

Figure 3-47. *Other members of your family should understand that although a heat gun looks like a hair dryer, appearances may be deceptive.*

Slide the tubing along until the joint is centered under it, hold it in front of your heat gun, and switch on the gun (keeping your fingers away from the blast of superheated air). Turn the wire so that you heat both sides. The tubing should shrink tight around the joint within half a minute. If you overheat the tubing, it may shrink so much that it splits, at which point you must remove it and start over. As soon as the tubing is tight around the wire, your job is done, and there's no point in making it any hotter. Figures 3-48 through 3-50 show the desired result. I used white tubing because it shows up well in photographs. Different colors of heat-shrink tubing all perform the same way.

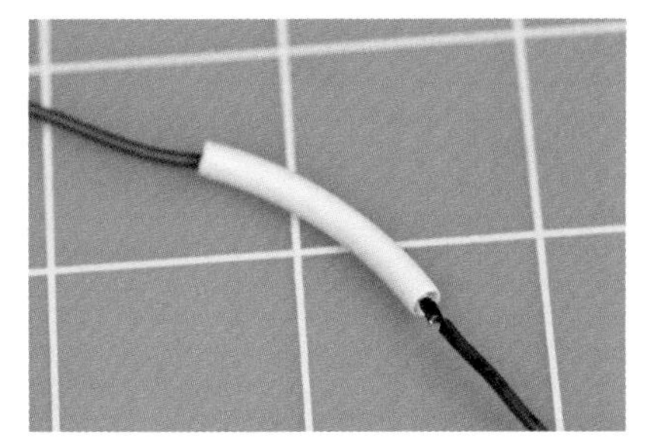

Figure 3-48. *Slip the tubing over your wire joint.*

Figure 3-49. *Apply heat to the tubing.*

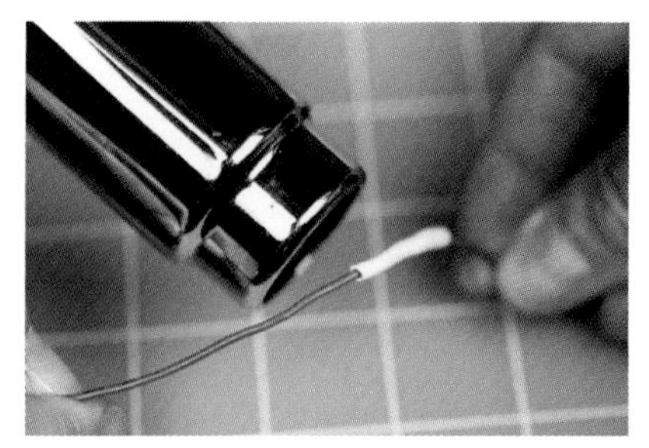

Figure 3-50. *Leave the heat on the tubing until it shrinks to firmly cover the joint.*

I suggest you next practice your soldering skills on a couple of practical projects. In the first one, you can add color-coded, solid-core wires to your AC adapter, and in the second one, you can shorten the power cord for a laptop power supply. You can use your larger soldering iron for both of these tasks, because neither of them involves any heat-sensitive components.

Modifying an AC Adapter

In the previous chapter, I mentioned the irritation of being unable to push the wires from your AC adapter into the holes of your breadboard. So, let's fix this right now:

1. Cut two pieces of solid-conductor 22-gauge wire—one of them red, the other black or blue. Each should be about 2 inches long. Strip a quarter-inch of insulation from both ends of each piece of wire.

2. Trim the wire from your AC adapter. You need to expose some fresh, clean copper to maximize your chance of getting the solder to stick.

   I suggest that you make one conductor longer than the other to minimize the chance of the bare ends touching and creating a short circuit. Use your meter, set to DC volts, if you have any doubt about which conductor is positive.

Solder the wires and add heat-shrink tubing as you did in the practice session. The result should look like Figure 3-51.

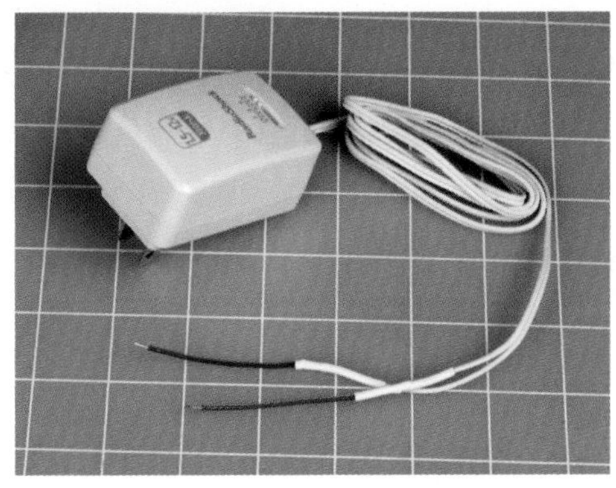

Figure 3-51. *Solid-core color-coded wires, soldered onto the wires from an AC adapter, provide a convenient way to feed power to a breadboard. Note that the wires are of differing lengths to reduce the risk of them touching each other.*

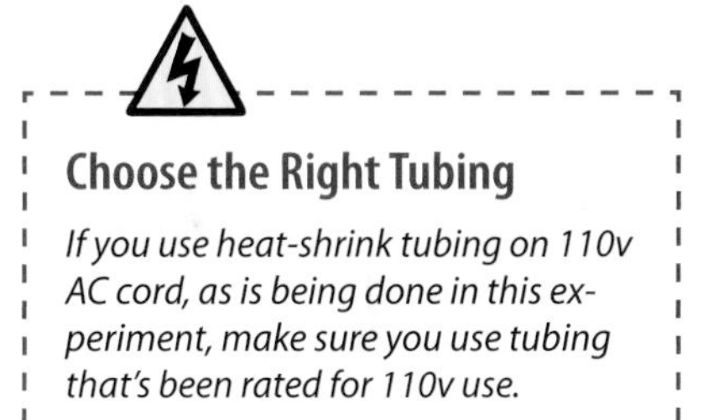

Choose the Right Tubing

If you use heat-shrink tubing on 110v AC cord, as is being done in this experiment, make sure you use tubing that's been rated for 110v use.

Shortening a Power Cord

When I travel, I like to minimize everything. It always annoys me that the power cord for the power supply of my laptop is 4 feet long. The thinner wire that connects the power supply to the computer is also 4 feet long, and I just don't need that much wire.

After searching exhaustively I couldn't find any laptop power cables shorter than 3 feet, so I decided to shorten one myself. If you feel no need to do this, you should try the following procedure on an old extension cord, just as an exercise. You do need to go through these steps to acquire some practice in soldering heavier, stranded wire and using heat-shrink tubing:

1. Use your wire cutters to chop the wire, and then a utility knife to split the two conductors, with one shorter than the other. When splicing a power cord or similar cable containing two or more conductors, it's good to avoid having the joints opposite each other. They fit more snugly if they are offset, and there's less risk of a short circuit if a joint fails.

Figure 3-52

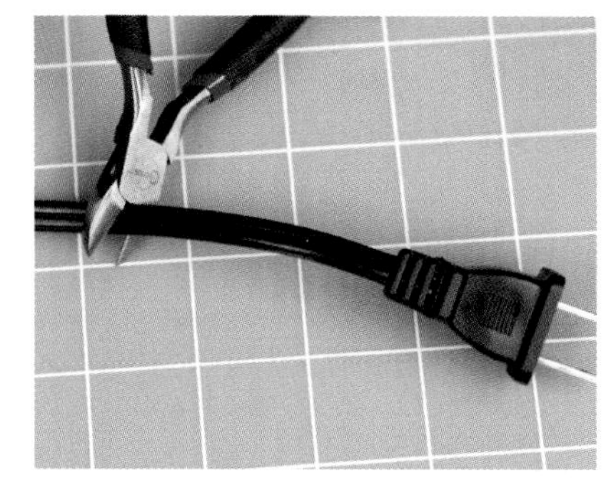

Figure 3-53

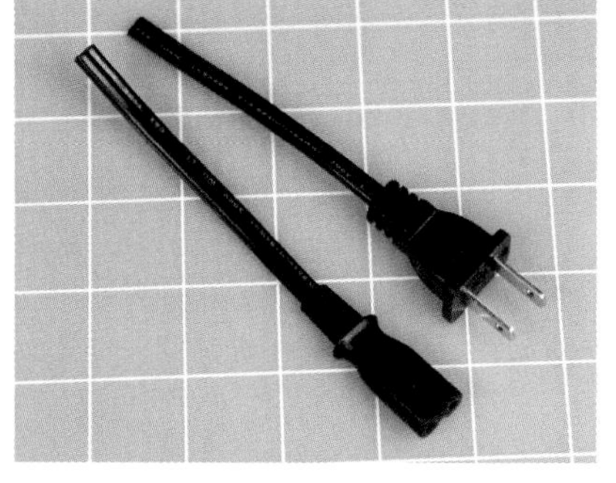

Figure 3-54

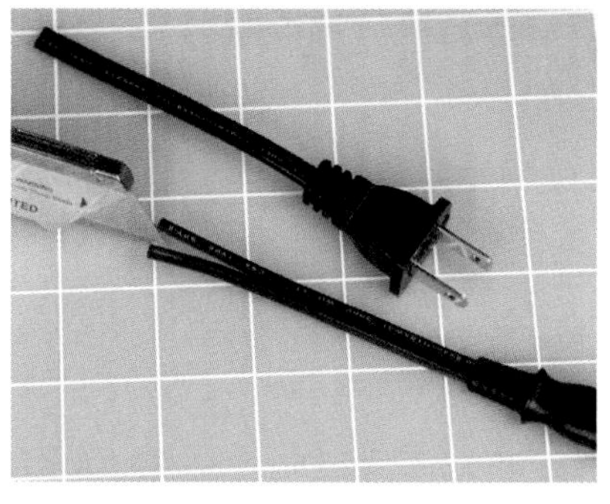

Figure 3-55

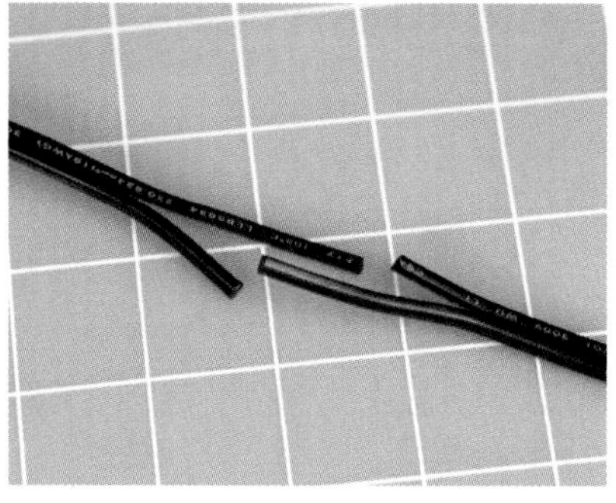

Figure 3-56

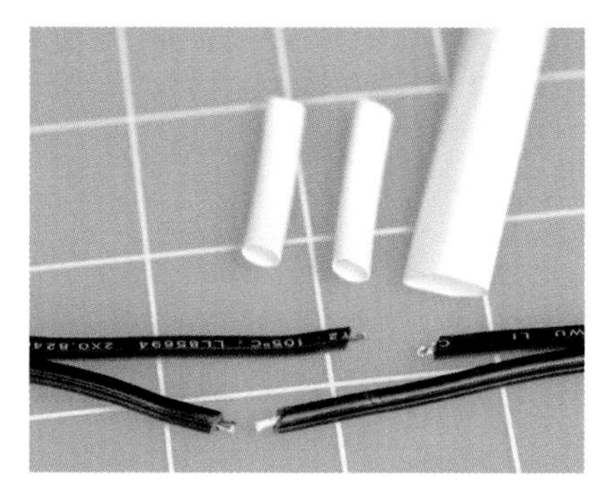

Figure 3-57

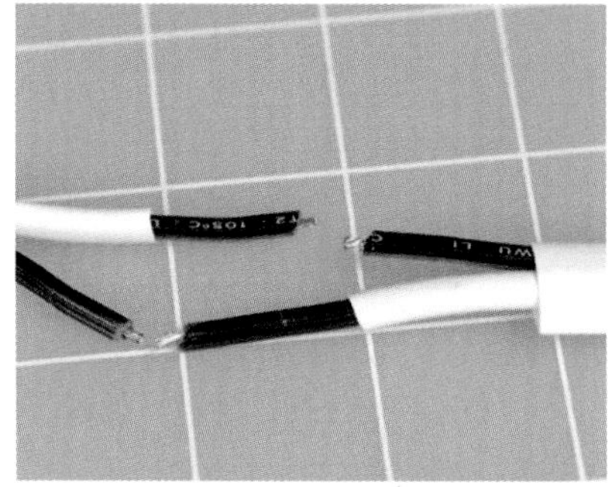

Figure 3-58. *Figures 3-52 through 3-58 illustrate the sequence of steps to prepare for making a shortened power cord for a laptop computer power supply.*

2. Strip off a minimal amount of insulation. One-eighth of an inch (3 mm) is sufficient. The automatic wire strippers that I mentioned in the shopping list in Chapter 1 are especially convenient, but regular wire strippers will do the job.

3. Cut two pieces of heat-shrink tubing, each 1 inch long, big enough to slide over the separate conductors in your cable. Cut a separate 2-inch piece of larger tubing that will slide over the entire joint when it's done. The steps described so far are illustrated in Figures 3-52 through 3-58.

4. Now for the most difficult part: activating your human memory. You have to remember to slide the tubing onto the wire *before* you make your solder joint, because the plugs on the ends of the wires will prevent you from adding any heat-shrink tubing later. If you're as impatient as I am, it's very difficult to remember to do this every time.

5. Use your helping hand to align the first joint. Push the two pieces of wire together so that the strands intermingle, and then squeeze them tight between finger and thumb, so that there are no little bits sticking out. A stray strand of wire can puncture heat-shrink tubing when the tubing is hot and soft and is shrinking around the joint.

6. The wire that you're joining is much heavier than the 22-gauge wire that you worked with previously, so it will suck up more heat, and you must touch the soldering iron to it for a longer time. Make sure that the solder flows all the way into the joint, and check the underside after the joint is cool. Most likely you'll find some bare copper strands there. The joint should become a nice solid, rounded, shiny blob. Keep the heat-shrink tubing as far away from the joint as possible while you're using the soldering iron, so that heat from the iron doesn't shrink the tubing prematurely, preventing you from sliding it over the joint later.

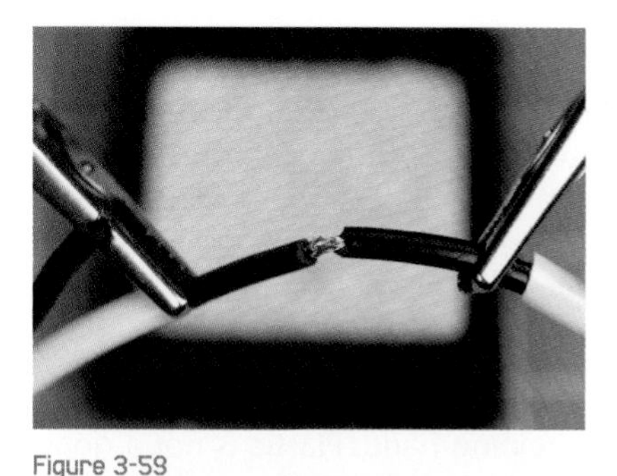

Figure 3-59

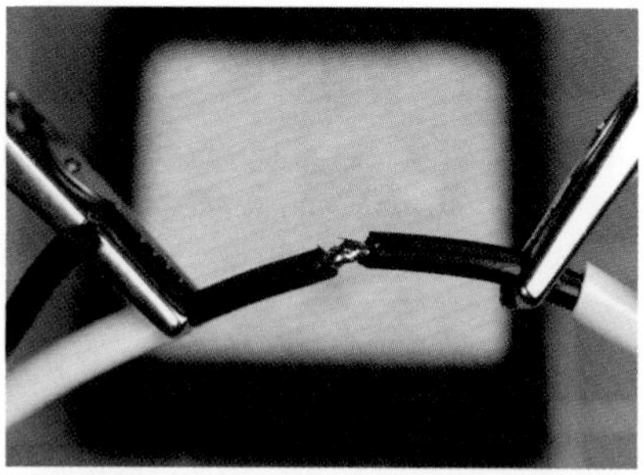

Figure 3-60

Figure 3-61

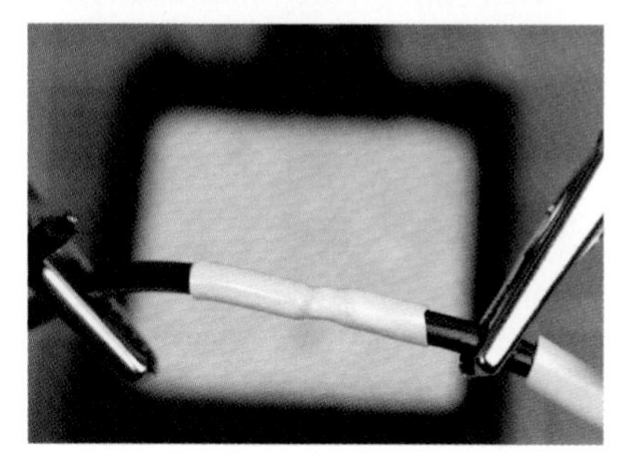

Figure 3-62

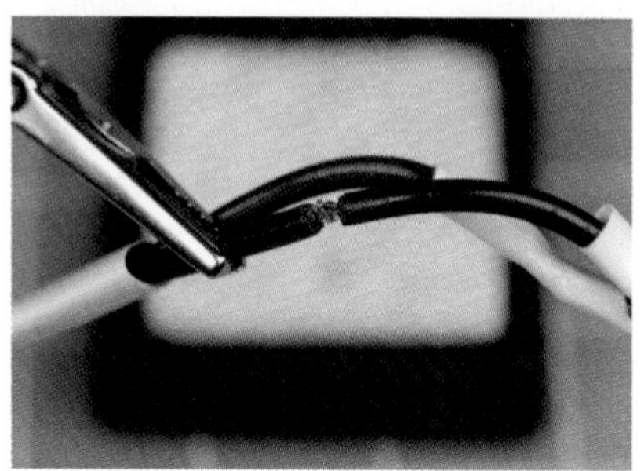

Figure 3-63

Figure 3-64

7. When the joint has cooled, slide the heat-shrink tubing over it, and apply the heat gun. Now repeat the process with the other conductor. Finally, slide the larger piece of tubing over the joint. You did remember to put the large tubing onto the wire at the beginning, didn't you?

Figures 3-59 through 3-65 show the steps all the way through to the end.

If you have completed the soldering exercises so far, you now have sufficient basic skills to solder your first electronic circuit. But first, I want you to verify the vulnerability of components to heat.

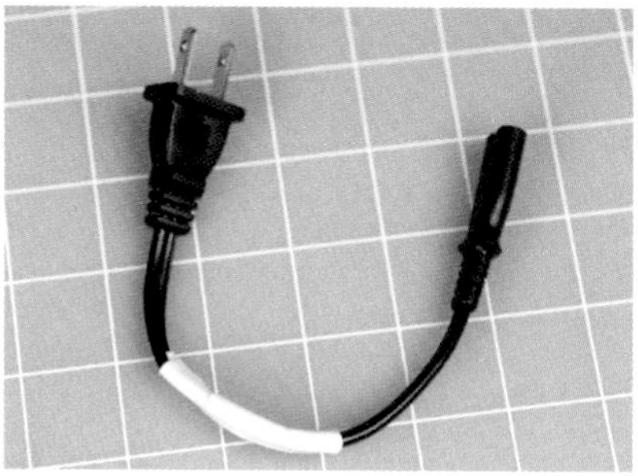

Figure 3-65. *Completion of the shortened power cord for a laptop power supply.*

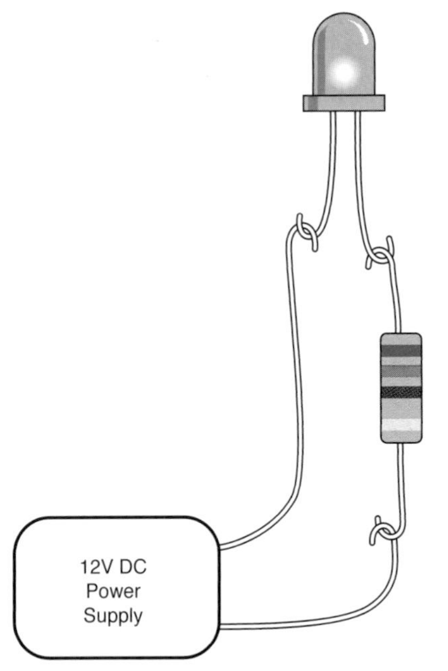

Figure 3-66. *By literally hooking together the leads from a resistor and a white-light LED, we minimize pathways for heat to escape during the subsequent test.*

Figure 3-67. *Applying heat with a 15-watt soldering iron. A typical LED should withstand this treatment for two or three minutes, but if you substitute a 30-watt soldering iron, the LED is likely to burn out in under 15 seconds.*

Experiment 13: Broil an LED

In Chapter 1, you saw how an LED can be damaged if too much current flows through it. The electricity caused heat, which melted the LED. Unsurprisingly, you can just as easily melt it by applying too much heat to one of its leads with a soldering iron. The question is: how much heat is too much? Let's find out.

You will need:

- 30-watt or 40-watt soldering iron
- 15-watt pencil-type soldering iron
- A couple of LEDs (that are expendable)
- 680Ω resistor
- Wire cutters and sharp-nosed pliers
- "Helping hand" gadget to hold your work

I don't want you to use alligator clips to join the LED to a power supply, because the alligator clip will divert and absorb some of the heat from your soldering iron. Instead, please use some sharp-nosed pliers to bend each of the leads from an LED into little hooks, and do the same thing with the wires on a 680Ω load resistor. Finally bend the new wires on your AC adapter so that they, too, are tiny hooks. Now you can put the hooks together like links in a chain, as shown in Figure 3-66.

Grip the plastic body of the LED in your helping hand. Plastic is not a good thermal conductor, so the helping hand shouldn't siphon too much heat away from our target. The resistor can dangle from one of the leads on the LED, and the wire from the AC adapter can hang from that, a little farther down. Gravity should be sufficient to make this work. Set your AC adapter to deliver 12 volts as before, plug it in, and your LED should be shining brightly. I used a white LED in this experiment, because it's easier to photograph.

Make sure your two soldering irons are really hot. They should have been plugged in for at least five minutes. Now take the pencil-style iron and hold its tip firmly against one of the leads on your glowing LED, while you check the time with a watch. Figure 3-67 shows the setup.

I'm betting that you can sustain this contact for a full three minutes without burning out the LED. This is why you use a 15-watt soldering iron for delicate electronics work—it doesn't endanger the components.

Allow your LED wire to cool, and then apply your more powerful soldering iron to the same piece of wire as before. Again, make sure it is completely hot, and I think you'll find that the LED will go dark after as little as 10 seconds (note, some LEDs can survive higher temperatures than others). This is why you *don't* use a 30-watt soldering iron for delicate electronics work.

The large iron doesn't necessarily reach a higher temperature than the small one. It just has a larger heat capacity. In other words, a greater quantity of heat can flow out of it, at a faster rate.

Throw away your burned-out LED. Substitute a new one, connected as before, but add a full-size copper alligator clip to one of the leads up near the body of the LED, as shown in Figure 3-68. Press the tip of your 30-watt or 40-watt soldering iron against the lead just *below* the alligator clip. This time, you should be able to hold the powerful soldering iron in place for a full two minutes without burning out the LED.

Figure 3-68. *When a copper alligator clip is used as a heat sink, you should be able to apply a 30-watt soldering iron (below the clip) without damaging the LED.*

Imagine the heat flowing out through the tip of your soldering iron, into the wire that leads to the LED—except that the heat meets the alligator clip along the way, as shown in Figure 3-69. The clip is like an empty vessel waiting to be filled. It offers much less resistance to heat than the remainder of the wire leading to the LED, so the heat prefers to flow into the copper clip, leaving the LED unharmed. At the end of your experiment, if you touch the clip, you'll find that it's hot, while the LED remains relatively cooler.

The alligator clip is known as a heat sink, and it should be made of copper, because copper is one of the best conductors of heat.

Because the 15-watt soldering iron failed to harm the LED, you may conclude that the 15-watt iron is completely safe, eliminating all need for a heat sink. Well, this may be true. The problem is, you don't really know whether some semiconductors may be more heat-sensitive than LEDs. Because the consequences of burning out a component are so exasperating, I suggest you should play it safe and use a heat sink in these circumstances:

- If you apply 15-watt iron extremely close to a semiconductor for 20 seconds or more.
- If you apply a 30-watt iron near resistors or capacitors for 10 seconds or more. (Never use it near semiconductors.)
- If you apply a 30-watt iron near anything meltable for 20 seconds or more. Meltable items include insulation on wires, plastic connectors, and plastic components inside switches.

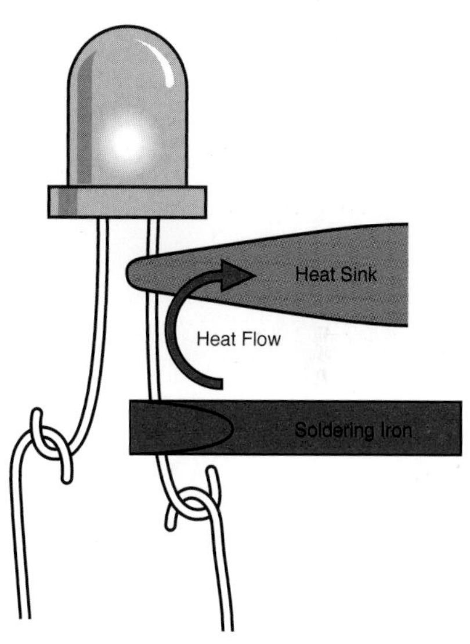

Figure 3-69. *The heat sink intercepts the heat, sucks it up, and protects the LED from damage.*

Rules for Heat Sinking

1. Full-size copper alligator clips do work better.
2. Clamp the alligator clip as close as possible to the component and as far as possible from the joint. (You don't want to suck too much heat away from the joint.)
3. Make sure there is a metal-to-metal connection between the alligator clip and the wire to promote good heat transfer.

FUNDAMENTALS

All about perforated board

For the remainder of this book, you'll be using perforated board whenever you want to create permanent, soldered circuits. There are three ways to do this:

1. Point-to-point wiring. You use perforated board that has no connections behind the holes. Either the board has no copper traces on it at all, as in Figure 3-70, or you will find a little circular copper circle around each hole, as in Figure 3-71. These circles are not connected with each other and are used only to stabilize the components that you assemble.

   Point-to-point wiring allows you to place the components in a convenient, compact layout that can be very similar to a schematic. Under the board you bend the wires to link the components, and solder them together, adding extra lengths of wire if necessary. The advantage of this system is that it can be extremely compact. The disadvantage is that the layout can be confusing, leading to errors.

2. Breadboard-style wiring. Use perforated board that is printed with copper traces in exactly the same pattern as the conductors inside a breadboard. Once your circuit works on the breadboard, you move the components over to the perfboard one by one, maintaining their exact same positions relative to each other. You solder the "legs" of the components to the copper traces, which complete the circuit. Then you trim off the surplus wire. The advantage of this procedure is that it's quick, requires very little planning, and minimizes the possibility for errors. The disadvantage is that it tends to waste space. A cheap example is shown in Figure 3-72.

3. You can etch your own circuit board with customized copper traces that link your components in a point-to-point layout. This is the most professional way to complete a project, but it requires more time, trouble, and equipment than is practical in this book.

   Point-to-point wiring is like working with alligator clips, on a much smaller scale. The first soldered project will use this procedure.

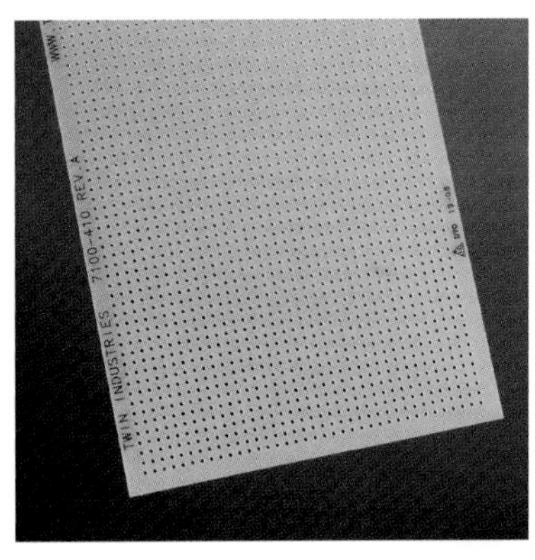

Figure 3-70

Figure 3-71. *Either this type of perforated board or the type in Figure 3-70 can be used for point-to-point wiring in Experiment 14.*

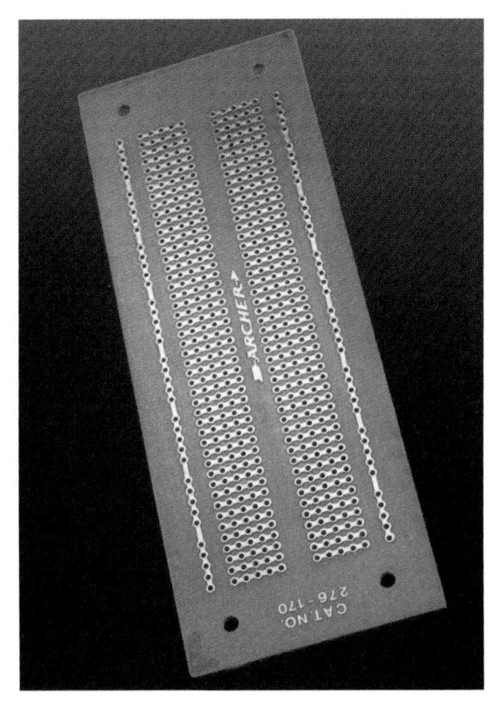

Figure 3-72. *Perforated board etched with copper in variants of a breadboard layout. This example is appropriate for Experiment 15.*

Experiment 14: A Pulsing Glow

You will need:

- Breadboard
- 15-watt pencil-type soldering iron
- Thin solder (0.022 inches or similar)
- Wire strippers and cutters
- Plain perforated board (no copper etching necessary)
- Small vise or clamp to hold your perforated board
- Resistors, various
- Capacitors, electrolytic, 100 µF and 220 µF, one of each
- Red LED, 5 mm, rated for 2 volts approximately
- 2N6027 programmable unijunction transistor

Your first circuit using a PUT was a slow-speed oscillator that made an LED flash about twice each second. The flashes looked very "electronic," by which I mean that the LED blinked on and off without a gradual transition between each state. I'm wondering if we can modify this circuit to make the LED pulse in a more gentle, interesting way, like the warning light on an Apple MacBook when it's in "sleep" mode. I'm thinking that something of this sort might be wearable as an ornament, if it's small enough and elegant enough.

I'm also thinking that this first soldering project will serve three purposes. It will test and refine your skill at joining wires together, will teach you point-to-point wiring with perfboard, and will give you some additional insight into the way that capacitors can be used to adjust timing.

Look back at the original circuit in Experiment 11, on page 82. Refresh your memory about the way it worked. The capacitor charges through a resistor until it has enough voltage to overcome the internal resistance in the PUT. Then the capacitor discharges through the PUT and flashes the LED.

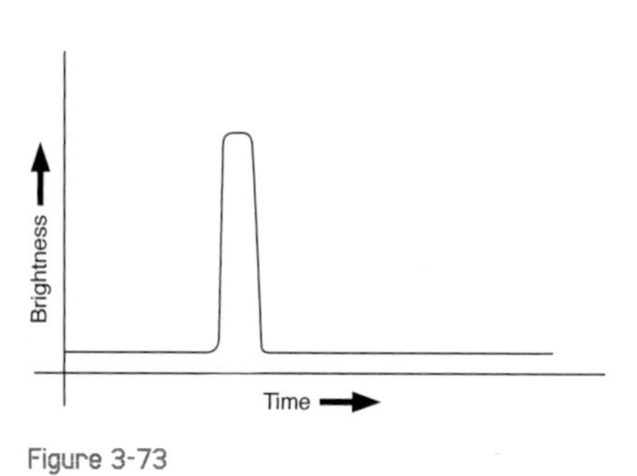

Figure 3-73

If you drew a graph of the light coming out of the LED, it would be a thin, square-shaped pulse, as shown in Figure 3-73. How can we fill it out to make it more like the curve in Figure 3-74, so that the LED fades gently on and off, like a heartbeat?

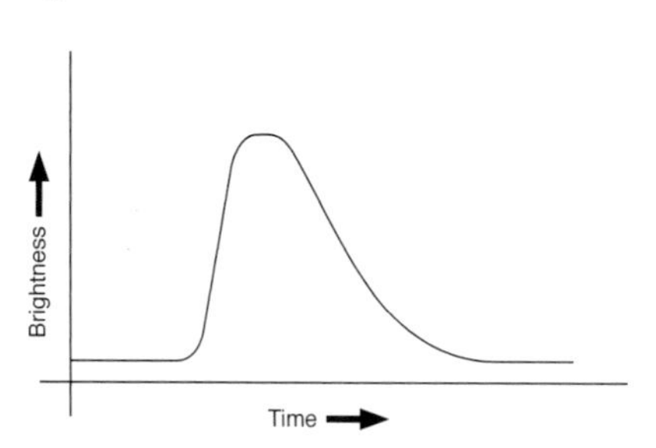

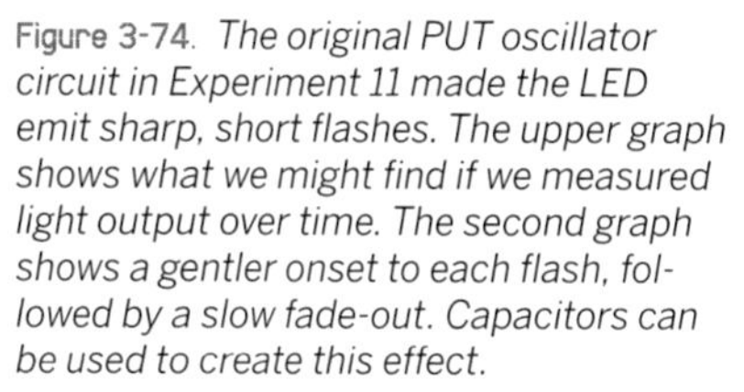

Figure 3-74. *The original PUT oscillator circuit in Experiment 11 made the LED emit sharp, short flashes. The upper graph shows what we might find if we measured light output over time. The second graph shows a gentler onset to each flash, followed by a slow fade-out. Capacitors can be used to create this effect.*

One thing is obvious: the LED is going to be emitting a greater total amount of light in each cycle. Therefore, it's going to need more power. This means that C1, in Figure 3-75, must be a larger capacitor.

When we have a larger capacitor, it takes longer to charge. To keep the flashes reasonably frequent, we'll need a lower-value resistor for R1 to charge the capacitor quickly enough. In addition, reducing the values of R2 and R3 will program the PUT to allow a longer pulse.

Most important, I want to discharge the capacitor through a resistor to make the onset of the pulse gradual instead of sudden. Remember, when you have a resistor in series with a capacitor, the capacitor not only charges more slowly, but discharges more slowly.

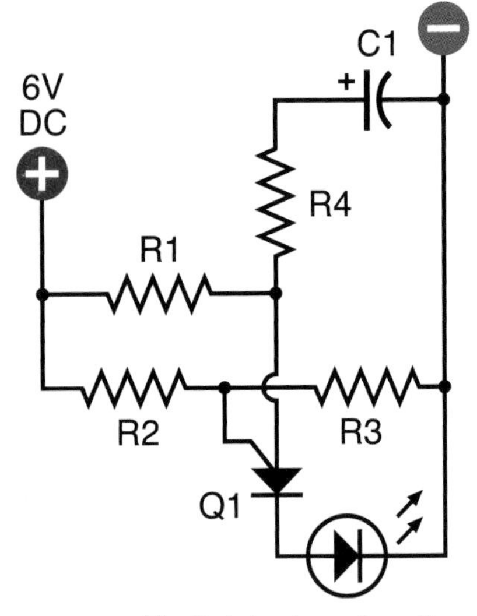

Figure 3-75. *The first step toward creating a gentler flashing effect is to use a larger capacitor for C1 and discharge it through a resistor, R4. Lower-value resistors are necessary to charge the capacitor rapidly enough.*

R1: 33K
R2: 1K
R3: 1K
R4: 1K
C1: 100 µF electrolytic
Q1: 2N6027

Figure 3-75 shows these features. Compare it with Figure 2-103 on page 85. R1 is now 33K instead of 470K. R2 and R3 are reduced to 1K. R4 also is 1K, so that the capacitor takes longer to discharge. And C1 is now 100 µF instead of 2.2 µF.

Assemble this circuit on a breadboard, and compare the results when you include R4 or bypass it with a plain piece of jumper wire. It softens the pulse a bit, but we can work on it some more. On the output side of the PUT, we can add another capacitor. This will charge itself when the pulse comes out of the PUT, and then discharge itself gradually through another resistor, so that the light from the LED dies away more slowly.

Figure 3-76 shows the setup. C2 is large—220 µF—so it sucks up the pulse that comes out of the PUT, and then gradually releases it through 330Ω resistor R5 and the LED. You'll see that the LED behaves differently now, fading out instead of blinking off. But the resistances that I've added have dimmed the LED, and to brighten it, you should increase the power supply from 6 volts to 9 volts.

Remember that a capacitor imposes a smoothing effect only if one side of it is grounded to the negative side of the power supply. The presence of the negative charge on that side of the capacitor attracts the positive pulse to the other side.

I like the look of this heartbeat effect. I can imagine a piece of wearable electronic jewelry that pulses in this sensual way, very different from the hard-edged, sharp-on-and-off of a simple oscillator circuit. The only question is whether we can squeeze the components into a package that is small enough to wear.

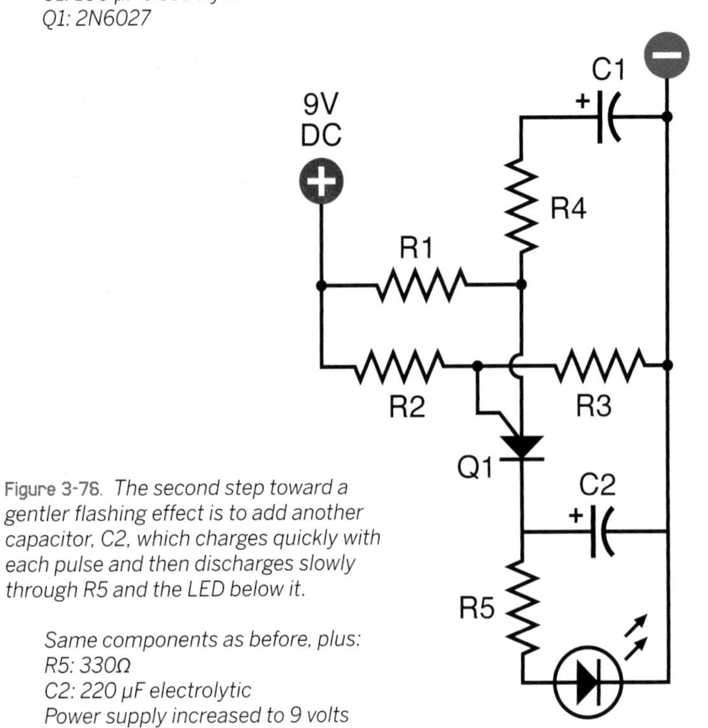

Figure 3-76. *The second step toward a gentler flashing effect is to add another capacitor, C2, which charges quickly with each pulse and then discharges slowly through R5 and the LED below it.*

Same components as before, plus:
R5: 330Ω
C2: 220 µF electrolytic
Power supply increased to 9 volts

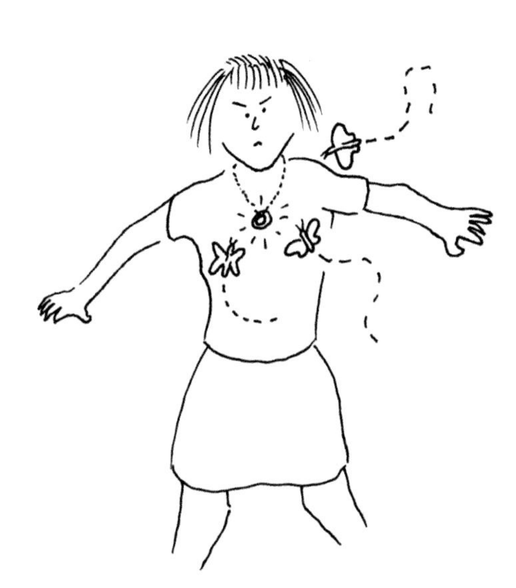

Figure 3-77. *On a dark night in a rural area, the heartbeat flasher may be attractive in unexpected ways.*

Resizing the Circuit

The first step is to look at the physical components and imagine how to fit them into a small space. Figure 3-78 shows a 3D view of a compact arrangement. Check this carefully, tracing all the paths through the circuit, and you'll see that it's the same as the schematic. The trouble is that if we solder the components together like this, they won't have much strength. All the little wires can bend easily, and there's no easy way to mount the circuit in something or on something.

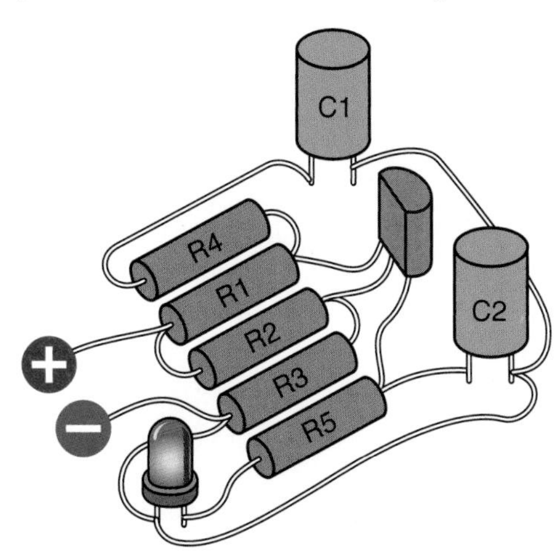

Figure 3-78. *This layout of components replicates their connections in the schematic diagram while squeezing them into a minimal amount of space.*

The answer is to put it on a substrate, which is one of those terms that people in the electronics field like to use, perhaps because it sounds more technical than "perfboard." But perforated board is what we need, and Figure 3-79 shows the components transferred onto a piece of board measuring just 1 inch by 0.8 inch.

The center version of this diagram uses dotted lines to show how the components will be connected with each other underneath the board. Mostly the leads that stick out from underneath the components will be long enough to make these connections.

Finally, the bottom version of the perfboard diagram shows the perfboard flipped left-to-right (notice the L and the R have been transposed to remind you, and I've used a darker color to indicate the underside of the board). Orange circles indicate where solder joints will be needed.

The LED should be unpluggable, because we may want to run it at some distance from the circuit. Likewise the power source should be unpluggable. Fortunately we can buy miniature connectors that fit right into the perforated board. You may have to go to large online retail suppliers such as Mouser.com for these. Some manufacturers call them "single inline sockets and headers," while others call them "boardmount sockets and pinstrip headers." Refer back to Figure 3-29 and check the shopping list for more details.

This is a very compact design that will require careful work with your pencil-style soldering iron. Because a piece of perforated board as small as this will tend to skitter around, I suggest that you apply your miniature vise to one end to anchor it with some weight while still allowing you to turn it easily.

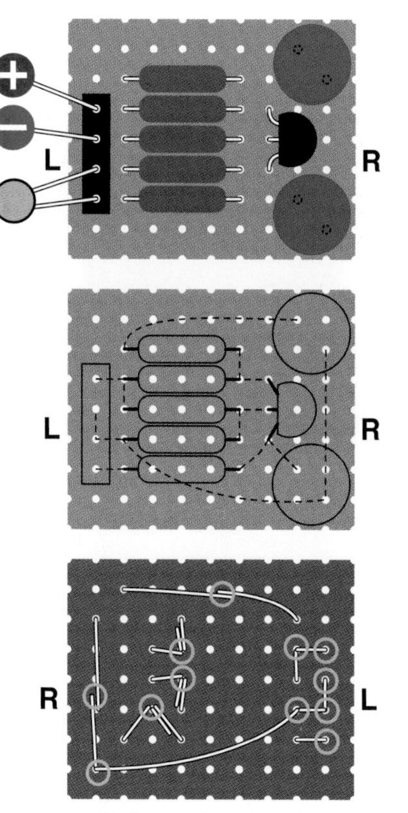

Figure 3-79. *Perforated board can be used to support the layout of components. Their leads are soldered together under the board to create the circuit. The middle diagram shows the wires under the board as dashed lines. The bottom diagram shows the board from underneath, flipped left to right. Orange circles indicate where solder joints will be necessary.*

When I'm working on this kind of project, I like to place it (with the vise attached) on a soft piece of polyurethane foam—the kind of slab that is normally used to make a chair cushion. The foam protects the components from damage when the board is upside-down, and again helps to prevent the work from sliding around unpredictably.

Step by Step

Here's the specific procedure for building this circuit:

1. Cut the small piece of perfboard out of a sheet that has no copper traces on it. You can cut the section using your miniature hobby saw, or you may be able to snap the board along its lines of holes, if you're careful. Alternatively, use a small ready-cut piece of perfboard with copper circles on it that are not connected to one another. You'll ignore the copper circles in this project. (In the next experiment, you'll deal with the additional challenge of making connections between components and copper traces on perforated board.)
2. Gather all the components and carefully insert them through holes in the board, counting the holes to make sure everything is in the right place. Flip the board over and bend the wires from the components to anchor them to the board and create connections as shown. If any of the wires isn't long enough, you'll have to supplement it with an extra piece of 22-gauge wire from your supply. You can remove all the insulation, as we'll be mounting the perfboard on a piece of insulating plastic.
3. Trim the wires approximately with your wire cutters.
4. Make the joints with your pencil soldering iron. Note that in this circuit, you are just joining wires to each other. The components are so close together that they'll prevent each other from wiggling around too much. If you are using board with copper pads (as I did), and some solder connects with them, that's OK—as long as it doesn't creep across to the neighboring component and create a short circuit.
5. Check each joint using a close-up magnifying glass, and wiggle it with pointed-nosed pliers. If there isn't enough solder for a really secure joint, reheat it and add more. If solder has created a connection that shouldn't be there, use a utility knife to make two parallel cuts in the solder, and scrape away the little section between them.

Figure 3-80. *Components mounted on a piece of perforated board.*

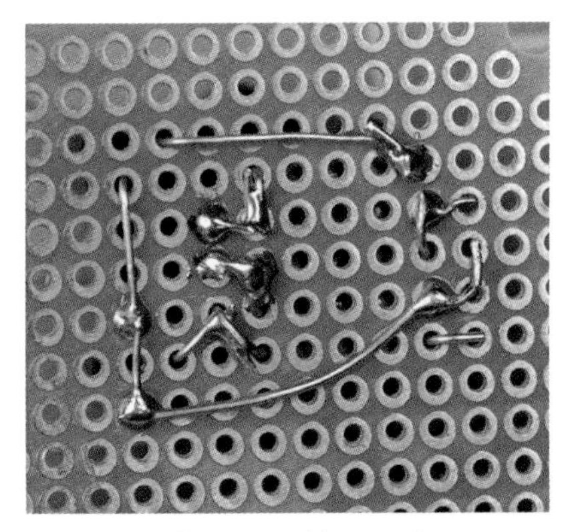

Figure 3-81. *The assembly seen from below. The copper circles around the holes are not necessary for this project. Some of them have picked up some solder, but this is irrelevant as long as no unintentional short circuits are created.*

Generally, I insert three or four components, trim the wires approximately, solder them, trim their wires finally, then pause to check the joints and the placement. If I solder too many components in succession, there's a greater risk of missing a bad joint, and if I make an error in placing a component, undoing it will be much more problematic if I have already added a whole lot more components around it.

Figures 3-80 and 3-81 show the version of this project that I constructed, before I trimmed the board to the minimum size.

Flying Wire Segments

The jaws of your wire cutters exert a powerful force that peaks and then is suddenly released when they cut through wire. This force can be translated into sudden motion of the snipped wire segment. Some wires are relatively soft, and don't pose a risk, but harder wires can fly in unpredictable directions at high speed, and may hit you in the eye. The leads of transistors are especially hazardous in this respect.

I think it's a good idea to wear safety glasses when trimming wires.

Finishing the Job

I always use bright illumination. This is not a luxury; it is a necessity. Buy a cheap desk lamp if you don't already have one. I use a daylight-spectrum fluorescent desk lamp, because it helps me identify the colored bands on resistors more reliably. Note that this type of fluorescent lamp emits quite a lot of ultraviolet light, which is not good for the lens in your eye. Avoid looking closely and directly at the tube in the lamp, and if you wear glasses, they will provide additional protection.

No matter how good your close-up vision is, you need to examine each joint with that close-up magnifier. You'll be surprised how imperfect some of them are. Hold the magnifier as close as possible to your eye, then pick up the thing that you want to examine and bring it closer until it comes into focus.

Finally, you should end up with a working circuit. You can insert the wires from your power supply into two of the tiny power sockets, and plug a red LED into the remaining two sockets. Remember that the two center sockets are negative, and the two outer sockets are positive, because it was easier to wire the circuit this way. You should color-code them to avoid mistakes.

So now you have a tiny circuit that pulses like a heartbeat. Or does it? If you have difficulty making it work, retrace every connection and compare it with the schematic. If you don't find an error, apply power to the circuit, attach the black lead from your meter to the negative side, and then go around the circuit with the red lead, checking the presence of voltage. Every part of this circuit should show at least some voltage while it's working. If you find a dead connection, you may have made a bad solder joint, or missed one entirely.

When you're done, now what? Well, now you can stop being an electronics hobbyist and become a crafts hobbyist. You can try to figure out a way to make this thing wearable.

First you have to consider the power supply. Because of the components that I used, we really need 9 volts to make this work well. How are you going to make this 9-volt circuit wearable, with a bulky 9-volt battery?

I can think of three answers:

1. You can put the battery in a pocket, and mount the flasher on the outside of the pocket, with a thin wire penetrating the fabric. Note that the tiny power connector on the perforated board will accept two 22-gauge wires if they are solid core, or if they are stranded (like the wires from a 9-volt battery connector) but have been thinly coated with solder.
2. You could mount the battery inside the crown of a baseball cap, with the flasher on the front.
3. You can put together three 3-volt button batteries in a stack, held in some kind of plastic clip. If you try this option, it may not be a good idea to try to solder wire to a battery. You will heat the liquid stuff inside the battery, which may not be good for it, and may not be good for you if the liquid starts boiling and the battery bursts open. Also, solder doesn't stick easily to the metallic finish on most battery terminals.

Most LEDs create a sharply defined beam of light, which you may want to diffuse to make it look nicer. One way to do this is to use a piece of transparent acrylic plastic, at least 1/4 inch thick, as shown in Figure 3-82. Sandpaper the front of the acrylic, ideally using an orbital sander that won't make an obvious pattern. Sanding will make the acrylic translucent rather than transparent.

Drill a hole slightly larger than the LED in the back of the acrylic. Don't drill all the way through the plastic. Remove all fragments and dust from the hole by blasting some compressed air into it, or by washing it if you don't have an air compressor. After the cavity is completely dry, get some transparent silicone caulking or mix some clear five-minute epoxy and put a drop in the bottom of the hole. Then insert the LED, pushing it in so that it forces the epoxy to ooze around it, making a tight seal. See Figure 3-82.

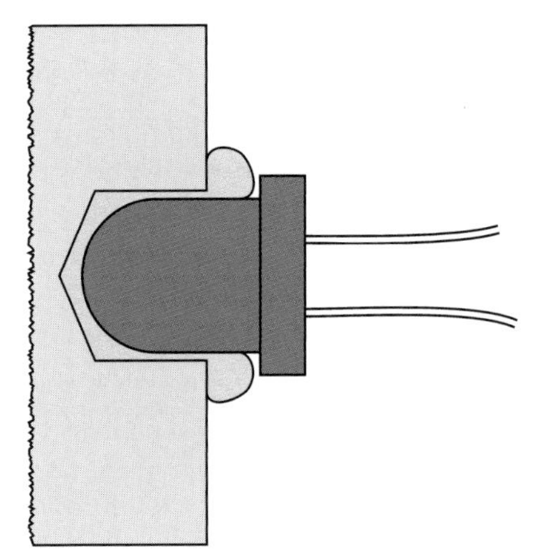

Figure 3-82. *This cross-sectional view shows a sheet of transparent acrylic in which a hole has been drilled part of the way from the back toward the front. Because a drill bit creates a hole with a conical shape at the bottom, and because the LED has rounded contours, transparent epoxy or silicone caulking can be injected into the hole before mounting the LED.*

Try illuminating the LED, and sand the acrylic some more if necessary. Finally, you can decide whether to mount the circuit on the back of the acrylic, or whether you want to run a wire to it elsewhere.

Because the LED will flash at about the speed of a human heart while the person is resting, it may look as if it's measuring your pulse, especially if you mount it on the center of your chest or in a strap around your wrist. If you enjoy hoaxing people, you can suggest that you're in such amazingly good shape, your pulse rate remains constant even when you're taking strenuous exercises.

To make a good-looking enclosure for the circuit, I can think of options ranging from embedding the whole thing in clear epoxy to finding a Victorian-style locket. I'll leave you to consider alternatives, because this is a book about electronics rather than handicrafts.

However, I will address one final issue: how long will this gadget continue flashing?

If you check the following section "Essentials: Battery life," you'll find that a regular alkaline 9-volt battery should keep the LED flashing for about 50 hours.

ESSENTIALS

Battery life

Any time you finish a circuit that you intend to run from a battery, you'll want to calculate the likely battery life. This is easily done, because manufacturers rate their batteries according to the "ampere hours" they can deliver. Keep the following in mind:

- The abbreviation for amp-hours is Ah, sometimes printed as AH. Milliampere-hours are abbreviated mAh.
- The rating of a battery in amp-hours is equal to the current, in amps, multiplied by the number of hours that the battery can deliver it.

Thus, in theory 1 amp-hour can mean 1 amp for 1 hour, or 0.1 amp for 10 hours, or 0.01 amp for 100 hours—and so on. In reality, it's not as simple as this, because the chemicals inside a battery become depleted more quickly when you draw a heavy current, especially if the battery gets hot. You have to stay within limits that are appropriate to the size of the battery.

For instance, if a small battery is rated for 0.5 amp-hours, you can't expect to draw 30 amperes from it for 1 minute. But you should be able to get 0.005 amps (i.e., 5 milliamps) for 100 hours without any trouble. Remember, though, that the voltage delivered by a battery will be greater than its rated voltage when the battery is fresh, and will diminish below its rated voltage while the battery is delivering power.

According to some test data that I trust (I think they are a little more realistic than the estimates supplied by battery manufacturers), here are some numbers for typical batteries:

- Typical 9 volt alkaline battery: 0.3 amp-hours, while delivering 100 mA.
- Typical AA size, 1.5-volt alkaline battery: 2.2 amp-hours, while delivering 100 mA.
- Rechargeable nickel-metal hydride battery: about twice the endurance of a comparably sized alkaline battery.
- Lithium battery: maybe three times the endurance of an alkaline battery.

BACKGROUND

Maddened by measurement

Throughout most of this book, I've mostly used measurements in inches, although sometimes I've digressed into the metric system, as when referring to "5-mm LEDs." This isn't inconsistency on my part; it reflects the conflicted state of the electronics industry, where you'll find inches and millimeters both in daily use, often in the very same data sheet.

The United States is the only major nation still using the old system of units that originated in England. (The other two holdouts are Liberia and Myanmar, according to the CIA's *World Factbook*.) Still, the United States has led many advances in electronics, especially the development of silicon chips, which have contacts spaced 1/10 inch apart. These standards became firmly established, and show no sign of disappearing.

To complicate matters further, even in the United States, you can encounter two incompatible systems for expressing fractions of an inch. Drill bits, for instance, are measured in multiples of 1/64 inch, while metal thicknesses may be measured in decimals such as 0.06 inch (which is approximately 1/16 inch).

The metric system is not necessarily more rational than the U.S. system. Originally, when the metric system was formally introduced in 1875, the meter was defined as being 1/10,000,000 of the distance between the North Pole and the equator, along a line passing through Paris—a quixotic, Francocentric conceit. Since then, the meter has been redefined three times, in a series of efforts to achieve greater accuracy in scientific applications.

As for the usefulness of a 10-based system, moving a decimal point is certainly simpler than doing calculations in 64ths of an inch, but the only reason we count in tens is because we happen to have evolved with that number of digits on our hands. A 12-based system would really be more convenient, as numbers would be evenly divisible by 2 and 3.

As we're stuck with the whimsical aspects of length measurement, I've created the charts in Figures 3-83 and 3-84 to assist you in going from one system to another. From these you will see that when you need to drill a hole for a 5 mm LED, a 3/16-inch drill bit is about right. (In fact, it results in a better, tighter fit than if you drill an actual 5 mm hole.)

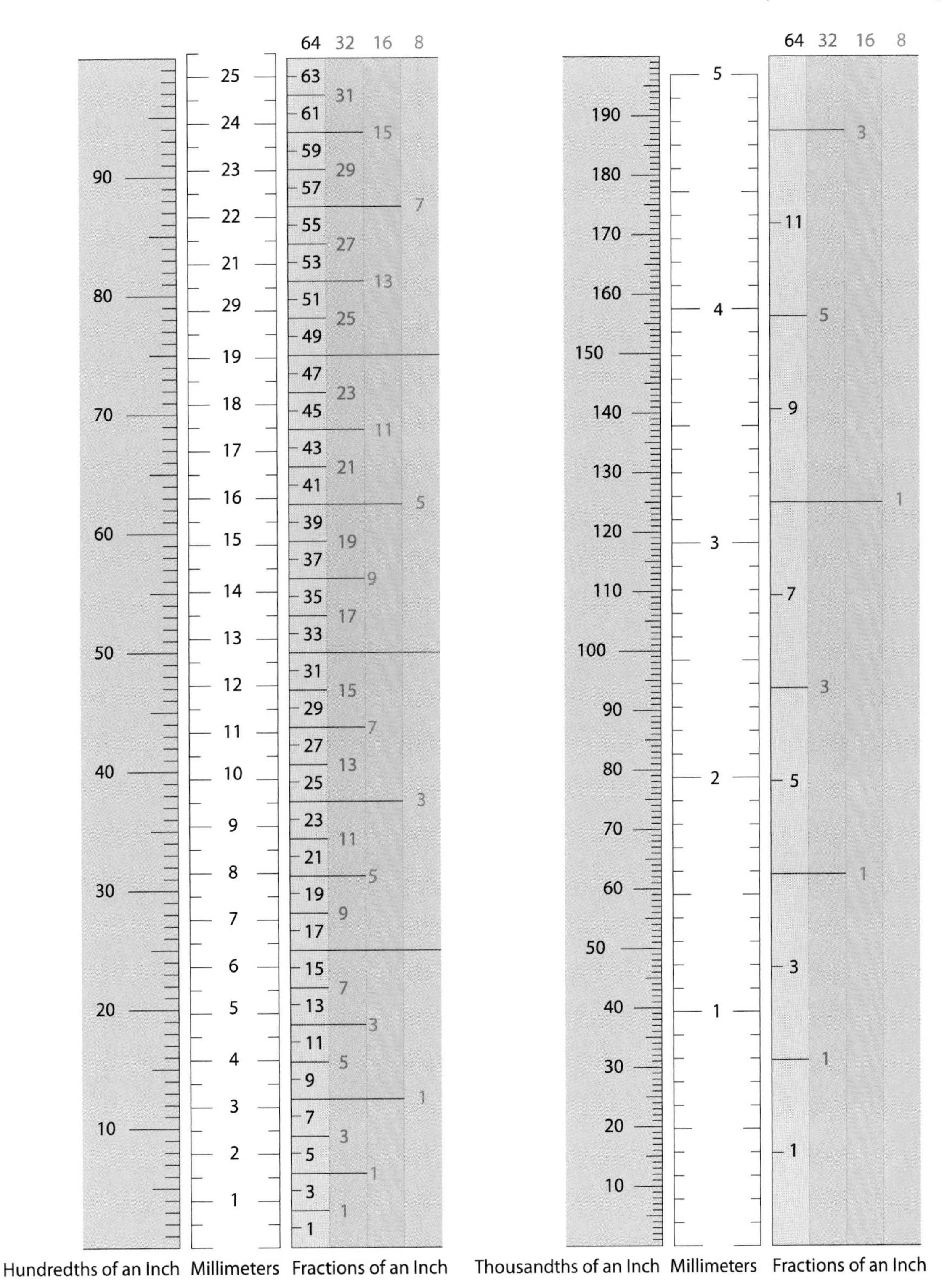

Figure 3-83. *Because units of measurement are not standardized in electronics, conversion is often necessary. The chart on the right is a 5x magnification of the bottom section of the chart on the left.*

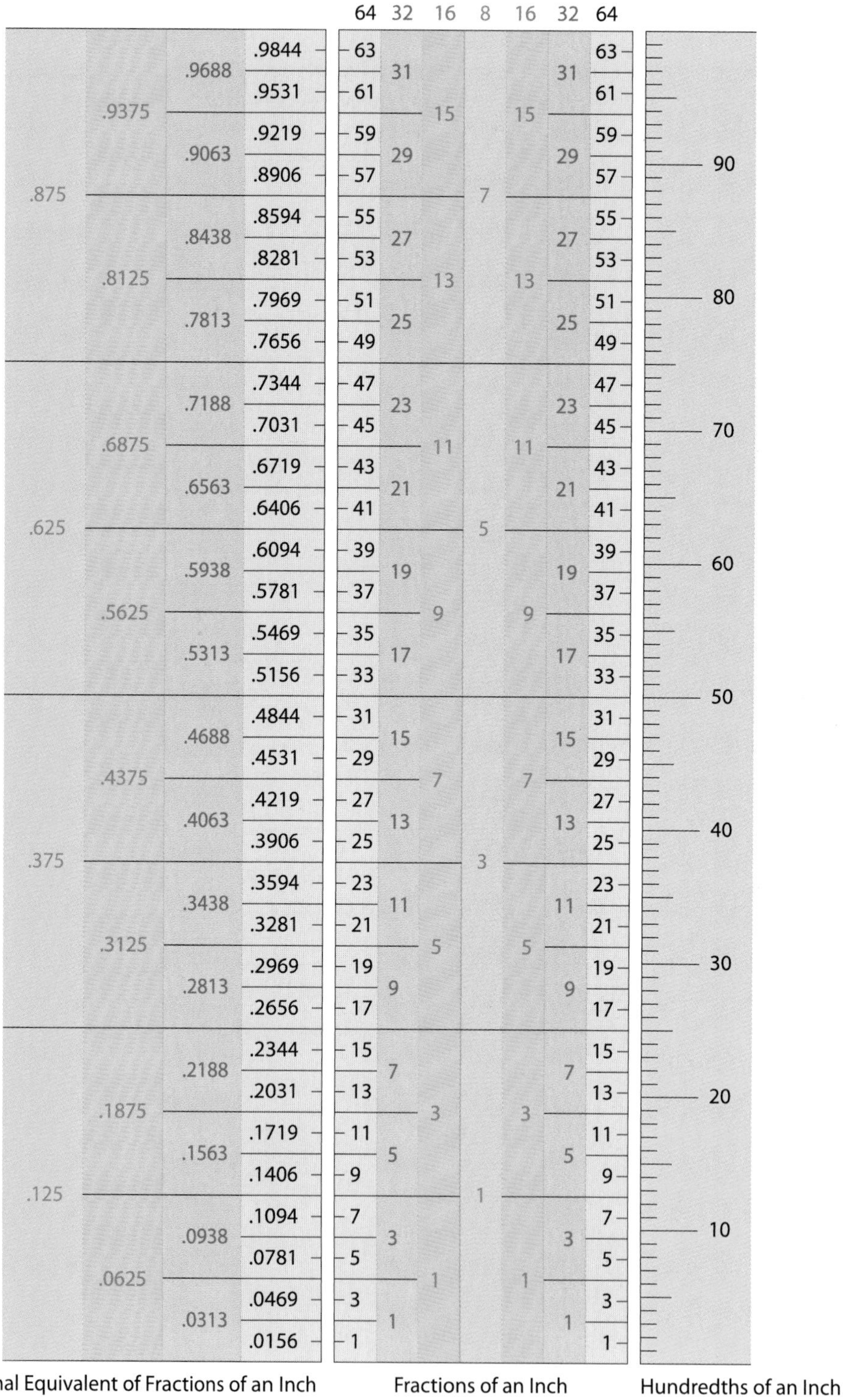

64	32	16	8	Decimal Equivalent
63				.9844
	31			.9688
61				.9531
		15		.9375
59				.9219
	29			.9063
57				.8906
			7	.875
55				.8594
	27			.8438
53				.8281
		13		.8125
51				.7969
	25			.7813
49				.7656
47				.7344
	23			.7188
45				.7031
		11		.6875
43				.6719
	21			.6563
41				.6406
			5	.625
39				.6094
	19			.5938
37				.5781
		9		.5625
35				.5469
	17			.5313
33				.5156
31				.4844
	15			.4688
29				.4531
		7		.4375
27				.4219
	13			.4063
25				.3906
			3	.375
23				.3594
	11			.3438
21				.3281
		5		.3125
19				.2969
	9			.2813
17				.2656
15				.2344
	7			.2188
13				.2031
		3		.1875
11				.1719
	5			.1563
9				.1406
			1	.125
7				.1094
	3			.0938
5				.0781
		1		.0625
3				.0469
	1			.0313
1				.0156

Hundredths of an Inch: 90, 80, 70, 60, 50, 40, 30, 20, 10

Decimal Equivalent of Fractions of an Inch — Fractions of an Inch — Hundredths of an Inch

Figure 3-84. *This chart allows conversion between hundredths of an inch, conventional U. S. fractions of an inch, and fractions expressed in thousandths of an inch.*

Experiment 15: Intrusion Alarm Revisited

Time now to add some of the enhancements to the intrusion alarm that I discussed at the end of Experiment 11. I'm going to show you how the alarm can be triggered if you install various detectors on windows and doors in your home. I'll also show how the alarm can be wired so that it locks itself on and continues to make noise even after a door or window is reclosed.

This experiment will demonstrate the procedure for transferring a project from a breadboard to a piece of perforated board that has copper connections laid out identically to the ones inside the breadboard, as shown earlier in Figure 3-72. And you'll mount the finished circuit in a project box with switches and connectors on the front.

When all is said and done, you'll be ready for wholesale circuit building. The explanations in the rest of this book will get gradually briefer, and the pace will increase.

You will need:

- 15-watt pencil-type soldering iron
- Thin solder (0.022 inches or similar)
- Wire strippers and cutters
- Perforated board etched with copper in a breadboard layout
- Small vise or clamp to hold your perforated board
- The same components that you used in Experiment 11, plus:
    - 2N2222 NPN transistor. Quantity: 1.
    - DPDT relay. Quantity: 1.
    - SPDT toggle switch. Quantity: 1.
    - 1N4001 diode. Quantity: 1.
    - Red and green 5mm LEDs. Quantity: 1 each.
    - Project box, 6 × 3 × 2 inches.
    - Power jack, type N, and matching power socket, type N.
    - Binding posts.
    - Stranded 22-gauge wire, three different colors.
    - Magnetic sensor switches, sufficient for your home.
    - Alarm network wiring, sufficient for your home.

Magnetic Sensor Switches

Figure 3-85. *In this simple alarm sensor switch, the lower module contains a magnet, which opens and closes a reed switch sealed into the upper module.*

A typical alarm sensor switch consists of two modules: the magnetic module and the switch module, as shown in Figures 3-85 and 3-86. The magnetic module contains a permanent magnet, and nothing else. The switch module contains a "reed switch," which makes or breaks a connection (like a contact inside a relay) under the influence of the magnet. When you bring the magnetic module close to the switch module, you may faintly hear the reed switch click as it flips from one state to the other.

Like all switches, reed switches can be normally open or normally closed. For this project, you want the kind of switch that is normally open, and closes when the magnetic module is close to it.

Attach the magnetic module to the moving part of a door or window, and attach the switch module to the window frame or door frame. When the window or door is closed, the magnetic module is almost touching the switch module. The magnet keeps the switch closed until the door or window is opened, at which point the switch opens.

The only question is: how do we use this component to trigger our alarm? As long as a small current flows through all our magnetic sensor switches, the alarm should be off, but if the flow of current stops, the alarm should switch on.

We could use a relay that is "always on" while the alarm is armed. When the circuit is interrupted, the relay relaxes and its other pair of contacts closes, which could power up the alarm noisemaker.

But I don't like this idea. Relays take significant power, and they can get hot. Most of them are not designed to be kept "always-on." I'd prefer to handle the task using a transistor.

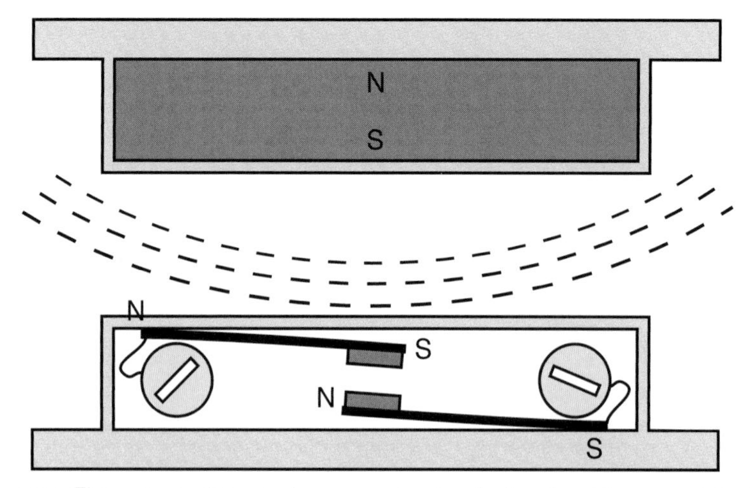

Figure 3-86. *This cutaway diagram shows a reed switch (bottom) and the magnet that activates it (top), inside an alarm sensor. The switch contains two flexible magnetized strips, the upper one with its south pole adjacent to an electrical contact, the lower one with its north pole adjacent to an electrical contact. When the south pole of the magnet approaches the switch, the magnetic force (shown as dashed lines) repels the south contact and attracts the north contact, causing them to snap together. Two screws on the outside of the casing are connected with the strips inside.*

A Break-to-Make Transistor Circuit

First, recall how an NPN transistor works. When the base is not sufficiently positive, the transistor blocks current between its collector and emitter, but when the base is relatively positive, the transistor passes current.

Take a look at the schematic in Figure 3-87, which is built around our old friend the 2N2222 NPN transistor. When the switch is closed, it connects the base of the transistor to the negative side of the power supply through a 1K resistor. At the same time, the base is connected with the positive side of the power supply through a 10K resistor. Because of the difference in resistances and the relatively high turn-on voltage for the LED, the base is forced below its turn-on threshold, and as a result, the transistor will not pass much current. The LED will glow dimly at best.

Now what happens when the switch is opened? The base of the transistor loses its negative power supply and has only its positive power supply. It becomes much more positive, above the turn-on threshold for the transistor, which tells the transistor to lower its resistance and pass more current. The LED now glows brightly. Thus, when the switch is turned off and breaks the connection, the LED is turned on.

This seems to be what we want. Imagine a whole series of switches instead of just one switch, as shown in Figure 3-88. The circuit will still work the same way, even if the switches are scattered all over your home, because the resistance in the wires connecting the switches will be trivial compared with the resistance of the 1K resistor.

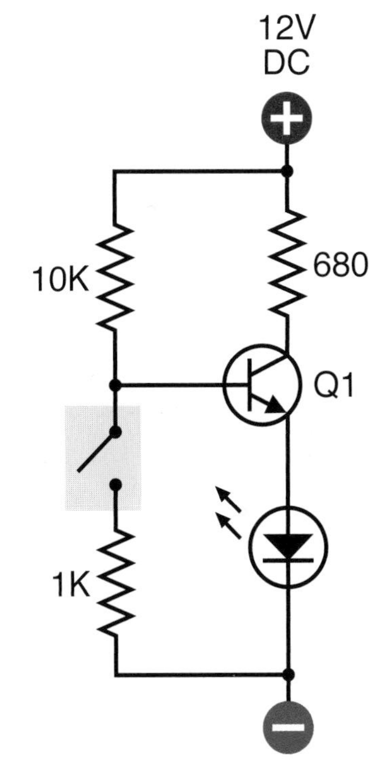

Figure 3-87. *In this demonstration circuit, when the switch is opened, it interrupts negative voltage to the base of the transistor, causing the transistor to lower its resistance, allowing current to reach the LED. Thus, when the switch is turned off, it turns on the LED.*

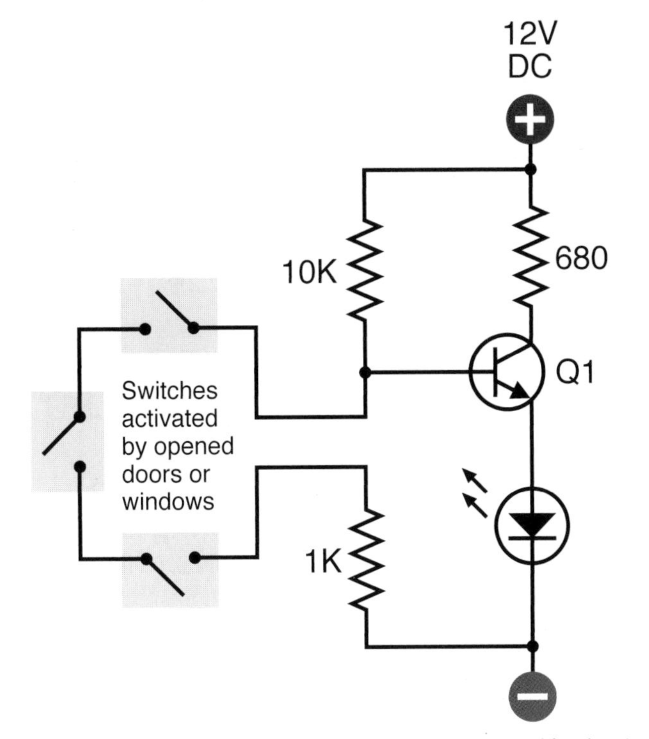

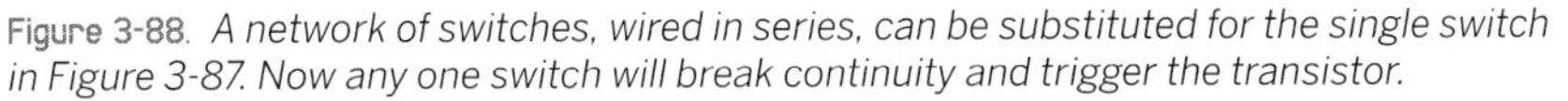
Figure 3-88. *A network of switches, wired in series, can be substituted for the single switch in Figure 3-87. Now any one switch will break continuity and trigger the transistor.*

I have shown the switches open, because that's the way the schematic for a switch is drawn, but imagine them all closed. The base of the transistor will now be supplied through the long piece of wire connecting all the closed switches, and the LED will stay dark. Now if just one switch is opened, or if anyone tampers with the wire linking them, the base of the transistor loses its connection to negative power, at which point the transistor conducts power and the LED lights up.

While all the switches remain closed, the circuit is drawing very little current—probably about 1.1 mA. So you could run it from a typical 12-volt alarm battery.

Now suppose we swap out the LED and put a relay in there instead, as shown in Figure 3-89. I don't mind using a relay in this location, because the relay will not be "always on." It will normally be off, and will draw power only when the alarm is triggered.

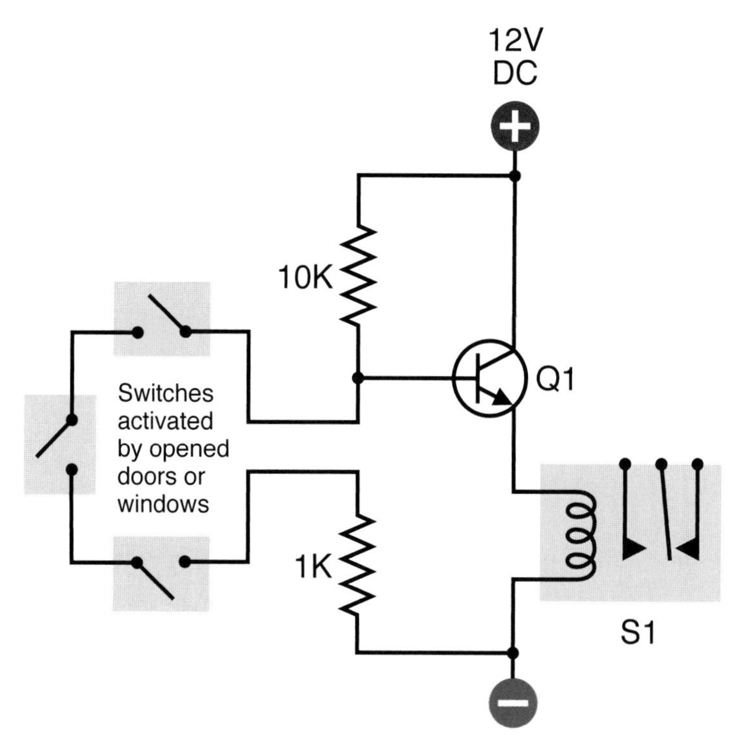

Figure 3-89. *If the LED and 680Ω resistor shown in Figure 3-88 are removed, and a relay takes their place, the relay will be activated when any switch in the sensor network is opened.*

Try one of the 12-volt relays that you used previously. You should find that when you open the switch, the relay is energized. When you close the switch, the relay goes back to sleep. Note that I eliminated the 680Ω resistor from the circuit, because the relay doesn't need any protection from the 12-volt power supply.

Self-Locking Relay

There's only one remaining problem: we want the alarm to continue making noise even after someone who has opened a door or window closes it again quickly. In other words, when the relay is activated, it must lock itself on.

One way to do this would be by using a latching relay. The only problem is that we would then need another piece of circuitry to unlatch it. I prefer to show you how you can make any relay keep itself switched on after it has received just one jolt of power. This idea will be useful to you later in the book as well.

The secret is to supply power to the relay coil through the two contacts inside the relay that are normally open. (Note that this is exactly opposite to the relay oscillator, which supplied power to its coil through the contacts that were normally closed. That setup caused the relay to switch itself off almost as soon as it switched itself on. This setup causes the relay to keep itself switched on, as soon as it has been activated.)

In Figure 3-90, the four schematics illustrate this. You can imagine them as being like frames in a movie, photographed microseconds apart. In the first picture, the switch is open, the relay is not energized, and nothing is happening. In the second, the switch has been closed to energize the coil. In the third, the coil has pulled the contact inside the relay, so that power now reaches the coil via two paths. In the fourth, the switch has been opened, but the relay is still powering its own coil through its contacts. It will remain locked in this state until the power is disconnected.

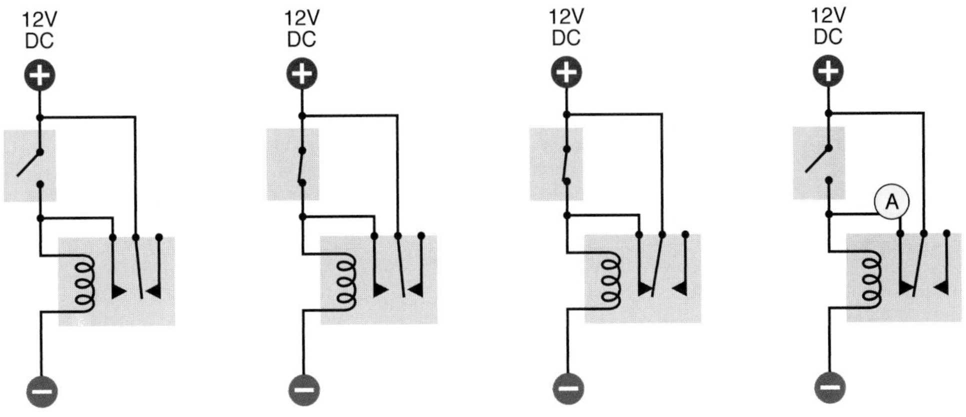

Figure 3-90. *This sequence of schematics shows the events that occur when a relay is energized. Initially, the switch is open. Then the switch is closed, activating the relay. The relay then powers itself through its own internal contacts. The relay remains energized even after the switch is opened again. Power switched by the relay can be taken from the circuit at point A.*

All we need to do, to make use of this idea, is to substitute the transistor for the on/off switch, and tap into the circuit at point A, running a wire from there to the noisemaking module.

Figure 3-91 shows how that would work. When the transistor is activated by any of the network of sensor switches, as previously explained, the transistor conducts power to the relay. The relay locks itself on, and the transistor becomes irrelevant.

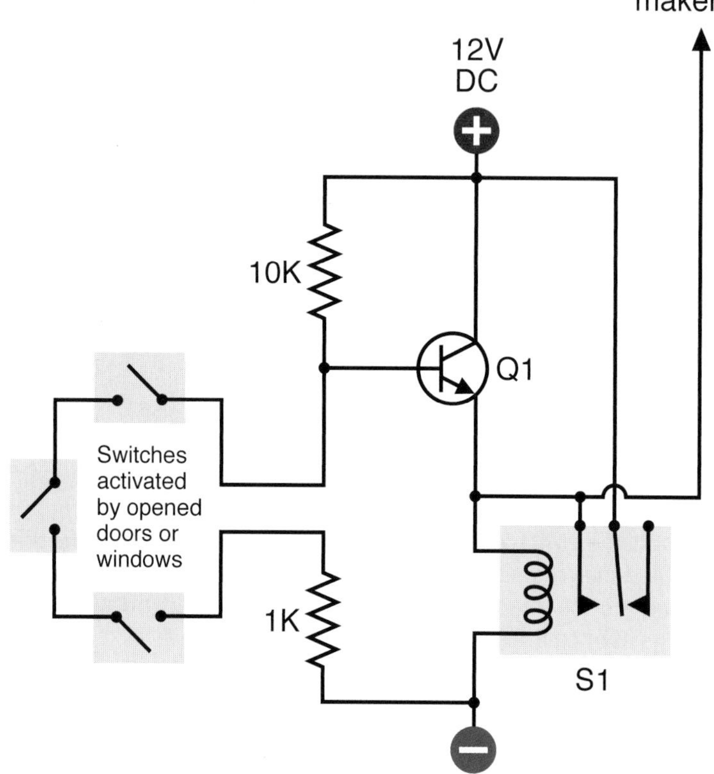

Figure 3-91. *The self-locking relay depicted in Figure 3-90 has been incorporated in the alarm circuit, so that if any switch in the network is opened, the relay will continue to power the noise maker even if the switch is closed again.*

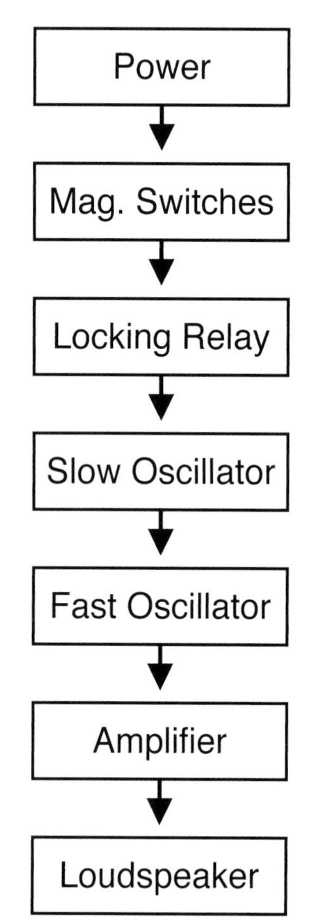

Figure 3-92. *This block diagram previously shown in Figure 2-112 on page 90 has been updated to include the magnetic-switch network and locking-relay control system.*

Because I've been adding pieces to the original alarm noisemaker circuit, I've updated the block diagram from Figure 2-112 to show that we can still break this down into modules with simple functions. The revised diagram is shown in Figure 3-92.

Blocking Bad Voltage

One little problem remains: in the new version of the circuit, if the transistor goes off while the relay is still on, current from the relay can flow back up the wire to the emitter of the transistor, where it will try to flow backward through the transistor to the base, which is "more negative," as it is linked through all the magnetic switches and the 1K resistor to the negative side of the power supply.

Applying power backward through a transistor is not a nice thing to do. Therefore the final schematic in this series shows one more component, which you have not seen before: a diode, labeled D1. See Figure 3-93. The diode looks like the heart of an LED, and indeed, that's pretty much what it is, although some diodes are much more robust. It allows electricity to flow in only one direction, from positive to negative, as shown by its arrow symbol. If current tries to flow in the opposite direction, the diode blocks it. The only price you pay for this service is that the diode imposes a small voltage drop on electricity flowing in the "OK" direction.

So now, positive flow can pass from the transistor, through the diode, to the relay coil, to get things started. The relay then supplies itself with power, but the diode prevents the positive voltage from getting back into the transistor the wrong way.

Perhaps a more elegant solution to the problem is to connect the NO (normally open) leg of the relay via a 10k resistor to the base connection. When the relay is not energized, the NO leg is inert and simply behaves as a parasitic capacitance on the node. When the relay becomes energized, the NO leg shunts +12V through the common terminal via a 10k resistor into the base of the transistor. In this circuit configuration, the transistor is never exposed to a potentially harmful voltage and you are not depending on leakage currents of non-ideal elements to protect devices.

However, I needed an opportunity to introduce you to the concept of diodes. You can check the following section "Essentials: All about diodes" to learn more.

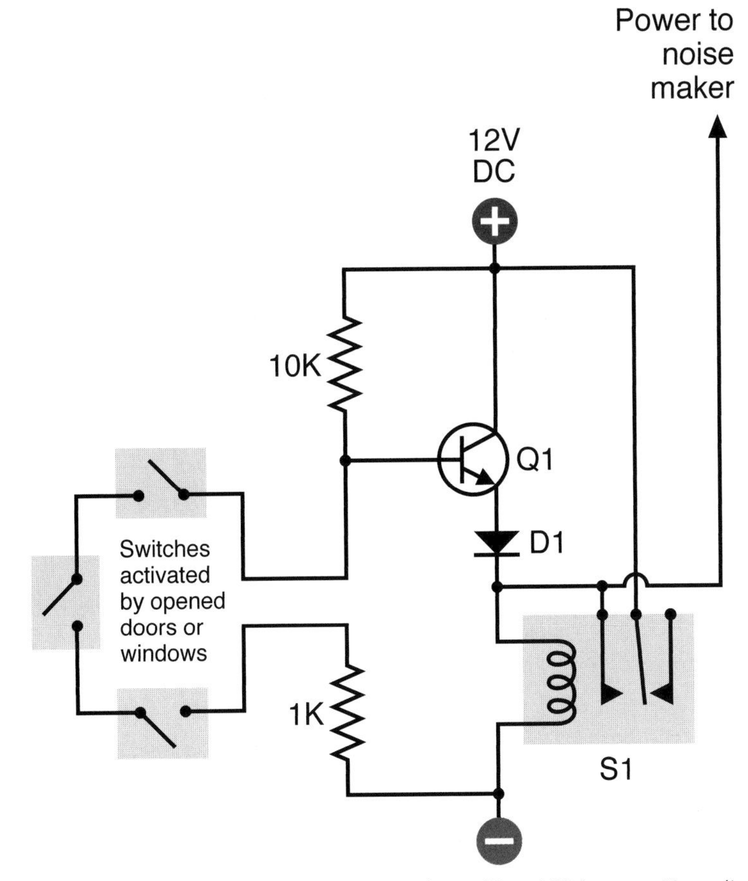

Figure 3-93. *Diode D1 has been added to protect the emitter of Q1 from positive voltage when the relay is energized.*

ESSENTIALS

All about diodes

A diode is a very early type of semiconductor. It allows electricity to flow in one direction, but blocks it in the opposite direction. (A light-emitting diode is a much more recent invention.) Like an LED, a diode can be damaged by reversing the voltage and applying excessive power, but most diodes generally have a much greater tolerance for this than LEDs. The end of the diode that blocks positive voltage is always marked, usually with a circular band, while the other end remains unmarked. Diodes are especially useful in logic circuits, and can also convert alternating current (AC) into direct current (DC).

A Zener diode is a special type that we won't be using in this book. It blocks current completely in one direction, and also blocks it in the other direction until a threshold voltage is reached—much like a PUT.

Signal diodes are available for various different voltages and wattages. The 1N4001 diode that I recommend for the alarm activation circuit is capable of handling a much greater load at a much higher voltage, but I used it because it has a low internal resistance. I wanted the diode to impose a minimal voltage drop, so that the relay would receive as much voltage as possible.

It's good practice to use diodes at less than their rated capacity. Like any semiconductor, they can overheat and burn out if they are subjected to mistreatment.

The schematic symbol for a diode has only one significant variant: sometimes the triangle is outlined instead of filled solid black (see Figure 3-94).

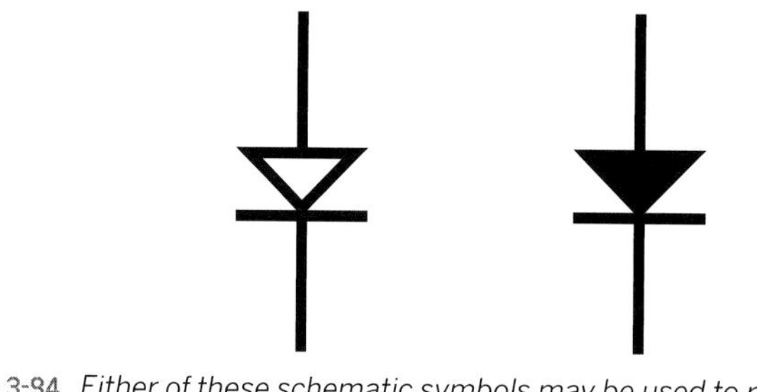

Figure 3-94. *Either of these schematic symbols may be used to represent a diode, but the one on the right is more common than the one on the left.*

Completing the Breadboard Alarm Circuit

It's time now to breadboard the control circuit for your alarm noisemaker. Figure 3-95 shows how this can be done. I am assuming that you still have the noisemaker, which functions as before. I'm assuming that you still have its relevant components mounted on the top half of the breadboard. To save space, I'm just going to show the additional components mounted on the bottom half of the same breadboard.

It's important to remember that you are not supplying power directly to the left and right "rails" on the breadboard anymore; you are supplying power to the relay-transistor section, and when the relay closes its contacts, the *relay* supplies power to the rails. These then feed the power up to the top half of the breadboard. So disconnect your power supply from the breadboard rails and reconnect it as shown in Figure 3-95.

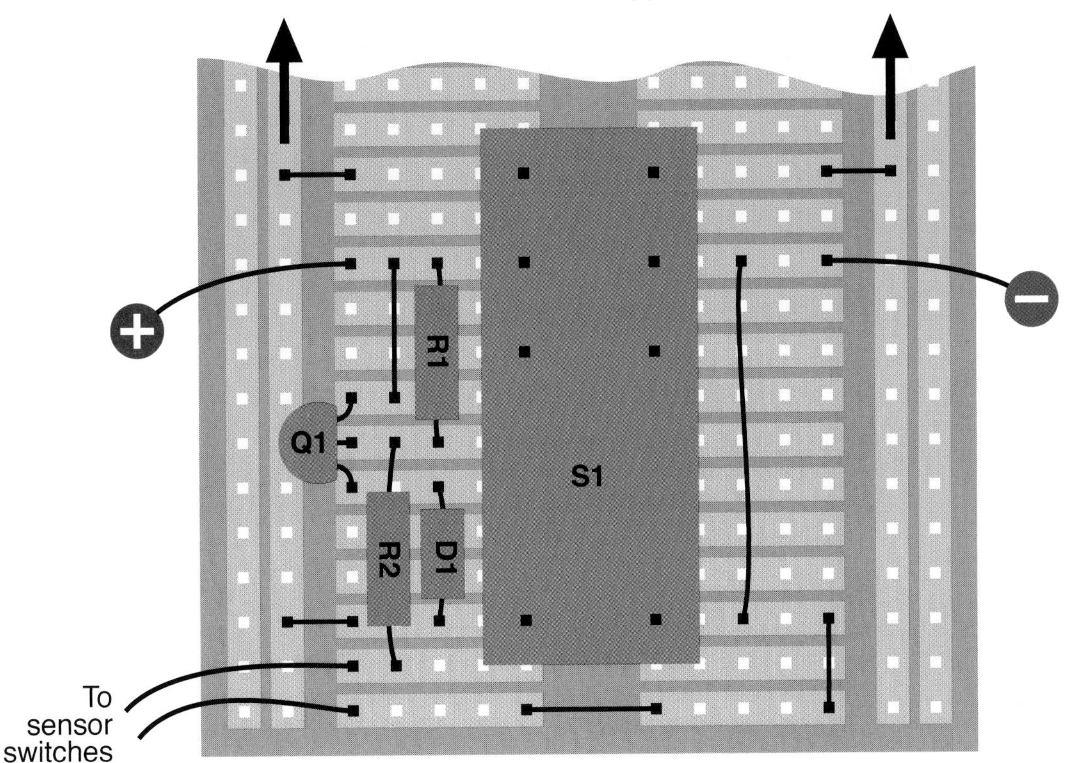

Figure 3-95. *The schematic that was developed in the previous pages can be emulated with components on a breadboard, as shown here. S1 is a DPDT relay. Wires to the sensor switch network and to the power supply must be added where shown.*

Because it's a double-pole relay, I am using it to switch negative as well as positive. This means that when the relay contacts are open, the noisemaking section of the circuit is completely isolated from the rest of the world.

The breadboarded relay circuit is exactly the same as the schematic in Figure 3-93. The components have just been rearranged and squeezed together so that they will fit alongside the relay. Two wires at the lower-left corner go to the network of magnetic sensor switches that will trip the alarm; for testing purposes, you can just hold the stripped ends of these two wires together to simulate all the switches being closed, and separate the wires to simulate a switch opening.

Two more wires bring power to the breadboard on either side of the relay. This is where you should connect your power supply during testing. The output from the relay, through its top pair of contacts, is connected with the rails of the breadboard by a little jumper wire at top left, and another at top right. Don't forget to include them! One more little wire at the lower-left corner (easily overlooked) connects the lefthand side rail to the lefthand coil terminal of the relay, so that when the relay is powering the noisemaker circuit, it powers itself as well.

When you mount the diode, remember that the end of it that is marked with a band around it is the end that blocks positive current. In this circuit, that's the lower end of the diode.

Try it to make sure that it works. Short the sensor wires together and then apply power. The alarm should remain silent. You can use your meter to check that no voltage exists between the side rails. Now separate the sensor wires, and the relay should click, supplying power to the side rails, which activates the noisemaker. Even if you bring the sensor wires back together, the relay should remain locked on. The only way to unlock it is to disconnect the power supply.

When the circuit is active, the transistor followed by the diode drops the voltage slightly, but the 12-volt relay should still work.

In my test circuit, trying three different relays, they drew between 27 and 40 milliamps at 9.6 volts. Some current still leaked through the transistor when it was in its "off" mode, but only a couple of milliamps at 0.5 volts. This low voltage was far below the threshold required to trip the relay.

Ready for Perfboarding

If the circuit works, the next step is to immortalize it on perforated board. Use the type of board that has a breadboard contact pattern etched on it in copper, as shown in Figure 3-72 on page 116. Check the following section, "Essentials: Perfboard procedure," for guidance on the best way to make this particular kind of solder joint—and the subsequent section for the most common problems.

ESSENTIALS

Perfboard soldering procedure

Carefully note the position of a component on your breadboard, and then move it to the same relative position on the perfboard, poking its wires through the little holes.

Turn the perfboard upside down, make sure that it's stable, and examine the hole where the wire is poking through, as shown in Figure 3-96. A copper trace surrounds this hole and links it with others. Your task is to melt solder so that it sticks to the copper and also to the wire, forming a solid, reliable connection between the two of them.

Take your pencil-style soldering iron in one hand and some solder in your other hand. Hold the tip of the iron against the wire and the copper, and feed some thin solder to their intersection. After two to four seconds, the solder should start flowing.

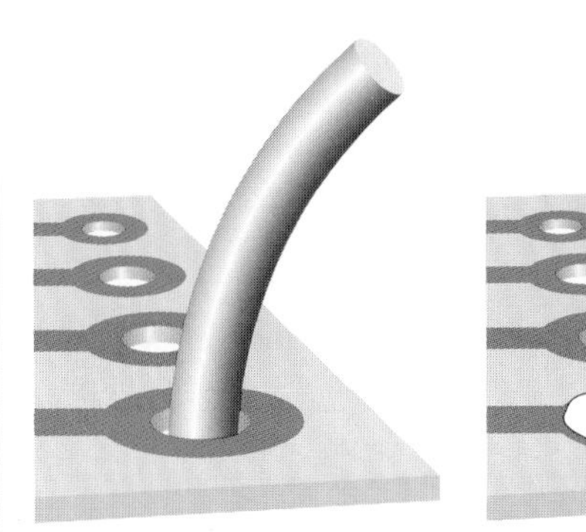

Figure 3-96

Figure 3-97

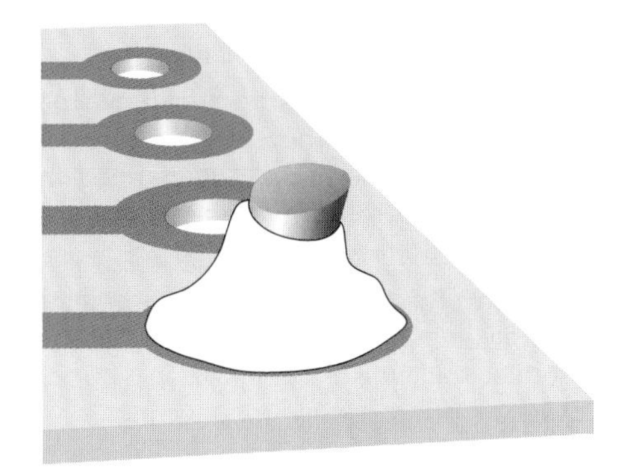

Figure 3-98. *To establish a connection between a section of wire and a copper trace on perforated board, the wire is pushed through the hole, and solder (shown in pure white for illustrative purposes) completes the connection. The wire can then be snipped short.*

Allow enough solder to form a rounded bump sealing the wire and the copper, as shown in Figure 3-97. Wait for the solder to harden thoroughly, and then grab the wire with pointed-nosed pliers and wiggle it to make sure you have a strong connection. If all is well, snip the protruding wire with your cutters. See Figure 3-98.

Because solder joints are difficult to photograph, I'm using drawings to show the wire before and after making a reasonably good joint, which is shown in pure white, outlined with a black line.

Actual soldered perfboard is shown in the photographs in Figures 3-99 and 3-100.

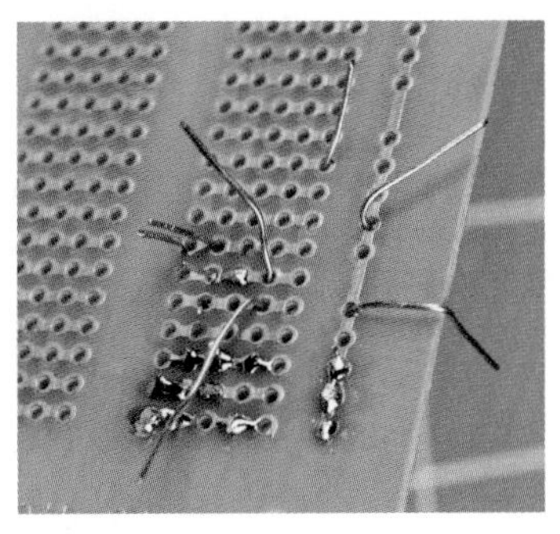

Figure 3-99. *This photograph was taken during the process of transferring components from breadboard to perforated board. Two or three components at a time are inserted from the other side of the board, and their leads are bent over to prevent them from falling out.*

Figure 3-100. *After soldering, the leads are snipped short and the joints are inspected under a magnifying glass. Another two or three components can now be inserted, and the process can be repeated.*

TOOLS

Four most common perfboarding errors

1. Too much solder

Before you know it, solder creeps across the board, touches the next copper trace, and sticks to it, as depicted in Figure 3-101. When this happens, you have to wait for it to cool, and then cut it with a utility knife. You can also try to remove it with a rubber bulb and solder wick, but some of it will tend to remain.

Even a microscopic trace of solder is enough to create a short circuit. Check the wiring under a magnifying glass while turning the perfboard so that the light strikes it from different angles (or use your solder wick to suck it away).

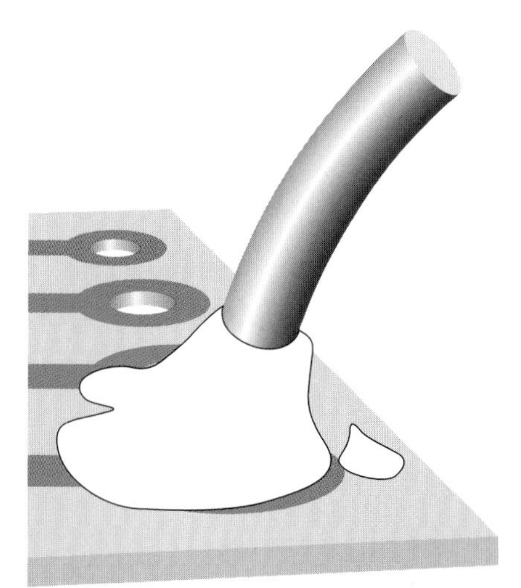

Figure 3-101. *If too much solder is used, it makes a mess and can create an unwanted connection with another conductor.*

2. Not enough solder

If the joint is thin, the wire can break free from the solder as it cools. Even a microscopic crack is sufficient to stop the circuit from working. In extreme cases, the solder sticks to the wire, and sticks to the copper trace around the wire, yet doesn't make a solid bridge connecting the two, leaving the wire encircled by solder yet untouched by it, as shown in Figure 3-102. You may find this undetectable unless you observe it with magnification.

You can add more solder to any joint that may have insufficient solder, but be sure to reheat the joint thoroughly.

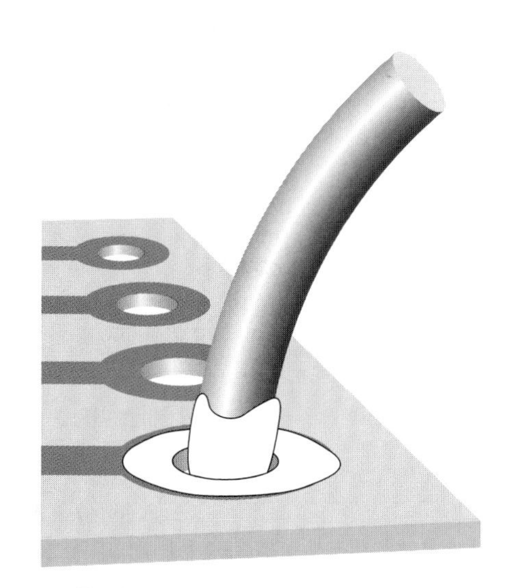

Figure 3-102. *Too little solder (or insufficient heat) can allow a soldered wire to remain separate from the soldered copper on the perforated board. Even a hair-thin gap is sufficient to prevent an electrical connection.*

3. Components incorrectly placed

It's very easy to put a component one hole away from the position where it should be. It's also easy to forget to make a connection.

I suggest that you print a copy of the schematic, and each time you make a connection on the perforated board, you eliminate that wire on your hardcopy, using a highlighter.

4. Debris

When you're trimming wires, the little fragments that you cut don't disappear. They start to clutter your work area, and one of them can easily get trapped under your perforated board, creating an electrical connection where you don't want it.

This is another reason for working with something soft, such as polyurethane foam, under your project. It tends to absorb or hold little pieces of debris, reducing the risk of you picking them up in your wiring.

Clean the underside of your board with an old (dry) toothbrush before you apply power to it, and keep your work area as neat as possible. The more meticulous you are, the fewer problems you'll have later.

Once again, be sure to check every joint with a magnifying glass.

Transferring components from the breadboard to the perforated board should be fairly simple, as long as you don't try to move too many at once. Follow the suggestions described in previous section "Essentials: Perfboard procedure," and pause frequently to check your connections. Impatience is almost always the cause of errors in this kind of work.

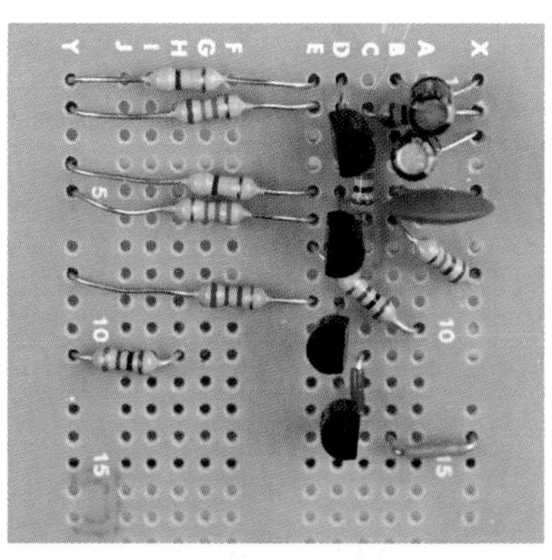

Figure 3-103. *The noisemaker circuit has been transplanted from breadboard to perforated board, with no additions or changes.*

Figure 3-103 shows the noisemaker section of the circuit on perfboard, with the components positioned to minimize wasted space. Figure 3-104 shows the perfboard with the relay and its associated components added. The two black wires will go to the loudspeaker, the black-and-red pair of wires will bring power to the board, and the green wires will go to the magnetic sensor switches. Each wire penetrates the board, and its stripped end is soldered to the copper beneath.

Test it now, in the same way that you tested the same circuit on the breadboard. If it doesn't work, check the following section, "Essentials: Real-world fault tracing." If it does work, you're ready to trim the board and mount it in a project box.

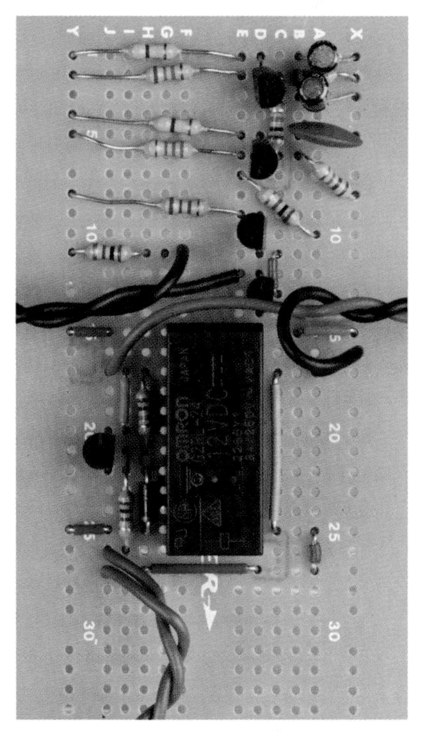

Figure 3-104. *The relay-transistor control circuit has been added. Wires to external devices have been stripped and poked into the perforated board, where they are soldered in place. The green wires connect with the sensor network, the black wires go to the loudspeaker, and the red-and-black wires supply power.*

ESSENTIALS

Real-world fault tracing

Here's a real-life description of the procedure for tracing a fault.

After I assembled the perfboard version of the combined noisemaker and relay circuit, I checked my work, applied power—and although the relay clicked, no sound came out of the loudspeaker. Of course, everything had worked fine on the breadboard.

First I looked at component placement, because this is the easiest thing to verify. I found no errors. Then I flexed the board gently while applying power—and the loudspeaker made a brief "beep." Any time this happens, you can be virtually certain that a solder joint has a tiny crack in it.

The next step was to anchor the black lead of my meter to the negative side of the power supply, and then switch on the power and go through the circuit point by point, from top to bottom, checking the voltage at each point with the red lead of the meter. In a simple circuit like this, every part should show at least some voltage.

But when I got to the second 2N2222 transistor, which powers the loudspeaker, its output was completely dead. Either I had melted the transistor while soldering it (unlikely), or there was a bad joint. I checked the perfboard beneath the transistor with the magnifying glass, and found that solder had flowed around one of the leads of the transistor without actually sticking to it. The gap must have been less than one-thousandth of an inch, but still, that was enough. Probably, the problem had been caused by dirt or grease.

This is the kind of patient inquiry you need to follow when a circuit doesn't work. Check whether your components are placed correctly, check your power supply, check the power on the board, check the voltage at each stage, and if you are persistent, you'll find the fault.

Switches and Inputs for the Alarm

Now you need to make the system easy to use. The block diagram in Figure 3-105 shows one additional box near the top of the sequence: User Controls. These will consist of switches, LEDs, and connections to the outside world. To plan this part of the job, first I have to summarize the way in which our alarm system works at this point in its evolution.

A full-featured home alarm system normally has two modes: in-home and away-from-home:

- Using the in-home mode, you switch on the alarm while you are at home so that it will alert you if an intruder opens a door or window.
- Using the away-from-home mode, typically you enter a code number, after which you have 30 seconds to leave and close the door behind you. When you return, you trigger the alarm by opening the door, but now you have 30 seconds to go to the control panel and enter your code number again to stop the alarm from sounding.

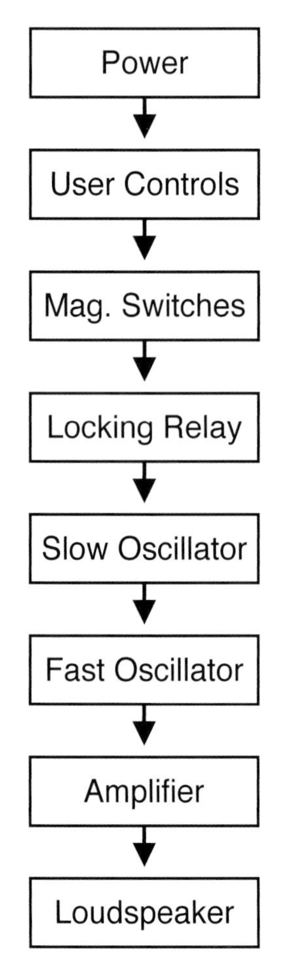

Figure 3-105. *The final block diagram for this phase of the project shows where user controls fit in the series of functions.*

So far, the alarm system that you've been building has only an in-home mode. Still, many people find this function useful and reassuring. Later in the book I'll suggest a way in which you can modify it to incorporate an away-from-home mode, but for now, making it practical for in-home use is enough of a challenge.

Consider how it should be used on an everyday basis. It should have an on/off switch, naturally. When it's on, any of the magnetic sensor switches should trigger the alarm. But what if you switch it on without realizing that you've left a window open? At that time it won't be appropriate for the alarm to sound. What you really need is a circuit-test feature, to tell you if all the doors and windows are closed. Then you can switch on the alarm.

I think a pushbutton would be useful to test the alarm circuit. When you press it, a green LED should light up to indicate that the circuit is good. After you see the green light, you let go of the pushbutton and turn on the power switch, which illuminates a red LED, to remind you that the alarm is now armed and ready.

One additional feature would be useful: an alarm noisemaker test feature, so that you can be sure that the system is capable of sounding its alert when required to do so.

The circuit shown in Figure 3-106 incorporates all of these features. S1 is a SPDT switch; S2 is a DPDT momentary pushbutton of ON-(ON) type. I spliced S2 into the circuit by first cutting the two green wires that connect the sensor switches with the rest of the relay circuit. I then attached one pair of those wires to one side of S2 and the other pair to the other side, as shown in the figure. The schematic shows it in its "relaxed" mode, when the button is not being pressed.

D1 is a red LED, D2 is a green LED, J1 is a power input jack (to be connected with an external 12-volt supply), and R1 is a 680Ω resistor to protect the LEDs. Note that J1 follows the usual practice of supplying positive voltage in its center contact, and negative in the circular shell around the center.

When S1 is in its Off position, it still supplies positive power through its upper contact to S2, the pushbutton. When the pushbutton is pressed, so that it goes into its "Test" position, the pole of S2 connects with the power and sends it out through the sensor switches on doors and windows. The wires to these switches will be attached via a couple of binding posts, shown here as two circles. If the sensor switches are all closed, power returns through the second binding post, passes through the lower set of contacts in S2, and lights D2, the green LED. Because S1 is not supplying power to the alarm circuit board, the alarm does not sound at this time.

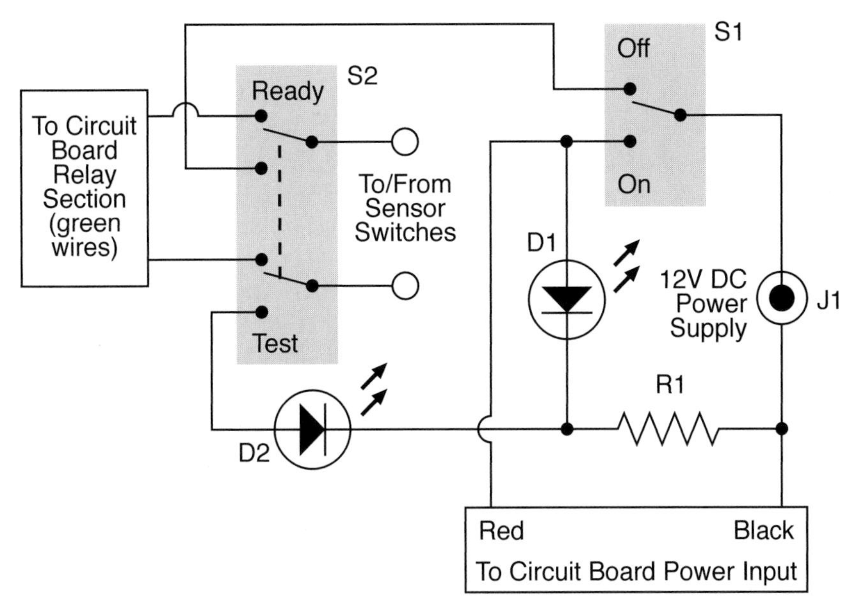

Figure 3-106. *This schematic suggests a convenient way to add an on/off switch, a continuity testing feature, and a noisemaker testing feature to the alarm.*

Now if S1 is turned to its On position, it sends power to the components on the perforated board. The relay section of the circuit sends power out along the green wires to S2, and as long as the button is not being pressed, the power goes out through the switch network and back through S2, to the relay section, just as before S2 was spliced in. So the alarm remains silent. But as soon as a sensor switch is opened, the circuit is broken and the alarm sounds. The only way to stop it will be by switching S1 into its Off position.

Finally, if you press S2, the pushbutton, while S1 is in its On mode, you interrupt the network of switches and activate the noisemaker. In this way, S2 does dual duty: when S1 is off, pushbutton S2 tests the sensor switches for continuity. When S1 is on, S2 tests the noisemaker to make sure it creates a noise. I think this is the simplest possible way to implement these features.

Installing the Switches

If you bought a project box from RadioShack, it may have come with two optional top panels: one made of metal, the other made of plastic. I'll assume that you'll use the plastic one, as you'll have more trouble drilling holes in metal. The type of plastic used by RadioShack is ABS, which is very easy to shape with the tools that I have recommended.

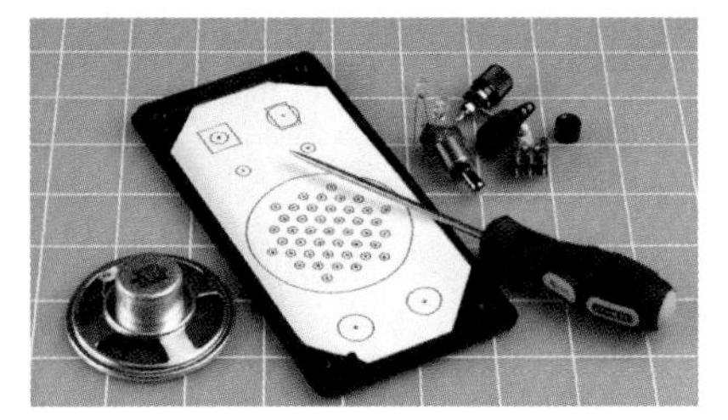

Figure 3-107. *A printed layout for the switches, LEDs, and other components has been taped to the underside of the lid of the project box. An awl is pressed through the paper to mark the center of each hole to be drilled in the lid.*

You have to choose a layout for the switches and other components that will share the top panel of the project box. I like a layout to be neat, so I take the trouble to draw it using illustration software, but a full-size pencil sketch is almost as good. Just make sure there's room for the components to fit together, and try to place them similarly to the schematic, to minimize the risk of confusion.

Tape your sketch to the inside of the top panel, as shown in Figure 3-107, and then use a sharp pointed tool, such as a pick, to press through and mark the plastic at the center of each hole. The indentations will help to center your bit when you drill the holes. Remember that you'll need to make multiple holes to vent the sound from the loudspeaker, which will be beneath the top panel of the box. The result is shown in Figure 3-108.

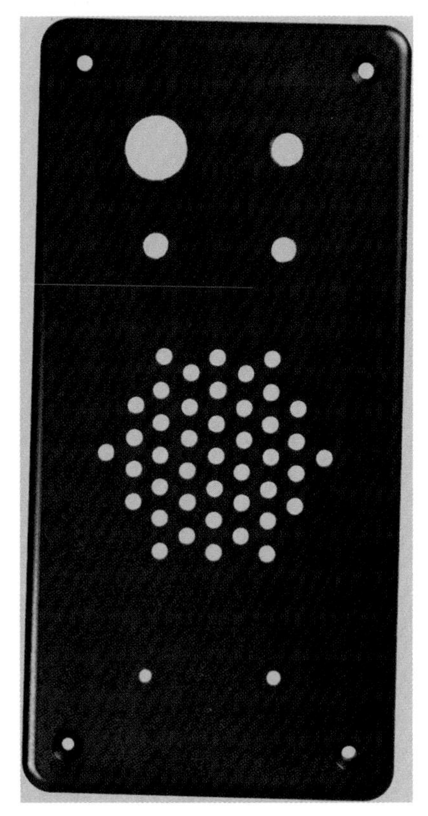

Figure 3-108. *The exterior of the panel after drilling. A small handheld cordless drill can create a neat result if the holes were marked carefully.*

I placed all the components on the top panel, with the exception of the power input jack, which I positioned at one end of the box. Naturally, each hole has to be sized to fit its component, and if you have calipers, they'll be very useful for taking measurements and selecting the right drill bit. Otherwise, make your best guess, too small being better than too large. A deburring tool is ideal for slightly enlarging a hole so that a component fits snugly. This may be necessary if you drill 3/16-inch holes for your 5-inch mm LEDs. Fractionally enlarge each hole, and the LEDs should push in very snugly.

If your loudspeaker lacks mounting holes, you'll have to glue it in place. I used five-minute epoxy to do this. Be careful not to use too much. You don't want any of the glue to touch the speaker cone.

Drilling large holes in the thin, soft plastic of a project box can be a problem. The drill bit tends to dig in and create a mess. You can approach this problem in one of three ways:

1. Use a Forstner drill bit if you have one. It creates a very clean hole.
2. Drill a series of holes of increasing size.
3. Drill a smaller hole than you need, and enlarge it with a deburring tool.

Regardless of which approach you use, you'll need to clamp or hold the top panel of the project box with its outside surface face-down on a piece of scrap wood. Then drill from the inside, so that your bit will pass through the plastic and into the wood.

Finally, mount the components in the panel, as shown in Figure 3-109, and turn your attention to the underneath part of the box.

The circuit board will sit on the bottom, held in place with four #4-size machine screws (bolts) with washers and nylon-insert locknuts. You need to use locknuts to eliminate the risk of a nut working loose and falling among components where it can cause a short circuit.

You'll have to cut the perfboard to fit, taking care not to damage any of the components on it. Also check the underside the board for loose fragments of copper traces after you finish cutting.

Drill bolt holes in the board, if necessary, taking care again not to damage any components. Then mark through the holes to the plastic bottom of the box, and drill the box. Countersink the holes (i.e., bevel the edges of a hole so that a flat-headed screw will fit into it flush with the surrounding surface), push the little bolts up from underneath, and install the circuit board. Be extremely careful not to attach the circuit board too tightly to the project box. This can impose bending stresses, which may break a joint or a copper trace on the board.

I like to include a soft piece of plastic under the board to absorb any stresses. Because you're using locknuts, which will not loosen, there's no need to make them especially tight.

Test the circuit again after mounting the circuit board, just in case.

Figure 3-109. *Components have been added to the control panel of the project box (seen from the underside). The loudspeaker has been glued in place. Spare glue was dabbed onto the LEDs, just in case. The SPDT on/off switch is at the top right, the DPDT pushbutton is at top left, and the binding posts, which will connect with the network of magnetic sensor switches, are at the bottom.*

Soldering the Switches

Figure 3-110 shows how the physical switches should be wired together. Remember that S1 is a toggle switch and S2 is a DPDT pushbutton. Your first step is to decide which way up they should be. Use your meter to find out which terminals are connected when the switch is flipped, and when the button is pressed. You'll probably want the switch to be on when the toggle is flipped upward. Be especially careful with the orientation of the pushbutton, because if you wire it upside-down, it will constantly have the alarm in "test" mode, which is not what you want.

Remember, the center terminal of any double-throw switch is almost always the pole of the switch, connecting with the terminals immediately above it and below it.

Stranded wire is appropriate to connect the circuit board with the components in the top panel, because the strands flex easily and impose less stress on solder joints. Twisting each pair of wires together helps to minimize the mess.

Remember to install the LEDs with their short, negative wires connected with the resistor. This will entail some wire-to-wire soldering. You may want to protect some of these bare leads and joints with thin heat-shrink tubing, to minimize the risk of short circuits when you push all the parts into the box.

When you connect wires or components with the lugs on the switches, your pencil-style soldering iron probably won't deliver enough heat to make good joints. You can use your higher-powered soldering iron in these locations, but you absolutely must apply a good heat sink to protect the LEDs when you attach them, and don't allow the iron to remain in contact with anything for more than 10 seconds. It will quickly melt insulation, and may even damage the internal parts of the switches.

In projects that are more complex than this one, it would be good practice to link the top panel with the circuit board more neatly. Multicolored ribbon cable is ideal for this purpose, with plug-and-socket connectors that attach to the board. For this introductory project, I didn't bother. The wires just straggle around, as shown in Figure 3-111.

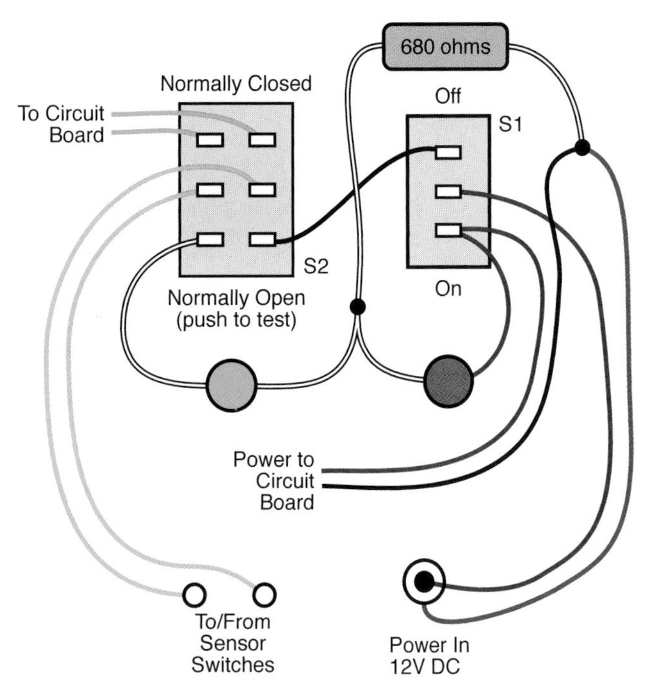

Figure 3-110. *The components can be wired together like this to replicate the circuit shown in The red and green circles are LEDs. Small, solid black circles indicate wire-to-wire solder joints.*

Figure 3-111. *The circuit board has been installed in the base of the project box, and the power input jack has been screwed into the end of the box. Twisted wire-pairs have been connected on a point-to-point basis, without much concern over neatness, as this is a relatively small project. The white insulation at the top-right corner of the front panel is heat-shrink tube that encloses a solder joint and the 680Ω load resistor. Soldering wires to the pushbutton switch requires care and precision, as the contacts are closely spaced.*

Final Test

When you've completed the circuit, test it! If you don't have your network of magnetic sensor switches set up yet, you can just use a piece of wire to connect the two binding posts. Make sure that S1 is in its Off position, then solder the appropriate plug to your 12-volt power source, and plug it into the power jack. When you press the button, the green LED should light up to show continuity between the two binding posts. Now disconnect the wire between the binding posts, press the button again, and the green LED should remain dark.

Reconnect the binding posts, flip S1 to its On position, and the red LED should light up. Press the button, and the alarm should start. Reset it by turning S1 off and then on again; then disconnect the wire between the binding posts. Again, the alarm should start, and it should continue even if you reconnect the wire.

If everything works the way it should, it's time to screw the top of the box in place, pushing the wires inside. Because you're using a large box, you should have no risk of metal parts touching each other accidentally, but still, proceed carefully.

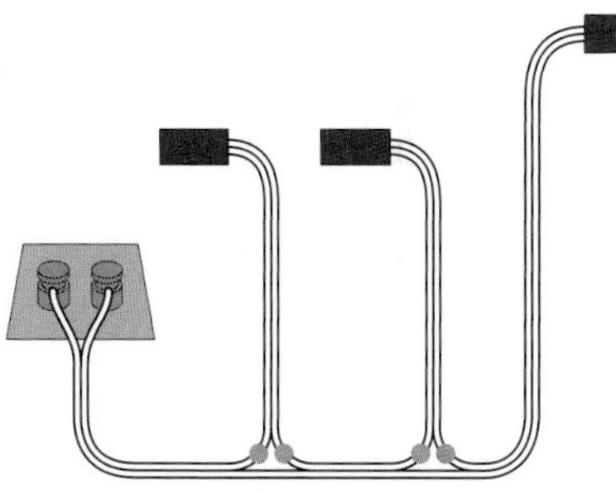

Figure 3-112. *Dual-conductor, white insulated wire can be used to connect the terminals on the alarm control box with magnetic sensors (shown in dark red). Because the sensors must be in series, the wire is cut and joined at positions marked with orange dots.*

Alarm Installation

Before you install your magnetic sensor switches, you should test each one by moving the magnetic module near the switch module and then away from it, while you use your meter to test continuity between the switch terminals. The switch should close when it's next to the magnet, and open when the magnet is removed.

Now draw a sketch of how you'll wire your switches together. Always remember that they have to be in series, not in parallel! Figure 3-112 shows the concept in theory. The two terminals are the binding posts on top of your control box (which is shown in green), and the dark red rectangles are the magnetic sensor switches on windows and doors. Because the wire for this kind of installation usually has two conductors, you can lay it as I've indicated but cut and solder it to create branches. The solder joints are shown as orange dots. Note how current flows through all the switches in series before it gets back to the control box.

Figure 3-113 shows the same network as you might actually install it in a situation where you have two windows and a door. The blue rectangles are the magnetic modules that activate the switch modules.

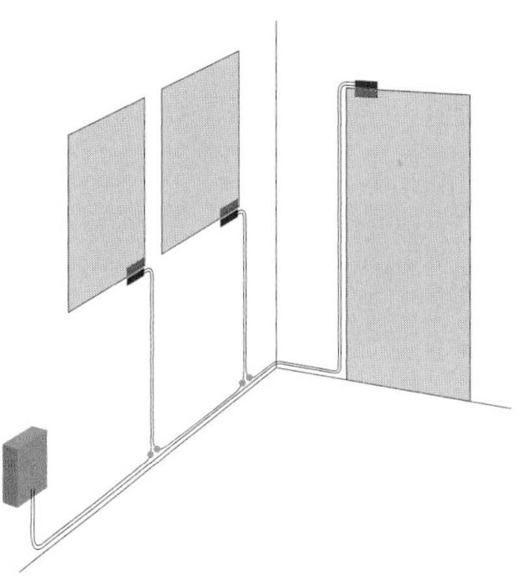

Figure 3-113. *In an installation involving two windows and a door, the magnetic components of the sensors (blue rectangles) could be placed as shown, while the switches (dark red) are located alongside them.*

You'll need a large quantity of wire, obviously. The type of white, stranded wire that is sold for doorbells or furnace thermostats is good. Typically, it is 20-gauge or larger.

After you install all the switches, clip your meter leads to the wires that would normally attach to the alarm box. Set your meter to test continuity, and open each window or door, one at a time, to check whether you're breaking the continuity. If everything is OK, attach the alarm wires to the binding posts on your project box.

Figure 3-114. *The intrusion alarm completed and in its project box.*

Now deal with the power supply. Use your AC adapter, set to 12 volts, hooked up to your type N power plug, or attach the power plug to a 12-volt alarm battery.

If you use a battery, be especially careful that the wire leading to the center terminal of your power plug is positive! A 12-volt battery can deliver substantial current, which can fry your components if you connect it the wrong way around. It would be a shame to destroy your entire project at the very last step.

The only remaining task is to label the switch, button, power socket, and binding posts on the alarm box. You know that the switch turns the power on and off, and the button tests the circuit and the noisemaker, but no one else knows, and you might want to allow a guest to use your alarm while you're away. For that matter, months or years from now, you may forget some details. Will you remember that the power source for this unit should be 12 volts?

Labeling really is a good idea. But as you can see in Figure 3-114, I haven't quite gotten around to it for the box that I built.

Conclusion

The alarm project has taken you through the basic steps that you will usually follow any time you develop something:

1. Draw a schematic and make sure that you understand it.
2. Modify it to fit the pattern of conductors on a breadboard.
3. Install components on the breadboard and test the basic functions.
4. Modify or enhance the circuit, and retest.
5. Transfer to perforated board, test, and trace faults if necessary.
6. Add switches, buttons, power jack, and plugs or sockets to connect the circuit with the outside world.
7. Mount everything in a box (and add labeling).

While going through this sequence, I hope you've learned the basics of electricity, along with some simple electrical theory, and fundamentals about electronic components. This knowledge should enable you to move on to the much more powerful realm of integrated circuit—which I'll cover in Chapter 4.

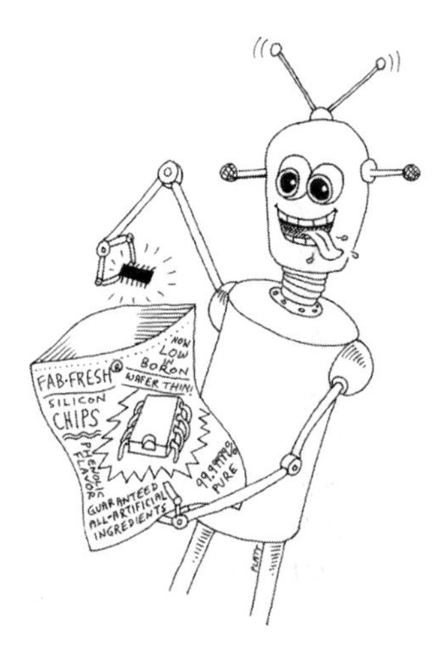

4

Chips, Ahoy!

Before I get into the fascinating topic of integrated circuit (IC) chips, I have to make a confession: some of the things I asked you to do in Chapter 3 could have been done a bit more simply. Does this mean you have been wasting your time? No, I firmly believe that by building circuits with old-fashioned components—capacitors, resistors, and transistors—you acquire the best possible understanding of the principles of electronics. Still, you are going to find that integrated circuit chips, containing dozens, hundreds, or even thousands of transistor junctions, will enable some shortcuts.

IN THIS CHAPTER

Shopping List: Experiments 16 Through 24

Tools

The only new tool that I recommend using in conjunction with chips is a logic probe. This tells you whether a single pin on a chip has a high or low voltage, which can be helpful in figuring out what your circuit is doing. The probe has a memory function so that it will light its LED, and keep it lit, in response to a pulse that may have been too quick for the eye to see.

Search online and buy the cheapest logic probe you can find. I don't have any specific brand recommendations. The one shown in Figure 4-1 is fairly typical.

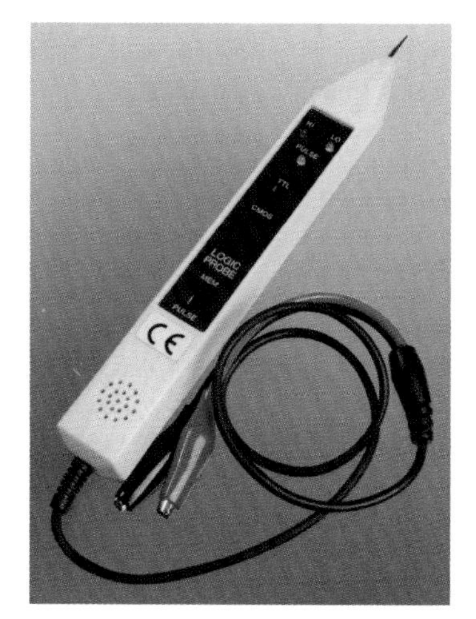

Figure 4-1. *A logic probe detects the high or low voltage on each pin of a chip, and reveals pulses that may occur too quickly for you to perceive them with the unaided eye.*

Supplies

Integrated circuit chips

If you buy everything on this shopping list, and you bought basic parts such as resistors and capacitors that were listed previously, you should have everything you need for all the projects in this chapter.

As chips are quite cheap (currently around 50 cents apiece), I suggest you buy extras. This way, if you damage one, you'll have some in reserve. You'll also have a stock for future projects.

Please read the next section, "Fundamentals: Choosing chips," before you begin chip-shopping. Chips should be easily obtainable from all the major electronics retail suppliers, and sometimes are found on eBay shops. Look in the appendix for a complete list of URLs.

FUNDAMENTALS

Choosing chips

Figure 4-2 shows what is often referred to as an integrated circuit (IC). The circuit is actually etched onto a tiny wafer or "chip" of silicon, embedded in a black plastic body, which is properly referred to as the "package." Tiny wires inside the package link the circuit with the row of pins on either side. Throughout this book, I will use the word "chip" to refer to the whole object, including its pins, as this is the most common usage.

Figure 4-2. *An integrated circuit chip in Plastic Dual-Inline Pin package, abbreviated PDIP, or, more often, DIP.*

The pins are mounted at intervals of 1/10 inch in two rows spaced 3/10 inch apart. This format is known as a Plastic Dual Inline Package, abbreviated PDIP, or, more often, just DIP. The chip in the photograph has four pins in each row; others may have many more. The first thing you need to know, when shopping for chips, is that you'll only be using the DIP package. This book will not be featuring the more modern type, known as "surface-mount," because they're much smaller, more difficult to handle, and require special tools that are relatively expensive. Figure 4-3 shows a size comparison between a 14-pin DIP package and a 14-pin surface-mount package. Many surface-mount chips are even smaller than the one shown.

Just about every chip has a part number printed on it. In Figure 4-2, the part number is KA555. In Figure 4-3, the DIP chip's part number is M74HC00B1, and the surface-mount chip is a 74LVC07AD. You can ignore the second line of numbers and/or letters on each chip, as they are not part of the part number.

Notice in Figure 4-3 that even though the chips look quite different from each other, they both have "74" in their part numbers. This is because both of them are members of the "7400" family of logic chips, which originally had part numbers from 7400 and upward (7400, 7401, 7402, 7403, and so on). Often they are now referred to as "74xx" chips, where "xx" includes all the members of the family. I'll be using this family a lot, so you need to know how to buy them. I'll give you some advice on that without going into details yet about what the chips actually do.

Figure 4-3. *The DIP chip, at the rear, has pins spaced 1/10 inch apart, suitable for insertion in a breadboard or perforated board. It can be soldered without special tools. The small-outline integrated circuit (SOIC) surface-mount chip (foreground) has solder tabs spaced at 1/20 inch. Other surface-mount chips have pins spaced at 1/40 inch or even less (these dimensions are often expressed in millimeters). Surface-mount chips are designed primarily for automated assembly and are difficult to work with manually. In this photo, the yellow lines are 1 inch apart to give you an idea of the scale.*

Take a look at Figure 4-4, which shows how to interpret a typical part number in a 74xx family member. The initial letters identify the manufacturer (which you can ignore, as it really makes no difference for our purposes). Skip the letters until you get to the "74." After that, you find two more letters, which are important. The 74xx family has evolved through many generations, and the letter(s) inserted after the "74" tell you which generation you're dealing with. Some generations have included:

- 74L
- 74LS
- 74C
- 74HC
- 74AHC

And there are more. Generally speaking, subsequent generations tend to be faster or more versatile than previous generations. In this book, for reasons I'll explain later, we are mostly using the HC generation.

FUNDAMENTALS

Choosing chips (continued)

After the letters identifying the generation, you'll find two (sometimes more) numerals. These identify the specific function of the chip. You can ignore any remaining letters and numerals. Looking back at Figure 4-3, the DIP chip part number, M74HC00B1 tells you that it is a chip in the 74xx family, HC generation, with its function identified by numerals 00. The surface-mount chip number, 74LVC07AD, tells you that it is in the 74xx family, LVC generation, with function identified by numerals 07. For convenience we could refer to the first chip as a "74HC00" and the second chip as a "74VC07" because, regardless of their different manufacturers and package sizes, the fundamental behavior of the circuit inside remains the same.

The purpose of this long explanation is to enable you to interpret catalog listings when you go chip shopping. You can search for "74HC00" and the online vendors are usually smart enough to show you appropriate chips from multiple manufacturers, even though there are letters preceding and following the term that you're searched for.

Suppose a circuit requires a 74HC04 chip. If you search for "74HC04" on the website of a parts supplier, you may find versions such as the CD74HC04M96 by Texas Instruments, the 74HC04N by NXP Semiconductors, or MM74HC04N by Fairchild Semiconductor. Because they all have "74HC04" in the middle, any of them will work.

Just be careful that you buy the larger DIP-style package, not the surface-mount package. If the part number has an "N" on the end, you can be sure that it's a DIP package. If there is no "N" on the end, it may or may not be a DIP package, and you will have to check a photo or additional description to make sure. If the part number begins with SS, SO, or TSS, it's absolutely definitely surface-mount, and you don't want it. Many catalogs show photographs of the chips to assist you in buying the right package style.

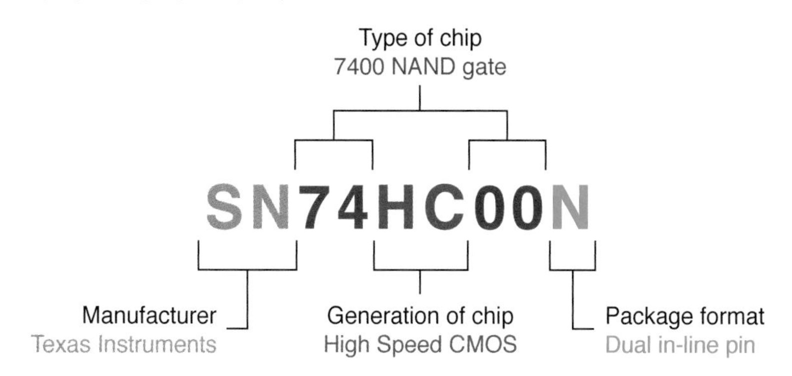

Figure 4-4. *Look for the chip family (74xx, in this case) with the correct generation (HC, in this case) embedded in the number. Make sure you buy the DIP version, not the surface-mount version. The manufacturer is unimportant.*

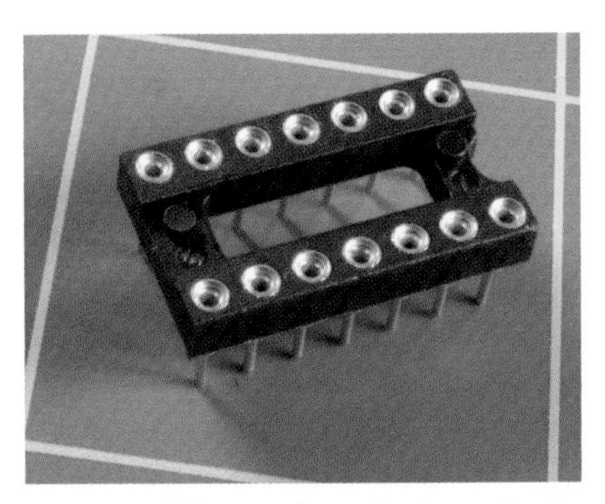

Figure 4-5. *When you're soldering a circuit onto perforated board, sockets eliminate the risk of overheating integrated circuit chips and reduce the risk of zapping them with static electricity, and enable easy replacement.*

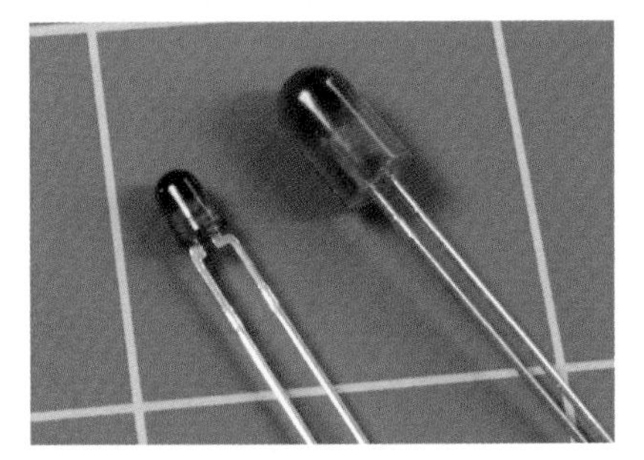

Figure 4-6. *An HC series logic chip is rated to deliver only 4mA at each pin. This is inadequate to drive a typical 5 mm LED (right), which is rated for 20mA forward current. Miniature, low-current LEDs (left) will use as little as 1mA in series with a suitable resistor, and are ideal for test circuits in which you want to see the output with a minimum of hassle.*

Figure 4-7. *Seven-segment displays are the simplest way to show a numeric output and can be driven directly by some CMOS chips. For finished projects, they are typically mounted behind transparent red acrylic plastic panels.*

Here's your chip list:

- 555 Timer. STMicroelectronics SA555N, Fairchild NE555D, RadioShack TLC555 (part number 276-1723), or similar. Do not get the "CMOS" version of this chip, or any fancy versions such as those of high precision. Buy the cheapest you can find. Quantity: 10. The chip in Figure 4-2 is a 555 timer.
- Logic chips types 74HC00, 74HC02, 74HC04, 74HC08, 74HC32, and 74HC86. Actual part numbers could be M74HC00B1, M74HC02B1, M74HC04B1, and so on, by STMicroelectronics, or SN74HC00N, SN74HC02N, SN74HC04N, and so on, by Texas Instruments. Any other manufacturers are acceptable.
- Remember, each part number should have "HC" in the middle of it, and you want the DIP or PDIP package, not surface-mount. Quantity of each: At least 4.
- 4026 Decade Counter (a chip that counts in tens). Texas Instruments CD-4026BE or similar. Quantity: 4 (you'll need 3, but because this is a CMOS chip sensitive to static electricity, you should have at least one in reserve). Any chip with "4026" in its part number should be OK.
- 74LS92 counter chip, 74LS06 open-collector inverter chip, and 74LS27 triple-input NOR chip. Quantity: 2 of each. Note the "LS" in these part numbers! There will be one experiment in which I want you to use the LS generation instead of the HC generation.

IC sockets

I suggest that you avoid soldering chips directly onto perforated board. If you damage them, they're difficult to remove. Buy some DIP sockets, solder the sockets onto the board, and then plug the chips into the sockets. You can use the cheapest sockets you can find (you don't need gold-plated contacts for our purposes). You will need 8-pin, 14-pin, and 16-pin DIP sockets, such as parts 276-1995, 276-1999, and 276-1998 from RadioShack. See Figure 4-5. Quantity of each: 5 minimum.

Low-power LEDs

The logic chips that you'll be using are not designed to deliver significant power. You'll need to add transistors to amplify their output if you want to drive bright LEDs or relays. Because adding transistors is a hassle, I suggest an alternative: Special low-power LEDs that will draw as little as 1mA, such as the Everlight model T-100 Low Current Red, part number HLMPK150. Figure 4-6 shows a size comparison with a regular 5mm LED. Quantity: 10 (at a minimum).

LED numeric displays

In at least one of our projects, you'll want to illuminate some seven-segment LED numerals. You'll need either three individual numerals, or one package containing three numerals, such as the Kingbright High Efficiency Red Diffused, part number BC56-11EWA, which will be specifically referred to in schematics in this book. If you buy a different seven-segment display, it must be an LED with a "common cathode." (Don't buy liquid-crystal LCD numerals; they require different electronics to drive them.) If you have a choice of power consumption, buy whichever product consumes the least current. See Figure 4-7.

Signal diode

1N4148 or similar. Quantity: 5 minimum.

Latching relays

You're going to need a 5-volt latching relay that has two coils, instead of one. The first coil flips the relay one way; the second coil flips it back. The relay consumes no additional power while remaining passively in each state. I suggest the Panasonic DS2E-SL2-DC5V relay. If you buy a different relay, it must be dual-coil latching to run off 5 volts DC, switching at least 1 amp, in a "2 form C" package, to fit your breadboard.

Potentiometers

You'll need 5K, 10K, and 100K linear potentiometers (one of each). Also, a 10K trimmer potentiometer (which you may find described just as a "trimmer"). The manufacturer is unimportant.

Voltage regulators

Because many logic chips require precisely 5 volts DC, you need a voltage regulator to deliver this. The LM7805 does the job. Here again, the chip number will be preceded or followed with an abbreviation identifying the manufacturer and package style, as in the LM7805CT from Fairchild. Any manufacturer will do, but the package should look like the one in Figure 4-8, and if you have a choice, buy a regulator that can deliver at least 1 amp.

Figure 4-8. *Many integrated circuit chips require a controlled power supply of 5 volts, which can be delivered by this regulator when you apply 7.5 to 9 volts to it. The lefthand pin is for positive input, the center pin is a common ground, and the righthand pin is the 5V output. For currents exceeding 250mA, you should bolt the regulator to a metal heat sink using the hole at the top.*

Tactile switches

These are SPST pushbuttons (momentary switches), usually with four legs. Look for the ALPS part number SKHHAKA010 or any similar item that has pins to fit your breadboard or perforated board. See Figure 4-9.

Figure 4-9. *A tactile switch delivers tactile feedback through your fingertip when you press it. They are almost always SPST pushbuttons designed for mounting in circuit boards with standard 1/10-inch hole spacing.*

12-key numeric keypad

Velleman "12 keys keyboard with common output" (no part number, but has been available through All Electronics under catalog code KP-12). Quantity: 1.

This type of keypad has the same layout as an old-fashioned touchtone phone. It should have at least 13 pins or contacts, 12 of which connect with individual pushbuttons, the thirteenth connecting with the other side of all the pushbuttons. In other words, the last pin is "common" to all of them, and this type of keypad is often described as having a "common output." The type of keypad that you don't want is "matrix-encoded," with fewer than 13 contacts, requiring additional external circuitry. See Figures 4-10 and 4-11. If you can't find the Velleman keypad that I suggest, look carefully at keypad descriptions and photographs to make sure that the one you buy is not matrix-encoded and has a common terminal.

Alternatively, you may substitute 12 cheap SPST NO pushbuttons and mount them in a small project box.

Figure 4-10. *When shopping for a numeric keypad, it should have 12 keys in "touch-tone phone" layout, and should have at least 13 contacts for input/output. The contacts are visible here along the front edge.*

Figure 4-11. *This keypad has insufficient pins and will not work in the circuit in this book.*

BACKGROUND

How chips came to be

The concept of integrating solid-state components into one little package originated with British radar scientist Geoffrey W. A. Dummer, who talked about it for years before he attempted, unsuccessfully, to build one in 1956. The first true integrated circuit wasn't fabricated until 1958 by Jack Kilby, working at Texas Instruments. Kilby's version used germanium, as this element was already in use as a semiconductor. (You'll encounter a germanium diode when I deal with crystal radios in the next chapter of this book.) But Robert Noyce, pictured in Figure 4-12, had a better idea.

Born in 1927 in Iowa, in the 1950s Noyce moved to California, where he found a job working for William Shockley. This was shortly after Shockley had set up a business based around the transistor, which he had coinvented at Bell Labs.

Noyce was one of eight employees who became frustrated with Shockley's management and left to establish Fairchild Semiconductor. While he was the general manager of Fairchild, Noyce invented a silicon-based integrated circuit that avoided the manufacturing problems associated with germanium. He is generally credited as the man who made integrated circuits possible.

Early applications were for military use, as Minuteman missiles required small, light components in their guidance systems. These applications consumed almost all chips produced from 1960 through 1963, during which time the unit price fell from around $1,000 to $25 each, in 1963 dollars.

In the late 1960s, medium-scale integration chips emerged, each containing hundreds of transistors. Large-scale integration enabled tens of thousands of transistors on one chip by the mid-1970s, and today's chips can contain as many as several billion transistors.

Robert Noyce eventually cofounded Intel with Gordon Moore, but died unexpectedly of a heart attack in 1990. You can learn more about the fascinating early history of chip design and fabrication at *http://www.siliconvalleyhistorical.org*.

Figure 4-12. *This picture of Robert Noyce, late in his career, is from the Wikimedia Commons.*

Experiment 16: Emitting a Pulse

I'm going to introduce you to the most successful chip ever made: the 555 timer. As you can find numerous guides to it online, you might question the need to discuss it here, but I have three reasons for doing so:

1. It's unavoidable. You simply have to know about this chip. Some sources estimate that more than 1 billion are still being manufactured annually. It will be used in one way or another in most of the remaining circuits in this book.
2. It provides a perfect introduction to integrated circuits, because it's robust, versatile, and illustrates two functions that we'll be dealing with later: comparators and a flip-flop.
3. After reading all the guides to the 555 that I could find, beginning with the original Fairchild Semiconductor data sheet and making my way through various hobby texts, I concluded that its inner workings are seldom explained very clearly. I want to give you a graphic understanding of what's happening inside it, because if you don't have this, you won't be in a good position to use the chip creatively.

You will need:

- 9-volt power supply.
- Breadboard, jumper wires, and multimeter.
- 5K linear potentiometer. Quantity: 1.
- 555 timer chip. Quantity: 1.
- Assorted resistors and capacitors.
- SPST tactile switches. Quantity: 2.
- LED (any type). Quantity: 1.

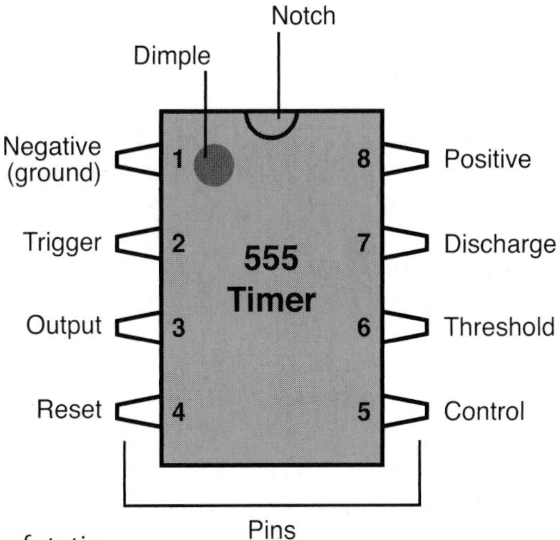

Figure 4-13. *The 555 timer chip, seen from above. Pins on chips are always numbered counterclockwise, from the top-left corner, with a notch in the body of the chip uppermost, or a circular indentation at top-left, to remind you which end is up.*

Procedure

The 555 chip is very robust, but still, in theory, you can zap it with a jolt of static electricity and kill it. Therefore, to be on the safe side, you should ground yourself before handling it. See the "Grounding yourself" warning on page 172 for details. Although this warning primarily refers to the type of chips known as CMOS, which are especially vulnerable, grounding yourself is always a sensible precaution.

Look for a small circular indentation, called the *dimple*, molded into the body of the chip, and turn the chip so that the indentation is at the top-left corner with the pins pointing down. Alternatively, if your chip is of the type with a notch at one end, turn the chip so that the notch is at the top.

The pins on chips are always numbered counterclockwise, starting from the top-left pin (next to the dimple). See Figure 4-13, which also shows the names of the pins on the 555 timer, although you don't need to know most of them just yet.

Insert the chip in your breadboard so that its pins straddle the channel down the center. Now you can easily feed voltages to the pins on either side, and read signals out of them. See Figure 4-14 for a precise guide to placement, in the first project. The timer is identified as "IC1," because "IC" is the customary abbreviation for "Integrated Circuit."

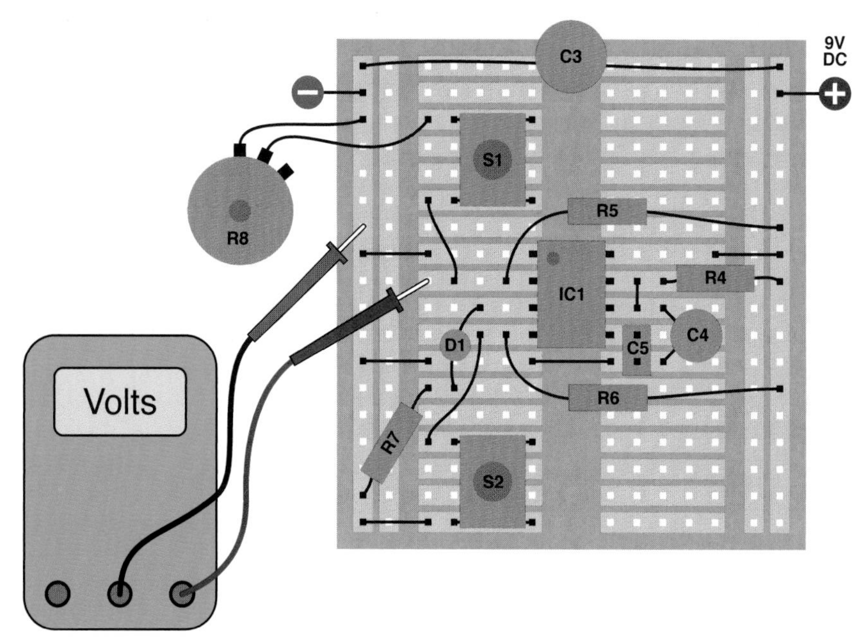

Figure 4-14. *This circuit allows you to explore the behavior of the 555 timer chip. Use your meter to monitor the voltage on pin 2 as shown. There are no resistors labeled R1, R2, or R3 and no capacitors labeled C1 or C2, because they'll be added in a later schematic. Component values in this schematic:*

R4: 100K
R5: 2K2
R6: 10K
R7: 1K
R8: 5K linear potentiometer
C3: 100 μF electrolytic
C4: 47 μF electrolytic
C5: 0.1 μF ceramic
IC1: 555 timer
S1, S2: SPST tactile switches (pushbuttons)
D1: Generic LED

R5 holds the trigger (pin 2) positive until S1 is pressed, which lowers the voltage depending on the setting of potentiometer R8. When the trigger voltage falls below 1/3 of the power supply, the chip's output (pin 3) goes high for a period determined by the values of R4 and C4. S2 resets (zeros) the timer, by reducing the voltage to pin 4, the Reset. C3 smoothes the power supply, and C5 isolates pin 5, the control, so that it won't interfere with the functioning of this test circuit. (We'll use the control pin in a future experiment.)

All integrated circuit chips require a power supply. The 555 is powered with negative voltage applied to pin 1 and positive to pin 8. If you reverse the voltage accidentally, this can permanently damage the chip, so place your jumper wires carefully.

Set your power supply to deliver 9 volts. It will be convenient for this experiment if you supply positive down the righthand side and negative down the lefthand side of the breadboard, as suggested in Figure 4-14. C3 is a large capacitor, at least 100 μF, which is placed across the power supply to smooth it out and provide a local store of charge to fuel fast-switching circuits, as well as to guard against other transient dips in voltage. Although the 555 isn't especially fast-switching, other chips are, and you should get into the habit of protecting them.

Begin with the potentiometer turned all the way counterclockwise to maximize the resistance between the two terminals that we're using, and when you apply the probe from your meter to pin 2, you should measure about 6 volts when you press S1.

Now rotate the potentiometer clockwise and press S1 again. If the LED doesn't light up, keep turning the potentiometer and pressing and releasing the button. When you've turned the potentiometer about two-thirds of the way, you should see the LED light up for just over 5 seconds when you press and release the button. Here are some facts that you should check for yourself:

- The LED will keep glowing after you release the button.
- You can press the button for any length of time (less than the timer's cycle time) and the LED always emits the same length of pulse.
- The timer is triggered by a fall in voltage on pin 2. You can verify this with your meter.
- The LED is either fully on or fully off. You can't see a faint glow when it's off, and the transition from off to on and on to off is very clean and precise.

Check Figure 4-16 to see how the components should look on your breadboard, and then look at the schematic in Figure 4-15 to understand what's happening. I will be adding more components later, which I will be labeling R1, R2, C1, and C2 to be consistent with data sheets that you may see for the 555 timer. Therefore, in this initial circuit the resistors are labeled R4 and up, and capacitors C3 and up.

When S1 (the tactile switch) is open, pin 2 of the 555 timer receives positive power through R5, which is 2K2. Because the input resistance of the timer is very high, the voltage on pin 2 is almost the full 9 volts.

When you press the button, it connects negative voltage through R8, the 5K potentiometer to pin 2. Thus, R8 and R5 form a voltage divider with pin 2 in the middle. You may remember this concept from when you were testing transistors. The voltage between the resistances will change, depending on the values of the resistances.

If R8 is turned up about halfway, it is approximately equal to R5, so the midpoint, connected to pin 2, has about half the 9-volt power supply. But when you turn the potentiometer so that its resistance falls farther, the negative voltage outweighs the positive voltage, so the voltage on pin 2 gradually drops.

If you have clips on your meter leads, you can hook them onto the nearest jumper wires and then watch the meter while you turn the potentiometer up and down and press the button.

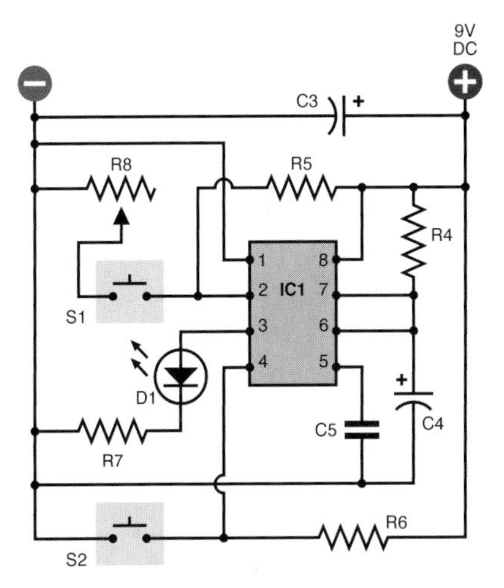

Figure 4-15. *A schematic view of the circuit shown in Figure 4-14. Throughout this chapter, the schematics will be laid out to emulate the most likely placement of components on a breadboard. This is not always the simplest layout, but will be easiest for you to build. Refer to Figure 4-14 for the values of the components.*

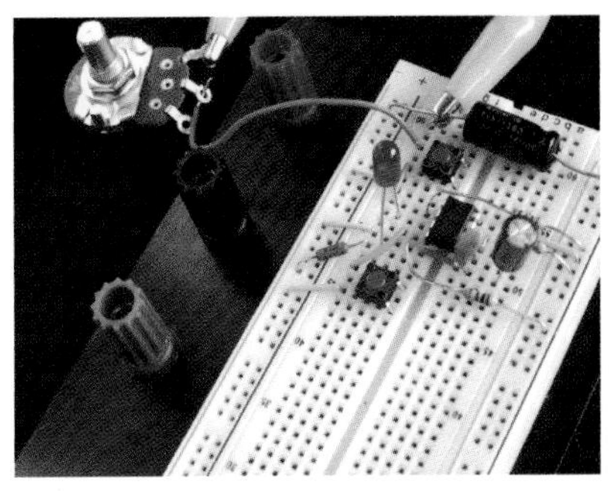

Figure 4-16. *This is how the components look when installed on the breadboard. The alligator clips are attached to a patch cord that links the 100 μF capacitor to the potentiometer. The power supply input is not shown.*

The graphs in Figure 4-17 illustrate what is happening. The upper graph shows the voltage applied to pin 2 by random button-presses, with the potentiometer turned to various values. The lower graph shows that the 555 is triggered if, and only if, the voltage on pin 2 actively drops from above 3 volts to below 3 volts. What's so special about 3 volts? It's one-third of our 9-volt power supply.

Here's the take-home message:

- The output of the 555 (pin 3) emits a *positive* pulse when the trigger (pin 2) drops *below* one-third of the supply voltage.
- The 555 delivers the *same duration* of positive pulse every time (so long as you don't supply a prolonged low voltage on pin 2).
- A *larger* value for R4 or for C4 will *lengthen* the pulse.
- When the output (pin 3) is high, the voltage is almost equal to the supply voltage. When the output goes low, it's almost zero.

The 555 converts the imperfect world around it into a precise and dependable output. It doesn't switch on and off absolutely instantly, but is fast enough to *appear* instant.

Now here's another thing to try. Trigger the timer so that the LED lights up. While it is illuminated, press S2, the second button, which grounds pin 4, the reset. The LED should go out immediately.

When the reset voltage is pulled *low*, the output goes *low*, regardless of what voltage you apply to the trigger.

There's one other thing I want you to notice before we start using the timer for more interesting purposes. I included R5 and R6 so that when you first switch on the timer, it is not emitting a pulse—but is ready to do so. These resistors apply a positive voltage to the trigger and the reset pin, to make sure that the 555 timer is ready to run when you first apply power to it.

As long as the *trigger* voltage is *high*, the timer *will not* emit a pulse. (It emits a pulse when the trigger voltage drops.)

As long as the *reset* voltage is *high*, the timer is *able* to emit a pulse. (It shuts down when the reset voltage drops.)

R5 and R6 are known as "pull-up resistors" because they pull the voltage up. You can easily overwhelm them by adding a direct connection to the negative side of the power supply. A typical pull-up resistor for the 555 timer is 10K. With a 9-volt power supply, it only passes 0.9mA (by Ohm's Law).

Finally, you may be wondering about the purpose of C5, attached to pin 5. This pin is known the "control" pin, which means that if you apply a voltage to it, you can control the sensitivity of the timer. I'll get to this in more detail a little later. Because we are not using this function right now, it's good practice to put a capacitor on pin 5 to protect it from voltage fluctuations and prevent it from interfering with normal functioning.

Make sure you become familiar with the basic functioning of the 555 timer before you continue.

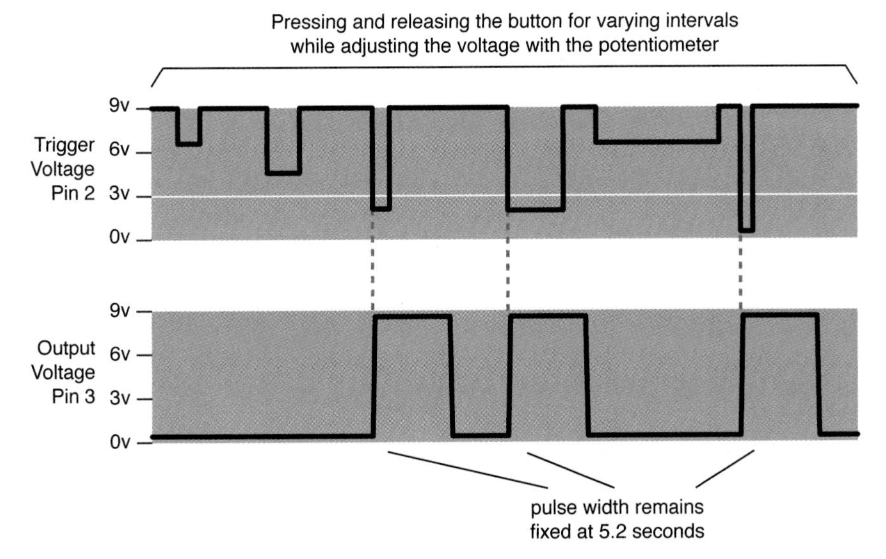

Figure 4-17. *The top graph shows voltage on the trigger (pin 2) when the pushbutton is pressed, for different intervals, at different settings of the potentiometer. The lower graph shows the output (pin 3), which rises until it is almost equal to the power supply, when the voltage on pin 2 drops below 1/3 the full supply voltage.*

FUNDAMENTALS

The following table shows 555 pulse duration in monostable mode:

- Duration is in seconds, rounded to two figures.
- The horizontal scale shows common resistor values between pin 7 and positive supply voltage.
- The vertical scale shows common capacitor values between pin 6 and negative supply voltage.

To calculate a different pulse duration, multiply resistance × capacitance × 0.0011 where resistance is in kilohms, capacitance is in microfarads, and duration is in seconds.

47 µF	0.05	0.11	0.24	0.52	1.1	2.4	5.2	11	24	52
22 µF	0.02	0.05	0.11	0.24	0.53	1.1	2.4	5.3	11	24
10 µF	0.01	0.02	0.05	0.11	0.24	0.52	1.1	2.4	5.2	11
4.7 µF		0.01	0.02	0.05	0.11	0.24	0.52	1.1	2.4	5.2
2.2 µF			0.01	0.02	0.05	0.11	0.24	0.53	1.1	2.4
1.0 µF				0.01	0.02	0.05	0.11	0.24	0.52	1.1
0.47 µF					0.01	0.02	0.05	0.11	0.24	0.52
0.22 µF						0.01	0.02	0.05	0.11	0.24
0.1 µF							0.01	0.02	0.05	0.11
0.047 µF								0.01	0.02	0.05
0.022 µF									0.01	0.02
0.01 µF										0.01
	1K	2K2	4K7	10K	22K	47K	100K	220K	470K	1M

THEORY

Inside the 555 timer: monostable mode

The plastic body of the 555 timer contains a wafer of silicon on which are etched dozens of transistor junctions in a pattern that is far too complex to be explained here. However, I can summarize their function by dividing them into groups, as shown in Figure 4-18. An external resistor and two external capacitors are also shown, labeled the same way as in Figure 4-15.

The negative and positive symbols inside the chip are power sources which actually come from pins 1 and 8, respectively. I omitted the internal connections to those pins for the sake of clarity.

The two yellow triangles are "comparators." Each comparator compares two inputs (at the base of the triangle) and delivers an output (from the apex of the triangle) depending on whether the inputs are similar or different. We'll be using comparators for other purposes later in this book.

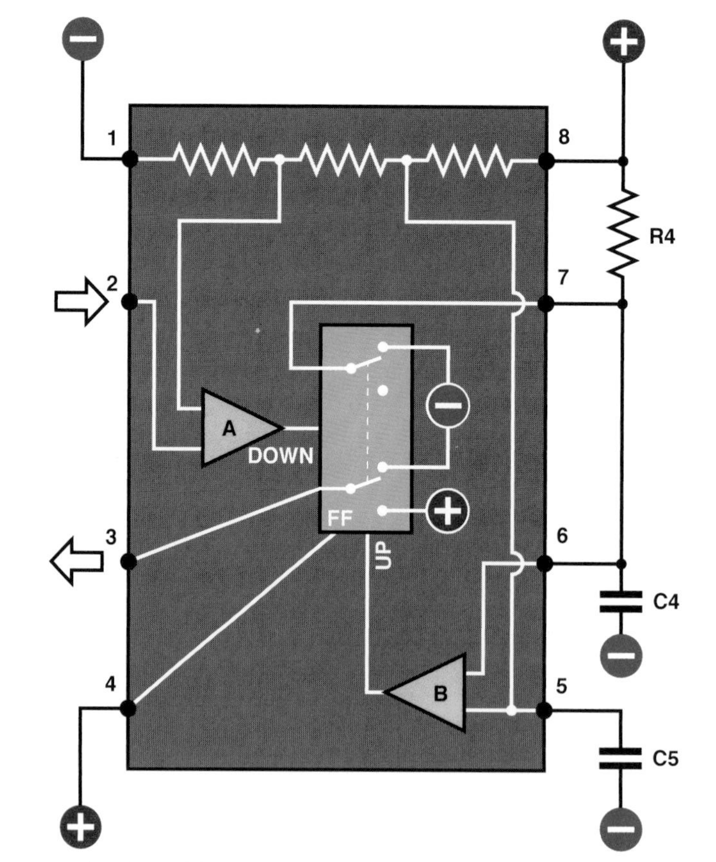

Figure 4-18. *Inside the 555 timer. White lines indicate connections inside the chip. A and B are comparators. FF is a flip-flop which can rest in one state or the other, like a double-throw switch. A drop in voltage on pin 2 is detected by comparator A, which triggers the flip-flop into its "down" position and sends a positive pulse out of pin 3. When C4 charges to 2/3 of supply voltage, this is detected by comparator B, which resets the flip-flop to its "up" position. This discharges C4 through pin 7.*

THEORY

Inside the 555 timer: monostable mode (continued)

The green rectangle, identified as "FF," is a "flip-flop." I have depicted it as a DPDT switch, because that's how it functions here, although of course it is really solid-state.

Initially when you power up the chip, the flip-flop is in its "up" position which delivers low voltage through the output, pin 3. If the flip-flop receives a signal from comparator A, it flips to its "down" state, and flops there. When it receives a signal from comparator B, it flips back to its "up" state, and flops there. The "UP" and "DOWN" labels on the comparators will remind you what each one does when it is activated.

Flip-flops are a fundamental concept in digital electronics. Computers couldn't function without them.

Notice the external wire that connects pin 7 with capacitor C4. As long as the flip-flop is "up," it sinks the positive voltage coming through R4 and prevents the capacitor from charging positively.

If the voltage on pin 2 drops to 1/3 of the supply, comparator A notices this, and flips the flip-flop. This sends a positive pulse out of pin 3, and also disconnects the negative power through pin 7. So now C4 can start charging through R4. While this is happening, the positive output from the timer continues.

As the voltage increases on the capacitor, comparator B monitors it through pin 6, known as the threshold. When the capacitor accumulates 2/3 of the supply voltage, comparator B sends a pulse to the flip-flop, flipping it back into its original state. This discharges the capacitor through pin 7, appropriately known as the discharge pin. Also, the flip-flop ends the positive output through pin 3 and replaces it with a negative voltage. This way, the 555 returns to its original state.

I'll sum up this sequence of events very simply:

1. Initially, the flip-flop grounds the capacitor and grounds the output (pin 3).
2. A drop in voltage on pin 2 to 1/3 the supply voltage or less makes the output (pin 3) positive and allows capacitor C4 to start charging through R4.
3. When the capacitor reaches 2/3 of supply voltage, the chip discharges the capacitor, and the output at pin 3 goes low again.

In this mode, the 555 timer is "monostable," meaning that it just gives one pulse, and you have to trigger it again to get another.

You adjust the length of each pulse by changing the values of R4 and C4. How do you know which values to choose? Check the table on page 157, which gives an approximate idea and also includes a formula so that you can calculate values of your own.

I didn't bother to include pulses shorter than 0.01 second in the table, because a single pulse of this length is usually not very useful. Also I rounded the numbers in the table to 2 significant figures, because capacitor values are seldom more accurate than that.

BACKGROUND

How the timer was born

Back in 1970, when barely a half-dozen corporate seedlings had taken root in the fertile ground of Silicon Valley, a company named Signetics bought an idea from an engineer named Hans Camenzind. It wasn't a huge breakthrough concept—just 23 transistors and a bunch of resistors that would function as a programmable timer. The timer would be versatile, stable, and simple, but these virtues paled in comparison to its primary selling point. Using the emerging technology of integrated circuits, Signetics could reproduce the whole thing on a silicon chip.

This entailed some trial and error. Camenzind worked alone, building the whole thing initially on a large scale, using off-the-shelf transistors, resistors, and diodes on a breadboard. It worked, so then he started substituting slightly different values for the various components to see whether the circuit would tolerate variations during production and other factors such as changes in temperature when the chip was in use. He made at least 10 different versions of the circuit. It took months.

Next came the crafts work. Camenzind sat at a drafting table and used a specially mounted X-Acto knife to scribe his circuit into a large sheet of plastic. Signetics then reduced this image photographically by a ratio of about 300:1. They etched it into tiny wafers, and embedded each of them in a half-inch rectangle of black plastic with the product number printed on top. Thus, the 555 timer was born.

It turned out to be the most successful chip in history, both in the number of units sold (tens of billions and counting) and the longevity of its design (unchanged in almost 40 years). The 555 has been used in everything from toys to spacecraft. It can make lights flash, activate alarm systems, put spaces between beeps, and create the beeps themselves.

Figure 4-19. *Hans Camenzind, inventor and developer of the 555 timer chip for Signetics.*

Today, chips are designed by large teams and tested by simulating their behavior using computer software. Thus, chips inside a computer enable the design of more chips. The heyday of solo designers such as Hans Camenzind is long gone, but his genius lives inside every 555 timer that emerges from a fabrication facility. (If you'd like to know more about chip history, see *http://www.semiconductormuseum.com/Museum_Index.htm*.)

FUNDAMENTALS

Why the 555 is useful

In its monostable mode (which is what you just saw), the 555 will emit a single pulse of fixed (but programmable) length. Can you imagine some applications? Think in terms of the pulse from the 555 controlling some other component. A motion sensor on an outdoor light, perhaps. When an infra-red detector "sees" something moving, the light comes on for a specific period—which can be controlled by a 555.

Another application could be a toaster. When someone lowers a slice of bread, a switch will close that triggers the toasting cycle. To change the length of the cycle, you could use a potentiometer instead of R4 and attach it to the external lever that determines how dark you want your toast. At the end of the toasting cycle, the output from the 555 would pass through a power transistor, to activate a solenoid (which is like a relay, except that it has no switch contacts) to release the toast.

Intermittent windshield wipers could be controlled by a 555 timer—and on earlier models of cars, they actually were.

And what about the burglar alarm that was described at the end of Chapter 3? One of the features that I listed, which has not been implemented yet, is that it should shut itself off after a fixed interval. We can use the change of output from a 555 timer to do that.

The experiment that you just performed seemed trivial, but really it implies all kinds of possibilities.

555 timer limits

1. The timer can run from a stable voltage source ranging from 5 to 15 volts.
2. Most manufacturers recommend a range from 1K to 1M for the resistor attached to pin 7.
3. The capacitor value can go as high as you like, if you want to time really long intervals, but the accuracy of the timer will diminish.
4. The output can deliver as much as 100mA at 9 volts. This is sufficient for a small relay or miniature loudspeaker, as you'll see in the next experiment.

Beware of Pin-Shuffling!

In all of the schematics in this book, I'll show chips as you'd see them from above, with pin 1 at top left. Other schematics that you may see, on websites or in other books, may do things differently. For convenience in drawing circuits, people shuffle the pin numbers on a chip so that pin 1 isn't necessarily shown adjacent to pin 2.

Look at the schematic in Figure 4-20 and compare it with the one in Figure 4-15. The connections are the same, but the one in Figure 4-20 groups pins to reduce the apparent complexity of the wiring.

"Pin shuffling" is common because circuit-drawing software tends to do it, and on larger chips, it is necessary for functional clarity of the schematic (i.e., logical groupings of pin names versus physical groupings on memory chips, for example). When you're first learning to use chips, I think it's easier to understand a schematic that shows the pins in their actual positions. So that's the practice I will be using here.

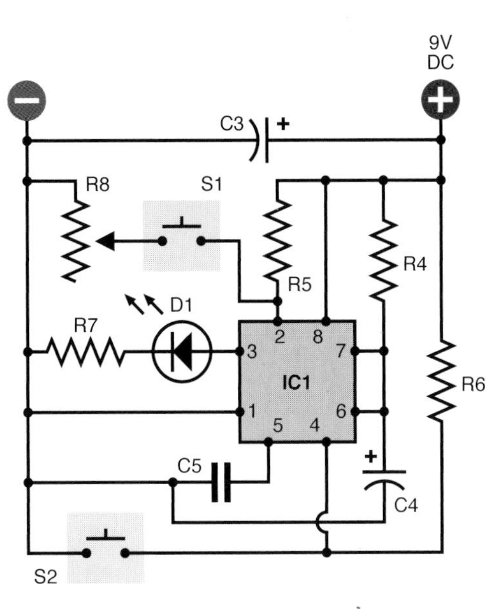

Figure 4-20. *Many people draw schematics in which the pin numbers on a chip are shuffled around to make the schematic smaller or simpler. This is not helpful when you try to build the circuit. The schematic here is for the same circuit as in Figure 4-15. This version would be harder to recreate on a breadboard.*

Experiment 17: Set Your Tone

I'm going to show you two other ways in which the 555 timer can be used.

You will need the same items as in Experiment 16, plus:

- Additional 555 timer chip. Quantity: 1.
- Miniature loudspeaker. Quantity: 1.
- 100K linear potentiometer. Quantity: 1.

Procedure

Leave the components from Experiment 16 where they are on the breadboard, and add the next section below them, as shown in Figures 4-21 and 4-22. Resistor R2 is inserted between pins 6 and 7, instead of the jumper wire that shorted the pins together in the previous circuit, and there's no external input to pin 2 anymore. Instead, pin 2 is connected via a jumper wire to pin 6. The easiest way to do this is by running the wire across the top of the chip.

I have omitted the smoothing capacitor from the schematic in Figure 4-22, because I'm assuming that you're running this circuit on the same breadboard as the first, where the previous smoothing capacitor is still active.

A loudspeaker in series with a 100Ω resistor (R3) has been substituted for the LED to show the output from the chip. Pin 4, the reset, is disabled by connecting it to the positive voltage supply, as I'm not expecting to use the reset function in this circuit.

Now what happens when you apply power? Immediately, you should hear noise through the loudspeaker. If you don't hear anything, you almost certainly made a wiring error.

Notice that you don't have to trigger the chip with a pushbutton anymore. The reason is that when C1 charges and discharges, its fluctuating voltage is connected via a jumper wire across the top of the chip to pin 2, the trigger. In this way, the 555 timer now triggers itself. I'll describe this in more detail in the next section "Theory: Inside the 555 timer: astable mode," if you want to see exactly what is going on.

In this mode, the chip is "astable," meaning that it is not stable, because it flips to and fro endlessly, sending a stream of pulses for as long as the power is connected. The pulses are so rapid that the loudspeaker reproduces them as noise.

In fact, with the component values that I specified for R1, R2, and C1, the 555 chip is emitting about 1,500 pulses per second. In other words, it creates a 1.5 KHz tone.

Check the table on page 166 to see how different values for R2 and C1 can create different pulse frequencies with the chip in this astable mode. Note that the table assumes a fixed value of 1K for R1!

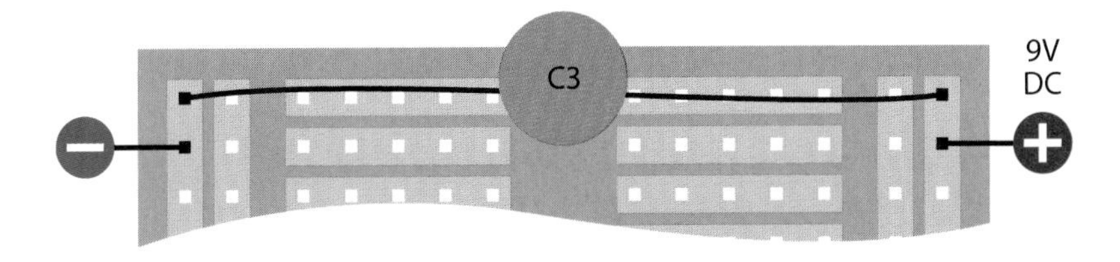

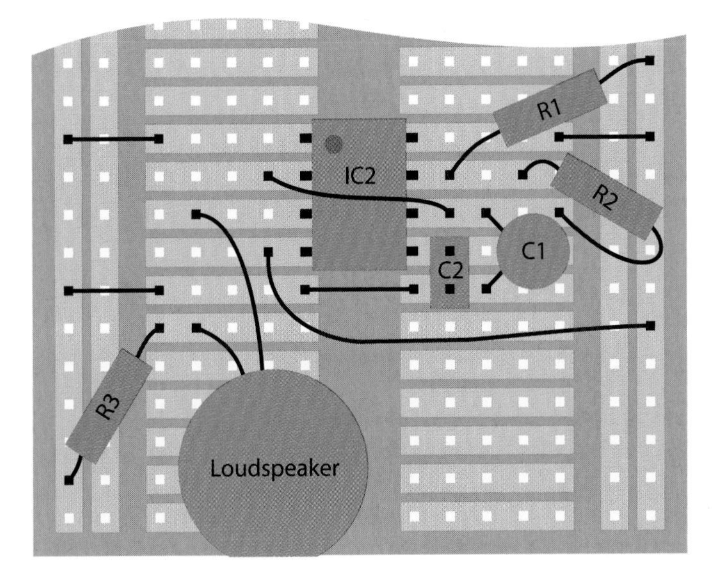

Figure 4-21. *These components should be added on the same breadboard below the components shown in Figure 4-14. Use the following values to test the 555 timer in its astable mode:*

R1: 1K
R2: 10K
R3: 100Ω
C1: 0.047 μF ceramic or electrolytic
C2: 0.1 μF ceramic
IC2: 555 timer

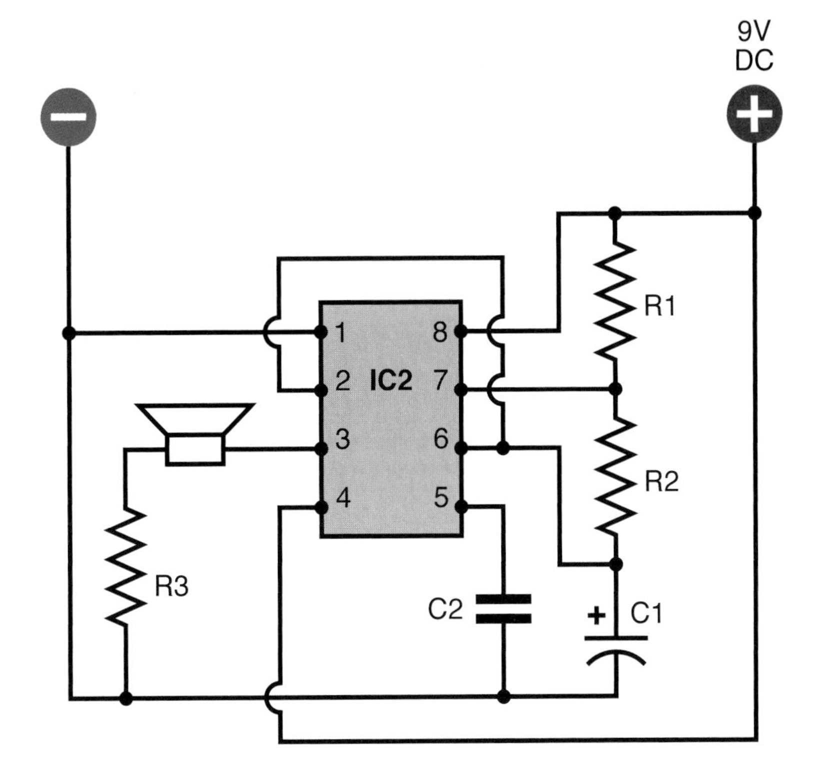

Figure 4-22. *This is the schematic version of the circuit shown in Figure 4-21. The component values are the same.*

THEORY

Inside the 555 timer: astable mode

Here's what is happening now, illustrated in Figure 4-23. Initially, the flip-flop grounds C1 as before. But now the low voltage on the capacitor is connected from pin 6 to pin 2 through an external wire. The low voltage tells the chip to trigger itself. The flip-flop obediently flips to its "on" position and sends a positive pulse to the loudspeaker, while removing the negative voltage from pin 6.

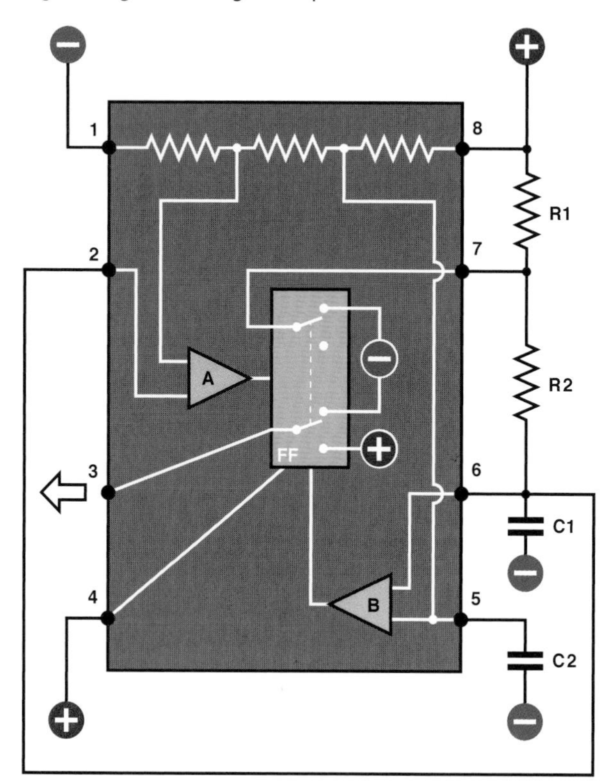

Figure 4-23. *When the 555 timer is used in astable mode, resistor R2 is placed between pin 6 and pin 7, and pin 6 is connected via an external wire to pin 2, so that the timer triggers itself.*

Now C1 starts charging, as it did when the timer was in monostable mode, except that it is being charged through R1 + R2 in series. Because the resistors have low values, and C1 is also small, C1 charges quickly. When it reaches 2/3 full voltage, comparator B takes action as before, discharging the capacitor and ending the output pulse from pin 3.

The capacitor takes longer to discharge than before, because R2 has been inserted between it and pin 7, the discharge pin. While the capacitor is discharging, its voltage diminishes, and is still linked to pin 2. When the voltage drops to 1/3 of full power or less, comparator A kicks in and sends another pulse to the flip-flop, starting the process all over again.

Summing up:

1. In astable mode, as soon as power is connected, the flip-flop pulls down the voltage on pin 2, triggering comparator A, which flips the flip-flop to its "down" position.
2. Pin 3, the output, goes high. The capacitor charges through R1 and R2 in series.
3. When the capacitor reaches 2/3 of supply voltage, the flip-flop goes "up" and the output at pin 3 goes low. The capacitor starts to discharge through R2.
4. When the charge on the capacitor diminishes to 1/3 of full voltage, the pull-down on pin 2 flips the flip-flop again and the cycle repeats.

Unequal on/off cycles

When the timer is running in astable mode, C1 charges through R1 and R2 in series. But when C1 discharges, it dumps its voltage through R2 only. This means that the capacitor charges more slowly than it discharges. While it is charging, the output on pin 3 is high; while it is discharging, the output on pin 3 is low. Consequently the "on" cycle is always longer than the "off" cycle. Figure 4-24 shows this as a simple graph.

If you want the on and off cycles to be equal, or if you want to adjust the on and off cycles independently (for example, because you want to send a very brief pulse to another chip, followed by a longer gap until the next pulse), all you need to do is add a diode, as shown in Figure 4-25.

Now when C1 charges, the electricity flows through R1 as before but takes a shortcut around R2, through diode D1. When C1 discharges, the diode blocks the flow of electricity in that direction, and so the discharge goes back through R2.

THEORY

Inside the 555 timer: astable mode (continued)

R1 now controls the charge time on its own, while R2 controls the discharge time. The formula for calculating the frequency is now:

Frequency = 1.44 / ((R1 + R2) × C1) or Frequency = 1.4 / ((R1 + R2) × C1)

If you set R1 = R2, you should get almost equal on/off cycles ("almost" because the diode itself imposes a small internal voltage drop of about 0.6V). The exact value depends primarily on the manufacturing process used to make the diode.

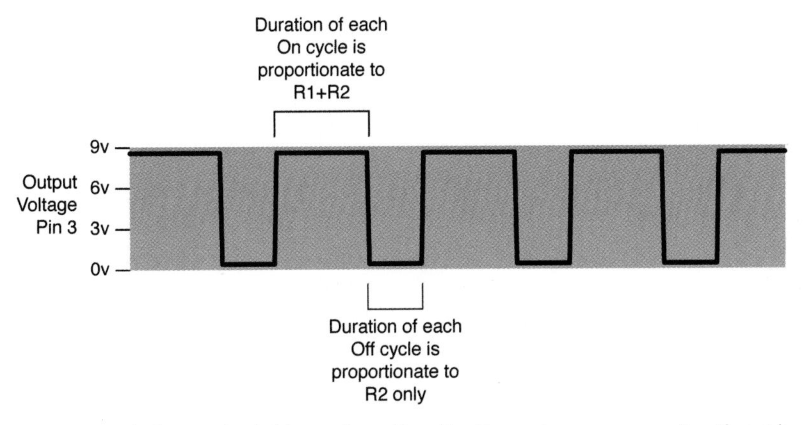

Figure 4-24. *In its usual astable configuration, the timer charges a capacitor through R1+R2 and discharges the capacitor through R2 only. Therefore its output on cycles are longer than its output off cycles.*

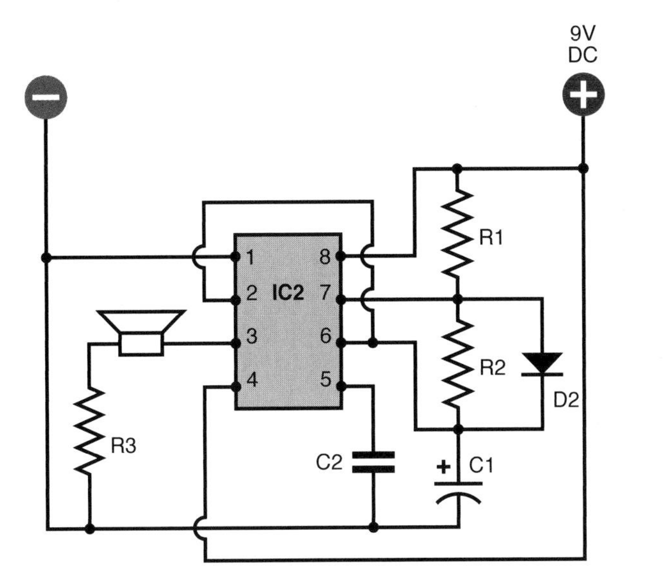

Figure 4-25. *This is a modification of the schematic shown in Figure 4-22. By adding a diode to a 555 timer running in astable mode, we eliminate R2 from the charging cycle of capacitor C1. Now we can adjust the output on cycle with the value of R1, and the output off cycle with the value of R2, so that the on and off durations are independent of each other.*

FUNDAMENTALS

The following table shows 555 timer frequency in astable mode:

- Frequency is in pulses per second, rounded to two figures.
- The horizontal scale shows common resistor values for R2.
- The vertical scale shows common capacitor values for C1. Resistor R1 is assumed to be 1K.
- Resistor R1 is assumed to be 1K.

To calculate a different frequency: double R2, add the product to R1, multiply the sum by C1, and divide the result into 1440. Like this:

$$\text{Frequency} = 1440 / ((R1 + 2R2) \times C1) \text{ cycles per second}$$

In this formula, R1 and R2 are in kilohms, C1 is in microfarads, and the frequency is in hertz (cycles per second). Note that the frequency is measured from the start of one pulse to the start of the next. The duration of each pulse is not the same as the length of time between each pulse. This issue is discussed in the previous section.

	1K	2K2	4K7	10K	22K	47K	100K	220K	470K	1M
47 µF	10	5.7	3.0	1.5	0.7	0.3	0.2	0.1		
22 µF	22	12	6.3	3.1	1.5	0.7	0.3	0.2	0.1	
10 µF	48	27	14	6.9	3.2	1.5	0.7	0.3	0.2	0.1
4.7 µF	100	57	30	15	6.8	3.2	1.5	0.7	0.3	0.2
2.2 µF	220	120	63	31	15	6.9	3.3	1.5	0.7	0.3
1.0 µF	480	270	140	69	32	15	7.2	3.3	1.5	0.7
0.47 µF	1,000	570	300	150	68	32	15	7	3.3	1.5
0.22 µF	2,200	1,200	630	310	150	69	33	15	7	3.3
0.1 µF	4,800	2,700	1,400	690	320	150	72	33	15	7.2
0.047 µF	10,000	5,700	3,000	1,500	680	320	150	70	33	15
0.022 µF	22,000	12,000	6,300	3,100	1,500	690	330	150	70	33
0.01 µF	48,000	27,000	14,000	6,900	3,200	1,500	720	330	150	72

Astable Modifications

In the circuits shown in Figures 4-22 or 4-25, if you substitute a 100K potentiometer for R2, you can adjust the frequency up and down by turning the shaft.

Another option is to "tune" the timer by using pin 5, the control, as shown in the Figure 4-26. Disconnect the capacitor that was attached to that pin and substitute the series of resistors shown. R9 and R11 are both 1K resistors, either side of R10, which is a 100K potentiometer. They ensure that pin 5 always has at least 1K between it and the positive and negative sides of the power supply. Connecting it directly to the power supply won't damage the timer, but will prevent it from generating audible tones. As you turn the potentiometer to and fro, the frequency will vary over a wide range. If you want to generate a very specific frequency, a trimmer potentiometer can be used instead.

A primary advantage of using pin 5 to adjust frequency is that you can control it remotely. Take the output from pin 3 of another 555 timer running slowly in astable mode, and pipe it through a 2K2 resistor to pin 5. Now you get a two-tone siren effect, as one timer controls the other. If, in addition, you add a 100 µF capacitor between pin 5 and ground, the charging and discharging of the capacitor will make the tone slide up and down instead of switching abruptly. I'll describe this in more detail shortly. This leads me to the whole topic of one chip controlling another chip, which will be our last variation on this experiment.

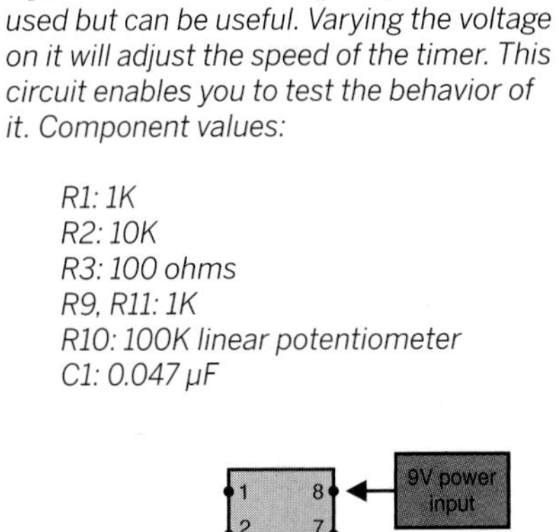

Figure 4-26. *The control (pin 5) is seldom used but can be useful. Varying the voltage on it will adjust the speed of the timer. This circuit enables you to test the behavior of it. Component values:*

R1: 1K
R2: 10K
R3: 100 ohms
R9, R11: 1K
R10: 100K linear potentiometer
C1: 0.047 µF

Chaining Chips

Generally speaking, chips are designed so that they can talk to each other. The 555 couldn't be easier in this respect:

- Pin 3, the output, from one 555 can be connected directly to pin 2, the trigger, of a second 555.
- Alternatively, the output can be sufficient to provide power to pin 8 of a second 555.
- The output is appropriate to control or power other types of chips too.

Figure 4-27 shows these options.

When the output from the first 555 goes high, it is about 70 to 80% of its supply voltage. In other words, when you're using a 9V supply, the high output voltage is at least 6 volts. This is still above the minimum of 5V that the second chip needs to trigger its comparator, so there's no problem.

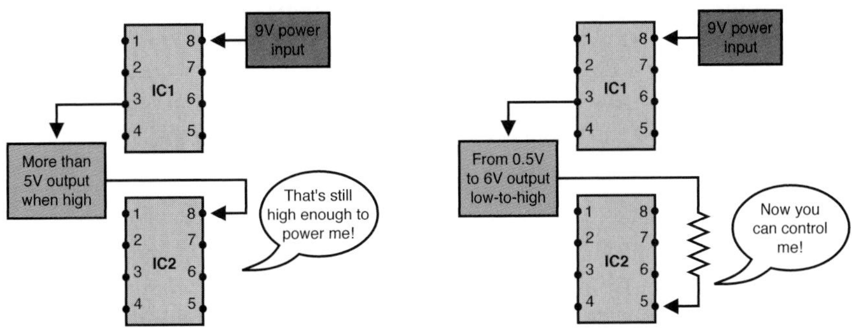

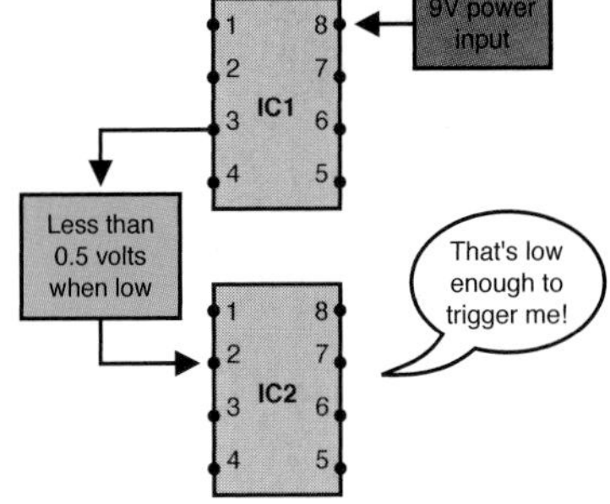

Figure 4-27. *Three ways to chain 555 timers together. The output of IC1 can power a second timer, or adjust its control voltage, or activate its trigger pin.*

You can chain together the two 555 timers that you already have on your breadboard. Figure 4-28 shows how to connect the two circuits that were shown previously in Figures 4-15 and 4-22. Run a wire from pin 3 (the output) of the first chip to pin 8 (the positive power supply) of the second chip, and disconnect the existing wire connecting pin 8 to your power supply. The new wire is shown in red. Now when you press the button to activate the first chip, its output powers the second chip.

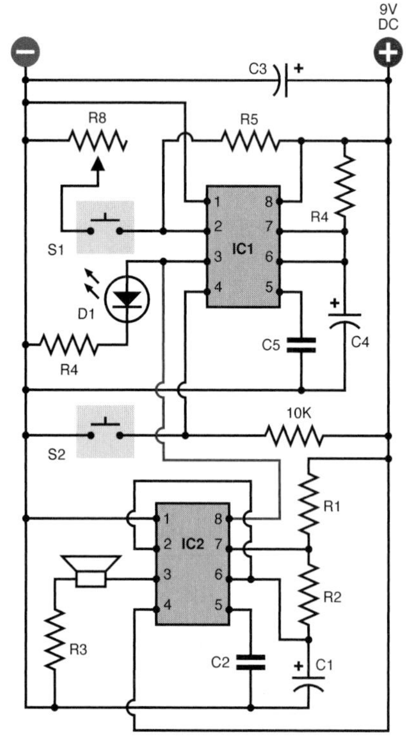

Figure 4-28. *You can combine the two circuits shown in Figures 4-15 and 4-22 simply by disconnecting the wire that provides power to pin 8 of the second timer, and running a substitute wire (shown in red).*

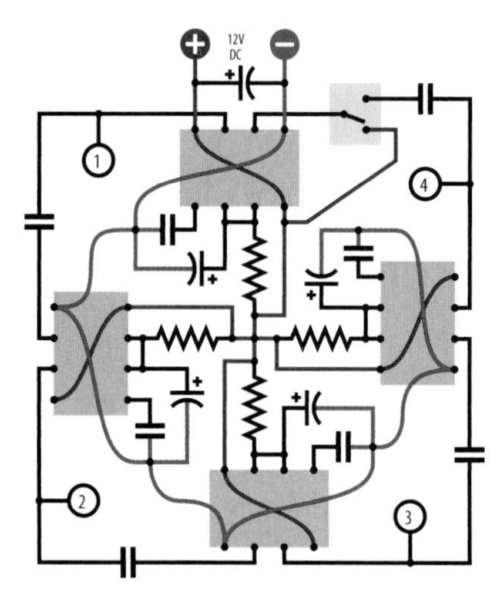

Figure 4-29. *Four 555 timers, chained together in a circle, can flash a series of four sets of LEDs in sequence, like Christmas lights or a movie marquee.*

You can also use the output from one chip to trigger another (i.e., you can connect pin 3 from the first chip to pin 2 of the second). When the output from the first chip is low, it's less than half a volt. This is well below the threshold that the second chip requires to be activated. Why would you want to do this? Well, you might want to have both timers running in monostable mode, so that the end of a high pulse from the first one triggers the start of a high pulse in the second one. In fact, you could chain together as many timers as you like in this way, with the last one feeding back and triggering the first one, and they could flash a series of LEDs in sequence, like Christmas lights. Figure 4-29 shows four timers linked this way, wired point-to-point on perforated board. Each output numbered 1 through 4 could drive up to 10 LEDs, using relatively high load resistors to limit their current. The four capacitors around the edges of the schematic should be 0.1µF, to isolate each timer from the next while allowing a triggering pulse to pass through.

Incidentally, you can reduce the chip count (the number of chips) by using two 556 timers instead of four 555 timers. The 556 contains a pair of 555 timers in one package. But because you have to make the same number of external connections (other than the power supply), I haven't bothered to use this variant.

You can even get a 558 timer that contains four 555 circuits, all preset to function in astable mode. I decided not to use this chip, because its output behaves differently from a normal 555 timer. But you can buy a 558 timer and play with it if you wish. It is ideal for doing the "chain of four timers" that I suggested previously. The data sheet even suggests this.

Lastly, going back to the idea of modifying the frequency of a 555 timer in astable mode, you can chain two timers, as shown in Figure 4-30. The red wire shows the connection from the output of the first timer to the control pin of the second. The first timer has now been rewired in astable mode, so that it creates an oscillating on/off output around once per second. This output flashes the LED (to give you a visual check of what's going on) and feeds through R7 to the control pin of the second timer.

But C2 is a large capacitor, which takes time to charge through R7. While this happens, the voltage detected by pin 5 slowly rises, so that the tone generated by IC2 gradually lowers in pitch. Then IC1 reaches the end of its on cycle and switches itself off, at which point C2 discharges and the pitch of the sound generated by IC2 falls again.

You can tweak this circuit to create all kinds of sounds, much more controllably then when you were using PUT transistors to do the same kind of thing. Here are some options to try:

- Double or halve the value of C2.
- Omit C2 completely, and experiment with the value of R7.
- Substitute a 10K potentiometer for R7.
- Change C4 to increase or decrease the cycle time of IC1.

- Halve the value of R5 while doubling the value of C4, so that the cycle time of IC1 stays about the same, but the On time becomes significantly longer than the Off time.
- Change the supply voltage in the circuit from 9 volts to 6 volts or 12 volts.

Remember, you can't damage a 555 timer by making changes of this kind. Just make sure that the negative side of your power supply goes to pin 1 and the positive side to pin 8.

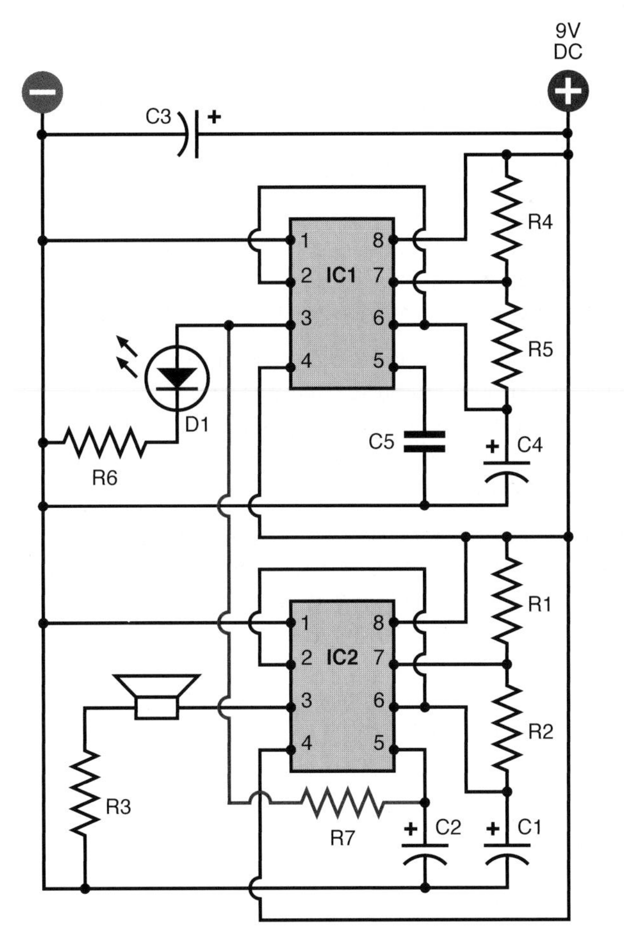

Figure 4-30. *When both timers are astable, but IC1 runs much more slowly than IC2, the output from IC1 can be used to modulate the tone generated by IC2. Note that as this is a substantial modification to the previous schematics, several components have been relabeled. To avoid errors, you may need to remove the old circuit from your breadboard and build this version from scratch. Try these values initially:*

R1, R4, R6, R7: 1K
R2, R5: 10K
R3: 100 ohms
C1: 0.047 µF
C2, C3: 100 µF
C4: 68 µF
C5: 0.1 µF

Figure 4-31. *After putting a 1K resistor between the common cathode of the display and the negative supply voltage, you can use the positive supply voltage to illuminate each segment in turn.*

Experiment 18: Reaction Timer

Because the 555 can easily run at thousands of cycles per second, we can use it to measure human reactions. You can compete with friends to see who has the fastest response—and note how your response changes depending on your mood, the time of day, or how much sleep you got last night.

Before going any further, I have to warn you that this circuit will have more connections than others you've tackled so far. It's not conceptually difficult, but requires a lot of wiring, and will only just fit on a breadboard that has 63 rows of holes. Still, we can build it in a series of phases, which should help you to detect any wiring errors as you go.

You will need:

- 4026 chip. Quantity: 4 (really you need only 3, but get another one in case you damage the others).
- 555 timers. Quantity: 3.
- Tactile switches (SPST momentary switches). Quantity: 3.
- Three numeric LEDs, or one 3-digit LED display (see the shopping list at the beginning of this chapter). Quantity: 1.
- Breadboard, resistors, capacitors, and meter, as usual.

Step 1: Display

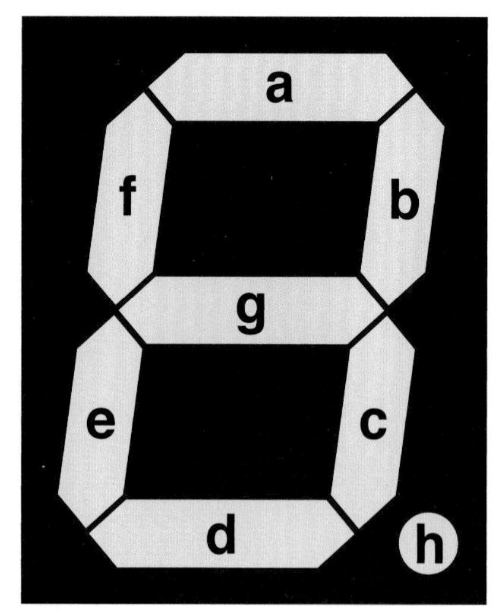

Figure 4-32. *The most basic and common digital numeral consists of seven LED segments identified by letters, as shown here, plus an optional decimal point.*

You can use three separate LED numerals for this project, but I suggest that you buy the Kingbright BC56-11EWA on the shopping list at the beginning of this chapter. It contains three numerals in one big package.

You should be able to plug it into your breadboard, straddling the center channel. Put it all the way down at the bottom of the breadboard, as shown in Figure 4-31. Don't put any other components on the breadboard yet.

Now set your power supply to 9 volts, and apply the *negative* side of it to the row of holes running up the breadboard on the *righthand* side. Insert a 1K resistor between that negative supply and each of pins 18, 19, and 26 of the Kingbright display, which are the "common cathode," meaning the negative connection shared by each set of LED segments in the display. (The pin numbers of the chip are shown in Figure 4-33. If you're using another model of display, you'll have to consult a data sheet to find which pin(s) are designed to receive negative voltage.)

Switch on the power supply and touch the free end of the positive wire to each row of holes serving the display on its left and right sides. You should see each segment light up, as shown in Figure 4-31.

Each numeral from 0 to 9 is represented by a group of these segments. The segments are always identified with lowercase letters *a* through *g*, as shown in Figure 4-32. In addition, there is often a decimal point, and although we won't be using it, I've identified it with the letter *h*.

Check Figure 4-33 showing the Kingbright display, and you'll see I have annotated each pin with its function. You can step down the display with the positive wire from your power supply, making sure that each pin lights an appropriate segment.

Incidentally, this display has two pins, numbered 3 and 26, both labeled to receive negative voltage for the first of the digits. Why two pins instead of one? I don't know. You need to use only one, and as this is a passive chip, it doesn't matter if you leave the unused one unconnected. Just take care not to apply positive voltage to it, which would create a short circuit.

A numeric display has no power or intelligence of its own. It's just a bunch of light-emitting diodes. It's not much use, really, until we can figure out a way to illuminate the LEDs in appropriate groups—which will be the next step.

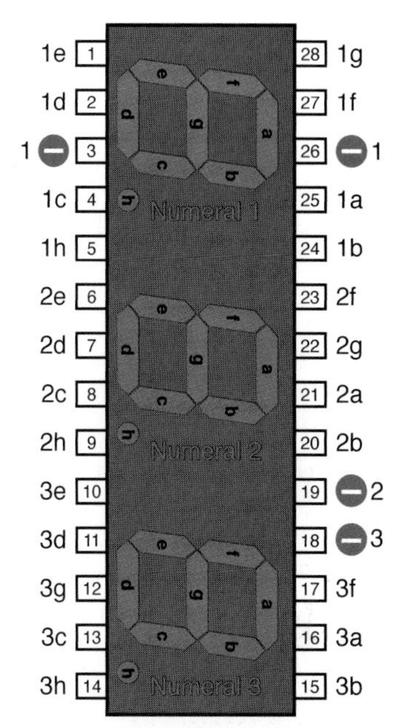

Figure 4-33. *This Kingbright unit incorporates three seven-segment numeric displays in one package, and can be driven by three chained 4026 decade counters. The pin numbers are shown close to the chip. Segments a through g of numeral 1 are identified as 1a through 1g. Segments a through g of numeral 2 are identified as 2a through 2g. Segments a through g of numeral 3 are identified as 3a through 3g.*

Step 2: Counting

Fortunately, we have a chip known as the 4026, which receives pulses, counts them, and creates an output designed to work with a seven-segment display so that it shows numbers 0–9. The only problem is that this is a rather old-fashioned CMOS chip (meaning, Complementary Metal Oxide Semiconductor) and is thus sensitive to static electricity. Check the caution on page 172 before continuing.

Switch off your power supply and connect its wires to the top of the breadboard, noting that for this experiment, we're going to need positive and negative power on both sides. See Figure 4-34 for details. If your breadboard does not already have the columns of holes color-coded, I suggest you use Sharpie markers to identify them, to avoid polarity errors that can fry your components.

The 4026 counter chip is barely powerful enough to drive the LEDs in our display when powered by 9 volts. Make sure you have the chip the right way up, and insert it into the breadboard immediately above your three-digit display, leaving just one row of holes between them empty.

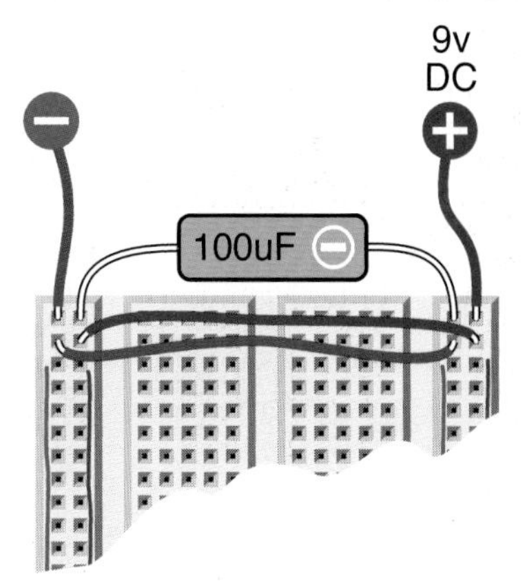

Figure 4-34. *When building circuits around chips, it's convenient to have a positive and negative power supply down each side of your breadboard. For the reaction timer circuit, a 9V supply with a 100 μF smoothing capacitor can be set up like this. If your breadboard doesn't color-code the columns of holes on the left and right sides, I suggest you do that yourself with a permanent marker.*

The schematic in Figure 4-35 shows how the pins of the 4026 chip should be connected. The arrows tell you which pins on the display should be connected with pins on the counter.

Figure 4-36 shows the "pinouts" (i.e., the functions of each pin) of a 4026 counter chip. You should compare this with the schematic in Figure 4-35.

Include a tactile switch between the positive supply and pin 1 of the 4026 counter, with a 10K resistor to keep the input to the 4026 counter negative until the button is pressed. Make sure all your positives and negatives are correct, and turn on the power. You should find that when you tap the tactile switch lightly, the counter advances the numeric display from 0 through 9 and then begins all over again from 0. You may also find that the chip sometimes misinterprets your button-presses, and counts two or even three digits at a time. I'll deal with this problem a little further on.

The LED segments will not be glowing very brightly, because the 1K series resistors deprive of them of the power they would really like to receive. Those resistors are necessary to avoid overloading the outputs from the counter.

Grounding Yourself

To avoid the frustration that occurs when you power up a circuit and nothing happens, be sure to take these precautions when you use the older generation of CMOS chips (which often have part numbers from 4000 upward, such as 4002, 4020, and so on):

Chips are often shipped with their legs embedded in black foam. This is electrically conductive foam, and you should keep the chips embedded in it until you are ready to use them.

If the chips are supplied to you in plastic tubes, you can take them out and poke their legs into pieces of conductive foam or, if you don't have any, use aluminum foil. The idea is to avoid one pin on a chip acquiring an electric potential that is much higher than another pin.

While handling CMOS components, grounding yourself is important. I find that in dry weather, I accumulate a static charge merely by walking across a plastic floor-protecting mat in socks that contain some synthetic fibers. You can buy a wrist strap to keep yourself grounded, or simply touch a large metal object, such as a file cabinet, before you touch your circuit board. I am in the habit of working with my socked foot touching a file cabinet, which takes care of the problem.

Never solder a CMOS chip while there is power applied to it.

Grounding the tip of your soldering iron is a good idea.

Better still, don't solder CMOS chips at all. When you're ready to immortalize a project by moving it from a breadboard into perforated board, solder a socket into your perforated board, then push the chip into the socket. If there's a problem in the future, you can unplug the chip and plug in another.

Use a grounded, conductive surface on your workbench. The cheapest way to do this is to unroll some aluminum foil and ground it (with an alligator clip and a length of wire) to a radiator, a water pipe, or a large steel object. I like to use an area of conductive foam to cover my workbench—the same type of foam that is used for packaging chips. However, this foam is quite expensive.

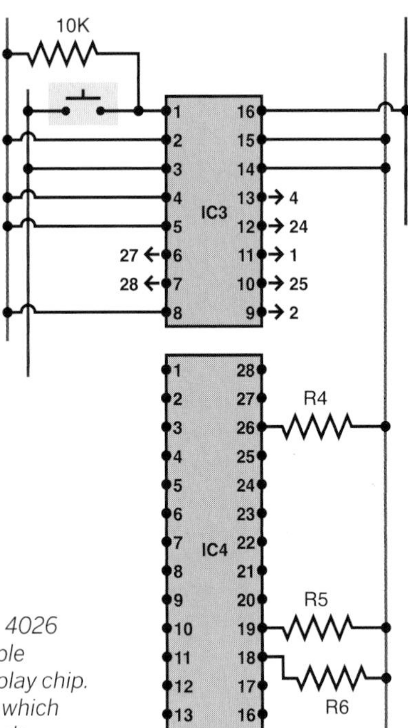

Figure 4-35. *IC3 is a 4026 counter. IC4 is a triple seven-segment display chip. The arrows tell you which pins on the LED display should be connected to the pins on the counter.*

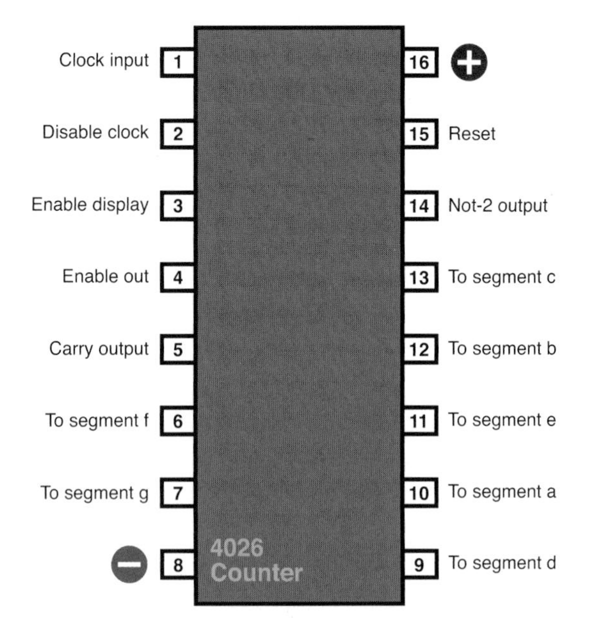

Figure 4-36. *The 4026 decade counter is a CMOS chip that accepts clock pulses on pin 1, maintains a running total from 0 to 9, and outputs this total via pins designed to interface with a seven-segment LED numeric display.*

FUNDAMENTALS

Counters and seven-segment displays

Most counters accept a stream of pulses and distribute them to a series of pins in sequence. The 4026 decade counter is unusual in that it applies power to its output pins in a pattern that is just right to illuminate the segments of a 7-segment numeric display.

Some counters create positive outputs (they "source" current) while others create negative outputs (they "sink" current). Some seven-segment displays require positive input to light up the numbers. These are known as "common cathode" displays. Others require negative input and are known as "common anode" displays. The 4026 delivers positive outputs and requires a common cathode display.

Check the data sheet for any counter chip to find out how much power it requires, and how much it can deliver. CMOS chips are becoming dated, but they are very useful to hobbyists, because they will tolerate a wide range of supply voltage—from 5 to 15 volts in the case of the 4026. Other types of chips are much more limited.

Most counters can source or sink only a few milliamps of output power. When the 4026 is running on a 9-volt power supply, it can source about 4mA of power from each pin. This is barely enough to drive a seven-segment display.

You can insert a series resistor between each output pin of the counter and each input pin of the numeric display, but a simpler, quicker option is to use just one series resistor for each numeral, between the negative-power pin and ground. The experiment that I'm describing uses this shortcut. Its disadvantage is that digits that require only a couple of segments (such as numeral 1) will appear brighter than those that use many segments (such as numeral 8).

If you want your display to look bright and professional, you really need a transistor to drive each segment of each numeral. An alternative is to use a chip containing multiple "op amps" to amplify the current.

When a decade counter reaches 9 and rolls over to 0, it emits a pulse from its "carry" pin. This can drive another counter that will keep track of tens. The carry pin on that counter can be chained to a third counter that keeps track of hundreds, and so on. In addition to decade counters, there are hexadecimal counters (which count in 16s), octal counters (in 8s), and so on.

Why would you need to count in anything other than tens? Consider that the four numerals on a digital clock each count differently. The rightmost digit rolls over when it reaches 10. The next digit to the left counts in sixes. The first hours digit counts to 10, gives a carry signal, counts to 2, and gives another carry signal. The leftmost hours digit is either blank or 1, when displaying time in 12-hour format. Naturally there are counters specifically designed to do all this.

Counters have control pins such as "clock disable," which tells the counter to ignore its input pulses and freeze the display, "enable display," which enables the output from the chip, and "reset," which resets the count to zero.

The 4026 requires a positive input to activate each control pin. When the pins are grounded, their features are suppressed.

To make the 4026 count and display its running total you must ground the "clock disable" and "reset" pins (to suppress their function) and apply positive voltage to the "enable display" pin (to activate the output). See Figure 4-36 to see these pins identified.

Assuming that you succeed in getting your counter to drive the numeric display, you're ready to add two more counters, which will control the remaining two numerals. The first counter will count in ones, the second in tens, and the third in hundreds.

In Figure 4-37, I've continued to use arrows and numbers to tell you which pins of the counters should be connected to which pins of the numeric display. Otherwise, the schematic would be a confusing tangle of wires crossing each other.

At this point, you can give up in dismay at the number of connections—but really, using a breadboard, it shouldn't take you more than half an hour to complete this phase of the project. I suggest you give it a try, because there's

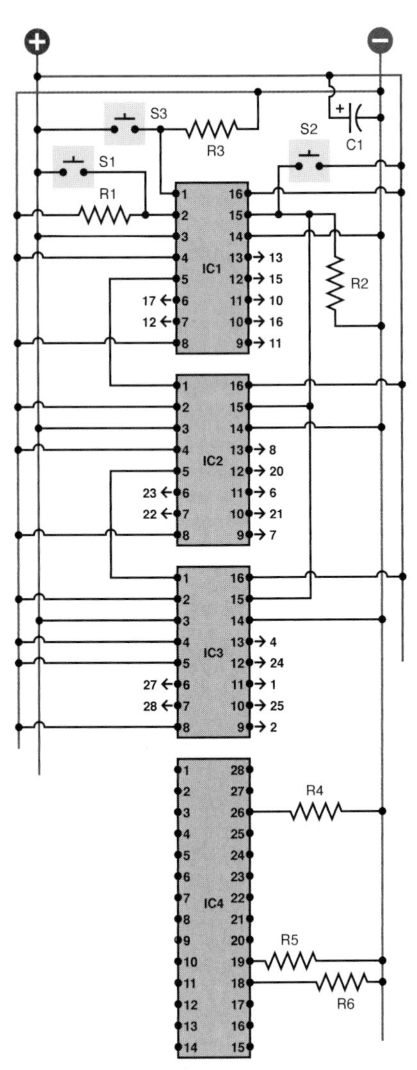

Figure 4-37. *This test circuit, laid out as you would be likely to place it on a breadboard, allows you to trigger a counter manually to verify that the display increments from 000 upward to 999.*

Component values:

All resistors are 1K.
S1, S2, S3: SPST tactile switches, normally open
IC1, IC2, IC3: 4026 decade counter chips
IC4: Kingbright 3-digit common-cathode display
C1: 100 μF (minimum) smoothing capacitor

Wire the output pins on IC1, IC2, and IC3 to the pins on IC4, according to the numbers preceded by arrows. The actual wires have been omitted for clarity. Check for the pinouts of IC4.

something magical about seeing a display count from 000 through 999 "all by itself," and I chose this project because it also has a lot of instructional value.

S1 is attached to the "clock disable" pin of IC1, so that when you hold down this button, it should stop that counter from counting. Because IC1 controls IC2, and IC2 controls IC3, if you freeze IC1, the other two will have to wait for it to resume. Therefore you won't need to make use of their "clock disable" features.

S2 is connected to the "reset" pins of all three counters, so that when you hold down this button, it should set them all to zero.

S3 sends positive pulses manually to the "clock input" pin of the first counter.

S1, S2, and S3 are all wired in parallel with 1K resistors connected to the negative side of the power supply. The idea is that when the buttons are not being pressed, the "pull-down" resistors keep the pins near ground (zero) voltage. When you press one of the buttons, it connects positive voltage directly to the chip, and easily overwhelms the negative voltage. This way, the pins remain either in a definitely positive or definitely negative state. If you disconnect one of these pull-down resistors you are likely to see the numeric display "flutter" erratically. (The numeric display chip has some unconnected pins, but this won't cause any problem, because it is a passive chip that is just a collection of LED segments.)

Always connect input pins of a CMOS chip so that they are either positive or negative. See the "No Floating Pins" warning on the next page.

I suggest that you connect all the wires shown in the schematic first. Then cut lengths of 22-gauge wire to join the remaining pins of the sockets from IC1, IC2, and IC3 to IC4.

Switch on the power and press S2. You'll see three zeros in your numeric display.

Each time you press S3, the count should advance by 1. If you press S2, the count should reset to three zeros. If you hold down S1 while you press S3 repeatedly, the counters should remain frozen, ignoring the pulses from S3.

FUNDAMENTALS

Switch bounce

When you hit S3, I think you'll find that the count sometimes increases by more than 1. This does not mean that there's something wrong with your circuit or your components; you are just observing a phenomenon known as "switch bounce."

On a microscopic level, the contacts inside a pushbutton switch do not close smoothly, firmly, and decisively. They vibrate for a few microseconds before settling; the counter chip detects this vibration as a series of pulses, not just one.

Various circuits are available to "debounce" a switch. The simplest option is to put a small capacitor in parallel with the switch, to absorb the fluctuations; but this is less than ideal. I'll come back to the topic of debouncing later in the book. Switch bounce is not a concern in this circuit, because we're about to get rid of S3 and substitute a 555 timer that generates nice clean bounceless pulses.

Pulse Generation

A 555 timer is ideal for driving a counter chip. You've already seen how to wire a 555 to create a stream of pulses that made noise through a loudspeaker. I'm reproducing the same circuit in Figure 4-38 in simplified form, using the positive and negative supply configuration in the current project. Also I'm showing the connection between pins 2 and 6 in the way that you're most likely to make it, via a wire that loops over the top of the chip.

For the current experiment, I'm suggesting initial component values that will generate only four pulses per second. Any faster than that, and you won't be able to verify that your counters are counting properly.

Install IC5 and its associated components on your breadboard immediately above IC1. Don't leave any gap between the chips. Disconnect S3 and R3 and connect a wire directly between pin 3 (output) of IC5 and pin 1 (clock) of IC1, the topmost counter. Power up again, and you should see the digits advancing rapidly in a smooth, regular fashion. Press S1, and while you hold it, the count should freeze. Release S1 and the count will resume. Press S2 and the counter should reset, even if you are pressing S1 at the same time.

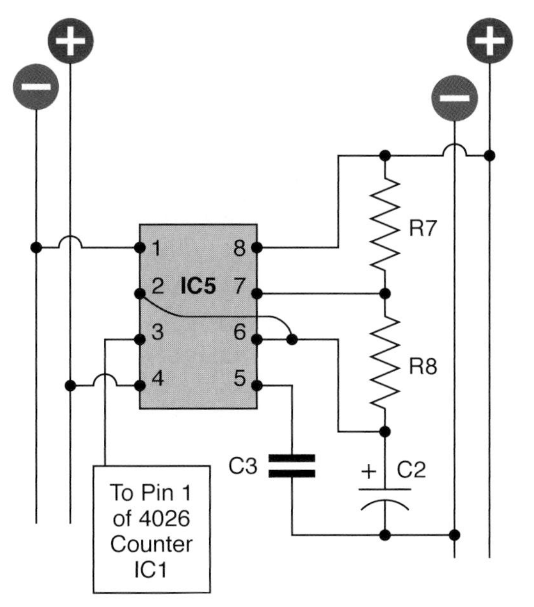

Figure 4-38. *A basic astable circuit to drive the decade counter in the previous schematic. Output is approximately 4 pulses per second.*

R7: 1K
R8: 2K2
C2: 68 µF
C3: 0.1 µF
IC5: 555 timer

No Floating Pins!

A CMOS chip is hypersensitive. Any pin that is not wired either to the supply voltage or to ground is said to be "floating" and may act like an antenna, sensitive to the smallest fluctuations in the world around it.

The 4026 counter chip has a pin labeled "clock disable." The manufacturer's data sheet helpfully tells you that if you give this pin a positive voltage, the chip stops counting and freezes its display. As you don't want to do that, you may just ignore that pin and leave it unconnected, at least while you test the chip. This is a very bad idea!

What the data sheet doesn't bother to tell you (presumably because "everyone knows" such things) is that if you want the clock to run normally, the clock-disable feature itself has to be disabled, by wiring it to negative (ground) voltage. If you leave the pin floating (and I speak from experience), the chip will behave erratically and uselessly.

All input pins must be either positively or negatively wired, unless otherwise specified.

Refinements

Now it's time to remember that what we really want this circuit to do is test a person's reflexes. When the user starts it, we want an initial delay, followed by a signal—probably an LED that comes on. The user responds to the signal by pressing a button as quickly as possible. During the time it takes for the person to respond, the counter will count milliseconds. When the person presses the button, the counter will stop. The display then remains frozen indefinitely, displaying the number of pulses that were counted before the person was able to react.

How to arrange this? I think we need a flip-flop. When the flip-flop gets a signal, it starts the counter running—and keeps it running. When the flip-flop gets another signal (from the user pressing a button), it stops the counter running, and keeps it stopped.

How do we build this flip-flop? Believe it or not, we can use yet another 555 timer, in a new manner known as bistable mode.

FUNDAMENTALS

The bistable 555 timer

Figure 4-39 shows the internal layout of a 555 timer, as before, but the external components on the righthand side have been eliminated. Instead, I'm applying a constant negative voltage to pin 6. Can you see the consequences? Suppose you apply a negative pulse to the trigger (pin 2). Normally when you do this and the 555 starts running, it generates a positive output while charging a capacitor attached to pin 6. When the capacitor reaches 2/3 of the full supply voltage, this tells the 555 to ends its positive output, and it flips back to negative.

Well, if there's no capacitor, there's nothing to stop the timer. Its positive output will just continue indefinitely. However, pin 4 (the reset) can still override everything, so if you apply negative voltage to pin 4, it flips the output to negative. After that, the output will stay negative indefinitely, as it usually does, until you trigger the timer by dropping the voltage to pin 2 again. This will flip the timer back to generating its positive output.

Here's a quick summary of the bistable configuration:

- A negative pulse to pin 2 turns the output positive.
- A negative pulse to pin 4 turns the output negative.
- The timer is stable in each of these states. Its run-time has become infinite.

It's OK to leave pins 5 and 7 of the timer unconnected, because we're pushing it into extreme states where any random signals from those pins will be ignored.

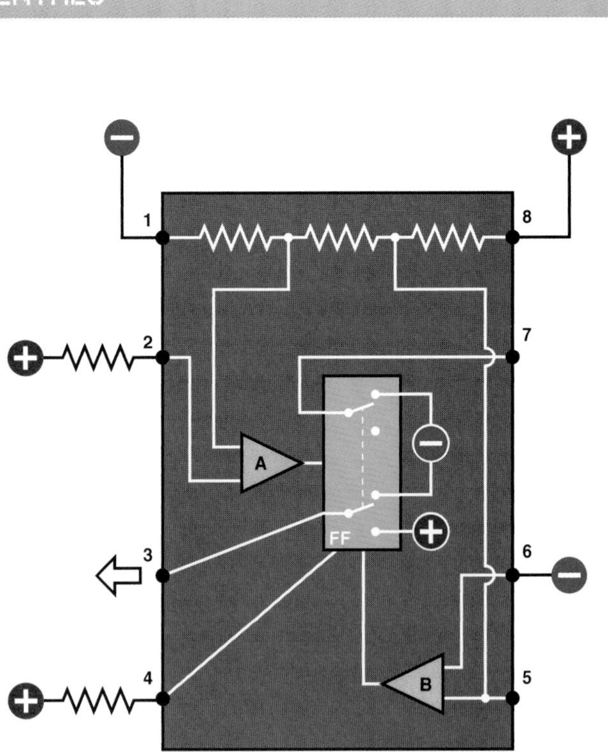

Figure 4-39. *In the bistable configuration, pin 6 of a 555 timer is perpetually negative, so the timer cycle never ends, unless you force it to do so by applying a negative pulse to pin 4 (the reset).*

In bistable mode, the 555 has turned into one big flip-flop. To avoid any uncertainty, we keep pins 2 and 4 normally positive via pull-up resistors, but negative pulses on those pins can overwhelm them when we want to flip the 555 into its opposite state. The schematic for running a 555 timer in bistable mode, controlled by two pushbuttons, is shown in Figure 4-40. You can add this above your existing circuit. Because you're going to attach the output from IC6 to pin 2 of IC1, the topmost counter, you can disconnect S1 and R1 from that pin. See Figure 4-41.

Now, power up the circuit again. You should find that it counts in the same way as before, but when you press S4 (in Figure 4-40), it freezes. This is because your bistable 555 timer is sending its positive output to the "clock disable" pin on the counter. The counter is still receiving a stream of pulses from the astable 555 timer, but as long as pin 2 is positive on the counter, the counter simply ignores the pulses.

Now press S5, which flips your bistable 555 back to delivering a negative output, at which point the count resumes.

We're getting close to a final working circuit here. We can reset the count to zero (with S3), start the count (with S5), and wait for the user to stop the count (with S4). The only thing missing is a way to start the count unexpectedly.

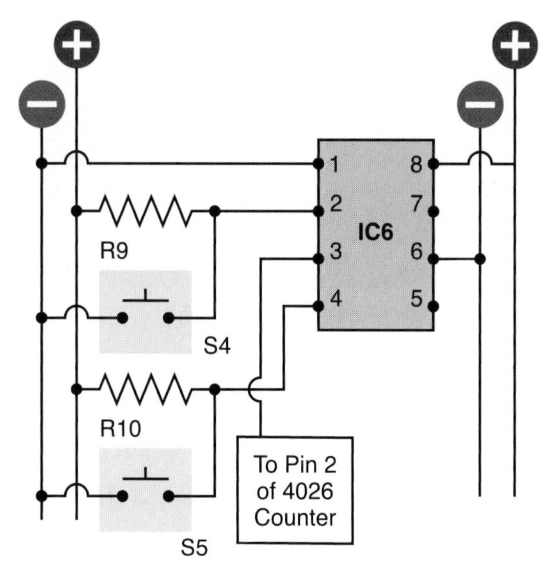

Figure 4-40. *Adding a bistable 555 timer to the reflex tester will stop the counter with a touch of a button, and keep it stopped.*

R9, R10: 1K
IC6: 555 timer

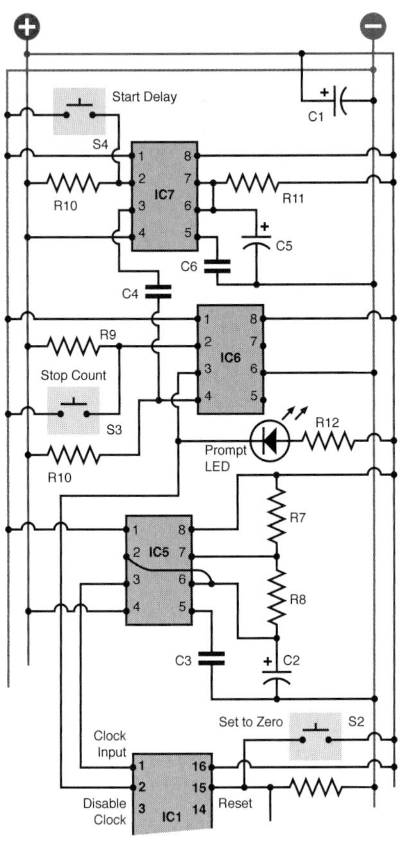

The completed control section of the circuit, to be added above these timers.

R7, R9, R12: 1K
R10 (both): 10K
R8: 2K2
R11: 330K
C1: 100 μF
C2: 68 μF
C3, C4, C6: 0.1 μF
C5: 10 μF
S1, S2, S3: tactile switches
IC5, IC6, IC7: 555 timers

The Delay

Suppose we set up yet another 555 in monostable mode. Trigger its pin 2 with a negative pulse, and the timer delivers a positive output that lasts for, say, 4 seconds. At the end of that time, its output goes back to being negative. Maybe we can hook that positive-to-negative transition to pin 4 of IC6. We can use this instead of switch S5, which you were pressing previously to start the count.

Check the new schematic in Figure 4-41 which adds another 555 timer, IC7 above IC6. When the output from IC7 goes from positive to negative, it will trigger the reset of IC6, flipping its output negative, which allows the count to begin. So IC7 has taken the place of the start switch, S4. You can get rid of S4, but keep the pull-up resistor, R9, so that the reset of IC6 remains positive the rest of the time.

The capacitor blocks DC voltage from IC7 until pin 3 transitions from high to low, at which point this transition is passed through as a pulse to pin 4 of IC6.

The final schematic in Figure 4-41 shows the three 555 timers all linked together, as you should insert them above the topmost counter, IC1. I also added an LED to signal the user. Figure 4-42 is a photograph of my working model of the circuit.

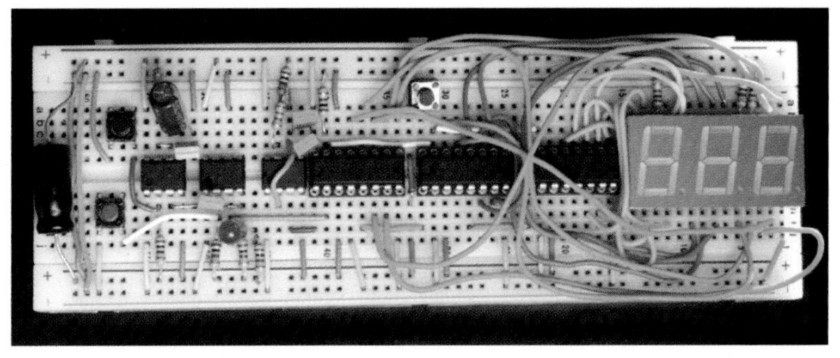

Figure 4-41. *The complete reaction-timer circuit barely fits on a 63-row breadboard.*

Because this circuit is complicated, I'll summarize the sequence of events when it's working. Refer to Figure 4-41 while following these steps:

1. User presses Start Delay button S4, which triggers IC7.
2. IC7 output goes high for a few seconds while C5 charges.
3. IC7 output drops back low.
4. IC7 communicates a pulse of low voltage through C4 to IC6, pin 4.
5. IC 6 output flips to low and flops there.
6. Low output from IC6 sinks current through LED and lights it.
7. Low output from IC6 also goes to pin 2 of IC1.

8. Low voltage on pin 2 of IC1 allows IC1 to start counting.
9. User presses S3, the "stop" button.
10. S3 connects pin 2 of IC6 to ground.
11. IC6 output flips to high and flops there.
12. High output from IC6 turns off the LED.
13. High output from IC6 also goes to pin 2 of IC1.
14. High voltage on pin 2 of IC1 stops it from counting.
15. After assessing the result, user presses S2.
16. S2 applies positive voltage to pin 15 of IC1, IC2, IC3.
17. Positive voltage resets counters to zero.
18. The user can now try again.
19. Meanwhile, IC5 is running continuously throughout.

In case you find a block diagram easier to understand, I've included that, too, in Figure 4-43.

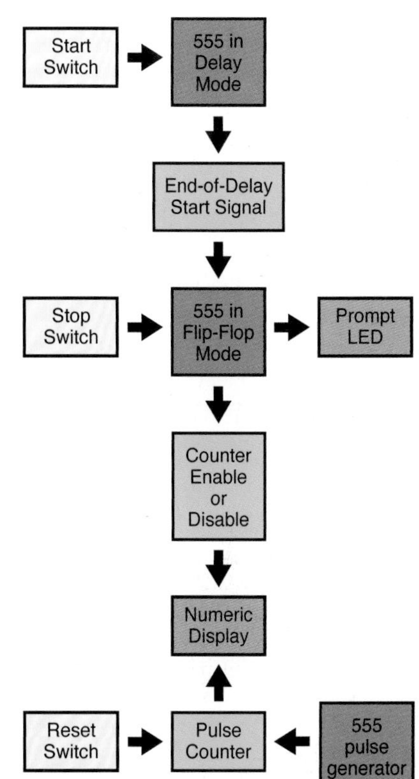

Figure 4-43. *The functions of the reflex tester, summarized as a block diagram.*

Using the Reflex Tester

At this point, you should be able to test the circuit fully. When you first switch it on, it will start counting, which is slightly annoying, but easily fixed. Press S3 to stop the count. Press S2 to reset to zero.

Now press S4. Nothing seems to happen—but that's the whole idea. The delay cycle has begun in stealth mode. After a few seconds, the delay cycle ends, and the LED lights up. Simultaneously, the count begins. As quickly as possible, the user presses S3 to stop the count. The numerals freeze, showing how much time elapsed.

There's only one problem—the system has not yet been calibrated. It is still running in slow-motion mode. You need to change the resistor and capacitor attached to IC5 to make it generate 1,000 pulses per second instead of just three or four.

Substitute a 10K trimmer potentiometer for R8 and a 0.1 µF capacitor for C2. This combination will generate about 690 pulses per second when the trimmer is presenting maximum resistance. When you turn the trimmer down to decrease its resistance, somewhere around its halfway mark the timer will be running at 1,000 pulses per second.

How will you know exactly where this point is? Ideally, you would attach an oscilloscope probe to the output from IC5. But, most likely, you don't have an oscilloscope, so here are a couple other suggestions.

First remove the 0.1 µF capacitor at C2 and substitute a 1 µF capacitor. Because you are multiplying the capacitance by 10, you will reduce the speed by 10. The leftmost digit in your display should now count in seconds, reaching 9 and rolling over to 0 every 10 seconds. You can adjust your trimmer potentiometer while timing the display with a stopwatch. When you have it right, remove the 1 µF capacitor and replace the 0.1 µF capacitor at C2.

The only problem is, the values of capacitors may be off by as much as 10%. If you want to fine-tune your reflex timer, you can proceed as follows.

Disconnect the wire from pin 5 of IC3, and substitute an LED with a 1K series resistor between pin 5 and ground. Pin 5 is the "carry" pin, which will emit a positive pulse whenever IC3 counts up to 9 and rolls over to start at 0 again. Because IC3 is counting tenths of a second, you want its carry output to occur once per second.

Now run the circuit for a full minute, using your stopwatch to see if the flashing LED drifts gradually faster or slower than once per second. If you have a camcorder that has a time display in its viewfinder, you can use that to observe the LED.

If the LED flashes too briefly to be easily visible, you can run a wire from pin 5 to another 555 timer that is set up in monostable mode to create an output lasting for around 1/10 of a second. The output from that timer can drive an LED.

Enhancements

It goes without saying that anytime you finish a project, you see some opportunities to improve it. Here are some suggestions:

1. No counting at power-up. It would be nice if the circuit begins in its "ready" state, rather than already counting. To achieve this you need to send a negative pulse to pin 2 of IC6, and maybe a positive pulse to pin 15 of IC1. Maybe an extra 555 timer could do this. I'm going to leave you to experiment with it.
2. Audible feedback when pressing the Start button. Currently, there's no confirmation that the Start button has done anything. All you need to do is buy a piezoelectric beeper and wire it between the righthand side of the Start button and the positive side of the power supply.
3. A random delay interval before the count begins. Making electronic components behave randomly is very difficult, but one way to do it would be to require the user to hold his finger on a couple of metal contacts. The skin resistance of the finger would substitute for R11. Because the finger pressure would not be exactly the same each time, the delay would vary. You'd have to adjust the value of C5.

Summing Up

This project demonstrated how a counter chip can be controlled, how counter chips can be chained together, and three different functions for 555 timers. It also showed you how chips can communicate with each other, and introduced you to the business of calibrating a circuit after you've finished building it.

Naturally, if you want to get some practical use from the circuit, you should build it into an enclosure with heavier-duty pushbuttons—especially the button that stops the count. You'll find that when people's reflexes are being tested, they are liable to hit the stop button quite hard.

Because this was a major project, I'll follow it up here with some quicker, easier ones as we move into the fascinating world of another kind of integrated circuit: logic chips.

Experiment 19: Learning Logic

You will need:

- Assorted resistors and capacitors.
- 74HC00 quad 2-input NAND chip, 74HC08 quad 2-input AND chip, and LM7805 voltage regulator. Quantity: 1 of each.
- Signal diode, 1N4148 or similar. Quantity: 1.
- Low-current LED. Quantity: 1.
- SPST tactile switches. Quantity: 2.

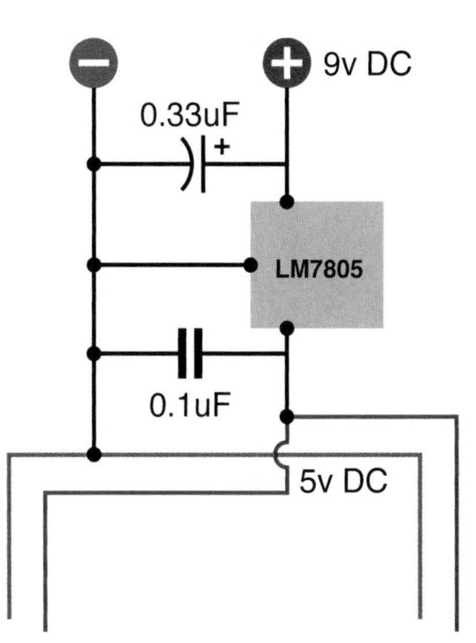

Figure 4-44. *This simple circuit is essential to provide a regulated 5V DC supply for logic chips.*

You're going to be entering the realm of pure digital electronics, using "logic gates" that are fundamental in every electronic computing device. When you deal with them individually, they're extremely easy to understand. When you start chaining them together, they can seem intimidatingly complex. So let's start with them one at a time.

Logic gates are much fussier than the 555 timer or the 4026 counter that you used previously. They demand an absolutely precise 5 volts DC, with no fluctuations or "spikes" in the flow of current. Fortunately, this is easy to achieve: just set up your breadboard with an LM7805 voltage regulator, as shown in the schematic in Figure 4-44 and the photograph in Figure 4-45. The regulator receives 9 volts from your usual voltage supply, and reduces it to 5 volts, with the help of a couple of capacitors. You apply the 9 volts to the regulator, and distribute the 5 volts down the sides of your breadboard instead of the unregulated voltage that you used previously. Use your meter to verify the voltage, and make sure you have the polarity clearly marked.

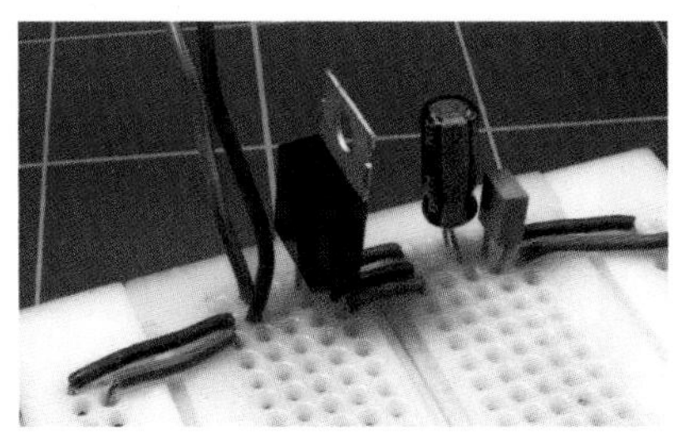

Figure 4-45. *The voltage regulator and its two capacitors can fit snugly at the top of a breadboard. Remember to apply the 9V input voltage at the left pin of the regulator, and distribute the 5V output down the sides of the breadboard.*

After installing your regulator, take a couple of tactile switches, two 10K resistors, a low-current LED, and a 1K resistor, and set them around a 74HC00 logic chip as shown in Figure 4-46. You may notice that many of the pins of the chip are shorted together and connected to the negative side of the power supply. I'll explain that in a moment.

FUNDAMENTALS

Voltage regulators

The simplest versions of these little semiconductors accept a higher DC voltage on one pin and deliver a lower DC voltage on another pin, with a third pin (usually in the middle) serving as a common negative, or ground. You should also attach a couple of capacitors to smooth the current, as shown in Figure 4-46.

Typically you can put a 7.5-volt or 9-volt supply on the "input" side of a 5-volt regulator, and draw a precise 5 volts from the "output" side. If you're wondering where the extra voltage goes, the answer is, the regulator turns the electricity into heat. For this reason, small regulators (such as the one in Figure 4-8) often have a metal back with a hole in the top. Its purpose is to radiate heat, which it will do more effectively if you bolt it to a piece of aluminum, since aluminum conducts heat very effectively. The aluminum is known as a heat sink, and you can buy fancy ones that have multiple cooling fins.

For our purposes, we won't be drawing enough current to require a heat sink.

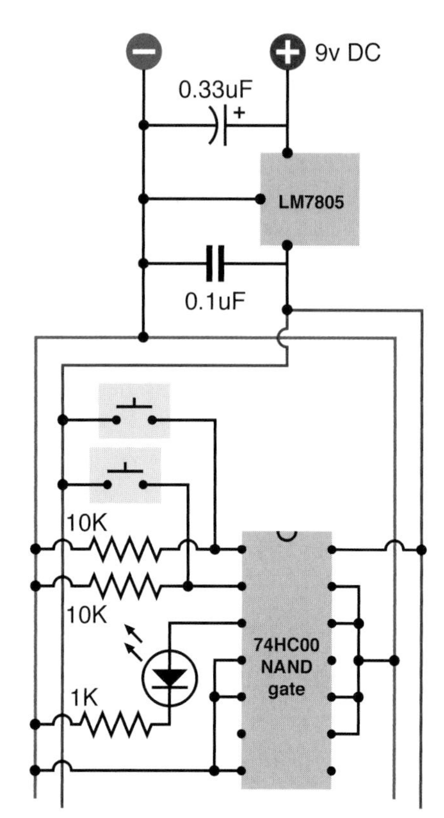

Figure 4-46. *By observing the LED when you press either, both, or neither of the buttons, you can easily figure out the logical function of the NAND gate.*

When you connect power, the LED should light up. Press one of the tactile switches, and the LED remains illuminated. Press the other tactile switch, and again the LED stays on. Now press both switches, and the light should go out.

Pins 1 and 2 are logic inputs for the 74HC00 chip. Initially they were held at negative voltage, being connected to the negative side of the power supply through 10K resistors. But each pushbutton overrides its pull-down resistor and forces the input pin to go positive.

The logic output from the chip, as you saw, is normally positive—but *not* if the first input *and* the second input are positive. Because the chip does a "Not AND" operation, it's known as a NAND logic gate. You can see the breadboard layout in Figure 4-47. Figure 4-48 is a simplified version of the circuit. The U-shaped thing with a circle at the bottom is the logic symbol for a NAND gate. No power supply is shown for it, but in fact all logic chips require a power supply, which enables them to put out more current than they take in. Anytime you see a symbol for a logic chip, try to remember that it has to have power to function.

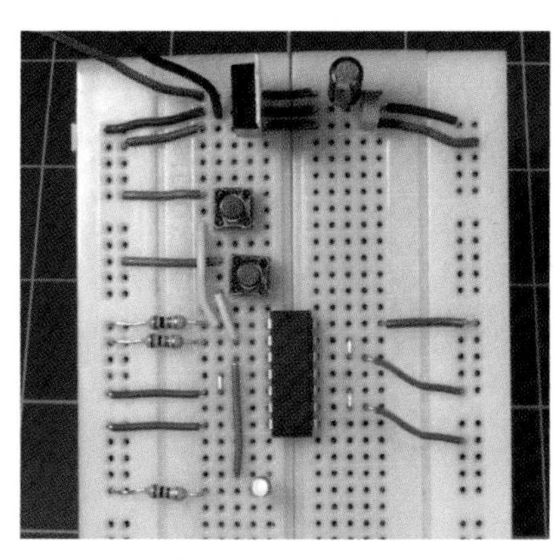

Figure 4-47. *This breadboard layout is exactly equivalent to the schematic in Figure 4-46.*

The 74HC00 actually contains four NAND gates, each with two logical inputs and one output. They are arrayed as shown in Figure 4-49. Because only one gate was needed for the simple test, the input pins of the unused gates were shorted to the negative side of the power supply.

Pin 14 supplies positive power for the chip; pin 7 is its ground pin. Almost all the 7400 family of logic chips use the same pins for positive and negative power, so you can swap them easily.

In fact, let's do that right now. First, disconnect the power. Carefully pull out the 74HC00 and put it away with its legs embedded in conductive foam. Substitute a 74HC08 chip, which is an AND chip. Make sure you have it the right way up, with its notch at the top. Reconnect the power and use the pushbuttons as you did before. This time, you should find that the LED comes on if the first input AND the second input are both positive, but it remains dark otherwise. Thus, the AND chip functions exactly opposite to the NAND chip. Its pinouts are shown in Figure 4-50.

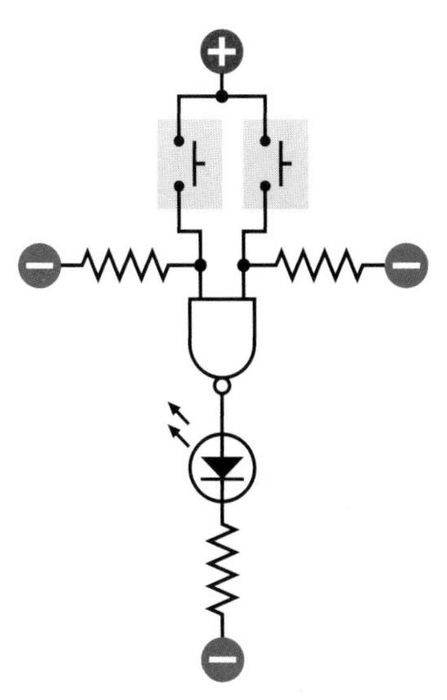

Figure 4-48. *The structure and function of the NAND gate is easier to visualize with this simplified schematic that omits the power supply for the chip and doesn't attempt to place the wires to fit a breadboard layout.*

You may be wondering why these things are useful. Soon you'll see that we can put logic gates together to do things such as create an electronic combination lock, or a pair of electronic dice, or a computerized version of a TV quiz show where users compete to answer a question. And if you were really insanely ambitious, you could build an entire computer out of logic gates.

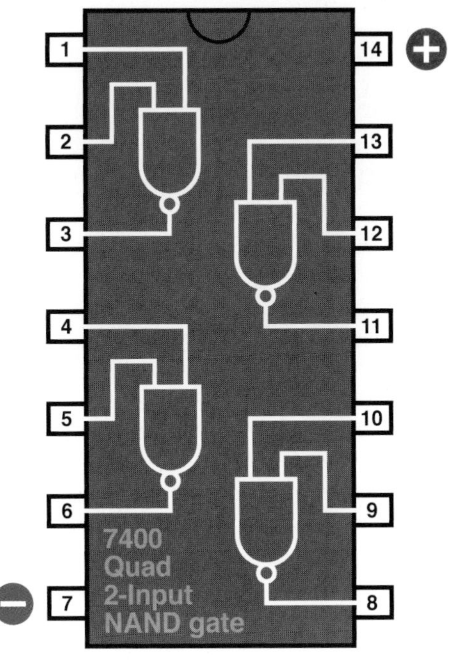

Figure 4-49. *The pinouts of the logic gates in a 74HC00chip.*

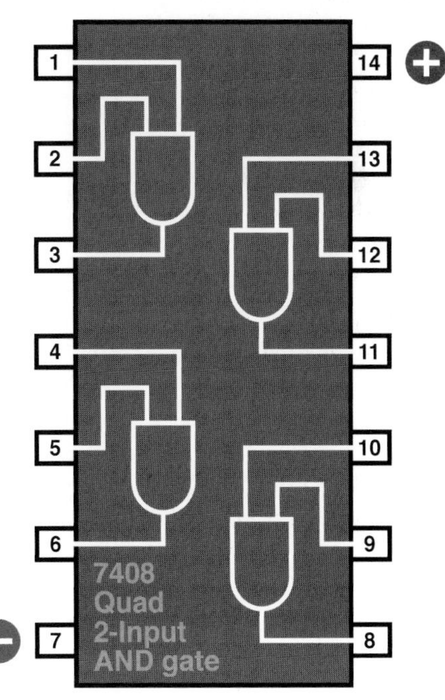

Figure 4-50. *The pinouts of the logic gates in a 74HC08chip.*

BACKGROUND

From Boole to Shannon

George Boole was a British mathematician, born in 1815, who did something that few people are ever lucky enough or smart enough to do: he invented an entirely new branch of mathematics.

Interestingly, it was not based on numbers. Boole had a relentlessly logical mind, and he wanted to reduce the world to a series of true-or-false statements which could overlap in interesting ways. For instance, suppose there is a couple named Ann and Bob who have so little money, they only own one hat. Clearly, if you happen to run into Ann and Bob walking down the street, there are four possibilities: neither of them may be wearing a hat, Ann may be wearing it, or Bob may be wearing it, but they cannot both be wearing it.

The diagram in Figure 4-51 illustrates this. All the states are possible except the one where the circles overlap. (This is known as a Venn diagram. I leave it to you to search for this term if it interests you and you'd like to learn more.) Boole took this concept much further, and showed how to create and simplify extremely complex arrays of logic.

Figure 4-51. *This slightly frivolous Venn diagram illustrates the various possibilities for two people, Ann and Bob, who own only one hat.*

Another way to summarize the hat-wearing situation is to make the "truth table" shown in Figure 4-52. The rightmost column shows whether each combination of propositions can be true. Now check the table in Figure 4-53. It's the same table but uses different labels, which describe the pattern you have seen while using the NAND gate.

Boole published his treatise on logic in 1854, long before it could be applied to electrical or electronic devices. In fact, during his lifetime, his work seemed to have no practical applications at all. But a man named Claude Shannon encountered Boolean logic while studying at MIT in the 1930s, and in 1938 he published a paper describing how Boolean analysis could be applied to circuits using relays. This had immediate practical applications, as telephone networks were growing rapidly, creating complicated switching problems.

BACKGROUND

From Boole to Shannon (continued)

Ann is wearing the hat	Bob is wearing the hat	This combination can be
NO	NO	TRUE
NO	YES	TRUE
YES	NO	TRUE
YES	YES	FALSE

Figure 4-52. *The hat-wearing possibilities can be expressed in a "truth table."*

Input A	Input B	Output
OFF	OFF	ON
OFF	ON	ON
ON	OFF	ON
ON	ON	OFF

Figure 4-53. *The truth-table from Figure 4-52 can be relabeled to describe the inputs and outputs of a NAND gate.*

A very simple telephone problem could be expressed like this. Suppose two customers in a rural area share one telephone line. If one of them wants to use the line, or the other wants to use it, or neither of them wants to use it, there's no problem. But they cannot both use it at once. You may notice that this is exactly the same as the hat-wearing situation for Ann and Bob.

We can easily draw a circuit using two normally closed relays that creates the desired outcome (see Figure 4-54), but if you imagine a telephone exchange serving many thousands of customers, the situation becomes very complicated indeed. In fact, in Shannon's time, no logical process existed to find the best solution and verify that it used fewer components than some other solution.

Shannon saw that Boolean analysis could be used for this purpose. Also, if you used an "on" condition to represent numeral 1 and an "off" condition to represent numeral 0, you could build a system of relays that could count. And if it could count, it could do arithmetic.

When vacuum tubes were substituted for relays, the first practical digital computers were built. Transistors took the place of vacuum tubes, and integrated circuit chips replaced transistors, leading to the desktop computers that we now take for granted today. But deep down, at the lowest levels of these incredibly complex devices, they still use the laws of logic discovered by George Boole. Today, when you use a search engine online, if you use the words AND and OR to refine your search, you're actually using Boolean operators.

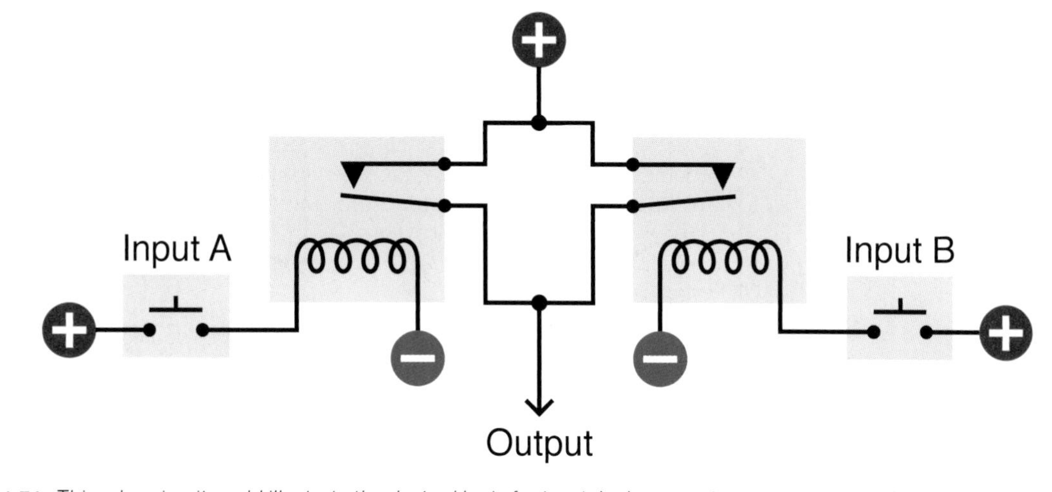

Figure 4-54. *This relay circuit could illustrate the desired logic for two telephone customers wanting to share one line, and its behavior is almost identical to that of the NAND schematic shown in Figure 4-48.*

Figure 4-55. *Ann and Bob attempt to overcome the limitations of Boolean logic.*

ESSENTIALS

Logic gate basics

The NAND gate is the most fundamental building block of digital computers, because (for reasons which I don't have space to explain here) it enables digital addition. If you want to explore more try searching online for topics such as "binary arithmetic" and "half-adder."

Generally, there are seven types of logic gates:

- AND
- NAND
- OR
- NOR
- XOR
- XNOR
- NOT

Of the six two-input gates, the XNOR is hardly ever used. The NOT gate has a single input, and simply gives a negative output when the input is positive or a positive output when the input is negative. The NOT is more often referred to as an "Inverter." The symbols for all seven gates are shown in Figure 4-56.

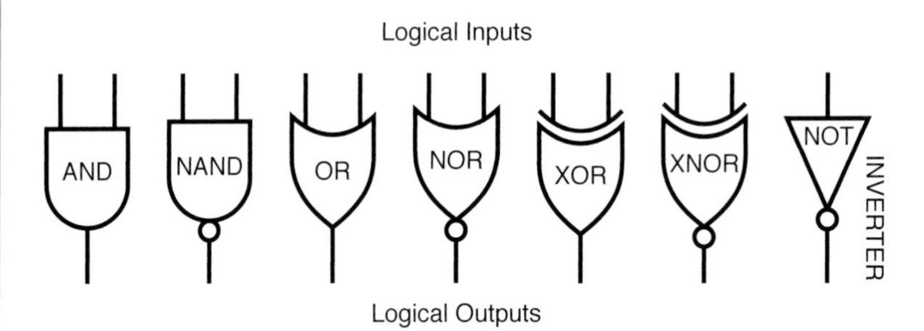

Figure 4-56. *American symbols for the six types of two-input logic gates, and the single-input inverter.*

I've shown the American symbols. Other symbols have been adopted in Europe, but the traditional symbols shown here are the ones that you will usually find, even being used by Europeans. I also show the truth tables, in Figure 4-57, illustrating the logical output (high or low) for each pair of inputs of each type of gate.

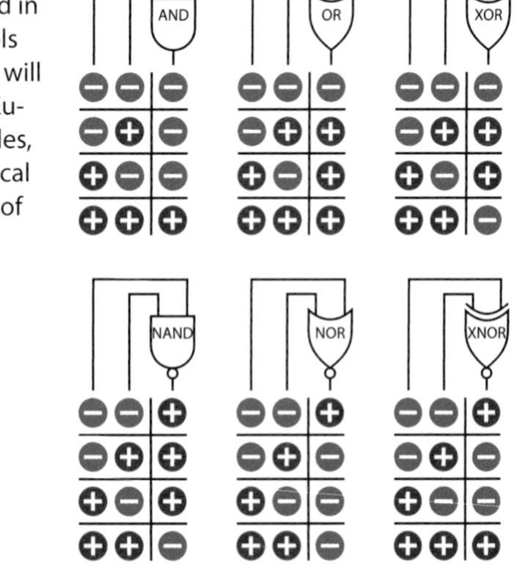

Figure 4-57. *Inputs and corresponding outputs for the six types of logic gates (note that the XNOR gate is seldom used). The minus signs indicate low voltage, close to ground potential. The plus signs indicate higher voltage, close to the positive potential of the power supply in the circuit. The exact voltages will vary depending on other components that may be actively connected.*

ESSENTIALS

Logic gate basics (continued)

If you have difficulty visualizing logic gates, a mechanical comparison may help. You can think of them as being like sliding plates with holes in them, in a bubblegum machine. Two people, A and B, can push the plates. These people are the two inputs, which are considered positive when they are doing something. (Negative logic systems also exist, but are uncommon, so I'm only going to talk about positive systems here.)

The flow of bubblegum represents a flow of positive current. The full set of possibilities is shown in Figures 4-58 through 4-63.

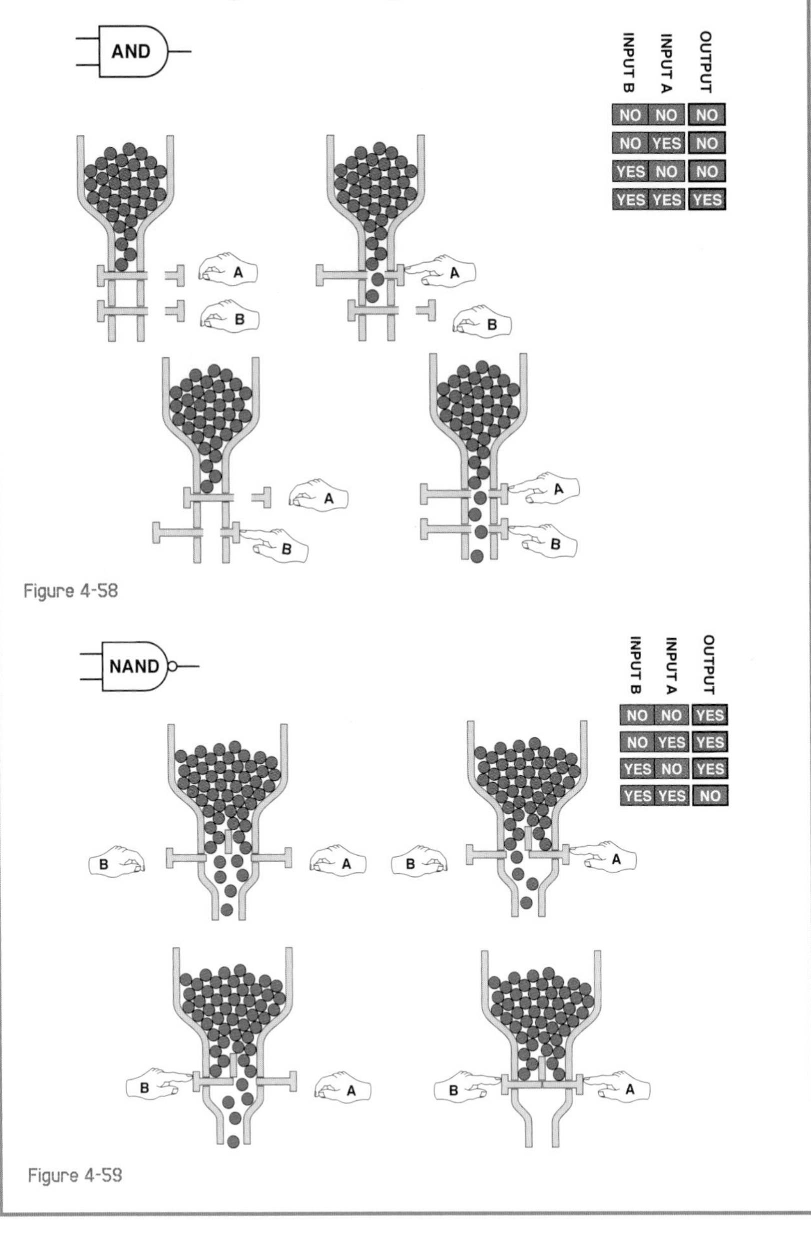

INPUT B	INPUT A	OUTPUT
NO	NO	NO
NO	YES	NO
YES	NO	NO
YES	YES	YES

Figure 4-58

INPUT B	INPUT A	OUTPUT
NO	NO	YES
NO	YES	YES
YES	NO	YES
YES	YES	NO

Figure 4-59

ESSENTIALS

Logic gate basics (continued)

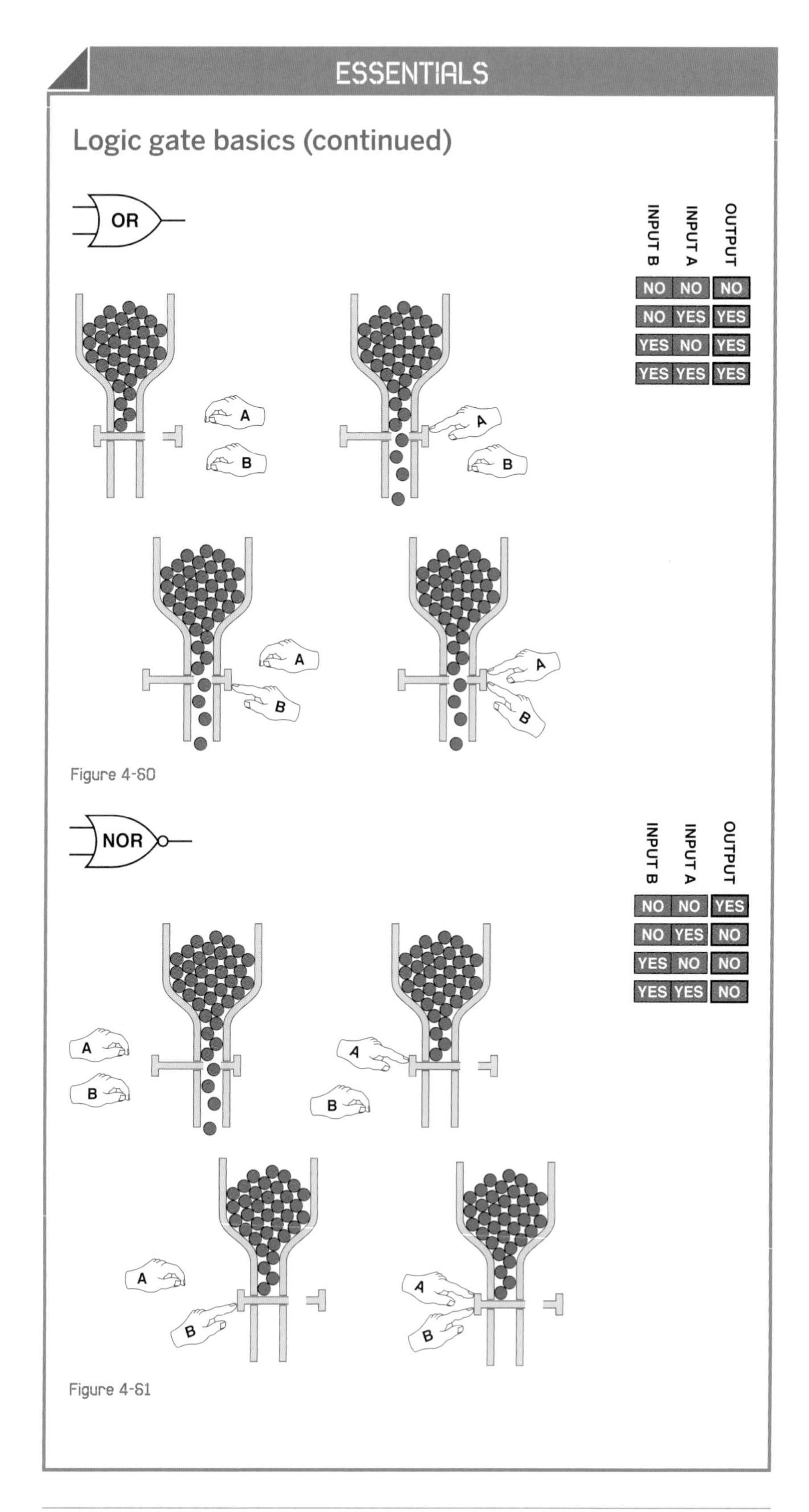

INPUT B	INPUT A	OUTPUT
NO	NO	NO
NO	YES	YES
YES	NO	YES
YES	YES	YES

Figure 4-60

INPUT B	INPUT A	OUTPUT
NO	NO	YES
NO	YES	NO
YES	NO	NO
YES	YES	NO

Figure 4-61

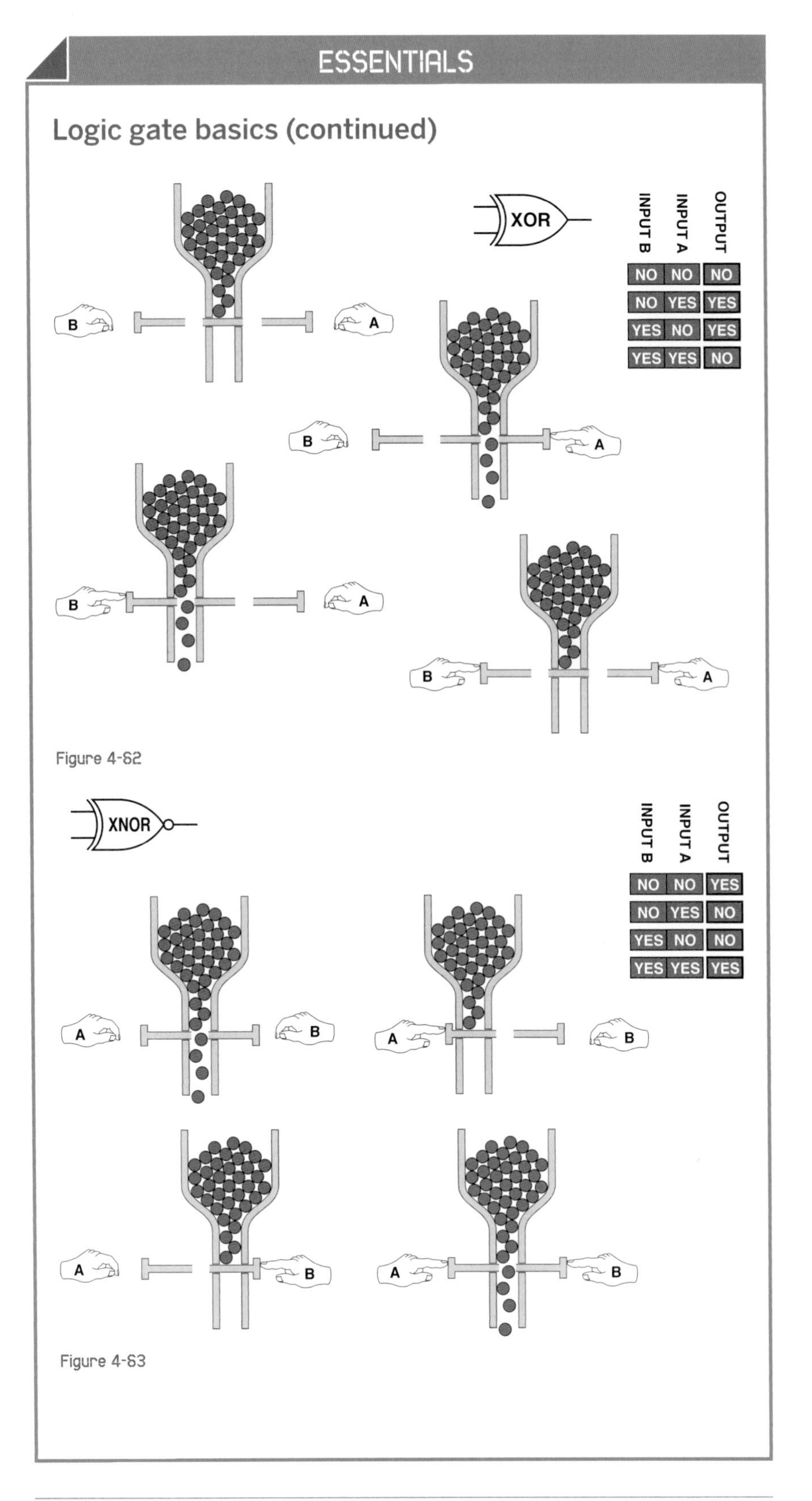

Figure 4-62

Figure 4-63

BACKGROUND

The confusing world of TTL and CMOS

Back in the 1960s, the first logic gates were built with Transistor-Transistor Logic, abbreviated TTL, meaning that tiny bipolar transistors were etched into a single wafer of silicon. Soon, these were followed by Complementary Metal Oxide Semiconductors, abbreviated CMOS. Each of these chips was a collection of metal-oxide field-effect transistors (known as MOSFETs). The 4026 chip that you used earlier is an old CMOS.

You may remember that bipolar transistors amplify current. TTL circuits are similar: they are sensitive to current, rather than voltage. Thus they require a significant flow of electricity, to function. But CMOS chips are like the programmable unijunction transistor that I featured previously. They are voltage-sensitive, enabling them to draw hardly any current while they are waiting for a signal, or pausing after emitting a signal.

The two families named TTL and CMOS still exist today. The table in Figure 4-64 summarizes their basic advantages and disadvantages. The CMOS series, with part numbers from 4000 upward, were easily damaged by static electricity but were valuable because of their meager power consumption. The TTL series, with part numbers from 7400 upward, used much more power but were less sensitive and very fast. So, if you wanted to build a computer, you used the TTL family, but if you wanted to build a little gizmo that would run for weeks on a small battery, you used the CMOS family.

From this point on everything became extremely confusing, because CMOS manufacturers wanted to grab market share by emulating the advantages of TTL chips. Newer generations of CMOS chips even changed their part numbers to begin with "74" to emphasize their compatibility, and the functions of pins on CMOS chips were swapped around to match the functions of pins on TTL chips. Consequently, the pinouts of CMOS and TTL chips are usually now identical, but the meaning of "high" and "low" states changed in each new generation, and the maximum supply voltages for CMOS chips were revised downward. Note I have included question marks beside two categories in the CMOS column, as modern CMOS chips have overcome those disadvantages—at least to some extent.

Here's a quick summary, which will be useful to you if you look at a circuit that you find online, and you wonder about the chips that have been specified.

Where you see a letter "x," it means that various numbers may appear in that location. Thus "74xx" includes the 7400 NAND gate, the 7402 NOR gate, the 74150 16-bit data selector, and so on. A combination of letters preceding the "74" identifies the chip manufacturer, while letters following the part number may identify the style of package, may indicate whether it contains heavy metals that are environmentally toxic, and other details.

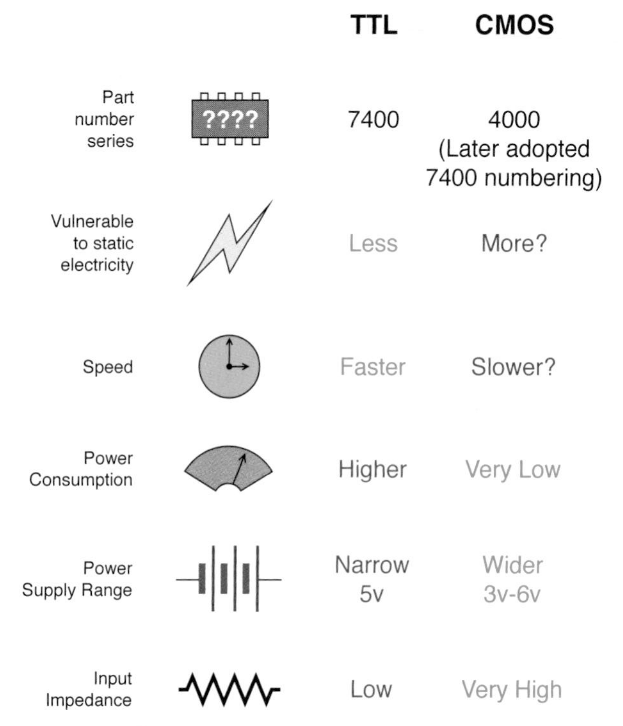

Figure 4-64. *The basic differences between the two families of logic chips. In successive generations, these differences have gradually diminished.*

TTL family:

74xx
: The old original generation, now obsolete.

74Sxx
: Higher speed "Schottky" series, now obsolete.

74LSxx
: Lower power Schottky series, still used occasionally.

74ALSxx
: Advanced low-power Schottky.

74Fxx
: Faster than the ALS series.

BACKGROUND

The confusing world of TTL and CMOS (continued)

CMOS family:

40xx
: The old original generation, now obsolete.

40xxB
: The 4000B series was improved but still susceptible to damage from static electricity. Many hobby circuits still use these chips because they will run from relatively high voltages and can power LEDs and even small relays directly.

74HCxx
: Higher-speed CMOS, with part numbers matching the TTL family, and pinouts matching the TTL family, but input and output voltages not quite the same as the TTL family. I've used this generation extensively in this book, because it's widely available, and the circuits here have no need for greater speed or power.

74HCTxx
: Like the HC series but matching the TTL voltages.

74ACxx
: Advanced version of HC series. Faster, with higher output capacity.

74ACTxx
: Like the AC series but with the same pin functions and voltages as TTL.

74AHCxx
: Advanced higher-speed CMOS.

74AHCTxx
: Like the AHC series but with the same pin functions and voltages as TTL.

74LVxx
: Lower voltage (3.3v) versions, including LV, LVC, LVT, and ALVC series.

As you can see, these days you have to interpret part numbers very carefully. But which family and generation of chips should you use? Well, that depends! Following are some guidelines.

What you don't need:

1. Speed differences are irrelevant from our point of view, as we're not going to be building circuits running at millions of cycles per second.
2. Price differences are so small as to be inconsequential.
3. Lower-voltage (LV) CMOS chips are not very interesting for our small experimental circuits.
4. Try to avoid mixing different families, and different generations of the same family, in the same circuit. They may not be compatible.
5. Some modern chip varieties may be only available in the surface-mount package format. Because they're so much more difficult to deal with, and their only major advantage is miniaturization, I don't recommend them.
6. In the TTL family, the LS and ALS series cannot handle as much output current as the S series and the F series. You don't need them.

What you should use:

1. The old 74LSxx series of TTL chips was so popular, you'll still find schematics that specify these chips. You should still be able to buy them from sources online, but if not, you can substitute the 74HCTxx chips, which are designed to function identically.
2. The old 4000B series of CMOS chips are still used by hobbyists because their willingness to tolerate high voltages is convenient. While TTL or TTL-compatible chips require a carefully regulated 5 volts, the 4000B chips would handle 15 volts—and also delivered enough power to energize LEDs or even very small relays. Some hobbyists also have a nostalgic affection for the 74Cxx series of chips, which had the same pin connections as the TTL chips but could still tolerate higher voltages and higher output current. The trouble is, some of the 74Cxx chips are almost extinct, and while the 4000B chips are still available, they are considered almost obsolete.

BACKGROUND

The confusing world of TTL and CMOS (continued)

Bottom line: I suggest you use the 4000B chips only if you want to replicate an old circuit, or if a modern equivalent is unavailable (which is why I specified the 4026B chip for the reaction timer—I could not find a modern equivalent that will drive seven-segment numeric displays directly, and I didn't want you to have to deal with more parts than necessary).

If you check online suppliers such as *Mouser.com* you'll find that the HC family is by far the most popular right now. They are all available in through-hole format (to fit your breadboard and perforated board). They have the high input impedance of CMOS (which is useful) and they have the same pin identities as the old 74LSxx series.

Abbreviations

When looking at data sheets, you are likely to encounter some or all of these abbreviations:

- VOH min: Minimum output voltage in high state
- VOL max: Maximum output voltage in low state
- VIH min: Minimum input voltage to be recognized as high
- VIL max: Maximum input voltage recognized as low

BACKGROUND

Logic gate origins

The 7400 family of integrated circuits was introduced by Texas Instruments, beginning with the 7400 NAND gate in 1962. Other companies had sold logic chips previously, but the 7400 series came to dominate the market. The Apollo lunar missions used a computer built with 7400 chips, and they were a mainstay of minicomputers during the 1970s.

RCA introduced its 4000 series of logic chips in 1968, built around CMOS transistors; Texas Instruments had chosen TTL. The CMOS chips used less power, thus generating much less heat and enabling flexible circuit design, as each chip could power many others. CMOS was also tolerant of wide voltage ranges (from 3 to 15 volts) but prohibited switching speeds faster than around 1MHz. TTL was 10 times faster.

Design tweaks gradually eradicated the speed penalty for CMOS, and TTL chips have become relatively rare. Still, some people retain a special nostalgic loyalty to "the logic gates that went to the moon." A hardcore enthusiast named Bill Buzbee has built an entire web server from TTL-type 7400 chips, currently online at *http://magic-1.org*. Figure 4-65 shows just one of the handmade circuit boards that Bill assembled to run his computer.

Figure 4-65. *Hobbyist Bill Buzbee built himself a web server entirely from 7400 series logic chips, the oldest of which was fabricated back in 1969. The web server can be found online at http://magic-1.org, displaying pictures of itself and details of its construction. The picture here that Bill took shows just one of the circuit boards of this remarkable project.*

FUNDAMENTALS

Common part numbers

Each 14-pin chip can contain four 2-input gates, three 3-input gates, two 4-input gates, one 8-input gate, or six single-input inverters, as shown in the following table.

	2 input	3 input	4 input	8 input
AND	7408	7411	7421	
NAND	7400	7410	7420	7430
OR	7432			744078*
NOR	7402	7427		744078*
XOR	7486			
XNOR	747266			
Inverter	(1 input) 7404			

*The 744078 has an OR output and a NOR output on the same chip.

Figures 4-66 through 4-74 show the internal connections of the logic chips that you are most likely to use. Note that the 7402 NOR gate has its logical inputs and outputs arranged differently from all the other chips.

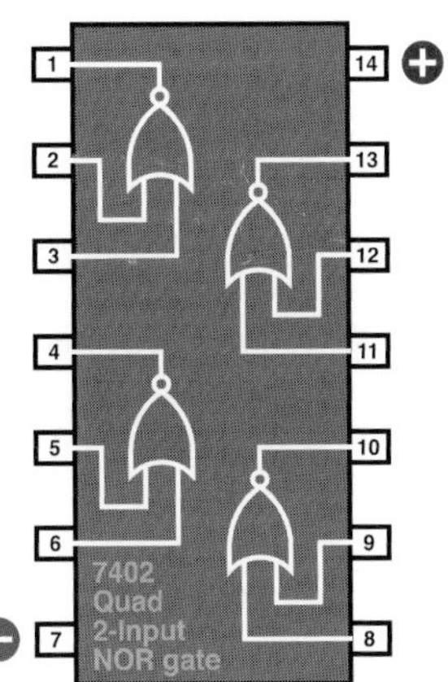

Figure 4-66. *Figures 4-66 through 4-74 show pinouts for some of the most widely used logic chips. Note that the inputs of the 7402 are reversed compared with the other chips.*

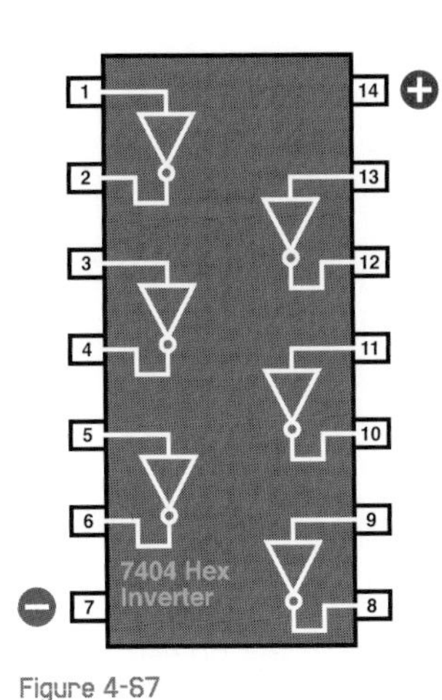

Figure 4-67

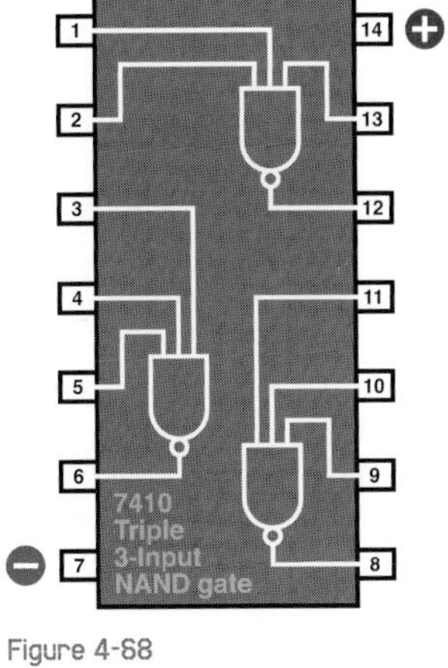

Figure 4-68

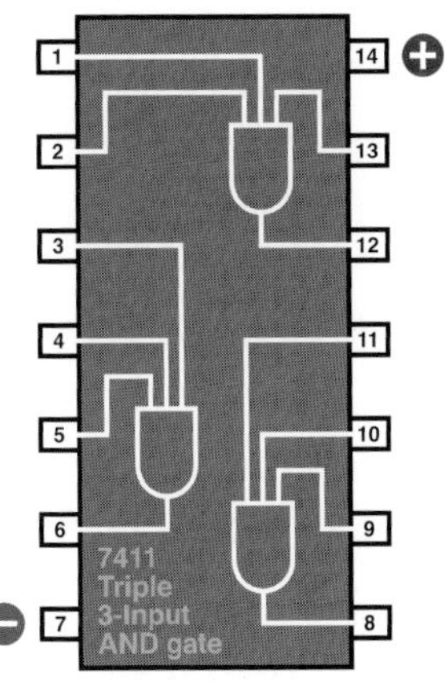

Figure 4-69

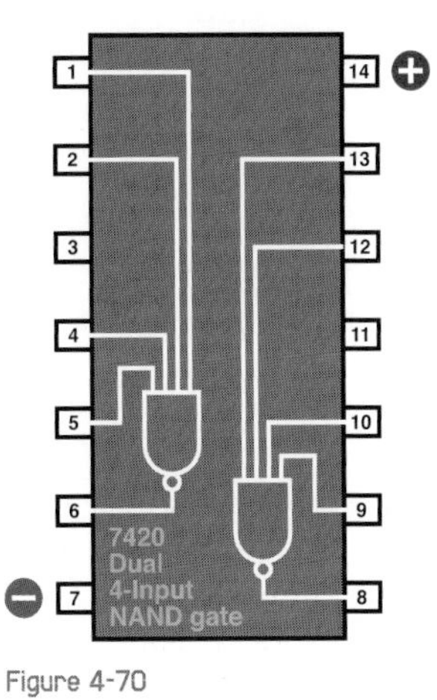

Figure 4-70

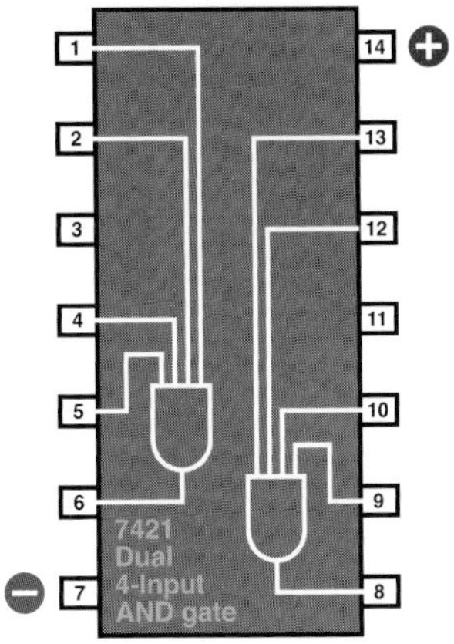

Figure 4-71

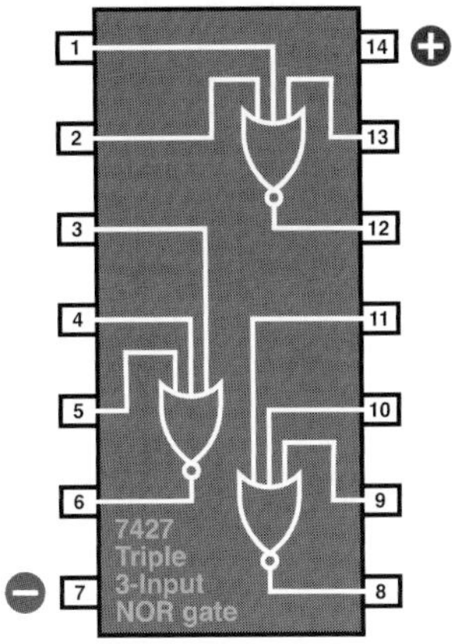

Figure 4-72

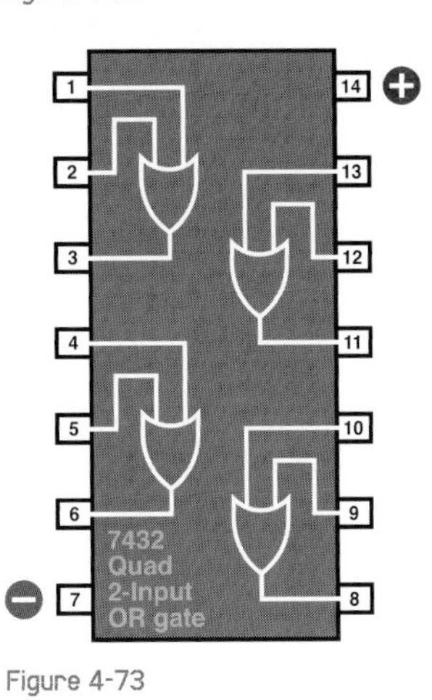

Figure 4-73

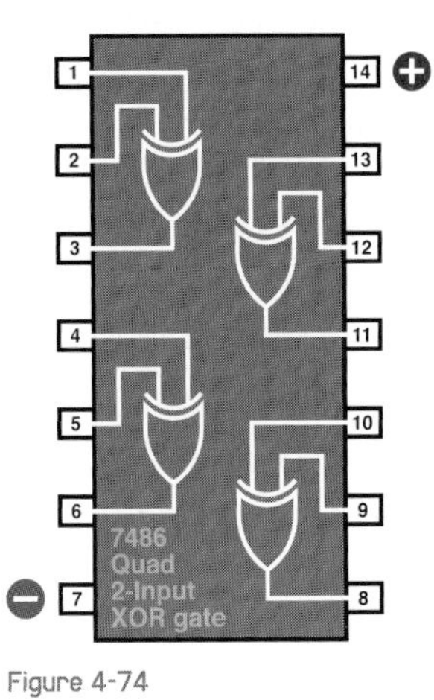

Figure 4-74

FUNDAMENTALS

Rules for connecting logic gates

Permitted:

- You can connect the input of a gate directly to your regulated power supply, either positive side or negative side.
- You can connect the output from one gate directly to the input of another gate.
- The output from one gate can power the inputs of many other gates (this is known as "fanout"). The exact ratio depends on the chip, but you can always power at least ten inputs with one logic output. The output from a logic chip can drive the trigger (pin 2) of a 555 timer. The output from the timer can then deliver 100mA, easily enough for half-a-dozen LEDs or a small relay.
- Low input doesn't have to be zero. A 74HCxx logic gate will recognize any voltage up to 1 volt as "low."
- High input doesn't have to be 5 volts. A 74HCxx logic gate will recognize any voltage above 3.5 volts as "high."

See Figures 4-75 and 4-76 for a comparison of permitted voltages on the input and output side of 74HCxx and 74LSxx chips.

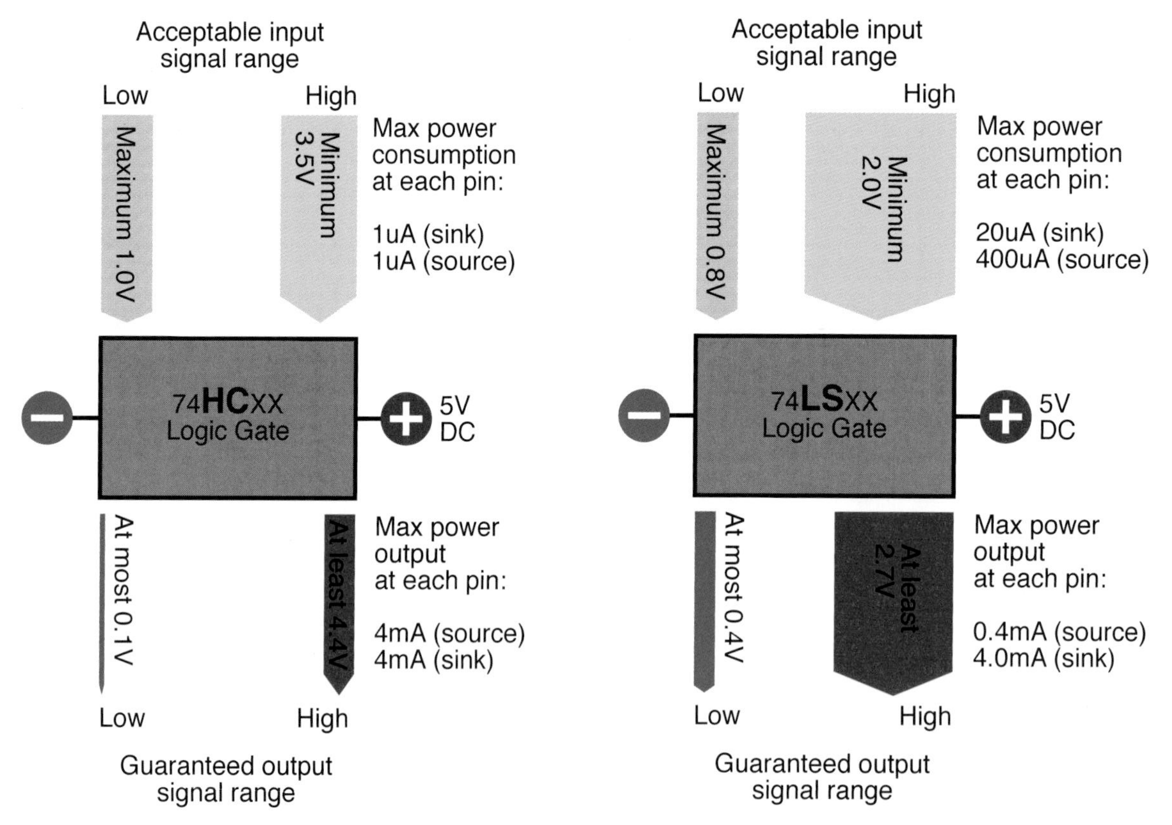

Figure 4-75. *Each family of logic chips, and each generation in each family, has different standards for input and output minimum and maximum voltages. This diagram shows the standards used by the HC generation of the CMOS family, which was chosen for most of the projects in this book. Note that the current required for input is minimal compared with the current available for output. The power supply to the chip makes up the difference.*

Figure 4-76. *Because the LS generation of the TTL family has such different tolerances for input voltages and different standards for output voltages, the LS generation of TTL chips should not be mixed in the same circuit as the HC generation of CMOS chips, unless pull-up resistors are used to bring the LS chips into conformance with standards expected by the HC chips. See Experiment 21 for a case study in using LS chips.*

FUNDAMENTALS

Rules for connecting logic gates (continued)

Not permitted:

- No floating-input pins! On CMOS chips such as the HC family, you must always connect all input pins with a known voltage, even if they supply a gate on the chip that you're not using. When you use a SPST switch to control an input, remember that in its "off" position, it leaves the input unconnected. Use a pull-up or pull-down resistor to prevent this situation. See Figure 4-77.
- Don't use an unregulated power supply, or more than 5 volts, to power 74HCxx or 74LSxx logic gates.
- Be careful when using the output from a logic gate to power even a low-current LED. Check how many milliamps are being drawn. Also be careful when "sharing" the output from a logic gate with the input of another gate, at the same time that it is driving an LED. The LED may pull down the output voltage, to a point where the other gate won't recognize it. Always check currents and voltages when modifying a circuit or designing a new one.
- Never apply a significant voltage or current to the output pin of a logic gate. In other words, don't force an input into an output.
- Never link the outputs from two or more logic gates. If they must share a common output wire, use diodes to protect them from each other. See Figure 4-78.

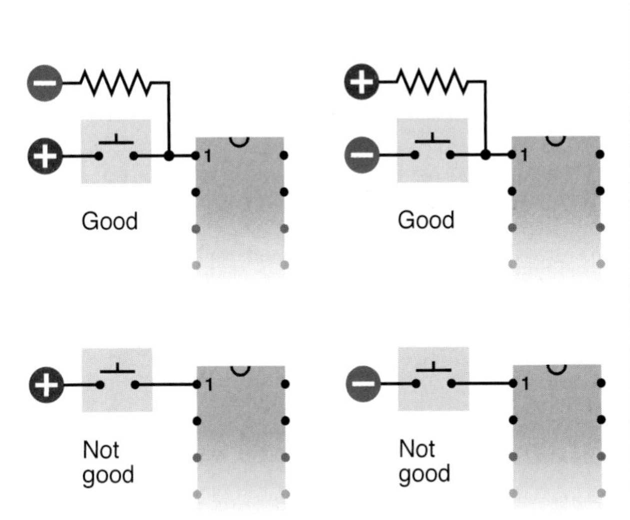

Figure 4-77. *Because a CMOS chip is so sensitive to input fluctuations, a logical input should never be left "floating," or unattached to a defined voltage source. This means that any single-throw switch or pushbutton should be used with a pull-up or pull-down resistor, so that when the contacts are open, the input is still defined.*

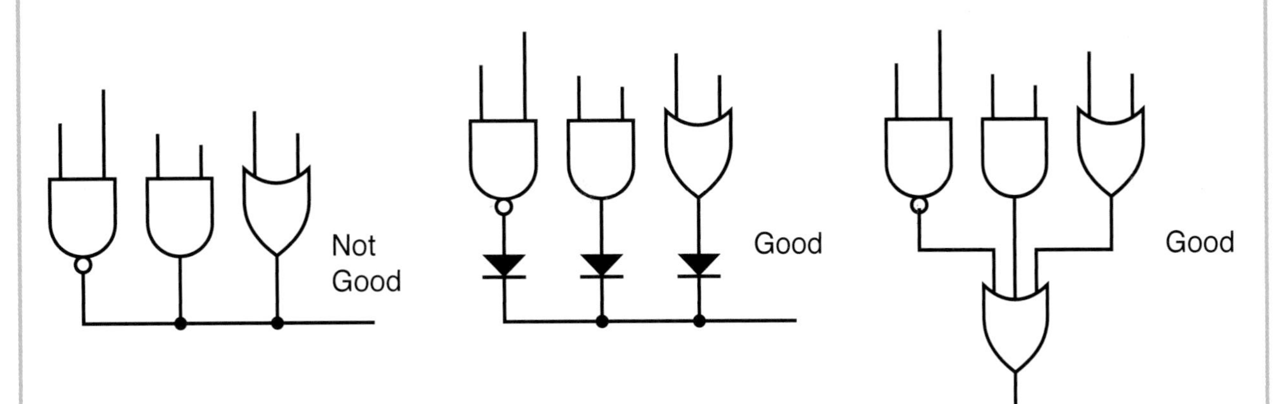

Figure 4-78. *The output from one logic gate must not be allowed to feed back into the output from another logic gate. Diodes can be used to isolate them, or they can be linked via another gate.*

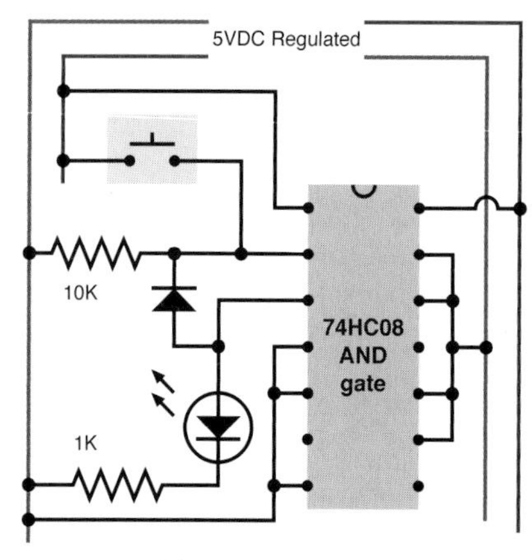

Figure 4-79. *Using a diode, the logical output from a gate can be allowed to feed back to one of its inputs, so that the gate latches after receiving a brief logical input pulse.*

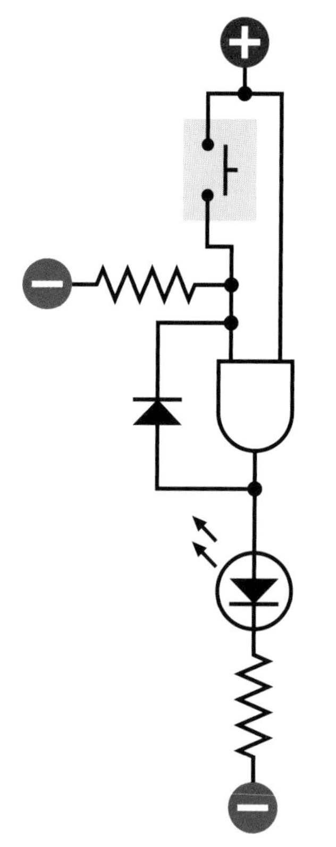

Figure 4-80. *The breadboard-format schematic in is simplified here to show more clearly the way in which a gate can latch itself after receiving an input pulse.*

In the 74HCxx logic family, each input of a logic gate consumes just a microamp, while the output can source 4 milliamps. This seems paradoxical: how can the chip give out more than it takes in? The answer is that it also consumes power from the power supply attached to pins 7 and 14. That's where the additional electricity comes from.

Because the logical output from a chip can be greater than the logical input, we can put the chip in a state where it keeps itself "switched on" in a way which is similar to the way the relay in the alarm project was wired to lock itself on. The simplest way to do this in a logic chip is by feeding some of the output back to one of the inputs.

Figure 4-79 shows an AND gate with one of its inputs wired to positive and its other input held low by a pull-down resistor, with a pushbutton that can make the input high. A signal diode connects the output of the chip back to the pushbutton-controlled input. Remember that the diode has a mark on it indicating the end which should be connected to the *negative* side of the power supply, which in this case will be the end of the 10K resistor.

The schematic in Figure 4-79 shows how the circuit should look in breadboard format. Figure 4-80 shows it in a simpler format.

From this point on, I won't bother to show the power regulator and the capacitors associated with it. Just remember to include them every time you see the power supply labeled as "5V DC Regulated."

When you switch on the power, the LED is dark, as before. The AND gate needs a positive voltage on both of its logical inputs, to create a positive output, but it now has positive voltage only on one of its inputs, while the other input is pulled down by the 10K resistor. Now touch the pushbutton, and the LED comes on. Let go of the pushbutton, and the LED stays on, because the positive output from the AND gate circulates back through the diode and is high enough to overcome the negative voltage coming through the pull-down resistor.

The output from the AND gate is powering its own input, so the LED will stay on until we disconnect it. This arrangement is a simple kind of "latch," and can be very useful when we want an output that continues after the user presses and releases a button.

You can't just connect the output from the gate to one of its inputs using an ordinary piece of wire, because this would allow positive voltage from the tactile switch to flow around and interfere with the output signal. Remember, you must never apply voltage to the output pin of a logic gate. The diode prevents this from happening.

If you've grasped the basics of logic gates, you're ready now to continue to our first real project, which will use all the information that I've set out so far.

Experiment 20: A Powerful Combination

Suppose you want to prevent other people from using your computer. I can think of two ways to do this: using software, or using hardware. The software would be some kind of startup program that intercepts the normal boot sequence and requests a password. You could certainly do it that way, but I think it would be more fun (and more relevant to this book) to do it with hardware. What I'm imagining is a numeric keypad requiring the user to enter a secret combination before the computer can be switched on.

You will need:

- Numeric keypad. As specified in the shopping list at the beginning of this chapter, it must have a "common terminal" or "common output." The schematic in Figure 4-82 shows what I mean. Inside the keypad, one conductor (which I have colored red to distinguish it from the others) connects with one side of every pushbutton. This conductor is "common" to all of them. It emerges from the keypad on an edge connector or set of pins at the bottom, which I've colored yellow.

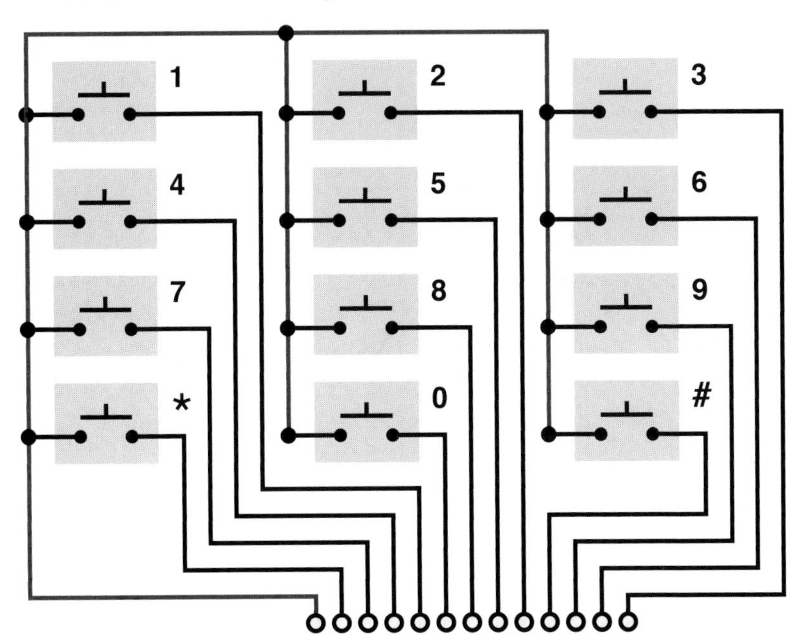

Figure 4-82. *A keypad of the type required for Experiment 20 incorporates a common terminal connected to one side of each of the 12 pushbuttons. The wire from the common terminal is shown red, here, to make it more easily identifiable.*

- Keypads that use "matrix encoding" won't work with the circuit that I'm going to describe. If the Velleman keypad, which I recommend, is unavailable, and you can't find another like it, you can use 12 separate SPST pushbuttons. Of course, that will cost a little more.
- 74HC08 logic chip containing four AND gates. Quantity: 1.
- 74HC04 logic chip containing six inverters. Quantity: 1.
- 555 timer chip. Quantity: 1.

The Warranty Issue

If you follow this project all the way to its conclusion, you'll open your desktop computer, cut a wire, and saw a hole in the cabinet. Without a doubt, this will void your warranty. If this makes you nervous, here are three options:

1. Breadboard the circuit for fun, and leave it at that.

2. Use the numeric keypad on some other device.

3. Use it on an old computer.

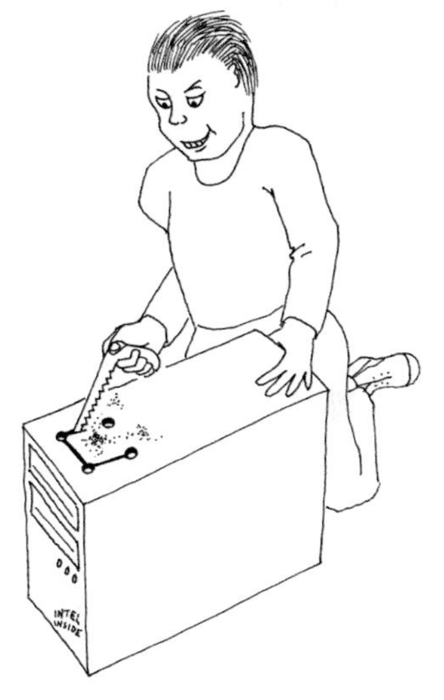

Figure 4-81. *Caution: This just might void your warranty.*

- Latching relay, 5 volt, DPST or DPDT, "2 form C" package, Panasonic DS2E-SL2-DC5V or similar. Must have two separate coils (one to latch, one to unlatch) with separate inputs. Quantity: 1.
- LEDs, 5mm generic, your choice of colors. Quantity: 3.
- Ribbon cable, with six conductors minimum, if you want to do a really neat job. You can use a cable of the type sold for hard drives, and split off the six conductors that you need, or shop around on eBay.
- Tools to open your computer, drill four holes, and make saw cuts between the holes, to create a rectangular opening for the keypad (if you want to take this project to its conclusion). Also, four small bolts to attach the keypad to the computer cabinet after you create the opening for it.

The Schematic

This time I'd like you to study the schematic before building anything. Let's start with the simplified version, shown in Figure 4-83.

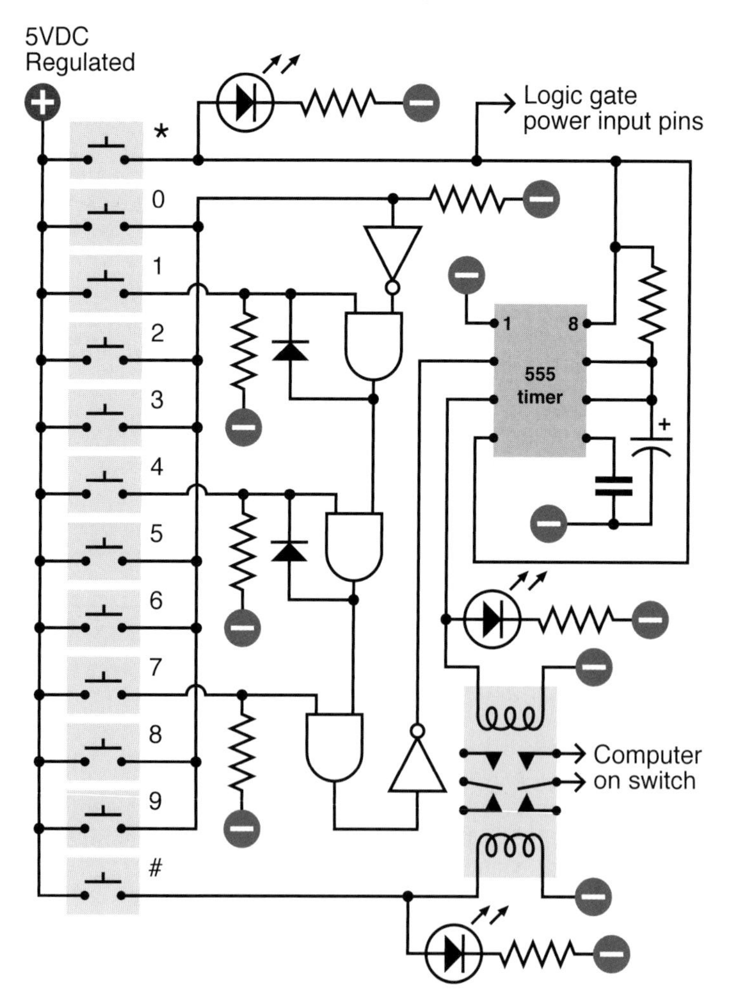

Figure 4-83. *A simplified schematic showing the basic structure of the combination lock circuit.*

I want this to be a battery-powered circuit, so that you don't have to run a separate power supply to it or (worse) try to tap into your computer's 5-volt bus. Battery power means that the circuit has to be "off" most of the time, to prevent the battery from running down. Because the keypad has two spare buttons (the asterisk and the pound sign), I'm going to use the asterisk as the "power on" button. When you press it, the LED at the top of the schematic lights up to confirm that everything's working, and the button sends power to the two logic chips and the 555 timer. You have to hold down the asterisk button while you punch in a three-digit code to unlock the computer.

Arbitrarily, I've chosen 1-4-7 as the three-digit code. Let's track what happens when you enter this sequence. (Naturally, if you build the circuit, you can wire it to choose any three digits you prefer.)

Pressing the 1 button sends positive power to one logical input of the first AND gate. The other logical input of this gate is also positive, because an inverter is supplying it, and the input of the inverter is being held negative by a pull-down resistor. When an inverter has a negative input, it gives a positive output, so pressing the 1 button activates the AND gate, and makes its output positive. The AND gate locks itself on, as its output cycles back to its switched input via a diode. So the gate output remains high even after you let go of the 1 button.

The output from the first AND gate also supplies one logical input of the second AND gate. When you press the 4 button, you send positive voltage to the other logical input of this AND gate, so its output goes high, and it locks itself on, just as the first gate did.

The second AND gate feeds the third AND gate, so when you press the 7 button, the third AND gate changes its output from low to high. This passes through an inverter, so the output from the inverter goes from high to low. This in turn goes to the trigger of a 555 timer wired in monostable mode.

When the trigger of a 555 timer goes from high to low, the timer emits a positive pulse through its output, pin 3. This runs down to the upper coil of the latching relay, and also flashes an LED to confirm that the code has been accepted and the relay has been activated.

Two of the contacts in the relay are wired into the power-up button of your computer. A little later in this description I'll explain why this should be safe with any modern computer.

Because we're using a latching relay, it flips into its "on" state and remains there, even when the power pulse from the timer ends. So now you can let go of the asterisk button to disconnect the battery power to your combination lock, and press the power-up button that switches on your computer.

At the end of your work session, you shut down your computer as usual, then press the pound button on your keypad, which flips the relay into its other position, reactivating the combination lock.

Incorrect Inputs

What happens if you enter the wrong code? If you press any button other than 1, 4, or 7, it sends positive voltage to the inverter near the top of the schematic. The positive voltage overwhelms the negative voltage being applied to the inverter through a pull-down resistor, and causes the inverter to output a negative voltage, which it applies to one of the logical inputs of the first AND gate. If the AND gate was locked on, the negative input will switch it off. If it was supplying the second AND gate, it'll switch that one off too.

Thus, any error when entering the first, second, or third digit of the secret code will reset the AND gates, forcing you to begin the sequence all over again.

What if you enter 1, 4, and 7 out of their correct sequence? The circuit won't respond. The third AND gate needs a high input supplied by the second AND gate, and the second AND gate needs a high input supplied by the first AND gate. So you have to activate the AND gates in the correct sequence.

Questions

Why did I use a 555 timer to deliver the pulse to the relay? Because the logical output from an AND gate cannot deliver sufficient power. I could have passed it through a transistor, but I liked the idea of a pulse of a fixed length to flip the relay and illuminate an LED for about 1 second, regardless of how briefly the user presses the 7 button.

Why do I need three LEDs? Because when you're punching buttons to unlock your computer, you need to know what's going on. The Power On LED reassures you that your battery isn't dead. The Relay Active LED tells you that the system is now unlocked, in case you are unable to hear the relay click. The System Relocked LED reassures you that you have secured your computer.

Because all the LEDs are driven either directly from the 5-volt supply or from the output of the 555 timer, they don't have to be low-current LEDs and can be used with 330Ω series resistors, so they'll be nice and bright.

How do you connect the keypad with the circuit? That's where your ribbon cable comes in. You carefully strip insulation from each of the conductors, and solder them to the contact strip or edge connector on your keypad. Push the conductors on the other end of the cable into your breadboard (when you're test-building the circuit) or solder them into perforated board (when you're building it permanently). Find a convenient spot inside your computer case where you can attach the perforated board, with double-sided adhesive or small bolts or whatever is convenient. Include a 9-volt battery carrier, and don't forget your power regulator to step the voltage down to 5 volts.

Breadboarding

No doubt you have realized by now that breadboards are very convenient as a quick way to push in some components and create connections, but the layout of their conductors forces you to put components in unintuitive configurations. Still, if you carefully compare the breadboard schematic in Figure 4-84 with the simplified schematic in Figure 4-83, you'll find that the connections are the same.

To help it make sense, I've shown the logic gates that exist inside the chips. I've also colored the power supply wires, as before, to reduce the risk of confusion. The positive side of the supply goes only to the common terminal on your keypad, and you have to press the asterisk key to send the power back down the ribbon cable, to supply the chips.

If you build the circuit and you can't understand why everything's dead, it's most likely because you forgot to hold down the asterisk button.

Note that the "wrong" numbers on the keypad are all shorted together. This will create some inconvenience if you want to change the combination in the future. I'll suggest a different option in the "enhancements" section that follows. For now, ideally, you should run a wire from every contact on your keypad, down to your circuit on its breadboard, and short the "wrong" keypad numbers together with jumper wires on the breadboard.

Also note that if you use a meter to test the inputs to the AND gates, and you touch your finger against the meter probe while doing so, this can be sufficient to trigger the sensitive CMOS inputs and give a false positive.

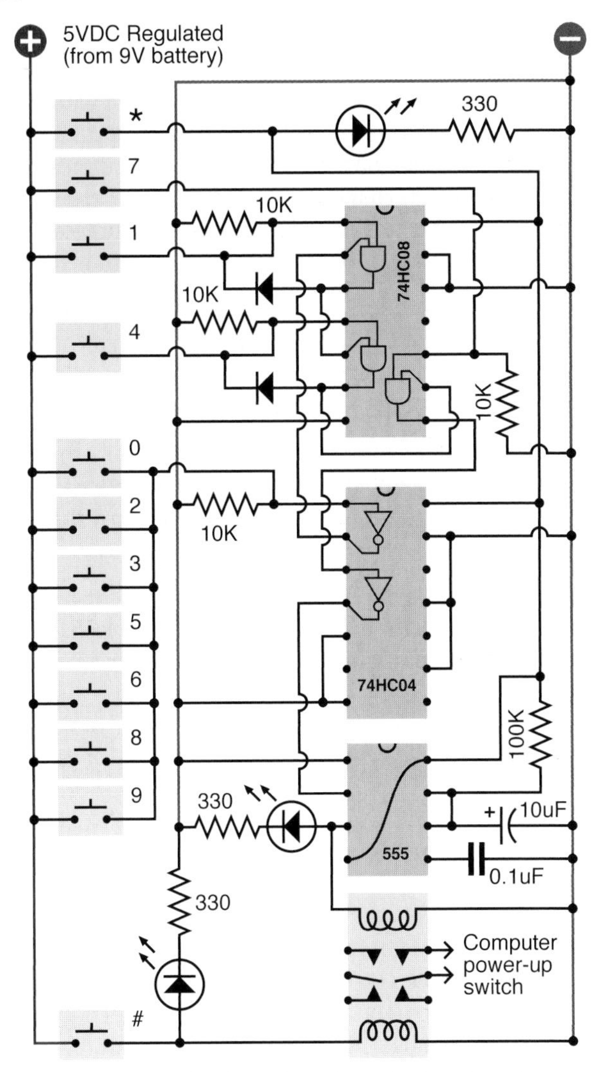

Figure 4-84. *The combination lock schematic redrawn to show how the components can be laid out on a breadboard.*

One Little Detail: The Computer Interface

Old computers used to have a big switch at the back, attached to the heavy metal box inside the computer, that transformed house current to regulated voltages that the computer needs. Most modern computers are not designed this way; you leave the computer plugged in, and you touch a little button on the box (if it's a Windows machine) or the keyboard (if it's a Mac), which sends a low-voltage pulse to the motherboard.

This is ideal from our point of view, because we don't have to mess with high voltages. Don't even think of opening that metal box with the fan mounted in it, containing the computer power supply. Just look for the wire (usually containing two conductors, on a Windows machine) that runs from the "power up" button to the motherboard.

To check that you found the right one, *make sure that your computer is unplugged*, ground yourself (because computers contain CMOS chips that are sensitive to static electricity) and very carefully snip just one of the two conductors in the wire. Now plug in your computer and try to use the "power up" button. If nothing happens, you've probably cut the right wire. Even if you cut the wrong wire, it still prevented your computer from booting, which is what you want, so you can use it anyway. Remember, we are not going to introduce any voltage to this wire. We're just going to use the relay as a switch to reconnect the conductor that you cut. You should have no problem if you maintain a cool and calm demeanor, and look for that single wire that starts everything. Check online for the maintenance manual for your computer if you're really concerned about making an error.

After you find the wire and cut just one of its conductors, unplug your computer again, and keep it unplugged during the next steps.

Find where the wire attaches to the motherboard. Usually there's a small unpluggable connector. First, mark it so that you know how to plug it back in the right way around, and then disconnect it while you follow the next couple of steps.

Strip insulation from the two ends of the wire that you cut, and solder an additional piece of two-conductor wire, as shown in Figure 4-85, with heat-shrink tube to protect the solder joints. (This is very important!)

Run your new piece of wire to the latching relay, making sure you attach it to the pair of contacts which close, inside the relay, when it is energized by the unlocking operation. You don't want to make the mistake of unlocking your computer when you think you're locking it, and vice versa.

Reconnect the connector that you disconnected from your motherboard, plug in your computer, and try to power it up. If nothing happens, this is probably good! Now enter the secret combination on your keypad (while holding down the asterisk button to provide battery power) and listen for the click as the relay latches. Now try the "power up" button again, and everything should work.

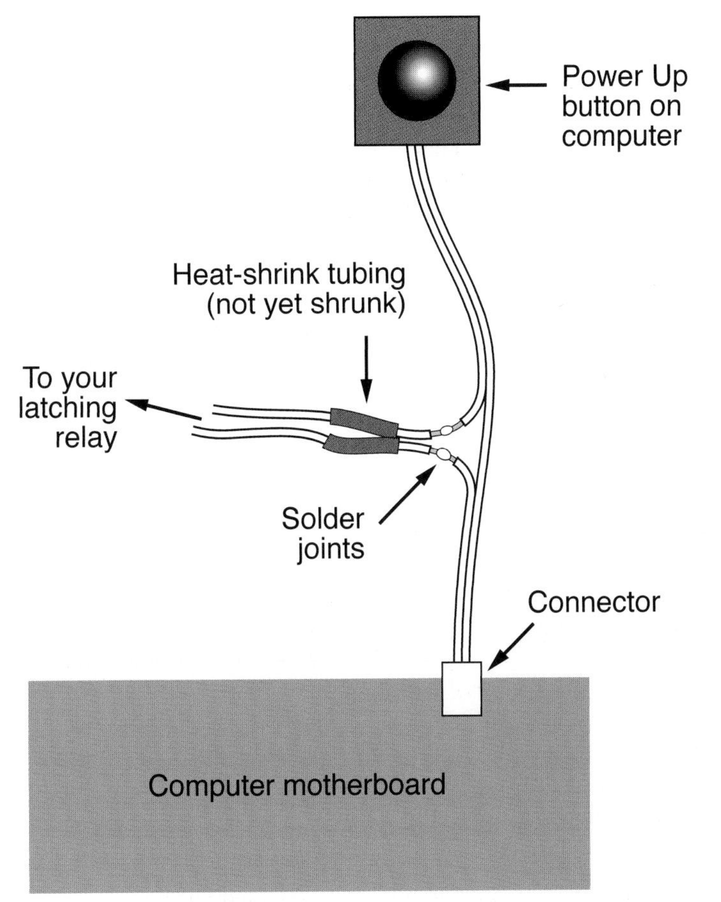

Figure 4-85. *The combination lock project can be interfaced with a typical desktop computer by cutting one conductor in the wire from the "power up" pushbutton, soldering an extension, and covering the joints with heat-shrink tube.*

Enhancements

At the end of any project, there's always more you can do.

To make this setup more secure, you could remove the usual screws that secure the case of the computer, and replace them with tamper-proof screws. Check any online source for "tamper-proof screws," such as *http://www.mcmaster.com*. Naturally, you will also need the special tool that fits the screws, so that you can install them (or remove them, if your security system malfunctions for any reason).

Another enhancement could be an additional 555 timer that is activated by the asterisk button, and delivers power to the other chips for, say, a limited period of 30 seconds, allowing you that much time to unlock the system. This would eliminate the need to hold down the asterisk button while you enter the unlocking code. A 555 timer can supply power to all the other chips, because they don't use very much. I omitted this feature for the sake of simplicity.

Yet another enhancement, if you are security-crazed, is to go for a four-button code. After all, the 74HC08 chip still has one unused AND gate. You could insert that into the chain of the existing AND gates and wire it to another keypad button of your choice.

Still another enhancement would be a way to change the code without unsoldering and resoldering wires. You can use the miniature sockets that I suggested in the heartbeat flasher project. These should enable you to swap around the ends of your wires from the keypad.

And for those who are absolutely, positively, totally paranoid, you could fix things so that entering a wrong code flips a second high-amperage relay which supplies a massive power overload, melting your CPU and sending a big pulse through a magnetic coil clamped to your hard drive, instantly turning the data to garbage (Figure 4-86). Really, if you want to protect information, messing up the hardware has major advantages compared with trying to erase data using software. It's faster, difficult to stop, and tends to be permanent. So, when the Record Industry Association of America comes to your home and asks to switch on your computer so that they can search for illegal file sharing, just accidentally give them an incorrect unlocking code, sit back, and wait for the pungent smell of melting insulation.

Figure 4-86. *For those who are absolutely, positively, totally paranoid: a meltdown/self-destruct system controlled by a secret key combination provides enhanced protection against data theft or intrusions by RIAA investigators asking annoying questions about file sharing.*

If you pursue this option, I *definitely* take no responsibility for the consequences.

On a more realistic level, no system is totally secure. The value of a hardware locking device is that if someone does defeat it (for instance, by figuring out how to unscrew your tamper-proof screws, or simply ripping your keypad out of the computer case with metal shears), at least you'll know that something happened—especially if you put little dabs of paint over the screws to reveal whether they've been messed with. By comparison, if you use password-protection software and someone defeats it, you may never know that your system has been compromised.

Experiment 21: Race to Place

The next project is going to get us deeper into the concept of feedback, where the output is piped back to affect the input—in this case, blocking it. It's a small project, but quite subtle, and the concepts will be useful to you in the future.

You will need:

- 74HC32 chip containing four OR gates. Quantity: 1.
- 555 timers. Quantity: 2.
- SPDT switch. Quantity: 1.
- SPST tactile switches. Quantity: 2.
- Various resistors.
- 5V supply using power regulator as before.

The Goal

On quiz shows such as *Jeopardy*, contestants race to answer each question. The first person who hits his answer button automatically locks out the other contestants, so that their buttons become inactive. How can we make a circuit that will do the same thing?

If you search online, you'll find several hobby sites where other people have suggested circuits to work this way, but they lack some features that I think are necessary. The approach I'm going to use here is both simpler and more elaborate. It's simpler because it has a very low chip count, but it's more elaborate in that it incorporates "quizmaster control" to make a more realistic game.

I'll suggest some initial ideas for a two-player version. After I develop that idea, I'll show how it could be expanded to four or even more players.

A Conceptual Experiment

I want to show how this kind of project grows from an idea to the finished version. By going through the steps of developing a circuit, I'm hoping I may inspire you to develop ideas of your own in the future, which is much more valuable than just replicating someone else's work. So join me in a conceptual experiment, thinking our way from a problem to a solution.

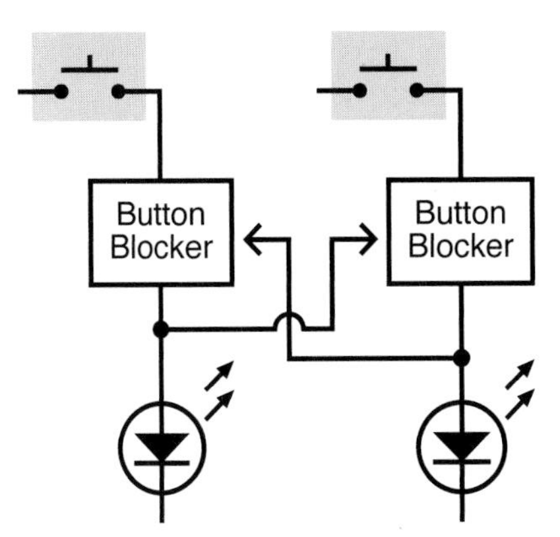

Figure 4-87. *The basic concept of the quiz project is that the output from one button should feed back to intercept the output of another button. At this point, the way in which the "button blocker" circuit works has not been figured out.*

First consider the basic concept: two people have two buttons, and whoever goes first locks out the other person. I always find it helps me to visualize this kind of thing if I draw a sketch, so that's where I'll begin. In Figure 4-87, the signal from each button passes through a component that I'll call a "button blocker," activated by the other person's button. I'm not exactly sure what the button blocker will be or how it will work, yet.

Now that I'm looking at it, I see a problem here. If I want to expand this to three players, it will get complicated, because each player must activate the "button blockers" of *two* opponents. Figure 4-88 shows this. And if I have four players, it's going to get even more complicated.

Anytime I see this kind of complexity, I think there has to be a better way.

Also, there's another problem. After a player lets his finger off the button, the other players' buttons will be unblocked again. I need a latch to hold the signal from the first player's button and continue to block the other players.

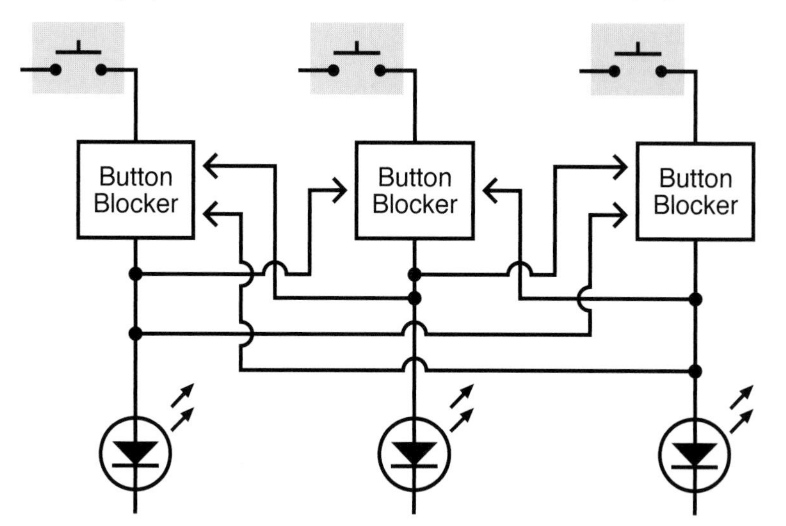

Figure 4-88. *The quiz concept becomes more complicated when an additional pushbutton is added. Now each button must block two other buttons. If a fourth button is added, the circuit will become unmanageably complex. There has to be a better way.*

This now sounds even more complicated. But wait a minute, if I have a latch which allows the winning player to take his finger off his button, I don't care if any of the buttons are being pressed anymore—including the button of the winning player. As soon as his signal is latched, *all* the buttons can be blocked. This makes things much simpler. I can summarize it as a sequence of events:

1. First player presses his button.
2. The signal is latched.
3. The latched signal feeds back and blocks all the buttons.

The new sketch in Figure 4-89 shows this. Now the configuration is modular, and can be expanded to almost any number of players, just by adding more modules.

There's something important missing, though: a reset switch, to put the system back to its starting mode after the players have had time to press their buttons and see who won. Also, I need a way to prevent players from pressing their buttons too soon, before the quizmaster has finished asking the question. Maybe I can combine this function in just one switch, which will be under the quizmaster's control. In its Reset position, the switch can reset the system and remove power to the buttons. In its Play position, the switch stops holding the system in reset mode, and provides power to the buttons. Figure 4-90 shows this. I've gone back to showing just two players, to minimize the clutter of lines and boxes, but the concept is still easily expandable.

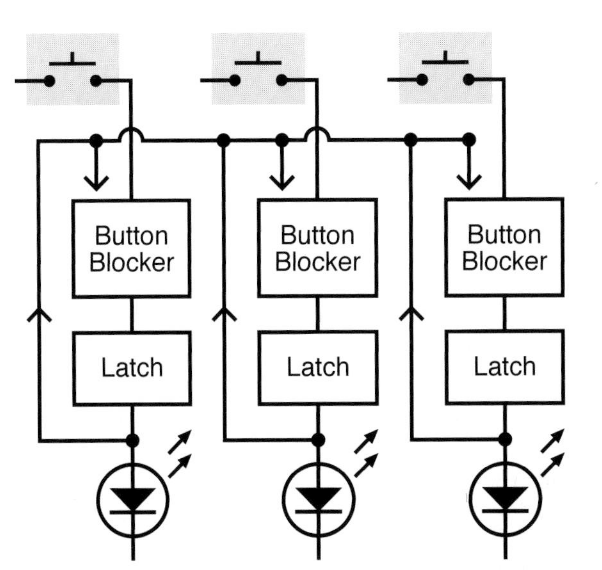

Figure 4-89. *If a latch is added below each button, it can retain one input and then block all inputs from all buttons. This simplifies the concept.*

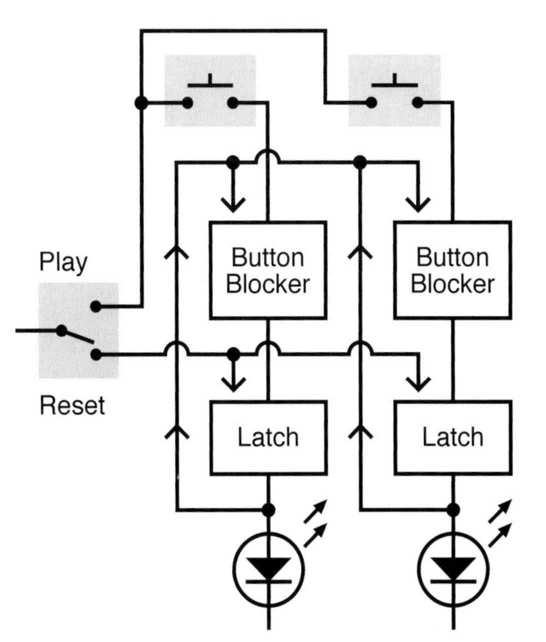

Figure 4-90. *A quizmaster switch will be needed to activate the buttons initially and then reset the circuit after a winning input has been recorded.*

Now I have to deal with a logic problem in the diagram. The way I've drawn it, after the output from the lefthand latch goes up to the "button blockers," it can also run down the wire to the other half of the circuit (against the direction to the arrows), because everything is joined together. In other words, if the lefthand LED lights up, the righthand LED will light up, too. How can I stop this from happening?

Well, I could put diodes in the "up" wires to block current from running down them. But I have a more elegant idea: I'll add an OR gate, because the inputs to an OR gate are separated from each other electrically. Figure 4-91 shows this.

Usually an OR gate has only two logical inputs. Will this prevent me from adding more players? No, because you can actually buy an OR that has eight inputs. If any one of them is high, the output is high. For fewer than eight players, I can short the unused inputs to ground, and ignore them.

Looking again at Figure 4-91, I'm getting a clearer idea of what the thing I've called a "button blocker" should actually be. I think it should be another logic gate. It should say, "If there's only one input, from a button, I'll let it through. But if there is a second input from the OR gate, I won't let it through."

That sounds like a NAND gate, but before I start choosing chips, I have to decide what the latch will be. I can buy an off-the-shelf flip-flop, which flips "on" if it gets one signal and "off" if it gets another, but the trouble is, chips containing flip-flops tend to have more features than I need for a simple circuit like this. Therefore I'm going to use 555 timers again, in flip-flop mode. They require very few connections, work very simply, and can deliver a good amount of current. The only problem with them is that they require a negative input at the trigger pin to create a positive output. But I think I can work with that.

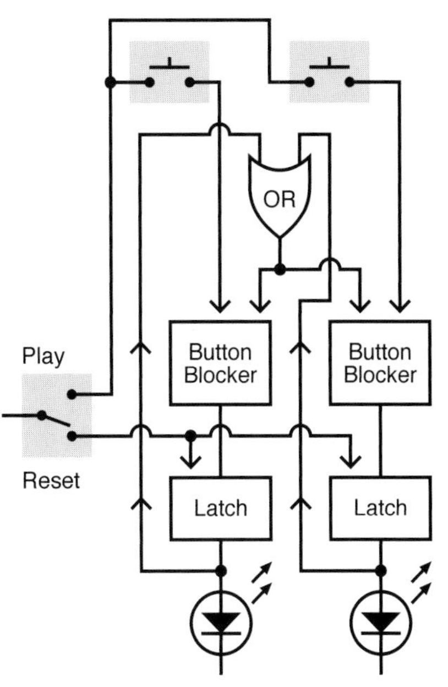

Figure 4-91. *To prevent the output from one latch feeding back around the circuit to the output from another latch, the outputs can be combined in an OR gate.*

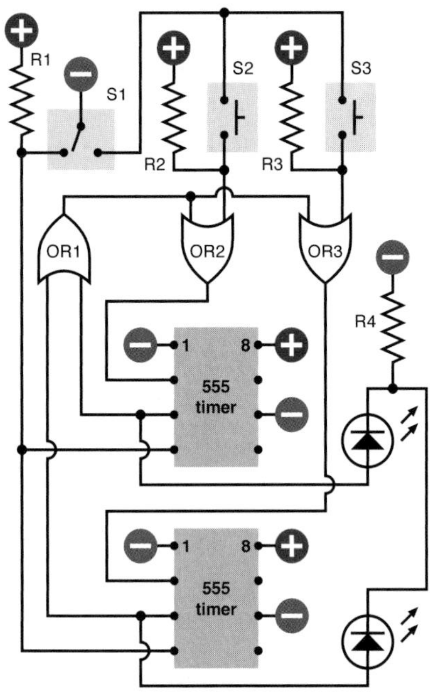

Figure 4-92. *Now that the basic concept of the quiz circuit has been roughed out, specific components can be inserted, with compatible inputs and outputs.*

So now, finally, here's a simplified schematic, in Figure 4-92. I like to show the pins of the 555 timers in their correct positions, so I had to move the components around a little to minimize wire crossovers, but you can see that logically, it's the same basic idea.

Before you try to build it, just run through the theory of it, because that's the final step, to make sure there are no mistakes. The important thing to bear in mind is that because the 555 needs a negative input on its trigger pin to create its output, when any of the players presses a button, the button has to create a negative "flow" through the circuit. This is a bit counterintuitive, so I'm including a three-step visualization in Figure 4-93, showing how it will work.

In Step 1, the quizmaster has asked a question and flipped his switch to the right, to supply (negative) power to the players' buttons. So long as no one presses a button, the pull-up resistors supply positive voltage to OR2 and OR3. An OR gate has a positive output if it has any positive input, so OR2 and OR3 keep the trigger inputs of the 555 timers positive. Their outputs remain low, and nothing is happening yet.

In Step 2, the lefthand player has pressed his button. Now OR2 has two negative inputs, so its output has gone low. But IC1 hasn't reacted yet.

In Step 3, just a microsecond later, IC1 has sensed the low voltage on its trigger, so its output from pin 3 has gone high, lighting the LED. Remember, this 555 timer is in flip-flop mode, so it locks itself into this state immediately. Meanwhile its high output also feeds back to OR1. Because OR1 is an OR gate, just one high input is enough to make a high output, so it feeds this back to OR2 and OR3. And now that they have high inputs, their outputs also go high, and will stay high, regardless of any future button-presses.

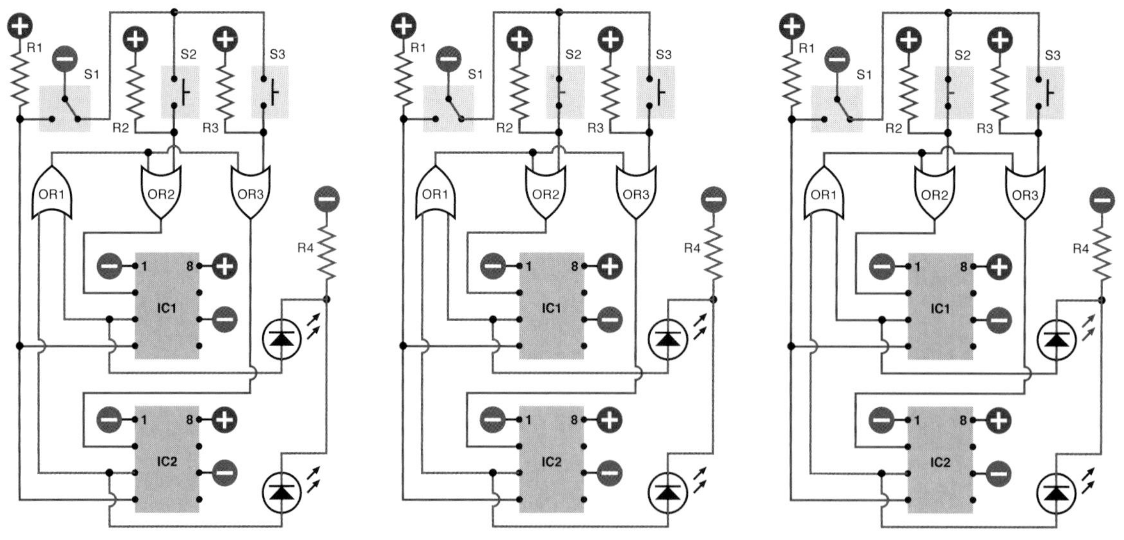

Figure 4-93. *These three schematics show the prevalence of higher and lower voltages (red and blue lines) through the quiz circuit when a pushbutton is pressed.*

Because OR2 and OR3 now have high inputs and outputs, IC1 and IC2 cannot be triggered. But IC1 is still locked into its "on" state, keeping the LED illuminated.

The only way to change IC1 is if the quizmaster flips his switch back to the left. That applies negative power to the reset pins of both the timers. Consequently their outputs go low, the LED goes out, and the circuit goes back into the same state as where it started. Having reset it, the quizmaster can ask another question, but the players' buttons are not activated until the quizmaster flips the switch back to the right again.

There's only one situation that I haven't addressed: what if both players press their buttons absolutely simultaneously? In the world of digital electronics, this is highly unlikely. Even a difference of a microsecond should be enough time for the circuit to react and block the second button. But if somehow both buttons are pressed at the same instant, both of the timers should react, and both of the LEDs will light up, showing that there has been a tie.

In case you feel a little uncertain about the way in which a two-player circuit can be upgraded to handle extra players, I've included a simplified three-player schematic in Figure 4-94.

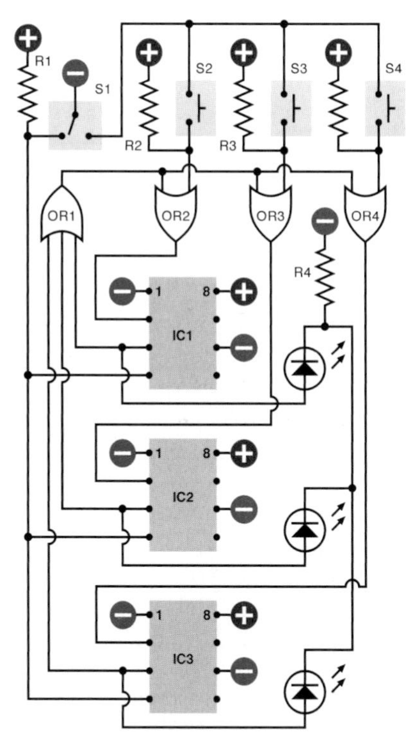

Figure 4-94. *The two-player schematic in can be easily upgraded to a three-player version, as shown here, provided the first OR gate can handle three inputs.*

Breadboarding It

Now it's time to create a schematic that's as close to the breadboard layout as possible, so that you can build this thing easily. The schematic is shown in Figure 4-95 and the actual components on a breadboard are in Figure 4-96. Because the only logic gates that I've used are OR gates, and there are only three of them, I just need one logic chip: the 74HC32, which contains four 2-input OR gates. (I've grounded the inputs to the fourth). The two OR gates on the left side of the chip have the same functions as OR2 and OR3 in my simplified schematic, and the OR gate at the bottom-right side of the chip works as OR1, receiving input from pin 3 of each 555 timer. If you have all the components, you should be able to put this together and test it quite quickly.

You may notice that I've made one modification of the previous schematic. A 0.01 μF capacitor has been added between pin 2 of each 555 timer (the Input) and negative ground. Why? Because when I tested the circuit without the capacitors, sometimes I found that one or both of the 555 timers would be triggered simply by flipping S1, the quizmaster switch, without anyone pressing a button.

At first this puzzled me. How were the timers getting triggered, without anyone doing anything? Maybe they were responding to "bounce" in the quizmaster switch. Sure enough, the small capacitors solved the problem. They may also slow the response of the 555 timers fractionally, but not enough to interfere with slow human reflexes.

As for the buttons, it doesn't matter if they "bounce," because each timer locks itself on at the very first impulse and ignores any hesitations that follow.

You can experiment building the circuit, disconnecting the 0.01 μF capacitors, and flipping S1 to and fro a dozen times. If you have a high-quality switch, you may not experience any problem. If you have a lower-quality switch, you may see a number of "false positives." I'm going to explain more about "bounce," and how to get rid of it, in the next experiment.

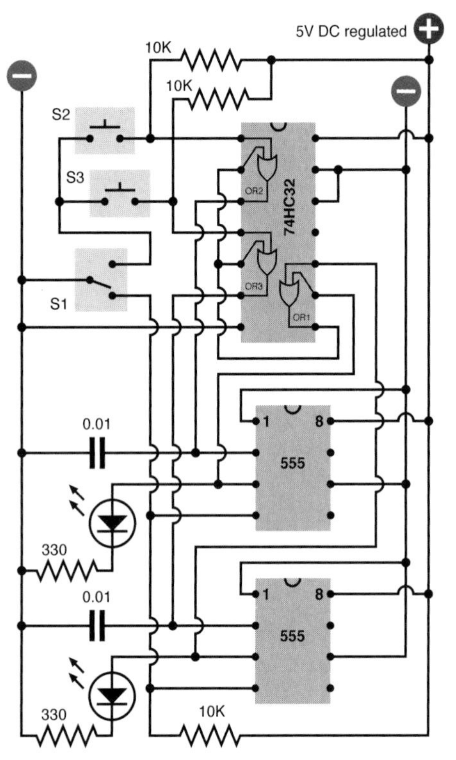

Figure 4-95. *Applying the simplified schematic to a breadboard inevitably entails a wiring layout that is less intuitively obvious and appears more complex. The connections are the same, though.*

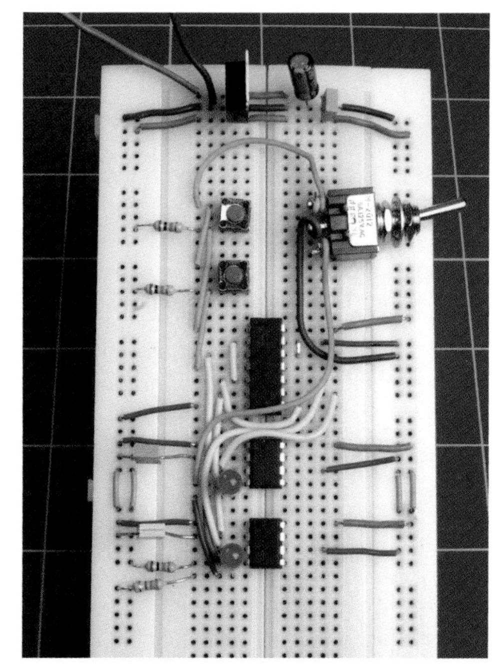

Figure 4-96. *The quiz schematic applied to a breadboard, to test the concept prior to full-scale implementation.*

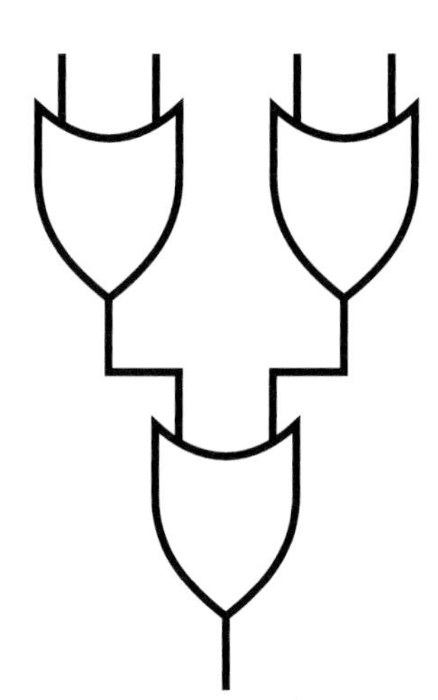

Figure 4-97. *Although a four-input OR gate is not manufactured, its functionality can be achieved easily by linking three 2-input OR gates together.*

Enhancements

After you breadboard the circuit, if you proceed to build a permanent version, I suggest that you expand it so that at least four players can participate. This will require an OR gate capable of receiving four inputs. The 74HC4078 is the obvious choice, as it allows up to eight. Just connect any unused inputs to negative ground.

Alternatively, if you already have a couple of 74HC32 chips and you don't want to bother ordering a 74HC4078, you can gang together three of the gates inside a single 74HC32 so that they function like a four-input OR. Look at the simple logic diagram in Figure 4-97 showing three ORs, and remember that the output from each OR will go high if at least one input is high.

And while you're thinking about this, see if you can figure out the inputs and output of three ANDs in the same configuration.

For a four-player game, you'll also need two additional 555 timers, of course, and two more LEDs, and two more pushbuttons.

As for creating a schematic for the four-player game—I'm going to leave that to you. Begin by sketching a simplified version, just showing the logic symbols. Then convert that to a breadboard layout. And here's a suggestion: pencil, paper, and an eraser can still be quicker, initially, than circuit-design software or graphic-design software, in my opinion.

Experiment 22: Flipping and Bouncing

I mentioned in the previous experiment that "bounce" from the buttons in the circuit wouldn't be a problem, because the buttons were activating 555 timers that were wired in bistable, flip-flop mode. As soon as the timer receives the very first pulse, it flips into its new state and flops there, ignoring any additional noise in the circuit. So can we debounce a switch or a button using a flip-flop? And as some 74HCxx chips are available containing flip-flops, can we use them?

The answers are yes, and yes, although it's not quite as simple as it sounds.

You will need:

- 74HC02 logic chip containing 4 NOR gates. 74HC00 logic chip containing 4 NAND gates. Quantity: 1 of each.
- SPDT switch. Quantity: 1.
- LEDs, low-current. Quantity: 2.
- 10K resistors and 1K resistors. Quantity: 2 of each.

Assemble the components on your breadboard, following the schematic shown in Figure 4-98. When you apply power (through your regulated 5-volt supply), one of the LEDs should be lit.

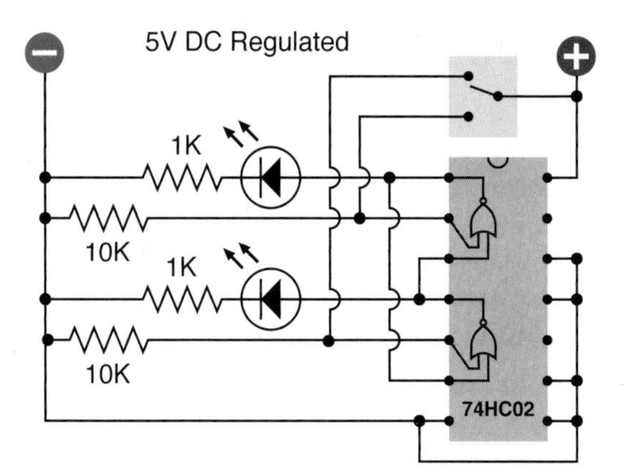

Figure 4-98. *A simple circuit to test the behavior of two NOR gates wired as a simple flip-flop that retains its state after an input pulse ceases.*

Now I want you to do something odd. Please disconnect the SPDT switch by taking hold of the wire that connects the positive power supply to the pole of the switch, and pulling the end of the wire out of the breadboard. When you do this, you may be surprised to find that the LED remains lit.

Push the wire back into the breadboard, flip the switch, and the first LED should go out, while the other LED should become lit. Once again, pull out the wire, and once again, the LED should remain lit.

Here's the take-home message:

- A flip-flop requires only an initial pulse.
- After that, it ignores its input.

How It Works

Two NOR gates or two NAND gates can function as a flip-flop:

- Use NOR gates when you have a positive input from a double-throw switch.
- Use NAND gates when you have a negative input from a double-throw switch.

Either way, you have to use a double-throw switch.

I've mentioned the double-throw switch three times (actually, four times if you count this sentence!) because for some strange reason, most introductory books fail to emphasize this point. When I first started learning electronics, I went crazy trying to understand how two NORs or two NANDs could debounce a simple SPST pushbutton—until finally I realized that they can't. The reason is that when you power up the circuit, the NOR gates (or NAND gates) need to be told in which state they should begin. They need an initial orientation, which comes from the switch being in one state or the other. So it has to be a double-throw switch. (Now I've mentioned it five times.)

I'm using another simplified multiple-step schematic, Figure 4-99, to show the changes that occur as the switch flips to and fro with two NOR gates. To refresh your memory, I've also included a truth table showing the logical outputs from NOR gates for each combination of inputs.

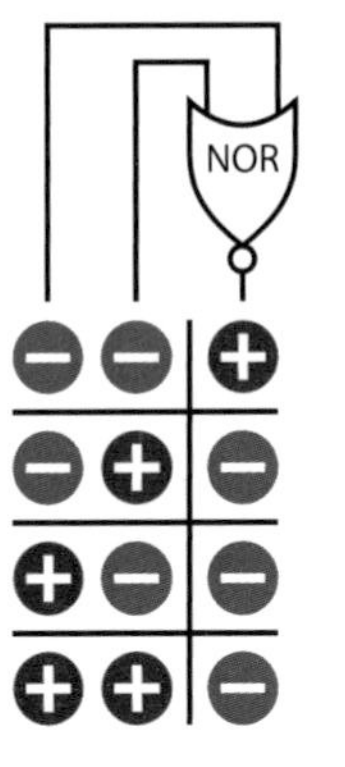

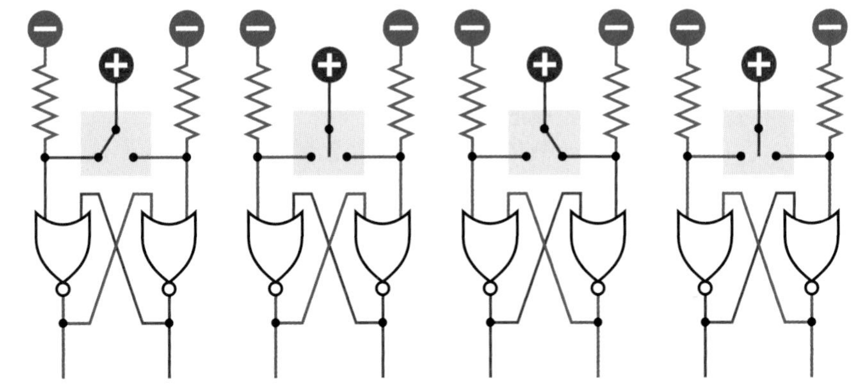

Figure 4-99. *Using two NOR gates in conjunction with a positive input through a SPDT switch, this sequence of four diagrams shows how a flip-flop circuit responds.*

Suppose that the switch is turned to the left. It sends positive current to the lefthand side of the circuit, overwhelming the negative supply from the pull-down resistor, so we can be sure that the NOR gate on the left has one positive logical input. Because any positive logical input will make the NOR give a negative output (as shown in the truth table), the negative output crosses over to the righthand NOR, so that it now has two negative inputs, which make it give a positive output. This crosses back to the lefthand NOR gate. So, in this configuration everything is stable.

Now comes the clever part. Suppose that you move the switch so that it doesn't touch either of its contacts. (Or suppose that the switch contacts are bouncing, and failing to make a good contact. Or suppose you disconnect the switch entirely.) Without a positive supply from the switch, the lefthand input of the left NOR gate goes from positive to negative, as a result of the pull-down resistor. But the righthand input of this gate is still positive, and one positive is all it takes to make the NOR maintain its negative output, so nothing changes. In other words, the circuit has "flopped" in this state.

Now if the switch turns fully to the right and supplies positive power to the righthand pin of the right NOR gate, quick as a flash, that NOR recognizes that it now has a positive logical input, so it changes its logical output to negative. That goes across to the other NOR gate, which now has two negative inputs, so its output goes positive, and runs back to the right NOR.

In this way, the output states of the two NOR gates change places. They flip, and then flop there, even if the switch breaks contact or is disconnected again. The second set of drawings in Figure 4-100 shows exactly the same logic, using a negatively powered switch and two NAND gates. You can use your 74HC00 chip, specified in the parts list for this experiment, to test this yourself.

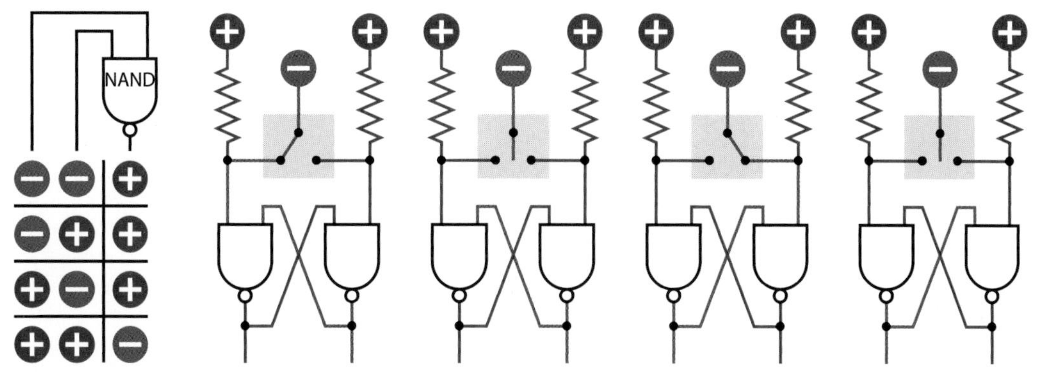

Figure 4-100. *The schematic from Figure 4-99 can be rewired with NAND gates and a negative switched input.*

Both of these configurations are examples of a jam-type flip-flop, so called because the switch forces it to respond immediately, and jams it into that state. You can use this circuit anytime you need to debounce a switch (as long as it's a double-throw switch).

A more sophisticated version is a clocked flip-flop, which requires you to set the state of each input first and then supply a clock pulse to make the flip-flop respond. The pulse has to be clean and precise, which means that if you supply it from a switch, the switch must be debounced—probably by using another jam-type flip-flop! Considerations of this type have made me reluctant to use clocked flip-flops in this book. They add a layer of complexity, which I prefer to avoid in an introductory text.

What if you want to debounce a single-throw button or switch? Well, you have a problem! One solution is to buy a special-purpose chip such as the 4490 "bounce eliminator," which contains digital delay circuitry. A specific part number is the MC14490PG from On Semiconductor. This contains six circuits for six separate inputs, each with an internal pull-up resistor. It's relatively expensive, however—more than 10 times the price of a 74HC02 containing NOR gates. Really, it may be simpler to use double-throw switches that are easily debounced as described previously.

Experiment 23: Nice Dice

This is the one experiment where I want you to use the 74LSxx generation of the TTL logic family, instead of the 74HCxx family of CMOS. Two reasons: first, I need to use the 7492 counter, which is unavailable in the HC family. And second, you should know the basic facts about the LS series of TTL chips, as they still crop up in circuits that you'll find in electronics books and online.

In addition, you'll learn about "open collector" TTL chips such as the 74LS06 inverter, which can be a convenient substitute for transistors when you want to deliver as much as 40mA of current.

The idea of this circuit is simple enough: run a 555 timer sending very fast pulses to a counter that counts in sixes, driving LEDs that are placed to imitate the spots on a die. (Note that the word "die" is the singular of "dice.") The counter runs so fast, the die-spots become a blur. When the user presses a button, the counter stops arbitrarily, displaying an unpredictable spot pattern.

Dice simulations have been around for many, many years, and you can even buy kits online. But this one will do something more: it will also demonstrate the principles of binary code.

So, if you're ready for the triple threat of TTL chips, open collectors, and binary, let's begin.

You will need:

- 74LS92 counter such as SN74LS92N by Texas Instruments. Quantity: 1 if you want to create one die, 2 to make two dice.
- 74LS27 three-input NOR gate such as SN74LS27N by Texas Instruments. Quantity: 1.
- 555 timers. Quantity: 1 if you want to make one die, 2 to make two dice.
- Signal diodes, 1N4148 or similar. Quantity: 4, or 8 to make two dice.

Seeing Binary

The counter that we dealt with before was unusual, in that its outputs were designed to drive seven-segment numerals. A more common type has outputs that count in binary code.

The 74LS92 pinouts are shown in Figure 4-101. Plug the chip into your breadboard and make connections as shown in Figure 4-102. Initially, the 555 timer will drive the counter in slow-motion, at around 1 step per second. Figure 4-103 shows the actual components on a breadboard.

Note that the counter has unusual power inputs, on pins 5 and 10 instead of at the corners. Also four of its pins are completely unused, and do not connect with anything inside the chip. Therefore, you don't need to attach any wire to them on the outside.

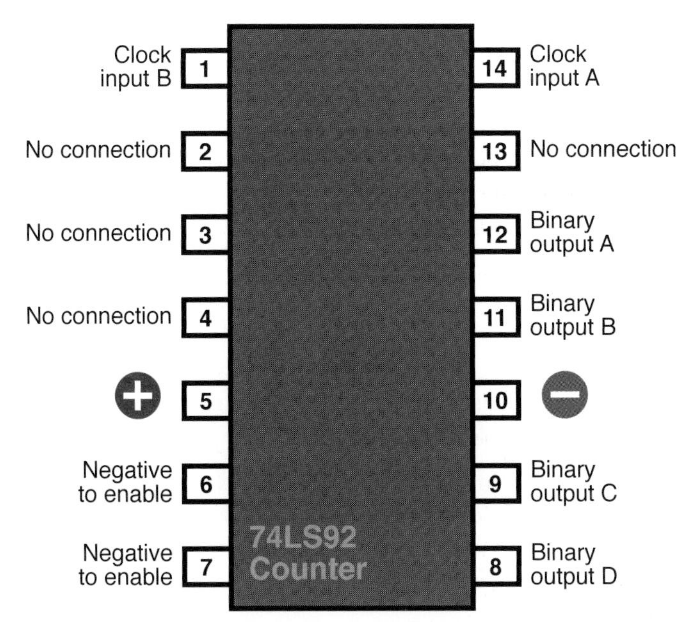

Figure 4-101. *The unusual pin assignments include four that have no connection of any kind inside the chip, and can be left unattached.*

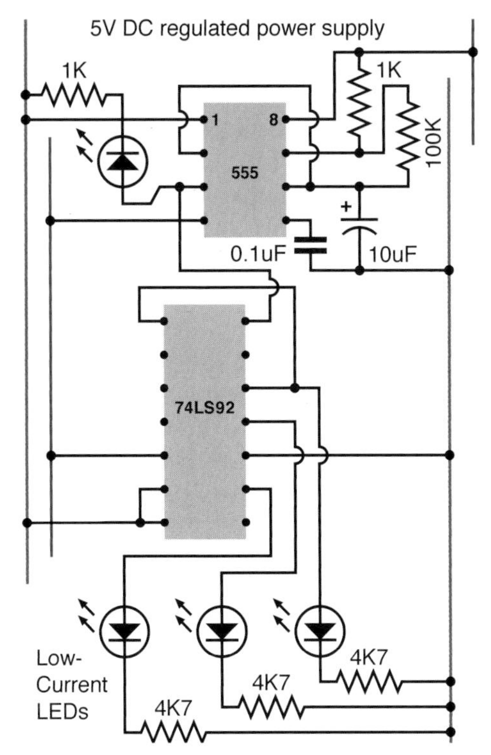

Figure 4-102. *This simple circuit uses a 555 timer running slowly to control the 74LS92 binary counter and display the succession of high states from its outputs.*

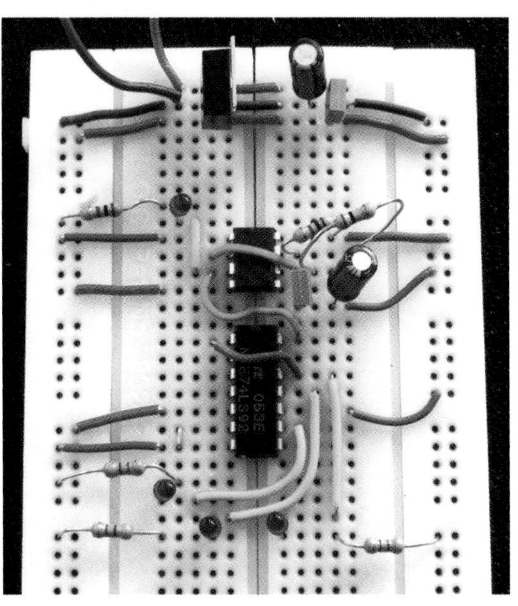

Figure 4-103. *The breadboard version of the schematic in Figure 4-102 to display the outputs from a 74LS92 counter.*

Now we come to the first new and difficult fact about the 74LSxx generation of TTL chips that makes them less desirable, for our purposes, than the 74HCxx generation of CMOS chips that I have recommended in previous projects. The modern and civilized HC chips will source 4mA or sink 4mA at each logical output, but the older LS generation is fussier. It will sink around 8mA into each output pin from a positive source, but when its output is high, it hardly gives you anything at all. This is a very basic principle:

- Outputs from TTL logic chips are designed to sink current.
- They are not designed to source significant current.

In fact, the 74LS92 is rated to deliver less than half a milliamp. This is quite acceptable when you're just connecting it with another logic chip, but if you want to drive an external device, it doesn't provide much to work with.

The proper solution is to say to the chip, "All right, we'll do it your way," and set things up with a positive source that flows through a load resistor to the LED that you want to use, and from there into the output from the chip. This is the "better" option shown in Figure 4-104.

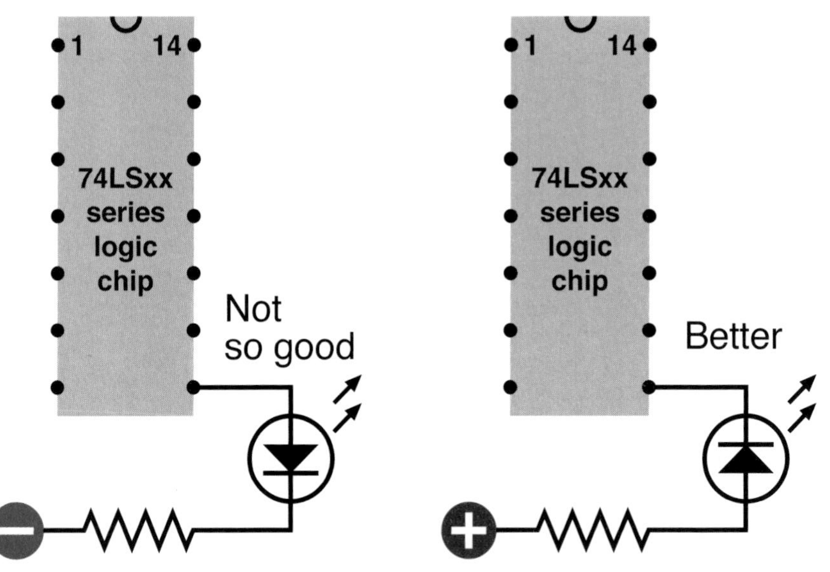

Figure 4-104. *Most TTL chips, including those in the LS generation, are unable to source much current from their logical output pins (left) and should usually be wired to sink current from a positive source (right).*

The only problem is that now the LED lights up when the counter's output is low. But the counter is designed to display its output in high pulses. So your LED is now off when it should be on, and on when it should be off.

You can fix this by passing the signal through an inverter, but already I'm getting impatient with this inconvenience. My way around the problem, at least for demo purposes, is to use the "Not so good" option in Figure 4-104 and make it work by connecting a very-low-current LED with a large 4K7 load resistor. This will enable us to "see" the output from the counter without asking it to give more than its rated limit, and if you want to create a more visibly powerful display for a finished version of the dice circuit, I'll deal with that later. According to my meter, the 4K7 resistor holds the current between 0.3mA and 0.4mA, which is the counter's rated maximum.

Set up your initial version of the circuit as shown in Figures 4-102 and 4-103. Be careful when you wire the positive and negative power supply to the counter chip, with its nonstandard pin assignments.

The 555 will run in astable mode, at about 1 pulse per second. This becomes the clock signal for the counter. The first three binary outputs from the counter then drive the three LEDs.

The counter advances when the input signal goes from *high* to *low*. So when the LED beside the 555 timer goes out, that's when the counter advances.

If you stare at the pattern generated by the outputs for long enough, you may be able to see the logic to it, bearing in mind that its zero state is when they are all off, and it counts up through five more steps before it repeats. The diagram in Figure 4-105 shows this sequence. If you want to know why the pattern works this way, check the following section, "Theory: Binary arithmetic."

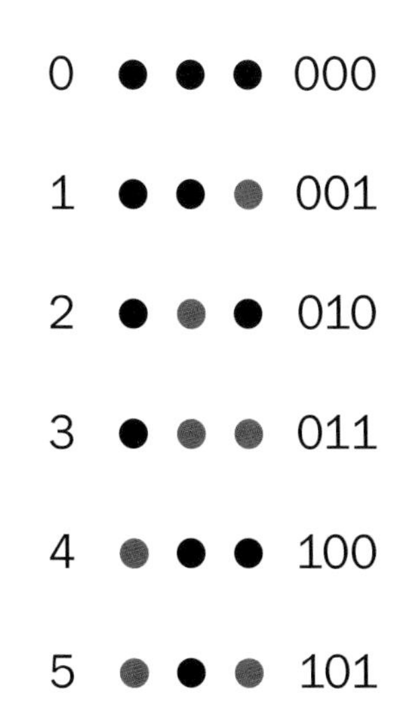

Figure 4-105. *The three output pins of the 74LS92 counter have high states shown by the red circles as the counter steps from 000 to 101 in binary notation.*

THEORY

Binary arithmetic

The rule for binary counting is just a variation of the rule that we normally use for everyday counting, probably without thinking much about it. In a 10-based system, we count from 0 to 9, then carry 1 over to the next position on the left, and go from 0 to 9 again in the right-most position. We repeat this procedure until we get to 99, then carry a 1 over to a new position to make 100, and continue counting.

In binary we do the same thing, except that we restrict ourselves to digits 0 and 1 only. So begin with 0 in the rightmost position, and count up to 1. As 1 is our limit, to continue counting we carry 1 over to the next place on the left, and start again from 0 in the right-most position. Count up to 1, then add 1 to the next place on the left—but, it already has a 1 in it, so it can't count any higher. So, carry 1 from there one space further, to the next place beside that—and so on.

If a glowing LED represents a 1, and a dark LED represents a 0, the diagram in Figure 4-105 shows how the 74LS92 counts up from 0 to (decimal) 5 or (binary) 101 in its inimitable fashion. I've also included a diagram in Figure 4-106 showing how a counter with four binary outputs would display decimal numbers from 0 through 15, again using the LEDs to represent 1s and 0s.

Here's a question for you: how many LEDs would you need to represent the decimal number 1024 in binary? And how many for 1023?

Obviously binary code is ideally suited to a machine full of logic components that either have a high or a low state. So it is that all digital computers use binary arithmetic (which they convert to decimal, just to please us).

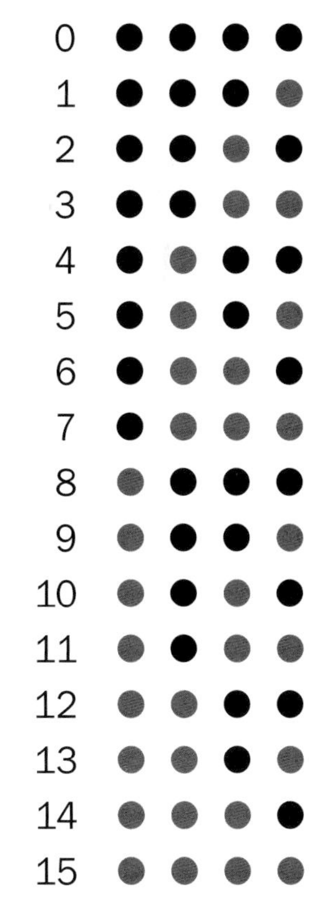

Figure 4-106. *A hexadecimal (16-based) binary counter would generate this succession of high states from its four output pins as it counts from 0 through 15 in decimal notation.*

Getting back to our project: I want to take the three binary outputs and make them create patterns like the spots on a die. How can I do this? Quite easily, as it turns out.

I'm assuming that I'll use seven LEDs to simulate the patterns of spots on a die. These patterns can be broken down into groups, which I have assigned to the three outputs from the counter in Figure 4-108. The first output (farthest to the right) can drive an LED representing the dot at the center of the die face. The second (middle) output can drive two more diagonal LEDs. The third output must switch on all four corner LEDs.

This will work for patterns 1 through 5, but won't display the die pattern for a 6. Suppose I tap into all three outputs from the counter with a three-input NOR gate. It has an output that goes high only when all three of its inputs are low, so it will only give a high output when the counter is beginning with all-low outputs. I can take advantage of this to make a 6 pattern.

Note that it's bad practice to mix the LS generation of TTL chips with the HC generation of CMOS chips, as their input and output ranges are different; so, the NOR chip has to be a 74LS27, not a 74HC27.

We're ready now for a simple schematic. In Figure 4-107 I've colored some of the wires just to make it easier for you to distinguish them. The colors have no other significance.

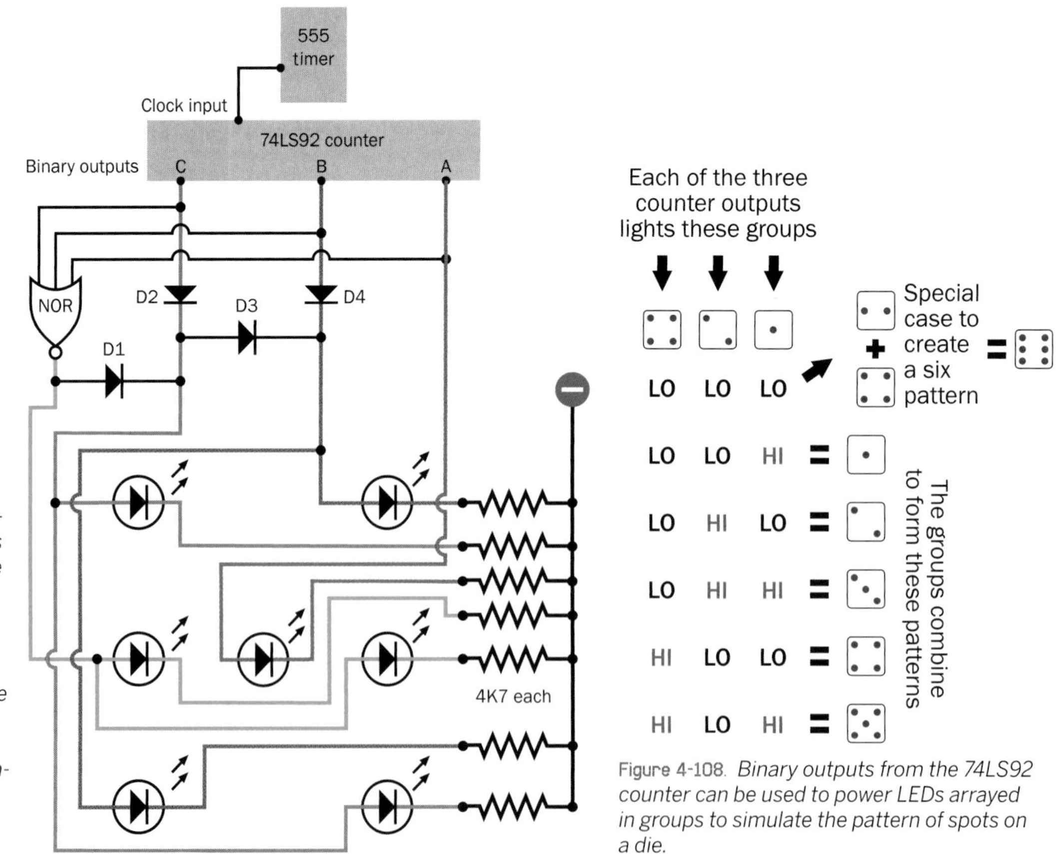

Figure 4-107. *A simplified schematic shows how outputs from the 74LS92 counter can be combined, with signal diodes and a single three-input NOR gate, to generate the spot patterns on a die. The wire colors have no special meaning and are used just to distinguish them from each other.*

Figure 4-108. *Binary outputs from the 74LS92 counter can be used to power LEDs arrayed in groups to simulate the pattern of spots on a die.*

Each of the LEDs is grounded through a separate 4K7 load resistor. Unfortunately, this means that when they are displaying the pattern for a 6, all of them are running in parallel from the output of the NOR gate, which overloads it. As long as you don't leave the display in this mode for very long periods, it shouldn't cause a problem. You could compensate by increasing the load resistors, or by running pairs of the LEDs through one resistor, but this will make them so dim that they'll be difficult to see, as they're so close to their lower limit for current already.

Notice how I have added four signal diodes, D1 through D4. When Output C goes high, it has to illuminate all four corner LEDs, and so its power goes into the brown wire as well as the gray wire. But we must never allow one output to feed back into another, so D4 is needed to protect Output B when Output C is high.

Because there is now a connection between B and C, we need D2 to protect Output C when Output B is high. And because Output B must only feed two of the corner LEDs, we also need D3 to stop it from illuminating the other two. And, we have to protect the output from the NOR gate when either Output C or Output B is high. This requires D1.

Figure 4-109 shows everything that I've described so far assembled in breadboard format, while Figure 4-110 shows the test version that I built. Note that the unused logical inputs on the 74LS27 chip are shorted together and connected to the *positive* side of the power supply. Here's the rule:

- When using CMOS chips (such as the HC series), connect unused logical inputs to the negative side of the power supply.
- When using TTL chips (such as the LS series), connect unused logical inputs to the positive side of the power supply.

I assume that you have had enough fun watching the LEDs count slowly, so I've changed the capacitor and resistor values for the 555 to increase its speed from approximately 1 pulse per second to about 50,000 pulses per second. The counter could run much faster than this, but I just want it to cycle fast enough so that when the user presses and releases a button, the count will stop at an unforeseeable number.

The button starts and stops the 555 timer by applying and releasing power to the timing circuit only. This is the equivalent of shaking and then throwing the die.

While the counter is running fast, the LEDs are flashing so fast that all of them will seem to be on at once. At the same time, the circuit charges a new 68 µF capacitor, which I have added between the pushbutton and ground. When you release the button, this capacitor discharges itself through the 1K timing resistor. As the charge dissipates, the timing capacitor will take longer and longer to charge, and discharge, and the frequency of the 555 will gradually diminish. Consequently the LED display will also flash slower, like the reel on a Las Vegas slot machine gradually coming to a stop. This increases the tension as players can see the die display counting to the number that they're hoping for—and maybe going one step beyond it.

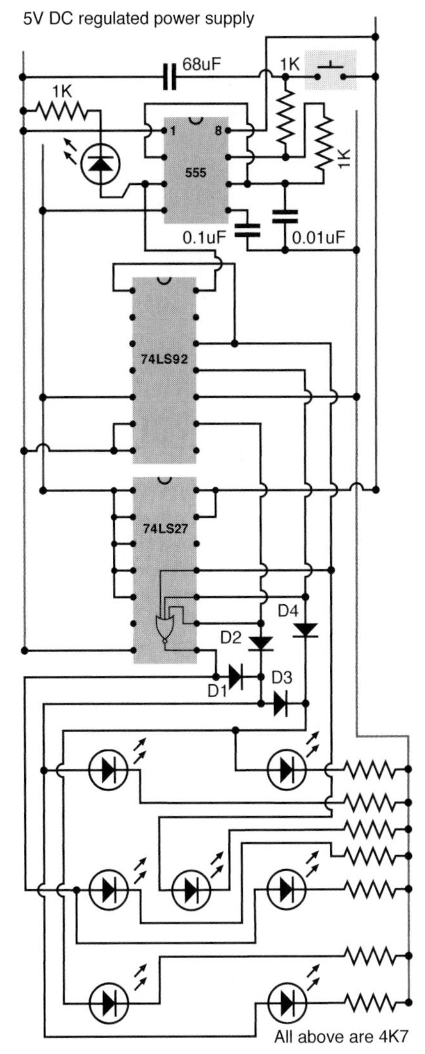

Figure 4-109. *With some extra components, the schematics from Figures 4-102 and 4-107 can be combined to make the working dice simulation.*

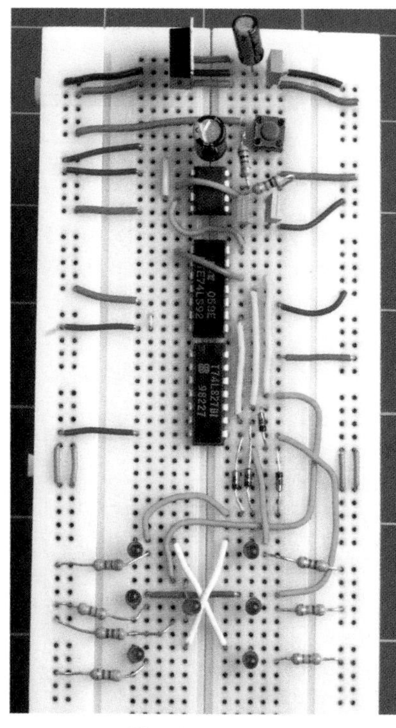

Figure 4-110. *The electronic dice schematic applied to a breadboard, with a pushbutton at the top to start and stop the counter, and 7 LEDs at the bottom to display the output.*

Note that to maximize this effect, the button has to be held down for a full second or more, so that the 68 µF capacitor becomes fully charged before the button is released.

So, this circuit now fulfills the original goal. But can it be better? Of course it can.

Enhancements

The main thing I want to improve is the brightness of the LEDs. I could add a transistor to amplify the current to each one, but there's a simpler alternative: a TTL "open collector" inverter.

I want to use an inverter because in the world of TTL, as I mentioned earlier, we can sink much more power into the output pin of a chip than we can source from it. So, I'm going to turn each LED the other way around and connect their load resistors to the positive side of the power supply. This way, they'll sink their power into the outputs of the inverter.

And the great advantage of an "open collector" version of the inverter chip is that it is designed to sink much more current than a normal TTL logic chip. It is rated for 40mA per pin. The only disadvantage is that it cannot source any current at all; instead of its output going high, it just behaves like an open switch. But that's OK for this circuit.

So the next and final schematic, in Figure 4-111, includes the 74LS06 inverter, which has also been added to the breadboarded version shown in Figure 4-112. I suggest that you put aside the little low-current LEDs and substitute some normal-size ones. Using Kingbright "standard" WP15031D 5mm LEDs, I find that each draws almost exactly 20mA with a voltage drop of about 2V with a 120 ohm series resistor. Because each output pin from the 74LS06 inverter powers no more than two LEDs at a time, this is exactly within its specification. I suggest that if you build this circuit, you check the consumption of your particular choice of LEDs and adjust the resistors if necessary.

Remember: to measure the voltage drop across an LED, simply touch the probes of your meter across it while it is illuminated. To measure the current, disconnect one side of the LED and insert the meter, in milliamp mode, between the leg of the LED and the contact that it normally makes in the circuit.

For a really dramatic display, you can get some 1 cm diameter LEDs (Figure 4-113). Check the specification, and you should find that many of these don't use more power than the usual 5 mm type. But whatever kind you use, don't forget to turn them around so that their negative sides face toward the inverter, and their positive sides face the resistors, which are connected to the positive side of the power supply.

One last detail: I had to add two 10K resistors to this version of the circuit. Can you see why? Diodes D1 through D4 are designed to transmit positive voltage through to the inverter when appropriate, but they prevent the inputs of the inverter from "seeing" the negative side of the power supply when the counter outputs are low. These inverter inputs require pull-down resistors to prevent them from "floating" and producing erroneous results.

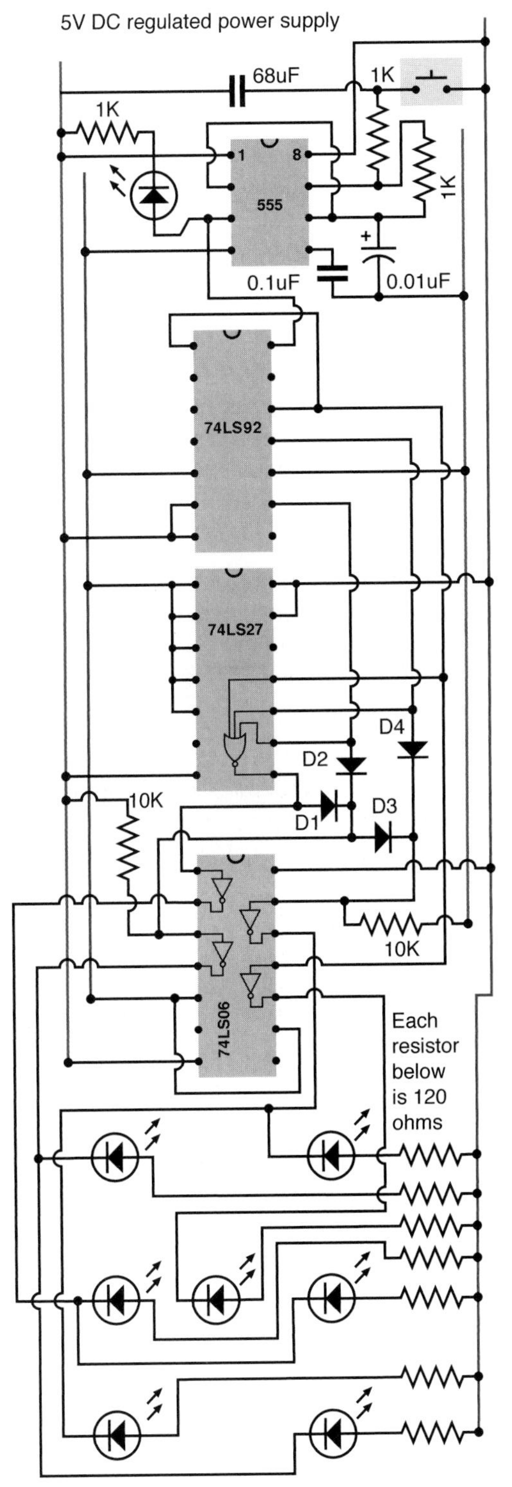

Figure 4-111. *If open-collector inverters are added to the dice schematic, it can drive full-size LEDs with up to 40mA, as long as the LEDs are turned around to sink current into the TTL output stage instead of trying to source current from it.*

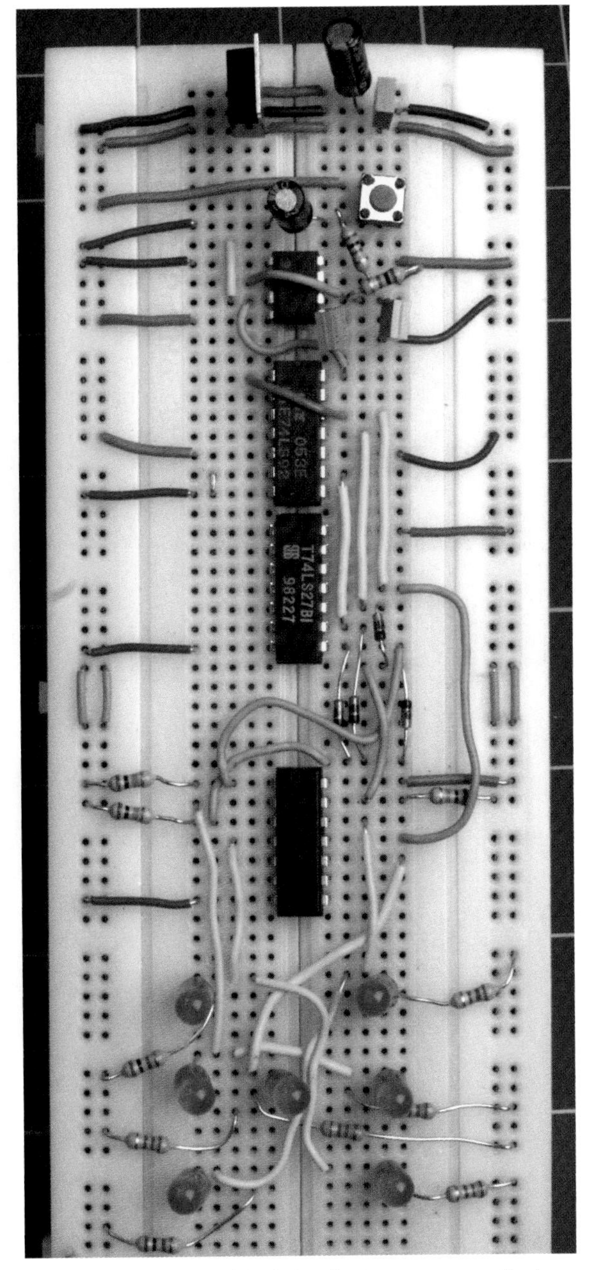

Figure 4-112. *The completed circuit using an open-collector inverter to drive full-size LEDs.*

The final enhancements are up to you. Most obviously, you can add a second die, as many games require two dice. The 74LS27 chip still has a couple of spare NOR gates in it, one of which you can make use of, but you will need an additional 555 timer, running at a significantly different speed to ensure randomness, and it will have to drive a second counter.

After you get your dice up and running, you may want to test them for randomness. Because the pulses from a 555 timer are of equal length, every number has an equal chance of coming up; but the longer you hold down the Start button, the better your odds are of interrupting the counting process at a truly random moment. Anyone using your electronic dice should be told that "shaking" them for a full second is mandatory.

Of course, I could have simulated dice more easily by writing a few lines of software to generate random numbers on a screen, but even a fancy screen image cannot have the same appeal as a well-made piece of hardware. Figure 4-113 shows white 1 cm LEDs mounted in a sanded polycarbonate enclosure for dramatic effect.

Most of all, I derived satisfaction from using simple, dedicated chips that demonstrate the binary arithmetic that is fundamental in every computer.

Figure 4-113. *The open-collector inverter chip in the dice circuit is sufficiently powerful to drive 1-cm white LEDs that draw about 20mA each, using a potential of 2V. In this finished version, the LEDs were embedded in cavities drilled from the underside of half-inch polycarbonate, which has been treated with an orbital sander to create a translucent finish.*

Experiment 24: Intrusion Alarm Completed

Now let me suggest how you can apply the knowledge from this chapter of the book to upgrade the burglar alarm project that was last modified in Experiment 15. You'll probably need to check Chapters 2 and 3 to refamiliarize yourself with some features of the alarm.

Upgrade 1: Delayed Activation

The biggest flaw in the alarm was that as soon as it was activated, it would immediately respond to any signal from the door and window sensors. It needed a feature to delay activation to give you a chance to exit from the building before the alarm armed itself. A 555 timer can provide this functionality, probably in conjunction with a relay. The power to the alarm should pass through the contacts of the relay, which are normally closed. When you press a button on the timer, it sends a positive pulse to the relay lasting for around 30 seconds, holding the relay open for that period. You could mount the timer in its own little box with a button on it, which you press when you're ready to leave the building. The 12-volt power supply to the burglar alarm passes through the box containing the delay circuit. For 30 seconds, the 555 interrupts power to the alarm, and then restores it, ready for action.

Upgrade 2: Keypad Deactivation

This is now really simple. You can substitute a latching relay instead of the switch, S1, on the alarm box (shown in Figure 3-110), and use a keypad to set and reset the relay in exactly the same way as in the combination lock in Experiment 20. You'll have to run an additional three wires from the relay, out of the alarm box, to the keypad (one supplying power to the "on" relay coil, another supplying power to the "off" coil, the third being a common ground). You can either use a 9-volt battery to power the electronics in association with the keypad, or run an additional fourth wire from the alarm box, to carry positive power to the logic chips, bearing in mind that you have to insert a voltage regulator at some point, to drop the 12 volts that the alarm uses to the 5 volts that the logic gates require. As the gates consume so little power, the 12-to-5 drop should be OK for the voltage regulator; it won't have to dissipate too much heat.

With this additional feature, you can use the alarm like this:

- Press the pound key on the keypad to flip the latching relay into its "on" mode, so that it passes power to the alarm box, which is now armed.
- If you want to leave the house, push the button on the delay unit to give you 30 seconds in which to do so.
- If the alarm is triggered, enter your secret code on the keypad to deactivate it by flipping the latching relay to its "off" position and cutting power to the alarm box.

These modifications are so simple that I think the block diagram in Figure 4-114 should be all you need. I don't think I need to give you any schematics. The only change you have to make to the existing alarm is to substitute the latching relay for the on/off switch.

But, there is still one obvious necessary enhancement needed: how can you get back into the house without instantly triggering the alarm?

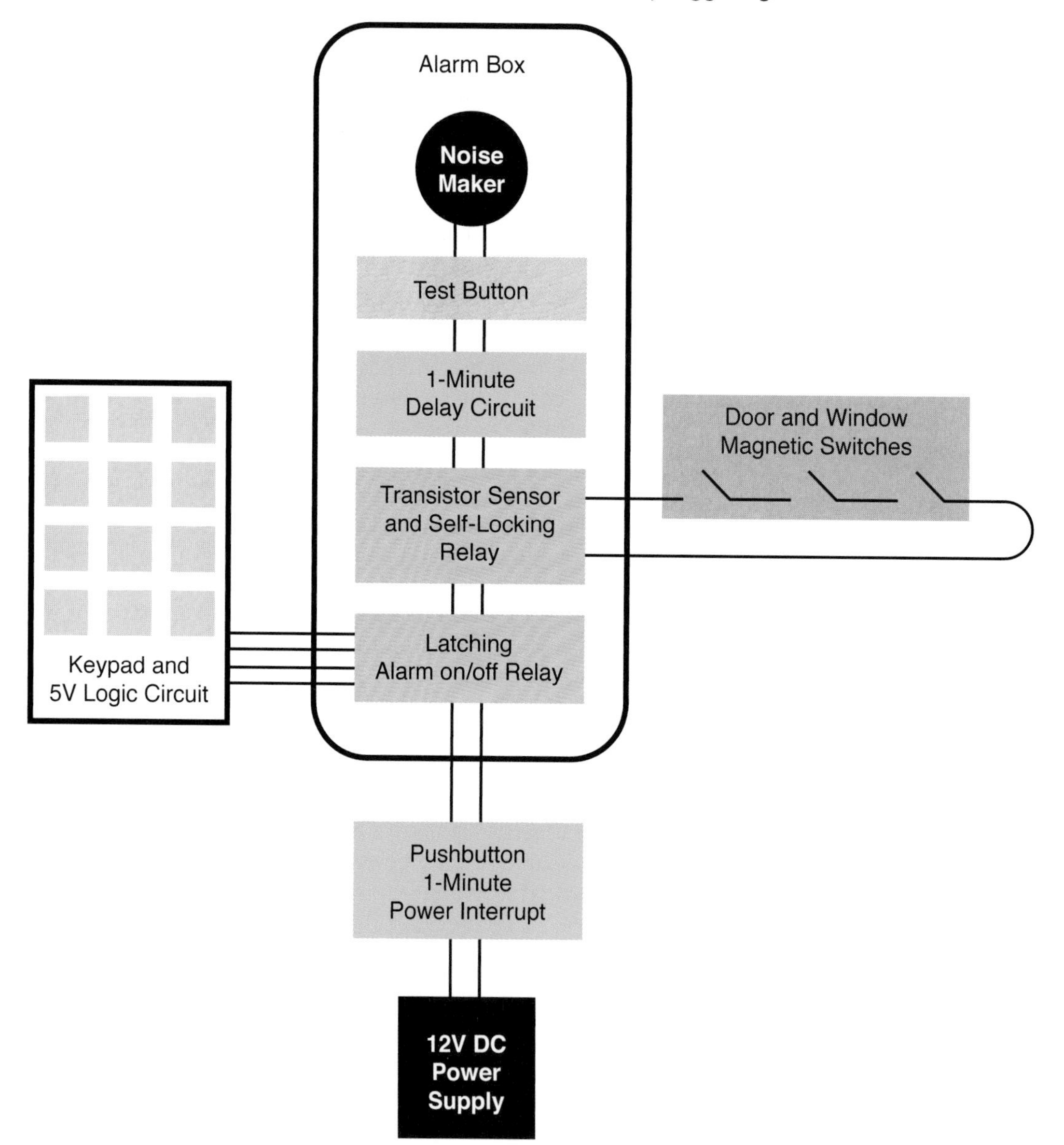

Figure 4-114. *This block diagram shows the relative placement of the old and new components. The pushbutton power interrupt (which allows you to leave the house before the alarm switches itself back on) goes between the power supply and everything else. The latching relay substitutes for the DPDT switch on the previous version of the alarm. The transistor and self-locking relay, connected with the door and window magnetic switches, remain unchanged. The new delay circuit is inserted between the self-locking relay and the noisemaker. The test button is wired with the latching relay in the same way that it was wired previously with the DPDT switch.*

Upgrade 3: Delay Before Deactivation

Typically, alarms include another delay feature. When you open a door on your way into the building and it triggers the alarm, you have 30 seconds to deactivate it, before it starts making a noise.

How can we implement this delay feature? If I try to use another 555 timer to generate a pulse to inhibit the noise, that won't work, because the output from either the transistor or the relay can continue indefinitely. The relay locks itself on, and the transistor can continue passing voltage for as long as someone leaves a door open. If either of these signals activates a timer in monostable mode, the pulse from the timer will never end, and it will suppress the alarm indefinitely.

I think what I have to do is use a resistor and a capacitor to create a delay. I'll power them through the existing relay, so that I can be sure that they'll receive the full voltage of the power supply, after beginning from zero. Gradually the capacitor will acquire voltage—but I can't connect this directly to the noisemaker, because the noisemaker will gradually get louder as the voltage increases.

I have to insert a device that will be triggered to give full voltage when the input rises past a certain point. To do this, I'll use a 555 timer that's wired in bistable mode. This kind of modification is generally known as a "kludge," because it's not elegant, uses too many components, and does not use them appropriately. What I really need is a comparator, but I don't have space to get into that topic. So, using the knowledge that you have so far, the schematic in Figure 4-115 shows how a delay can be added to the alarm—not elegantly, but reliably.

The only problem is that if you power up a 555 timer in bistable mode, there's a 50-50 chance that the timer starts itself with its output high or low. So I need to pull the voltage low on the reset pin (to start the timer with its output inhibited) and gradually let it become positive (to permit the output). At the same time, I want to start with the voltage high on the trigger pin and gradually lower it, until it falls below 1/3 of the power supply and triggers the output.

So there are two timing circuits. The one for the reset pin works faster than the one on the trigger pin, so that at the point when the timer is triggered, the reset won't stop it.

The schematic shows component values that will do this. The 10 μF capacitor starts low but is charged through the 10K resistor in a couple of seconds. The timer is now ready to be triggered. But the 68 μF capacitor starts high (being connected with the positive side of the power supply) and takes a full minute to be pulled down to 1/3 of supply voltage through the 1M resistor. At that point, its voltage is low enough to trigger the 555. The timer output goes high and supplies the noisemaker.

You should be able to insert this little delay module in your alarm box, between the output from the relay and the input to the noisemaker without too much trouble. And if you want to adjust the delay, just use a higher or lower value resistor than 1M.

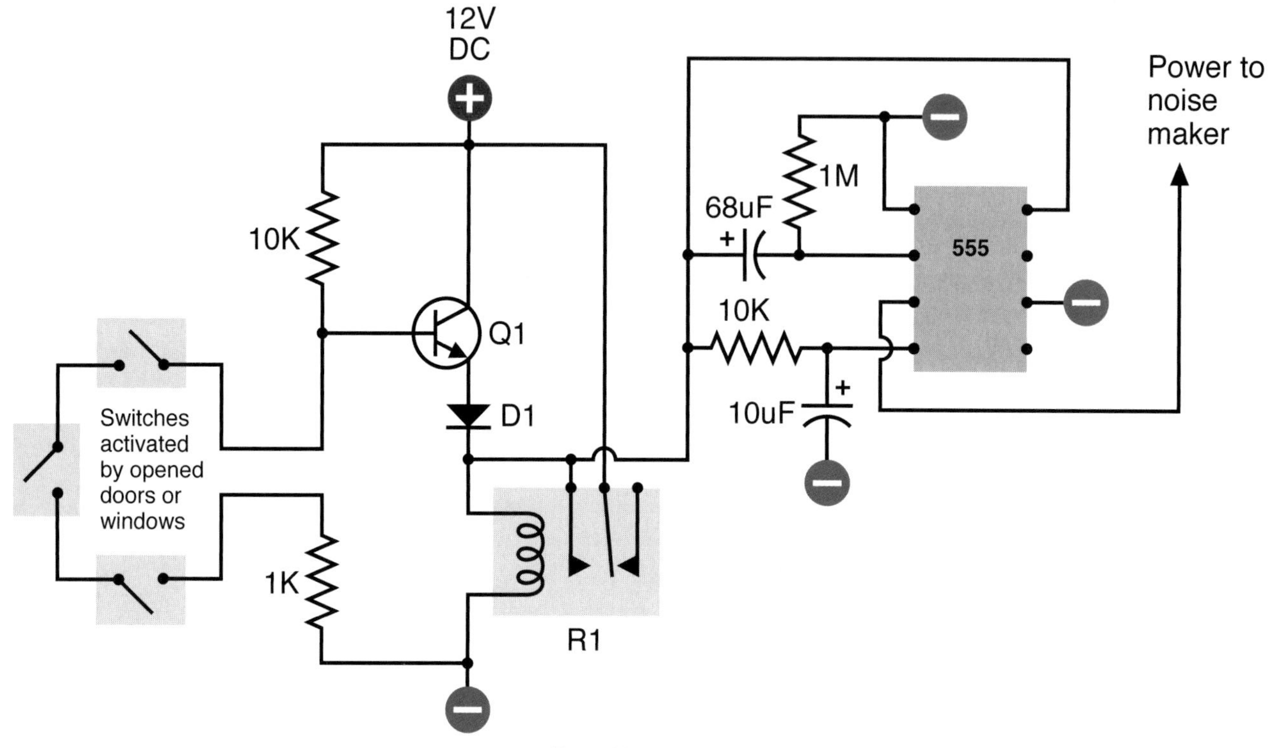

Figure 4-115. *This addition to the original alarm circuit imposes a one-minute delay before the alarm sounds. The 555 timer (wired in bistable mode) receives power through relay R1. The lower timing circuit initially applies negative voltage to the reset, ensuring that the 555 powers up with its output suppressed. This voltage quickly rises. Meanwhile the upper timing circuit applies a voltage to the trigger that gradually diminishes as the 68 μF capacitor equalizes its charge through the 1M resistor. When the voltage diminishes to 1/3 of the supply, the timer's output goes high and starts the noisemaker. If the power to the circuit is interrupted at any time before this, the relay relaxes, the capacitors gradually discharge, and the alarm does not sound.*

The Wrap-Up

If you add these three enhancements, your alarm will have all the features on my original wish list. Of course, if I were designing it from scratch, with all the information that has been added in this chapter of the book, it could be more elegant. But the modifications have not entailed making destructive changes to our original project, and all the design goals have been met.

What Next?

5

At this point, we can branch out in numerous directions. Here are some possibilities:

Audio electronics
: This is a field in itself, including hobby projects, such as simple amplifiers and "stomp boxes," to modify guitar sound.

Radio-frequency devices
: Anything that receives or transmits radio waves, from an ultra-simple AM radio to remote controllers.

Motors
: The field of robotics has encouraged the growth of many online sites selling stepper motors, gear motors, synchronous motors, servo motors, and more.

Programmable microcontrollers
: These are tiny computers on a single chip. You write a little program on your desktop computer, which will tell the chip to follow a series of procedures, such as receiving input from a sensor, waiting for a fixed period, and sending output to a motor. Then you download your program onto the chip, which stores it in nonvolatile memory. Popular controllers include the PICAXE, BASIC Stamp, Arduino, and many more. The cheapest ones retail for a mere $5 each.

Obviously, I don't have space to develop all of these topics fully, so what I'm going to do is introduce you to them by describing just one or two projects in each category. You can decide which interests you the most, and then proceed beyond this book by reading other guides that specialize in that interest.

I'm also going to make some suggestions about setting up a productive work area, reading relevant books, catalogs, and other printed sources, and generally proceeding further into hobby electronics.

IN THIS CHAPTER

Shopping List: Experiments 25 Through 36

Tools

You won't need any new tools for this section of the book.

Supplies and Components

As we have progressed to the point where you may want to pick and choose which projects you attempt, I will list the supplies and components at the beginning of each experiment.

Customizing Your Work Area

At this point, if you're getting hooked on the fun of creating hardware but haven't allocated a permanent corner to your new hobby, I have some suggestions. Having tried many different options over the years, my main piece of advice is this: don't build a workbench!

Many hobby electronics books want you to go shopping for 2×4s and plywood, as if a workbench has to be custom-fabricated to satisfy strict criteria about size and shape. I find this puzzling. To me, the exact size and shape of a bench is not very important. I think the most important issue is storage.

I want tools and parts to be easily accessible, whether they're tiny transistors or big spools of wire. I certainly don't want to go digging around on shelves that require me to get up and walk across the room.

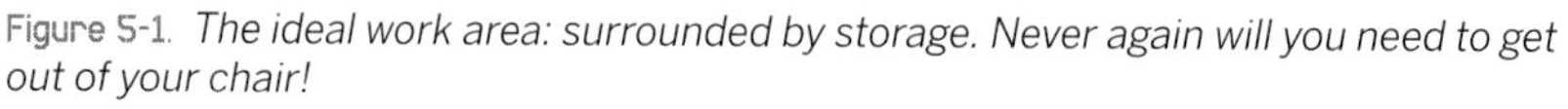

Figure 5-1. *The ideal work area: surrounded by storage. Never again will you need to get out of your chair!*

This leads me to two conclusions:

1. You need storage above the workbench.
2. You need storage below the workbench.

Many DIY workbench projects allow little or no storage underneath. Or, they suggest open shelves, which will be vulnerable to dust. My minimum configuration would be a pair of two-drawer file cabinets with a slab of 3/4-inch plywood or a Formica-clad kitchen countertop placed across them. File cabinets are ideal for storing all kinds of objects, not just files.

Of all the workbenches I've used, the one I liked best was an old-fashioned steel office desk—the kind of monster that dates back to the 1950s. They're difficult to move (because of their weight) and don't look beautiful, but you can buy them cheaply from used office furniture dealers, they're generous in size, they withstand abuse, and they last forever. The drawers are deep and usually slide in and out smoothly, like good file-cabinet drawers. Best of all, the desk has so much steel in it that you can use it to ground yourself before touching components that are sensitive to static electricity. If you use an anti-static wrist strap, you can simply attach it to a sheet-metal screw that you drive into one corner of the desk.

What will you put in the deep drawers of your desk or file cabinets? Some paperwork may be useful, perhaps including the following documents:

- Product data sheets
- Parts catalogs
- Sketches and plans that you draw yourself

The remaining capacity of each drawer can be filled with plastic storage boxes. The boxes can contain tools that you don't use so often (such as a heat gun or a high-capacity soldering iron), and larger-sized components (such as loudspeakers, AC adapters, project boxes, and circuit boards). You should look for storage boxes that measure around 11 inches long, 8 inches wide, and 5 inches deep, with straight sides. Boxes that you can buy at Wal-Mart will be cheaper, but they often have tapering sides (which are not space-efficient).

Figure 5-2. *Akro-Grid boxes contain grooves allowing them to be partitioned into numerous compartments for convenient parts storage.*

The boxes that I like best are Akro-Grids, made by Akro-Mils (see Figures 5-2 and 5-3). These are very rugged, straight-sided, with optional transparent snap-on lids. You can download the full Akro-Mills catalog from *http://www.akro-mils.com* and then search online for retail suppliers. You'll find that Akro-Mils also sells an incredible variety of parts bins, but I don't like open bins because their contents are vulnerable to dust and dirt.

Figure 5-3. *Lids are sold separately for Akro-Grid boxes to keep the contents dust-free. The height of the box in Figure 5-2 allows three to be stacked in a typical file-cabinet drawer. The box shown here allows two to be stacked.*

For medium-size components, such as potentiometers, power connectors, control knobs, and toggle switches, I like storage containers measuring about 11 inches long, 8 inches wide, and 2 inches deep , divided into four to six sections. You can buy these from Michaels (the craft store), but I prefer to shop online for the Plano brand, as they seem more durably constructed. The Plano products that are most suitable for medium-size electronic parts are classified as fishing-tackle boxes, and you'll see them at *http://www.planomolding.com/tackle/products.asp*.

Figure 5-4. *This Plano brand box is undivided, making it useful for storing spools of wire or medium-size tools. When stacked upright on its long edge, three will fit precisely in a file-cabinet drawer.*

For undivided, flat-format storage boxes, the Prolatch 23600-00 is ideally sized to fit a file-cabinet drawer, and the latches are sufficiently secure that you could stack a series of them on their long edges. See Figure 5-4.

Plano also sells some really nicely designed toolboxes, one of which you can place on your desktop. It will have small drawers for easy access to screwdrivers, pliers, and other basics. Because you need a work area that's only about three feet square for most electronics projects, surrendering some desk space to a toolbox is not a big sacrifice.

If you have a steel desk with relatively shallow drawers, one of them can be allocated for printed catalogs. Don't underrate the usefulness of hard copy, just because you can buy everything online. The Mouser catalog, for instance, has an index, which is better in some ways than their online search feature, and the catalog is divided into helpful categories. Many times I've found useful parts that I never knew existed, just by browsing, which is much quicker than flipping through PDF pages online, even with a broadband connection. Currently, Mouser is still quite generous about sending out their catalogs, which contain over 2,000 pages. McMaster-Carr will also send you a catalog, but only after you've ordered from them, and only once a year.

Now, the big question: how to store all those dinky little parts, such as resistors, capacitors, and chips? I've tried various solutions to this problem. The most obvious is to buy a case of small drawers, each of which is removable, so you can place it on your desk while you access its contents. But I don't like this system, for two reasons. First, for very small components, you need to subdivide the drawers, and the dividers are never secure. And second, the removability of the drawers creates the risk of accidentally emptying the contents on the floor. Maybe you're too careful to allow this to happen, but I'm not!

Figure 5-5. *Darice Mini-Storage boxes are ideal for components such as resistors, capacitors, and semiconductors. The boxes can be stacked stably or stored on shelves, with their ends labeled. The brand sticker is easily removed after being warmed with a heat gun.*

My personal preference is to use Darice Mini-Storage boxes, shown in Figure 5-5. You can find these at Michaels in small quantities, or buy them more economically in bulk online from suppliers such as *http://www.craftamerica.com*. The blue boxes are subdivided into five compartments that are exactly the right size and shape for resistors. The yellow boxes are subdivided into ten compartments, which are ideal for semiconductors. The purple boxes aren't divided at all, and the red boxes have a mix of divisions.

The dividers are molded into the boxes, so you don't have the annoyance associated with removable dividers that slip out of position, allowing components to mix together. The box lids fit tightly, so that even if you drop one of the boxes, it probably won't open. The lids have metal hinges, and a ridge around the edge that makes the boxes securely stackable.

I keep my little storage boxes on a set of shelves above the desk, with a gap of 3 inches between one shelf and the next, allowing two boxes to be stacked on each shelf. If I want to work with a particular subset of boxes, I shift them onto the desktop and stack them there.

Labeling

No matter which way you choose to store your parts, labeling them is essential. Any ink-jet printer will produce neat-looking labels, and if you use peelable (nonpermanent) labels, you'll be able to reorganize your parts in the future, as always seems to become necessary. I use color-coded labels for my collection of resistors, so that I can compare the stripes on a resistor with the code on the label, and see immediately if the resistor has been put in the wrong place. See Figure 5-6.

Figure 5-6. *To check that resistors are not placed in the wrong compartments, print the color code on each label.*

Even more important: you need to place a second (non-adhesive) label inside each compartment with the components. This label tells you the manufacturer's part number and the source, so that reordering is easy. I buy a lot of items from Mouser, and whenever I open their little plastic bags of parts, I snip out the section of the bag that has the identifying label on it, and slide it into the compartment of my parts box before I put the parts on top of it. This saves frustration later.

If I were *really* well organized, I would also keep a database on my computer listing everything that I buy, including the date, the source, the type of component, and the quantity. But I'm not that well organized.

On the Bench

Some items are so essential that they should sit on the bench or desktop on a permanent basis. These include your soldering iron(s), helping hands with magnifier, desk lamp, breadboard, power strip, and power supply. For a desk lamp, I prefer the type that has a daylight-spectrum fluorescent bulb, because it spreads a uniform light and helps me to distinguish colors of adjacent stripes on resistors.

The power supply is a matter of personal preference. If you're serious about electronics, you can buy a unit that delivers properly smoothed current at a variety of properly regulated and calibrated voltages. Your little wall-plug unit from RadioShack cannot do any of these things, and its output may vary depending on how heavily you load it. Still, as you've seen, it is sufficient for basic experiments, and when you're working with logic chips, you need to mount a 5-volt regulator on your breadboard anyway. Overall, I consider a good power supply optional.

Another optional item is an oscilloscope. This will show you, graphically, the electrical fluctuations inside your wires and components, and by applying probes at different points, you can track down errors in your circuit. It's a neat gadget to own, but it will cost a few hundred dollars, and for our tasks so far, it has not been necessary. If you plan to get seriously into audio circuits, an oscilloscope becomes far more important, because you'll want to see the shapes of the waveforms that you generate.

You can try to economize on an oscilloscope by buying a unit that plugs into the USB port of your computer and uses your computer monitor to display the signal. I have tried one of these, and was not entirely happy with the results. It worked, but did not seem accurate or reliable for low-frequency signals. Maybe I was unlucky; I decided not to try any other brands.

The surface of your desk or workbench will undoubtedly become scarred by random scuffs, cut marks, and drops of molten solder. I use a piece of half-inch plywood, two feet square, to protect my primary work area, and I clamp a miniature vise to its edge. To reduce the risk of static electricity when working with sensitive components, I cover the plywood with a square of conductive foam. This is not cheap, but offers advantages in addition to protecting chips from being zapped. Instead of scattering stray components, I can stick them into the foam, like plants growing in a garden. And like a garden, I can divide it into sections, with resistors on one side, capacitors on the other, and chips straight ahead.

Inevitably, during your work you'll create a mess. Little pieces of bent wire, stray screws, fasteners, and fragments of stripped insulation tend to accumulate, and can be a liability. If metal parts or fragments get into a project that you're building, they can cause short circuits. So you need a trash container. But it has to be easy to use. I use a full-size garbage pail, because it's so big that I can't miss it when I throw something toward it, and I can never forget that it's there.

Last, but most essential: a computer. Now that all data sheets are available online, and all components can be ordered online, and many sample circuits are placed online by hobbyists and educators, I don't think anyone can work efficiently without quick Internet access. To avoid wasting space, I suggest you use a small, cheap laptop that has a minimal footprint. A possible workbench configuration, using a steel desk, is shown in Figure 5-7.

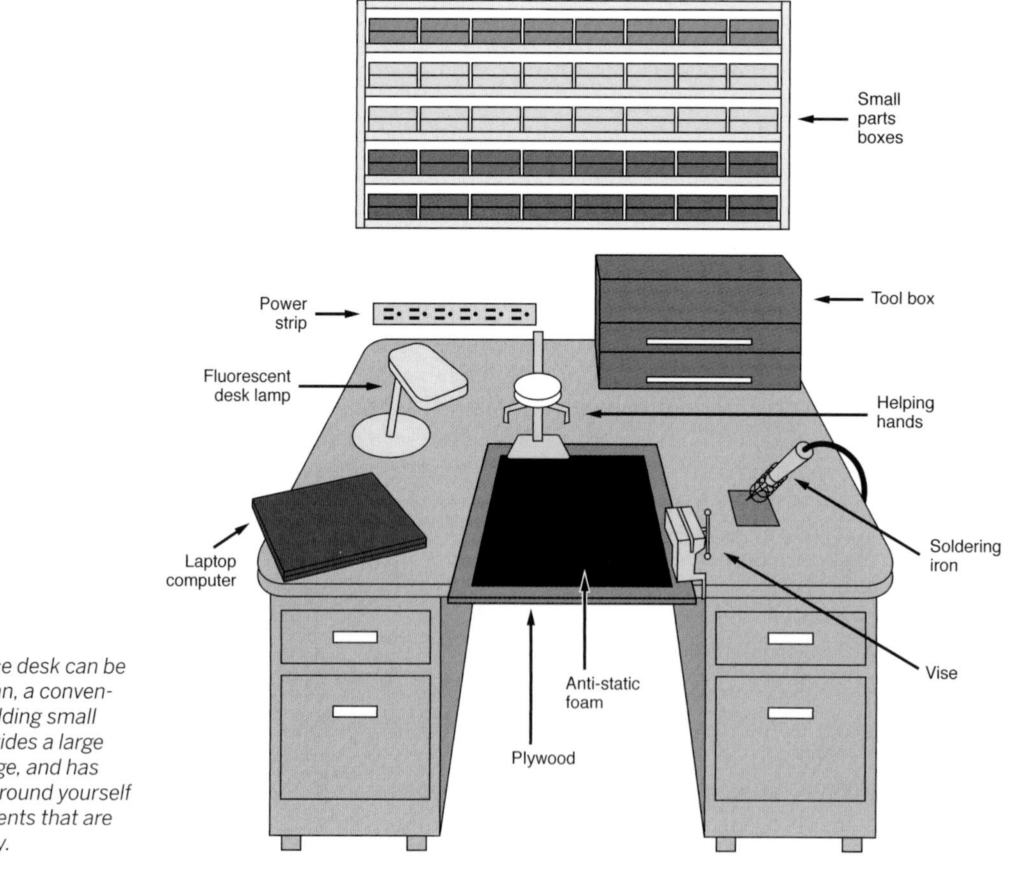

Figure 5-7. *An old steel office desk can be as good as, if not better than, a conventional workbench when building small electronics projects. It provides a large work area and ample storage, and has sufficient mass for you to ground yourself when dealing with components that are sensitive to static electricity.*

Reference Sources

Online

My favorite educational and reference site is Doctronics (*http://www.doctronics.co.uk*). I like the way they draw their schematics, and I like the way they include many illustrations of circuits on breadboards (which most sites don't bother to do). They also sell kits, if you're willing to pay and wait for shipping from the UK. Part of a page from the doctronics website is reproduced in Figure 5-8.

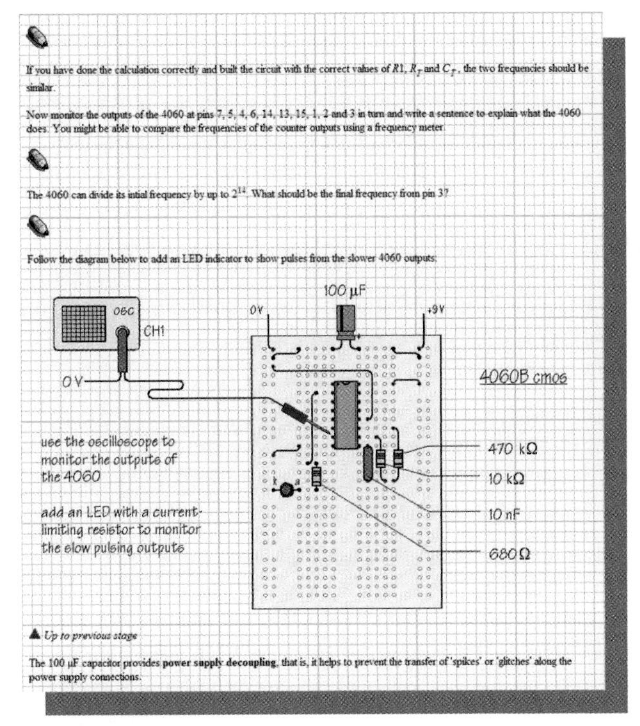

If you have done the calculation correctly and built the circuit with the correct values of $R1$, R_T and C_T, the two frequencies should be similar.

Now monitor the outputs of the 4060 at pins 7, 5, 4, 6, 14, 13, 15, 1, 2 and 3 in turn and write a sentence to explain what the 4060 does. You might be able to compare the frequencies of the counter outputs using a frequency meter.

The 4060 can divide its intial frequency by up to 2^{14}. What should be the final frequency from pin 3?

Follow the diagram below to add an LED indicator to show pulses from the slower 4060 outputs:

▲ *Up to previous stage*

The 100 µF capacitor provides **power supply decoupling**, that is, it helps to prevent the transfer of 'spikes' or 'glitches' along the power supply connections.

Figure 5-8. *A sample page from http://www.doctronics.co.uk shows their detailed instructional approach. This is a valuable free online resource.*

My next favorite hobby site is also British-based: the Electronics Club (*http://www.kpsec.freeuk.com*). It's not as comprehensive as Doctronics, but very friendly and easy to understand.

For a more theory-based approach, try *http://www.electronics-tutorials.ws*. This will go a little farther than the theory sections I've included here.

For an idiosyncratic selection of electronics topics, try Don Lancaster's Guru's Lair (*http://www.tinaja.com*). Lancaster wrote *The TTL Cookbook* more than 30 years ago, which opened up electronics to at least two generations of hobbyists and experimenters. He knows what he's talking about, and isn't afraid of getting into some fairly ambitious areas such as writing his own PostScript drivers and creating his own serial-port connections. You'll find a lot of ideas there.

Books

Yes, you do need books. As you're already reading this one, I won't recommend other beginners' guides. Instead, in keeping with the orientation of this chapter, I'll suggest some titles that will take you farther in various directions, and can be used for reference. I own all of these myself, and find them valuable:

Practical Electronics for Inventors, by Paul Scherz (McGraw-Hill, Second Edition, 2007)

This is a massive, comprehensive book, well worth the $40 cover price. Despite its title, you won't need to invent anything to find it useful. It's my primary reference source, covering a wide range of concepts, from the basic properties of resistors and capacitors all the way to some fairly high-end math. If you buy only one book (in addition to this one, of course!), this would be my recommendation.

Getting Started with Arduino, by Massimo Banzi (Make: Books, 2009)

If you enjoy the simplicity and convenience of the PICAXE programmable microcontroller that I describe later in this chapter, you'll find that the Arduino can do a lot more. *Getting Started* is the simplest introduction around, and will help to familiarize you with the Processing language used in Arduino (similar to the C language, not much like the version of BASIC used by the PICAXE).

Making Things Talk, by Tom Igoe (Make: Books, 2007)

This ambitious and comprehensive volume shows how to make the most of the Arduino's ability to communicate with its environment, even getting it to access sites on the Internet.

TTL Cookbook, by Don Lancaster (Howard W. Sams & Co, 1974)

The 1974 copyright date is not a misprint! You may be able to find some later editions, but whichever one you buy, it will be secondhand and possibly expensive, as this title now has collectible value. Lancaster wrote his guide before the 7400 series of chips was emulated on a pin-for-pin basis by CMOS versions, but it's still a good reference, because the concepts and part numbers haven't changed, and his writing is so accurate and concise.

CMOS Sourcebook, by Newton C. Braga (Sams Technical Publishing, 2001)

This book is entirely devoted to the 4000 series of CMOS chips, not the 74HC00 series that I've dealt with primarily here. The 4000 series is older and must be handled more carefully, because it's more vulnerable to static electricity than the generations that came later. Still, the chips remain widely available, and their great advantage is their willingness to tolerate a wide voltage range, typically from 5 to 15 volts. This means you can set up a 12-volt circuit that drives a 555 timer, and use output from the timer to go straight into CMOS chips (for example). The book is well organized in three sections: CMOS basics, functional diagrams (showing pinouts for all the main chips), and simple circuits showing how to make the chips perform basic functions.

The Encyclopedia of Electronic Circuits, by Rudolf F. Graf (Tab Books, 1985)

A totally miscellaneous collection of schematics, with minimal explanations. This is a useful book to have around if you have an idea and want to see how someone else approached the problem. Examples are often more valuable than general explanations, and this book is a massive compendium of examples. Many additional volumes in the series have been published, but start with this one, and you may find it has everything you need.

The Circuit Designer's Companion, by Tim Williams (Newnes, Second Edition, 2005)

Much useful information about making things work in practical applications, but the style is dry and fairly technical. May be useful if you're interested in moving your electronics projects into the real world.

The Art of Electronics, by Paul Horowitz and Winfield Hill (Cambridge University Press, Second Edition, 1989)

The fact that this book has been through 20 printings tells you two things: (1) Many people regard it as a fundamental resource; (2) Secondhand copies should be widely available, which is an important consideration, as the list price is over $100. It's written by two academics, and has a more technical approach than *Practical Electronics for Inventors*, but I find it useful when I'm looking for backup information.

Getting Started in Electronics, by Forrest M. Mims III (Master Publishing, Fourth Edition, 2007)

Although the original dates back to 1983, this is still a fun book to have. I think I have covered many of its topics here, but you may benefit by reading explanations and advice from a completely different source, and it goes a little farther than I have into some electrical theory, on an easy-to-understand basis, with cute drawings. Be warned that it's a brief book with eclectic coverage. Don't expect it to have all the answers.

Figure 5-9. *These books from MAKE provide guidance if you want to go beyond basic microcontrollers into the more exotic realms of the Arduino chip.*

Figure 5-10. *A sun-damaged copy of the Don Lancaster's classic TTL Cookbook, a 2,000-page catalog from the Mouser Electronics supply company, and two comprehensive reference books that can provide years of additional guidance in all areas of electronics.*

Experiment 25: Magnetism

This experiment should be a part of any school science class, but even if you remember doing it, I suggest that you do it again, because setting it up takes only a matter of moments, and it's going to be our entry point to a whole new topic: the relationship between electricity and magnetism. Quickly this will lead us into audio reproduction and radio, and I'll describe the fundamentals of self-inductance, which is the third and final basic property of passive components (resistance and capacitance being the other two). I left self-inductance until last because it's not very relevant to the experiments that you've done so far. But as soon as we start dealing with analog signals that fluctuate, it becomes essential.

You will need:

- Large screwdriver.
- 22-gauge wire (or thinner). Quantity: 6 feet.
- AA battery.

FUNDAMENTALS

A two-way relationship

Every electric motor that was ever made uses some aspect of the relationship between electricity and magnetism. It's absolutely fundamental in the world around us. Remember that electricity can create magnetism:

When electricity flows through a wire, it creates a magnetic force around the wire.

The principle works in reverse: magnetism can create electricity.

When a wire moves through a magnetic field, it creates a flow of electricity in the wire.

This second principle is used in power generation. A diesel engine, or a water-powered turbine, or a windmill, or some other source of energy either turns coils of wire through a powerful magnetic field, or turns magnets amid some massive coils of wire. Electricity is induced in the coils. In the next experiment, you'll see a dramatic mini-demo of this effect.

Procedure

This couldn't be simpler. Wind the wire around the shaft of the screwdriver, near its tip. The turns should be neat and tight and closely spaced, and you'll need to make 100 of them, within a distance of no more than 2 inches. To fit them into this space, you'll have to make turns on top of previous turns. If the final turn tends to unwind itself (which will happen if you're using stranded wire), secure it with a piece of tape. See Figure 5-11.

Now apply a battery, as shown in Figure 5-12. At first sight, this looks like a very bad idea, because you're going to short out your battery just as you did in Experiment 2. But by passing the current through a wire that's coiled instead of straight, we'll get some work out of it before the battery expires.

Put a small paper clip near the screwdriver blade, on a soft, smooth surface that will not present much friction. A tissue works well. Because many screwdrivers are already magnetic, you may find that the paper clip is naturally attracted to the tip of the blade. If this happens, move the clip just outside the range of attraction. Now apply the 1.5 volts to the circuit, and the clip should jump to the tip of the screwdriver. Congratulations: you just made an electromagnet.

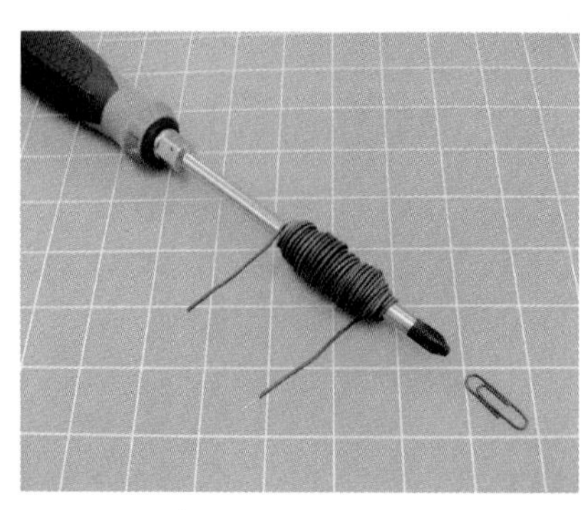

Figure 5-11. *Anyone who somehow missed this most basic childhood demo of electromagnetism should try it just for the fun of proving that a single AA battery can move a paper clip.*

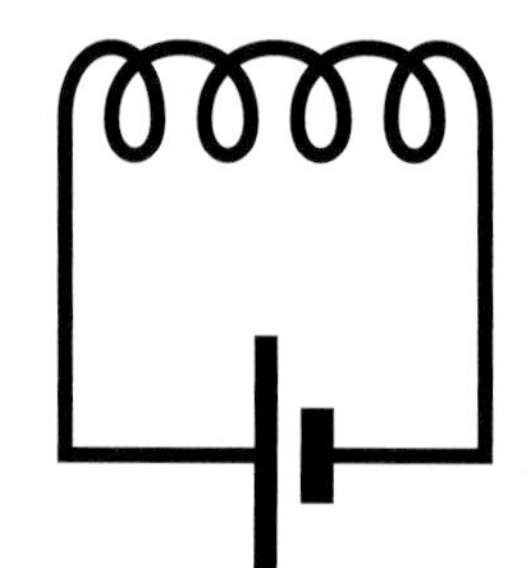

Figure 5-12. *A schematic can't get much simpler than this.*

THEORY

Inductance

When electricity flows through a wire, it creates a magnetic field around the wire. Because the electricity "induces" this effect, it is known as *inductance*. The effect is illustrated in Figure 5-13.

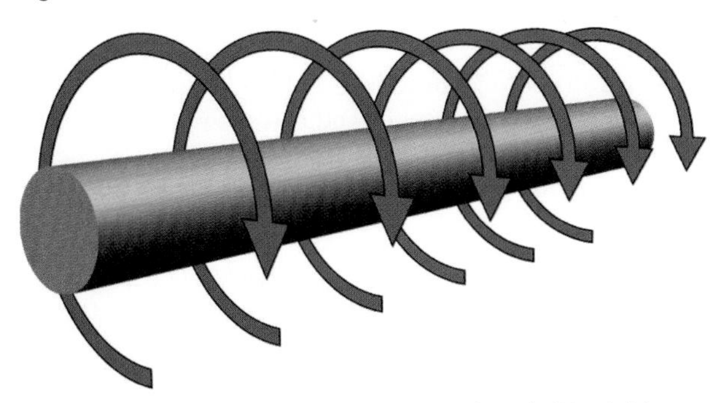

Figure 5-13. *When the flow of electricity is from left to right along this conductor, it induces a magnetic force shown by the green arrows.*

The field around a straight wire is very weak, but if we bend the wire into a circle, the magnetic force starts to accumulate, pointing through the center of the circle, as shown in Figure 5-14. If we add more circles, to form a coil, the force accumulates even more. And if we put a magnetic object (such as a screwdriver) in the center of the coil, the effectiveness increases further.

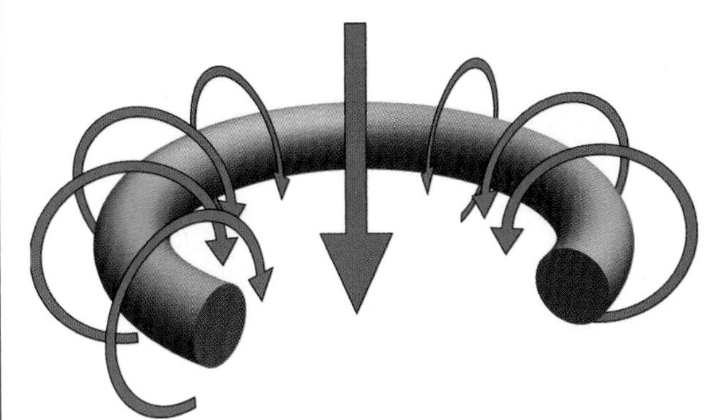

Figure 5-14. *When the conductor is bent to form a circle, the cumulative magnetic force acts through the center of the circle, as shown by the large arrow.*

Here's an approximated formula showing the relationship between the diameter of the coil, the width of the coil from end to end, the number of turns, and its inductance. The letter L is the symbol for inductance, even though the unit is the Henry, named after an American electrical pioneer named Joseph Henry:

L (in microHenrys) =
$[(D \times D) \times (N \times N)] / [(18 \times D) + (40 \times W)]$
(Approximately)

In this formula, D is the diameter of the coil, N is the number of turns, and W is the width of the coil from end to end. See Figure 5-15. Here are three simple conclusions from this formula:

- Inductance increases with the diameter of the coil.
- Inductance increases with the square of the number of turns. (In other words, three times as many turns create nine times the inductance.)
- If the number of turns remains the same, inductance is lower if you wind the coil so that it's slender and long, but is higher if you wind it so that it's fat and short.

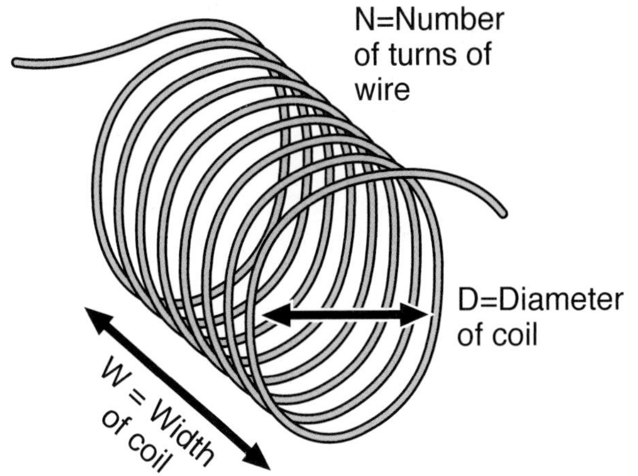

Figure 5-15. *The inductance of a coil increases with its diameter and with the square of its number of turns. If all other parameters remain the same, reducing the width (the distance from end to end) by packing the turns more tightly will increase the inductance.*

FUNDAMENTALS

Coil schematics and basics

Check the schematic symbols for coils in Figure 5-16. Note that if a coil has an iron core, this is shown with an extra couple of lines (sometimes only one line). If it has a ferrite core, the line is sometimes shown with dashes.

An iron core will add to the inductance of a coil, because it increases the magnetic effect.

A coil in isolation does not generally have any polarity. You can connect it either way around, but the magnetic force will be reversed accordingly (coils that interact with stuff—such as in transformers and solenoids—do have polarity).

Perhaps the most widespread application of coils is in transformers, where alternating current in one coil induces alternating current in another, often sharing the same iron core. If the primary (input) coil has half as many turns as the secondary (output) coil, the voltage will be doubled, at half the current—assuming hypothetically that the transformer is 100% efficient.

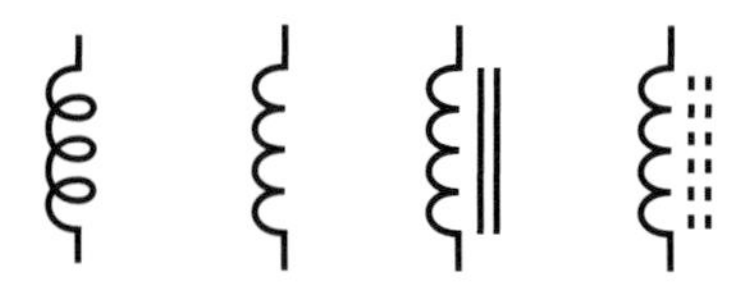

Figure 5-16. *Schematic diagrams represent coils. At far left is the older style. The third and fourth symbols indicate that the coil is wound around a solid or powdered magnetic core, respectively.*

BACKGROUND

Joseph Henry

Born in 1797, Joseph Henry was the first to develop and demonstrate powerful electromagnets. He also originated the concept of "self-inductance," meaning the "electrical inertia" that is a property of a coil of wire.

Henry started out as the son of a day laborer in Albany, New York. He worked in a general store before being apprenticed to a watchmaker, and was interested in becoming an actor. Friends persuaded him to enroll at the Albany Academy, where he turned out to have an aptitude for science. In 1826, he was appointed Professor of Mathematics and Natural Philosophy at the Academy, even though he was not a college graduate and described himself as being "principally self-educated." Michael Faraday was doing similar work in England, but Henry was unaware of it.

Henry was appointed to Princeton in 1832, where he received $1,000 per year and a free house. When Morse attempted to patent the telegraph, Henry testified that he was already aware of its concept, and indeed had rigged a system on similar principles to signal his wife, at home, when he was working in his laboratory at the Philosophical Hall.

Henry taught chemistry, astronomy, and architecture, in addition to physical science, and because science was not divided into strict specialties as it is now, he investigated phenomena such as phosphorescence, sound, capillary action, and ballistics. In 1846, he headed the newly founded Smithsonian Institution as its secretary.

Figure 5-17. *Joseph Henry was an American experimenter who pioneered the investigation of electromagnetism. This photograph is archived in Wikimedia Commons.*

Experiment 26: Tabletop Power Generation

If you have just three components, you can see magnetism generating electricity right in front of you, right now.

You will need:

- Cylindrical neodymium magnet, 3/4-inch diameter, axially magnetized. Quantity: 1. (Obtainable online at sites such as *http://www.kjmagnetics.com*.)
- Spool of hookup wire, 26-gauge, 100 feet. Quantity: 1.
- Spool of magnet wire, quarter-pound, 26-gauge, about 350 feet. Quantity: 1. (Search online for sources for "magnet wire.")
- Generic LED. Quantity: 1.
- 100 μF electrolytic capacitor. Quantity: 1.
- Signal diode, 1N4001 or similar. Quantity: 1.
- Jumper wires with alligator clips on the ends. Quantity: 2.

Procedure

You may be able to make this experiment work with the spool of hookup wire, depending on the size of the spool relative to the size of your magnet, but as the results are more likely to be better with the magnet wire, I'll assume that you're using that—initially, at least. The advantage of the magnet wire is that its very thin insulation allows the coils to be closely packed, increasing their inductance.

Figure 5-18. *An everyday 100-foot spool of hookup wire is capable of demonstrating the inductive power of a coil.*

First peek into the hollow center of the spool to see if the inner end of the wire has been left accessible, as is visible in Figures 5-18 and 5-19. If it hasn't, you have to unwind the wire onto any large-diameter cylindrical object, then rewind it back onto the spool, this time taking care to leave the inner end sticking out.

Scrape the transparent insulation off each end of the magnet wire with a utility knife or sandpaper, until bare copper is revealed. To check, attach your meter, set to measure ohms, to the free ends of the wire. If you make a good contact, you should measure a resistance of 30 ohms or less.

Place the spool on a nonmagnetic, nonconductive surface such as a wooden, plastic, or glass-topped table. Attach the LED between the ends of the wire using jumper wires. The polarity is not important. Now take a cylindrical neodymium magnet of the type shown in Figure 5-20 and push it quickly down into the hollow core, then pull it quickly back out. See Figure 5-21. You should see the LED blink, either on the down stroke or the up stroke.

Figure 5-19. *Magnet wire has thinner insulation than hookup wire, allowing the turns to be more densely packed, inducing a more powerful magnetic field.*

The same thing may or may not happen if you use 100 feet of 26-gauge hookup wire. Ideally, your cylindrical magnet should fit fairly closely in the hollow center of the spool. If there's a big air gap, this will greatly reduce the effect of the magnet. Note that if you use a weaker, old-fashioned iron magnet instead of a neodymium magnet, you may get no result at all.

Blood Blisters and Dead Media

Neodymium magnets can be hazardous. They're brittle and can shatter if they slam against a piece of magnetic metal (or another magnet). For this reason, many manufacturers advise you to wear eye protection.

Because a magnet pulls with increasing force as the distance between it and another object gets smaller, it closes the final gap very suddenly and powerfully. You can easily pinch your skin and get blood blisters.

If there's an object made of iron or steel anywhere near a neodymium magnet, the magnet will find it and grab it, with results that may be unpleasant, especially if the object has sharp edges and your hands are in the vicinity. When using a magnet, create a clear area on a nonmagnetic surface, and watch out for magnetic objects underneath the surface. My magnet sensed a steel screw embedded in the underside of a kitchen countertop, and slammed itself into contact with the countertop unexpectedly.

Be aware that magnets create magnets. When a magnetic field passes across an iron or steel object, the object picks up some magnetism of its own. Be careful not to magnetize your watch!

Don't use magnets anywhere near a computer, a disk drive, credit cards with magnetic stripes, cassettes of any type, and other media. Also keep magnets well away from TV screens and video monitors (especially cathode-ray tubes). Last but not least, powerful magnets can interfere with the normal operation of cardiac pacemakers.

Figure 5-20. *Three neodymium magnets, 1/4-, 1/2-, and 3/4-inch in diameter. I would have preferred to photograph them standing half-an-inch apart, but they refused to permit it.*

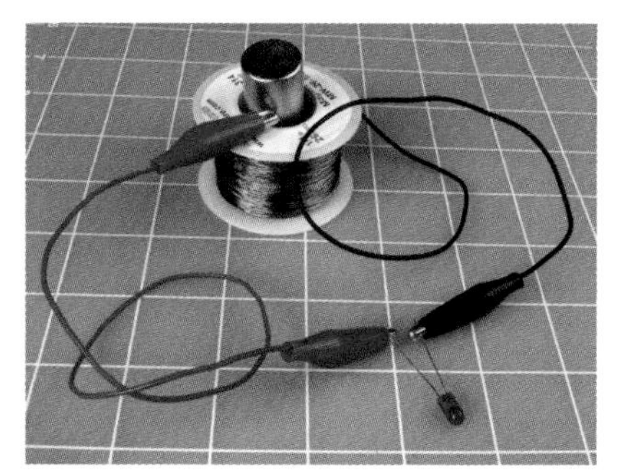

Figure 5-21. *By moving a magnet vigorously up and down through the center of a coil, you generate enough power to make the LED flash brightly.*

Here's another thing to try. Disconnect the LED and connect a 100 µF electrolytic capacitor in series with signal diode, as shown in Figure 5-23. Attach your meter, measuring volts, across the capacitor. If your meter has a manual setting for its range, set it to 20V DC. Make sure the positive (unmarked) side of the diode is attached to the negative (marked) side of the capacitor, so that positive voltage will pass through the capacitor and then through the diode.

Now move the magnet vigorously up and down in the coil. The meter should show that the capacitor is accumulating charge, up to about 10 volts. When you stop moving the magnet, the voltage reading will gradually decline, mostly because the capacitor discharges itself through the internal resistance of your meter.

This experiment is more important than it looks. Bear in mind that when you push the magnet into the coil, it induces current in one direction, and when you pull it back out again, it induces current in the opposite direction. You are actually generating alternating current.

The diode only allows current to flow one way through the circuit. It blocks the opposite flow, which is how the capacitor accumulates its charge. If you jump to the conclusion that diodes can be used to change alternating current to direct current, you're absolutely correct. We say that the diode is "rectifying" the AC power.

Experiment 25 showed that voltage can create a magnet. Experiment 26 has shown that a magnet can create voltage. We're now ready to apply these concepts to the detection and reproduction of sound.

Figure 5-22. *Because inductance increases with the diameter of a coil and with the square of the number of turns, your power output from moving a magnet through the coil can increase dramatically with scale. Those wishing to live "off the grid" may consider this steam-powered configuration, suitable for powering a three-bedroom home.*

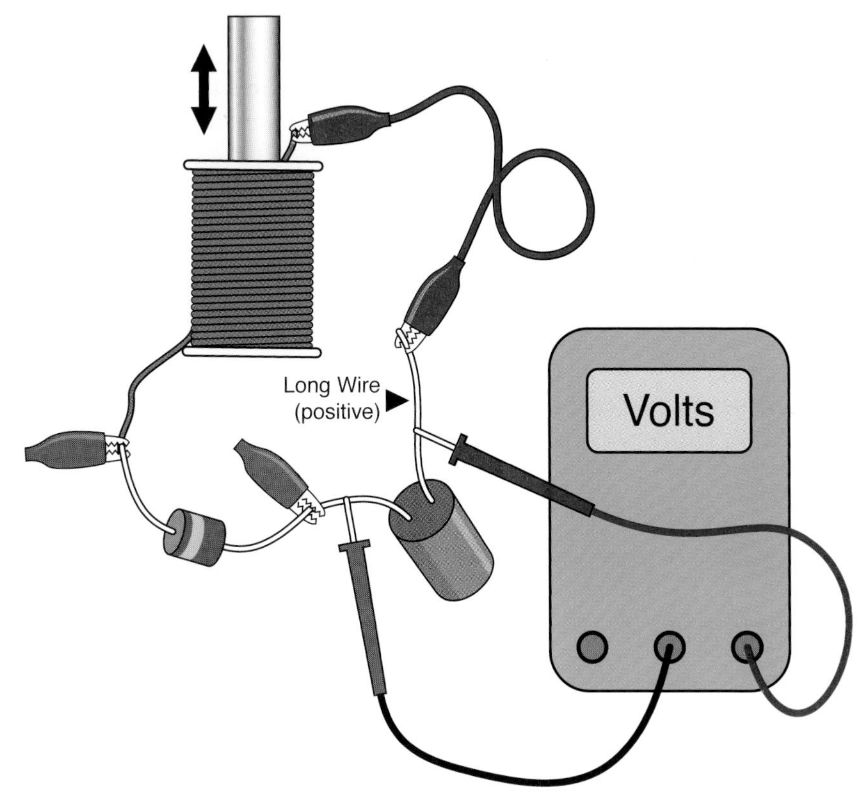

Figure 5-23. *Using a diode in series with a capacitor, you can charge the capacitor with the pulses of current that you generate by moving the magnet through the center of the coil. This demo illustrates the principle of rectifying alternating current.*

Figure 5-24. *A 2-inch loudspeaker can be instructively destroyed with a utility knife or X-Acto blade.*

Experiment 27: Loudspeaker Destruction

I'd like you to sacrifice a 2-inch loudspeaker, even though it means wasting the $5 or so that it probably costs. Actually, I don't consider this a waste, because if you want to learn how a component works, there's no substitute for actually seeing inside it. You might also have such a speaker already, part of a piece of cast-off personal electronics or toy you have in your basement.

You will need:

- Cheapest possible 2-inch loudspeaker. Quantity: 1. Figure 5-24 shows a typical example.

Procedure

Turn the loudspeaker face-up (as shown in Figure 5-25) and cut around the edge of its cone with a sharp utility knife or X-Acto blade. Then cut around the circular center and remove the ring of black paper that you've created. The result should look like Figure 5-26: you should see the flexible neck of the loudspeaker, which is usually made from a yellow weave. If you cut around its edge, you should be able to pull up the hidden paper cylinder, which has the copper coil of the loudspeaker wound around it. In Figure 5-27, I've turned it over so that it is easily visible. The two ends of this copper coil normally receive power through two terminals at the back of the speaker. When it sits in the groove visible between the inner magnet and the outer magnet, the coil reacts to voltage fluctuations by exerting an up-and-down force in reaction to the magnetic field. This vibrates the cone of the loudspeaker and creates sound waves.

Large loudspeakers in your stereo system work exactly the same way. They just have bigger magnets and coils that can handle more power (typically, as much as 100 watts).

Whenever I open up a small component like this, I'm impressed by the precision and delicacy of its parts, and the way it can be mass-produced for such a low cost. I imagine how astonished the pioneers of electrical theory (such as Faraday and Henry) would be, if they could see the components that we take for granted today. Henry spent days and weeks winding coils by hand to create electromagnets that were far less efficient than this cheap little loudspeaker.

Figure 5-25. *Loudspeaker ready for creative destruction.*

Figure 5-26. *The cone has been removed.*

Figure 5-27. *The neck of the cone has been pulled out. Note the coil of copper wire, which fits precisely in the groove between two magnets in the base of the speaker.*

BACKGROUND

Origins of loudspeakers

Loudspeakers utilize the fact that if you run a varying electrical current through a coil situated in a magnetic field, the coil will move in response to the current. This idea was introduced in 1874 by Ernst Siemens, a prolific German inventor. (He also built the world's first electrically powered elevator in 1880.) Today, Siemens AG is one of the largest electronics companies in the world.

When Alexander Graham Bell patented the telephone in 1876, he used Siemen's concept to create audible frequencies in the earpiece. From that point on, sound-reproduction devices gradually increased in quality and power, until Chester Rice and Edward Kellogg at General Electric published a paper in 1925 establishing basic principles that are still used in loudspeaker design today.

At *http://www.radiolaguy.com/Showcase/Gallery-HornSpkr.htm* you'll find photographs of very beautiful early loudspeakers, which used a horn design to maximize efficiency. As sound amplifiers became more powerful, speaker efficiency became less important compared with quality reproduction and low manufacturing costs. Today's loudspeakers convert only about 1% of electrical energy into acoustical energy.

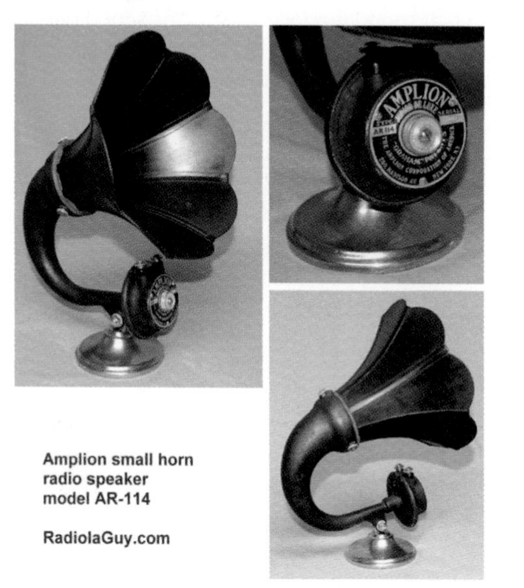

Figure 5-28. *This beautiful Amplion AR-114x illustrates the efforts of early designers to maximize efficiency in an era when the power of audio amplifiers was very limited. Photos by "Sonny, the RadiolaGuy." Many early speakers are illustrated at www.radiolaguy.com. Some are for sale.*

THEORY

Sound, electricity, and sound

Time now to establish a clear idea of how sound is transformed into electricity and back into sound again.

Suppose someone bangs a gong with a stick. The flat metal face of the gong vibrates in and out, creating sound waves. A sound wave is a peak of higher air pressure, followed by a trough of lower air pressure.

The wavelength of the sound is the distance (usually ranging from meters to millimeters) between one peak of pressure and the next peak.

The frequency of the sound is the number of waves per second, usually expressed as hertz.

Suppose we put a very sensitive little membrane of thin plastic in the path of the pressure waves. The plastic will flutter in response to the waves, like a leaf fluttering in the wind. Suppose we attach a tiny coil of very thin wire to the back of the membrane so that it moves with the membrane, and let's position a stationary magnet inside the coil of wire. This configuration is like a tiny, ultra-sensitive loudspeaker, except that instead of electricity producing sound, it is configured so that sound produces electricity. Sound pressure waves make the membrane move to and fro along the axis of the magnet, and the magnetic field creates a fluctuating voltage in the wire.

This is known as a *moving-coil* microphone. There are other ways to build a microphone, but this is the configuration that is easiest to understand. Of course, the voltage that it generates is very small, but we can amplify it using a transistor, or a series of transistors. Then we can feed the output through the coil around the neck of a loudspeaker, and the loudspeaker will recreate the pressure waves in the air. Figures 5-29 through 5-32 illustrate this sequence.

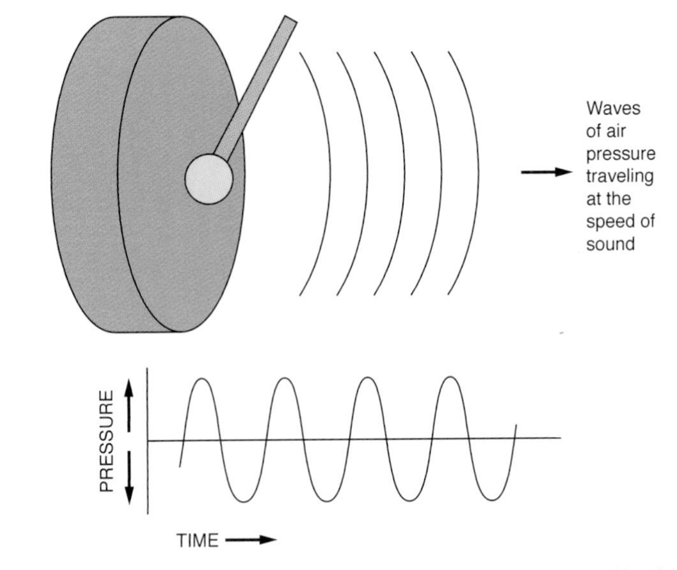

Figure 5-29. *Step 1 in the process of converting sound to electricity, and back again. When the hammer hits the gong, the face of the gong vibrates, creating pressure waves that travel through the air.*

THEORY

Sound, electricity, and sound (continued)

Somewhere along the way, we may want to record the sound and then replay it. But the principle remains the same. The hard part is designing the microphone, the amplifier, and the loudspeaker so that they reproduce the waveforms *accurately* at each step. It's a significant challenge, which is why accurate sound reproduction can be elusive.

Time now to think about what happens inside the wire when it generates a magnetic field. Obviously, some of the power in the wire is being transformed into magnetic force. But just what exactly is going on?

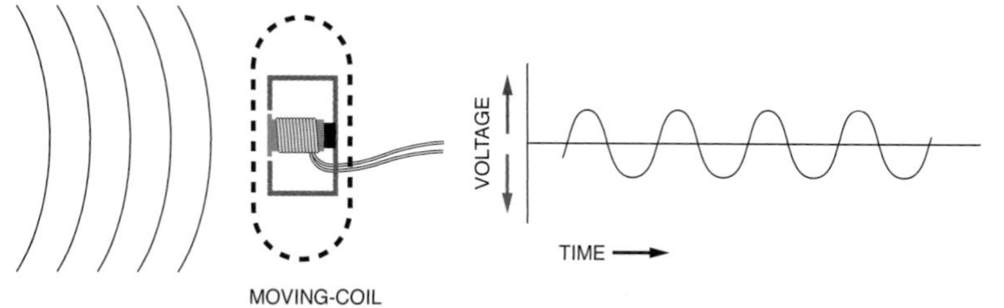

Figure 5-30. *Step 2: the pressure waves penetrate the perforated shell of a microphone and cause a diaphragm to vibrate in sympathy. The diaphragm has a coil attached to it. When the coil vibrates to and fro, a magnet at its center induces alternating current.*

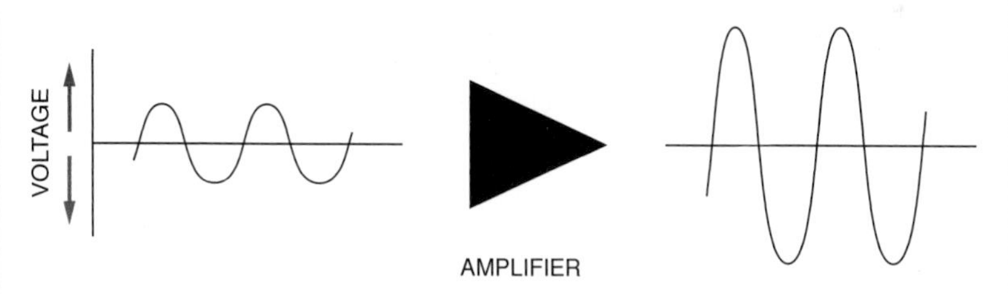

Figure 5-31. *Step 3: the tiny signals from the microphone pass through an amplifier, which enlarges their amplitude while retaining their frequency and the shape of their waveform.*

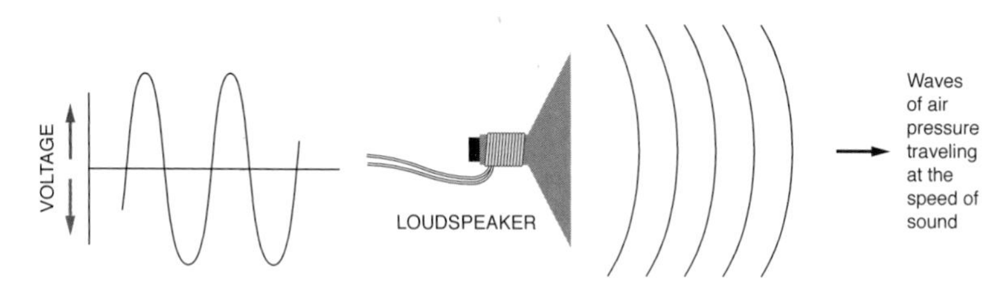

Figure 5-32. *Step 4: the amplified electrical signal is passed through a coil around the neck of a loudspeaker cone. The magnetic field induced by the current causes the cone to vibrate, reproducing the original sound.*

Experiment 28: Making a Coil React

A capacitor will absorb some DC current until it is fully charged, at which point it blocks the flow. There's another phenomenon that I haven't mentioned so far, which is the exact opposite of capacitance. It's known as *self-inductance*, and you find it in any coil of wire. Initially it blocks DC current (it reacts against it), but then its opposition gradually disappears. Here are a few definitions:

Resistance
: Constrains current flow and drops voltage.

Capacitance
: Allows current to flow initially and then blocks it. This behavior is properly known as capacitive reactance.

Self-Inductance
: Blocks the flow of current initially and then allows it. This is also often referred to as *inductive reactance*. In fact, you may find the term "reactance" used as if it means the same thing, but since self-inductance is the correct term, I'll be using it here.

In this experiment, you'll see self-inductance in action.

You will need:

- LEDs, low-current type. Quantity: 2.
- Spool of hookup wire, 26-gauge, 100 feet. Quantity: 1.
- Resistor, 220Ω, rated 1/4 watt or higher. Quantity: 1.
- Capacitor, electrolytic, 2,000 µF or larger. Quantity: 1.
- SPST tactile switch. Quantity: 1.

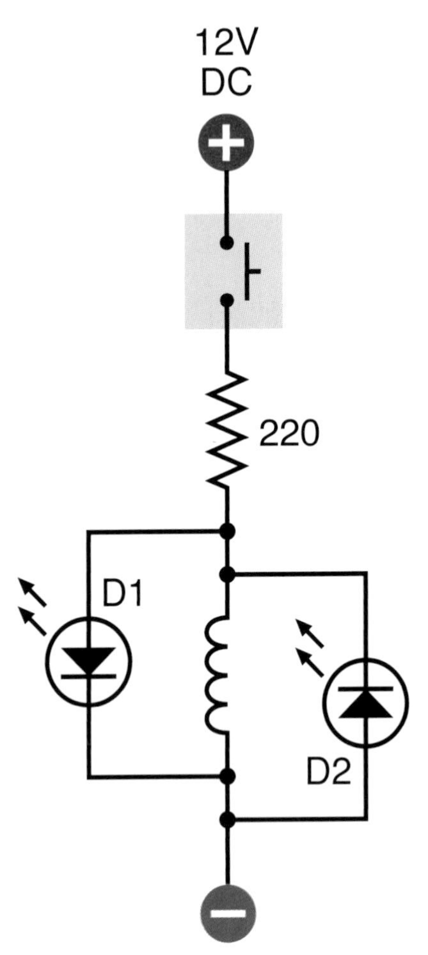

Figure 5-33. *In this demonstration of self-inductance, D1 and D2 are light-emitting diodes. When the switch is closed, D1 flashes briefly because the coil obstructs the initial flow of electricity. When the switch is opened, D2 flashes as the collapsing magnetic field induced by the coil releases another short burst of current.*

Procedure

Take a look at the schematic in Figure 5-33. At first it may not make much sense. The curly symbol is a coil of wire—nothing more than that. So apparently the voltage will pass through the 220Ω resistor, and then through the coil, ignoring the two LEDs, because the coil obviously has a much lower resistance than either of them (and one of them is upside-down anyway).

Is that what will happen? Let's find out. The coil can be a spool of 100 feet of 26-gauge (or smaller) hookup wire, although the spool of magnet wire listed in Experiment 25 will work better, if you have that. Once again, you will need access to both ends of the wire, and if the inner end is inaccessible, you'll need to rewind the coil, leaving the end sticking out.

Now that you have a coil, you can hook it up on your breadboard as shown in Figure 5-34, where the green circle is a tactile switch and the two circular red objects are LEDs. Make sure that you use low-current LEDs (otherwise, you may not see anything) and make sure that one of them is negative-side-up, positive-side-down and the other is positive-side-up, negative-side-down. Also, the 220Ω resistor should be rated at 1/4 watt or higher, if possible (see the following caution).

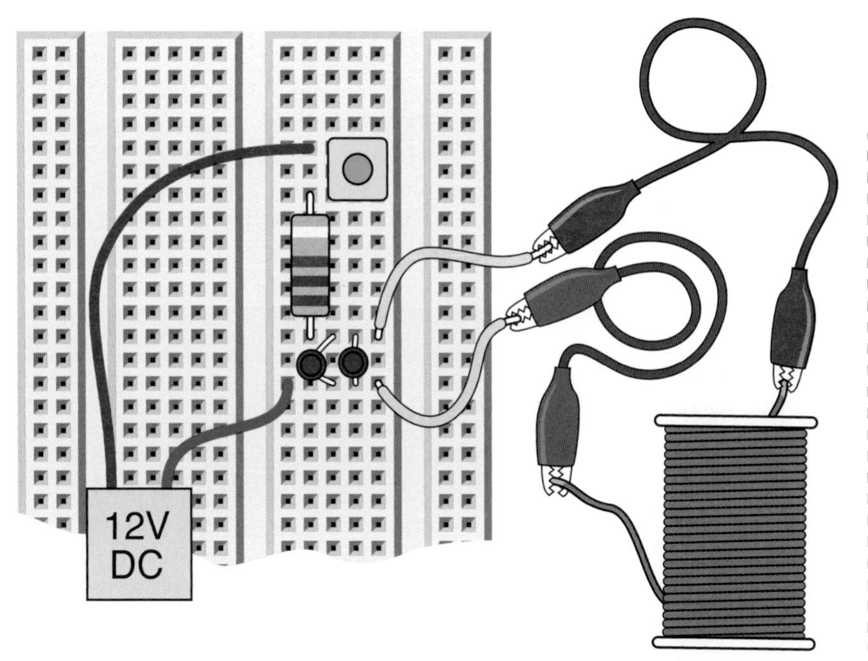

Figure 5-34. *The breadboarded version of the schematic in Figure 5-33 shows a simple way to set it up for a quick demo. The green button is a tactile switch. The two red LEDs should be placed so that the polarity of one is opposite to the polarity of the other.*

Hot Resistors

You'll be passing about 50mA through the 220Ω resistor, while the current is flowing. At 12 volts, this works out at 0.6 watts. If you use a 1/8-watt resistor, you will be overloading it, and it will get quite hot and may burn out. If you use a 1/4-watt resistor, it will still get hot, but is unlikely to burn out, as long as you don't press the button for more than a second or two.

Don't run the circuit without the coil of wire; you'll be trying to pass more than 50mA through the LEDs.

When you press the button, one LED should flash briefly. When you release the button, the other LED should flash.

What's happening here? The coil possesses self-inductance, which means that it reacts against any sudden change in the flow of electricity. First it fights it, and during that brief moment, it blocks most of the current. Consequently, the current looks for an alternative path and flows through D1, the lefthand LED in the schematic. (D2 doesn't respond, because it can pass current only in the opposite direction.)

Meanwhile, the voltage pressure overcomes the coil's self-inductance. When the self-inductance disappears, the resistance of the coil is no more than 10 ohms—so now the electricity flows mostly through the coil, and because the LED receives so little, it goes dark.

When you disconnect the power, the coil reacts again. It fights *any* sudden changes. After the flow of electricity stops, the coil stubbornly sustains it for a moment, because as the magnetic field collapses, it is turned back into electricity. This residual flow of current depletes itself through D2, the LED on the right.

In other words, the coil stores some energy in its magnetic field. This is similar to the way a capacitor stores energy between two metal plates, except that the coil blocks the current initially and then lets it build up, whereas the capacitor sucks up current initially, and then blocks it.

The more turns of wire you have in your coil, the more self-inductance the coil will have, causing your LEDs to flash more brightly.

Here's one last variation on this experiment to test your understanding of electrical fundamentals. Remove the 220Ω resistor, and substitute a 1K resistor (to protect your LED from sustained current). Remove the coil, and substitute a very large capacitor—ideally, about 4,700 µF. (Be careful to get its polarity the right way around.) What will you see when you press the button? Note that you will have to hold it down for a couple of seconds to get a result. And what will you see when you release the button? Remember: the behavior of capacitance is opposite to the behavior of self-inductance.

THEORY

Alternating current concepts

Here's a simple thought experiment. Suppose you set up a 555 timer to send a stream of pulses through a coil. This is a primitive form of alternating current.

We might imagine that the self-inductance of the coil will interfere with the stream of pulses, depending how long each pulse is, and how much inductance the coil has. If the pulses are too short, the self-inductance of the coil will tend to block them. Maybe if we can time the pulses exactly right, they'll synchronize with the time constant of the coil. In this way, we can "tune" a coil to allow a "frequency" to pass through it.

What happens if we substitute a capacitor? If the pulses are too long, compared with the time constant of the capacitor, it will tend to block them, because it will have enough time to become fully charged. But if the pulses are shorter, the capacitor can charge and discharge in rhythm with the pulses, and will seem to allow them through.

I don't have space in this book to get deeply into alternating current. It's a vast and complicated field where electricity behaves in strange and wonderful ways, and the mathematics that describe it can become quite challenging, involving differential equations and imaginary numbers. However, we can easily demonstrate the audio filtering effects of a loudspeaker and a coil.

Experiment 29: Filtering Frequencies

In this experiment, you'll see how self-inductance and capacitance can be used to filter audio frequencies. You're going to build a crossover network: a simple circuit that sends low frequencies to one place and high frequencies to another.

You will need:

- Loudspeaker, 8Ω, 5 inches in diameter. Quantity: 1. Figure 5-35 shows a typical example.
- Audio amplifier, STMicroelectronics TEA2025B or similar. Quantity: 1. See Figure 5-36.

Figure 5-35. *To hear the effects of audio filters using coils and capacitors, you'll need a loudspeaker capable of reproducing lower frequencies. This 5-inch model is the minimum required.*

Figure 5-36. *This single chip contains a stereo amplifier capable of delivering a total of 5 watts into an 8Ω speaker when the two channels are combined.*

Figure 5-37. *A nonpolarized electrolytic capacitor, also known as a bipolar capacitor, looks just like an electrolytic capacitor, except that it will have "NP" or "BP" printed on it.*

- Nonpolarized electrolytic capacitors (also known as bipolar). 100 µF, Quantity 2, and 10 µF, Quantity 1. A sample is shown in Figure 5-37. They should have "NP" or "BP" printed on them to indicate "nonpolarized" or "bipolar." (Because you'll be working with audio signals that alternate between positive and negative, you can't use the usual polarized electrolytic capacitors. If you want to avoid the trouble and expense of ordering nonpolarized capacitors, you can substitute two regular electrolytics in series, facing in opposite directions, with their negative sides joined in the middle. Just remember that when you put capacitors in series, their total capacitance is half that of each individual component. Therefore, you would need two 220 µF electrolytics in series to create 110 µF of capacitance. See Figure 5-38.)
- Regular electrolytic capacitors. 100uF. Quantity: 3.
- Coil, for crossover network. Quantity: 1. You can search a source such as eBay for keywords "crossover" and "coil," but if you can't find one at a reasonable price, you can make do with a spool of 100 feet of 20-gauge hookup wire.
- Plastic shoebox. Quantity: 1.

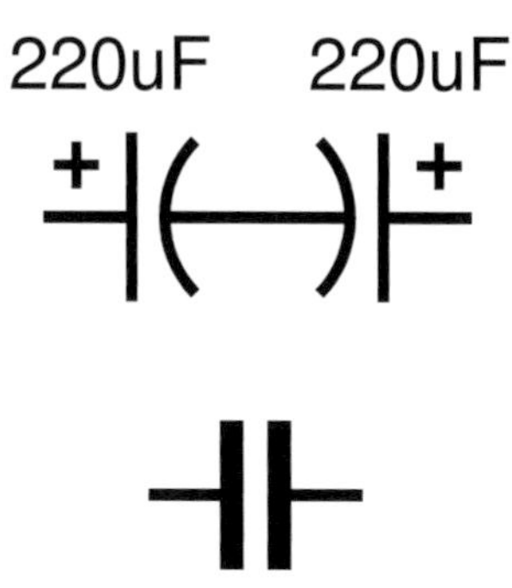

Figure 5-38. *You can make a nonpolarized electrolytic capacitor by putting two regular electrolytics in series. (In fact, that's what you'd find if you opened a real nonpolarized capacitor.) The symbol at the bottom is roughly equivalent to the pair of symbols at the top; bear in mind that two capacitors in series have a total capacitance that is half that of each of them.*

Figure 5-39. *A resonant enclosure is necessary if you want to hear some bass (lower frequencies) from your speaker. A cheap plastic shoebox is sufficient for demo purposes.*

Figure 5-40. *Drill some half-inch holes in the bottom of the box, then bolt the speaker in place, running a wire out through a hole in one end. Snap on the lid, and you're ready for not-quite-high-fidelity audio.*

Procedure

The purpose of the audio amplifier chip is to provide enough power to get a decent amount of sound out of your loudspeaker. The purpose of using a 5-inch speaker is to enable you to hear lower-frequency sounds than the baby speakers that we have used previously. Bass notes have long wavelengths that small speakers are not able to generate effectively.

Maybe you remember from building the intrusion alarm that a speaker makes much more noise if you prevent the sound waves from the back of the cone from cancelling the sound waves from the front of the cone. The obvious way to achieve this is by enclosing the speaker in a box. I suggest a plastic box, because they're cheap, and we don't care too much about sound quality as long as we can hear at least some of the low frequencies. Figure 5-39 shows the speaker bolted into the bottom of a plastic box, and Figure 5-40 shows the box turned upside-down after snapping its lid into place.

Normally, a speaker should be mounted in a cabinet of heavy, thick material that has a very low resonant frequency—below the limits of human hearing. To minimize the resonance of the shoebox, you can put some soft, heavy fabric inside it before you snap the lid on. A hand towel or some socks should be sufficient to absorb some of the vibration.

Adding an Amplifier

Back in the 1950s, you needed vacuum tubes, transformers, and other power-hungry heavyweight components to build an audio amplifier. Today, you can buy a chip for about $1 that will do the job, if you add a few capacitors around it, and a volume control. The TEA2025B that I'm recommending is intended for use in cheap portable cassette players and CD players, and can work in stereo or mono mode, from a power supply ranging from 3 to 9 volts. With 9 volts and the two sides of the chip bridged together to drive one 8Ω speaker, it can generate 5 watts of audio power. That doesn't sound much compared with a typical home theater system rated at 100 watts per channel, but because loudness is a logarithmic scale, 5 watts will be quite enough to irritate any family members in the same room—and possibly even in other rooms.

If you can't find the TEA2025B chip, you can use any alternative listed as an audio amplifier. Try to find one that is designed to drive an 8Ω speaker with up to 5 watts in mono mode. Check the manufacturer's data sheet to see where you attach capacitors around it. Note carefully whether some of the capacitors have no polarity marked, even though they have fairly high values, such as 100 µF. These capacitors must function regardless of which way the alternating current is flowing, and I've marked them "NP" in my schematic in Figure 5-41, meaning "nonpolarized." (You may find them identified as "bipolar" or "BP" in parts catalogs.) As noted in the shopping list, you can put two 220 µF capacitors in series, negative-to-negative, to get the same effect as a single 100 µF nonpolarized capacitor.

For this project, it's essential to include the regular 100 µF electrolytic smoothing capacitor across the power supply. Otherwise, the amplifier will pick up and—yes, amplify—small voltage spikes in the circuit.

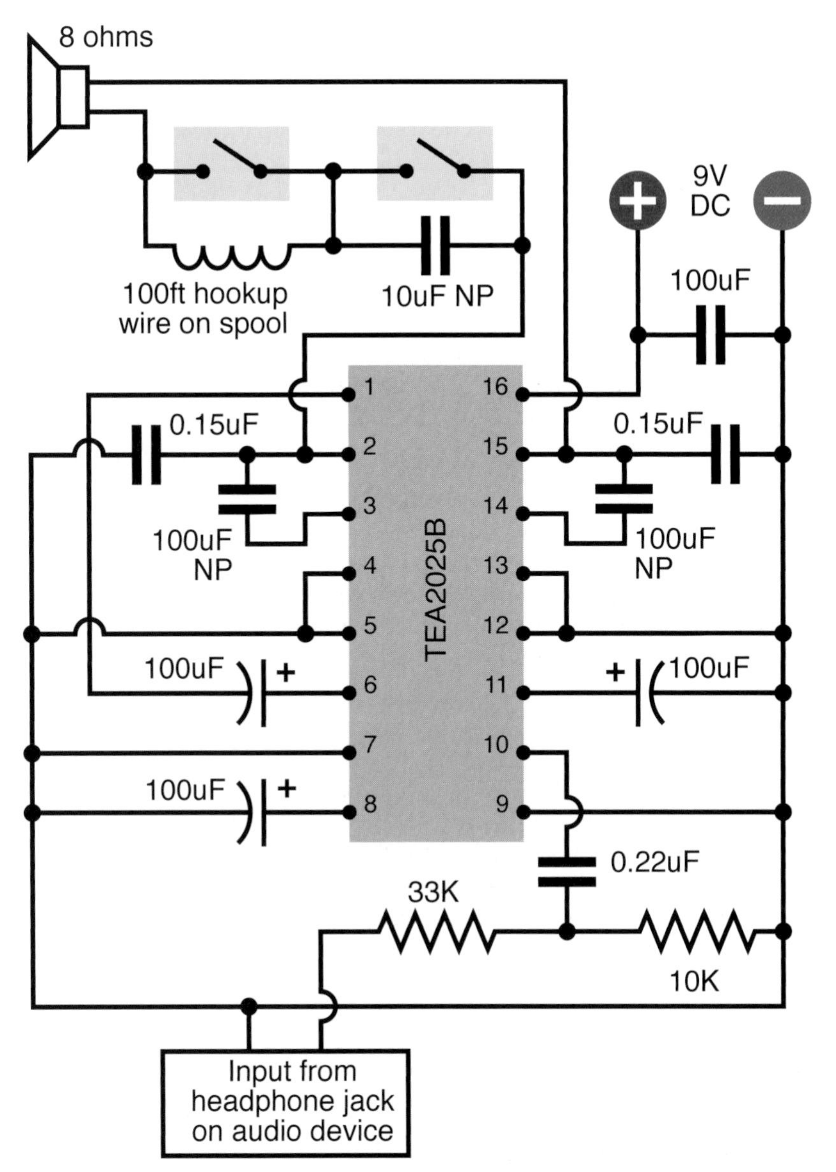

Figure 5-41. *The audio amplifier chip should be wired with capacitors around it as shown, "NP" denoting the ones that are not polarized. The acronym "BP," meaning bipolar, is also often used to mean the same thing. The output from pins 2 and 15 of the chip can be passed through a coil or a 10 μF capacitor to demonstrate audio filtering.*

The input shown in the schematic can receive a signal from a typical media player, such as a portable MP3 player, CD player, or cassette player. To connect its headphone jack to the breadboard, you can use an adapter that converts it to a pair of RCA-type audio jacks, and then stick a wire into one of them as shown in Figure 5-42. The wire will connect to the 33K resistor on the breadboard circuit. The chromed neck of the RCA jack (which is sometimes gold-plated, or at least gold-colored) *must* be connected with the negative side of your power supply on the breadboard; otherwise, you won't hear anything. You can ignore the second output on the adapter, because we're working in mono, here, not stereo.

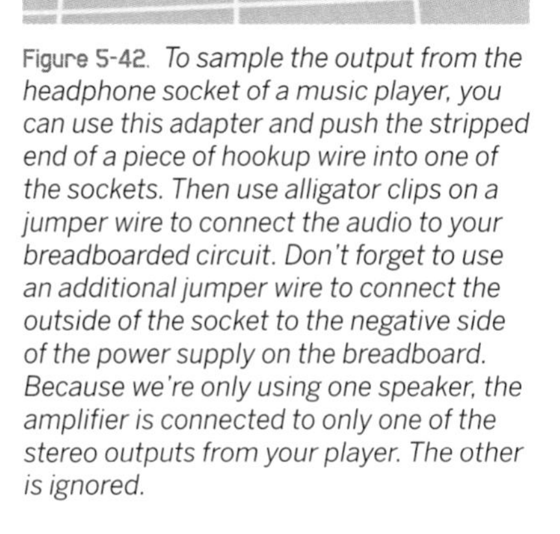

Figure 5-42. *To sample the output from the headphone socket of a music player, you can use this adapter and push the stripped end of a piece of hookup wire into one of the sockets. Then use alligator clips on a jumper wire to connect the audio to your breadboarded circuit. Don't forget to use an additional jumper wire to connect the outside of the socket to the negative side of the power supply on the breadboard. Because we're only using one speaker, the amplifier is connected to only one of the stereo outputs from your player. The other is ignored.*

The 33K resistor is necessary to protect the amplifier from being overdriven. If you don't get enough volume using your music player, decrease the 33K value. If the music is too loud and distorted, increase the value. You can also try omitting or increasing the 10K resistor next to it, which is included in an effort to reduce background hum noise.

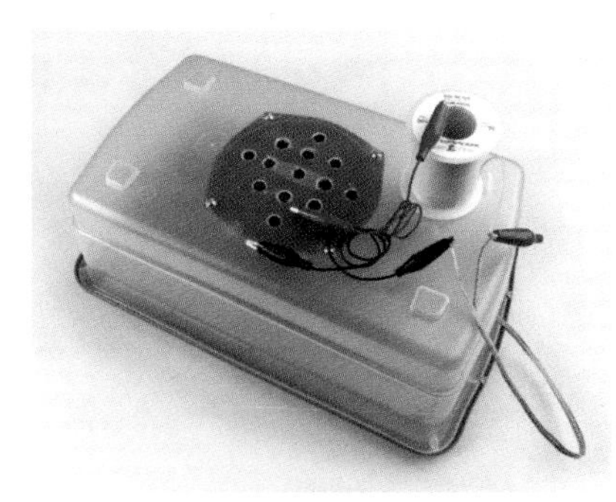

Figure 5-43. *The red and black alligator clips, lying on top of the shoebox, should connect with the output from your amplifier chip. The red jumper wire passes the signal through a coil of hookup wire on its way to the speaker. Note the change in sound when you short out the coil.*

I've shown two switches at the top of the schematic: one to bypass a coil, the other to bypass a capacitor. You can use alligator clips instead, as long as you can easily compare the sound when each of the components is inserted into the circuit.

Figure 5-43 shows a coil consisting of a spool of hookup wire being used. The red and black alligator clips resting loose on top of the shoebox will go to the output from the chip (on pins 2 and 15). There is no polarity; it doesn't matter which clip goes to which pin.

Begin by turning the volume control on your music source *all the way down* before you apply power. Don't be surprised if you hear humming or crackling noises when you activate the amplifier; it will pick up any stray voltages, because in this simple experiment, I haven't suggested that you should shield the input, and the amplifier circuit can pick up noise, as its wires can act like antennas.

Note that you may also get additional unwanted sound if you use the amplifier on a conductive desktop surface. Remove any aluminum foil or conductive foam for this project.

Make sure that your player is playing music, and slowly turn up its volume control until you hear it. If you don't hear anything, you'll have to check for circuit errors.

Now comes the interesting part. Insert the 100-foot spool of hookup wire between one output from the amplifier, and one input of the speaker (it doesn't matter which one), or if you used switches, open the switch that bypasses the coil. You should find that the music loses all its high-end response. By comparison, if you disconnect the coil and substitute a 10 µF capacitor, you should find that the music sounds "tinny," meaning that it loses all its low range, leaving only the high frequencies.

You've just tested two very simple filters. Here's what they are doing:

- The coil is a low-pass filter. It passes low frequencies but blocks high frequencies, because brief audio cycles don't have time to overcome the coil's self-inductance. A bigger coil eliminates a wider range of frequencies.
- The capacitor is a high-pass filter. It passes high frequencies and blocks low frequencies because longer audio cycles can fill the capacitance, at which point the capacitor stops passing current. A smaller capacitor eliminates a wider ranger of frequencies.

You can go a lot farther into filter design, using complex combinations of coils and capacitors to block frequencies at any point in the audible spectrum. Search online for audio filter schematics—you'll find hundreds of them.

Crossover Networks

In a traditional audio system, each speaker cabinet contains two drivers—one of them a small speaker called a *tweeter*, which reproduces high frequencies, the other a large speaker known as a *woofer*, which reproduces low frequencies. (Modern systems often remove the woofer and place it in a separate box of its own that can be positioned almost anywhere, because the human ear has difficulty sensing the direction of low-frequency sounds.)

The schematic that you just looked at and may have constructed is known as a "crossover network," and truly hardcore audiophiles have been known make their own (especially for use in car systems) to go with speakers of their choice in cabinets that they design and build themselves.

If you want to make a crossover network, you should use high-quality polyester capacitors (which have no polarity, last longer than electrolytics, and are better made) and a coil that has the right number of turns of wire and is the right size to cut high frequencies at the appropriate point. Figure 5-44 shows a polyester capacitor.

Figure 5-45 shows an audio crossover coil that I bought on eBay for $6. I was curious to find out what was inside it, so I bought two of them, and took one apart.

First I peeled away the black vinyl tape that enclosed the coil. Inside was some typical magnet wire—copper wire thinly coated with shellac or semitransparent plastic, as shown in Figure 5-46. I unwound the wire and counted the number of turns. Then I measured the length of the wire, and finally used a micrometer to measure the diameter of the wire, after which I checked online to find a conversion from the diameter in mils (1/1,000 of an inch) to American wire gauge.

As for the spool, it was plain plastic with an air core—no iron or ferrite rod in the center. Figure 5-47 shows the spool and the wire.

Figure 5-44. *Some nonelectrolytic capacitors have no polarity, such as this high-quality polyester film capacitor. However, they tend to be much more expensive, and are hard to find in values higher than 10 μF.*

Figure 5-45. *What exotic components may we find inside this high-end audio component that's used with a subwoofer to block high frequencies?*

Figure 5-46. *The black tape is removed, revealing a coil of magnet wire.*

Figure 5-47. *The audio crossover coil consists of a plastic spool and some wire. Nothing more.*

So here's the specification for this particular coil in an audio crossover network. Forty feet of 20-gauge copper magnet wire, wrapped in 200 turns around a spool of 1/16-inch-thick plastic with a hub measuring 7/8 inch in length between the flanges and 1/2-inch external diameter. Total retail cost of materials if purchased separately: probably about $1, assuming you can find or make a spool of the appropriate size.

Conclusion: there's a lot of mystique attached to audio components. They are frequently overpriced, and you can make your own coil if you start with these parameters and adjust them to suit yourself.

Suppose you want to put some thumping bass speakers into your car. Could you build your own filter so that they only reproduce the low frequencies? Absolutely—you just need to wind a coil, adding more turns until it cuts as much of the high frequencies as you choose. Just make sure the wire is heavy enough so that it won't overheat when you push 100 or more audio watts through it.

Here's another project to think about: a color organ. You can tap into the output from your stereo and use filters to divide audio frequencies into three sections, each of which drives a separate set of colored LEDs. The red LEDs will flash in response to bass tones, yellow LEDs in response to the mid-range, and green LEDs in response to high frequencies (or whatever colors you prefer). You can put signal diodes in series with the LEDs to rectify the alternating current, and series resistors to limit the voltage across the LEDs to, say, 2.5 volts (when the music volume is turned all the way up). You'll use your meter to check the current passing through each resistor, and multiply that number by the voltage drop across the resistor, to find the wattage that it's handling, to make sure the resistor is capable of dissipating that much power without burning out.

Audio is a field offering all kinds of possibilities if you enjoy designing and building your own electronics.

THEORY

Waveforms

If you blow across the top of a bottle, the mellow sound that you hear is caused by the air vibrating inside the bottle, and if you could see the pressure waves, they would have a distinctive profile.

If you could slow down time and draw a graph of the alternating voltage in any power outlet in your house, it would have the same profile.

If you could measure the speed of a pendulum swinging slowly to and fro in a vacuum, and draw a graph of the speed relative to time, once again it would have the same profile.

That profile is a *sine wave*, so called because you can derive it from basic trigonometry. In a right-angled triangle, the sine of an angle is found by dividing the length of the side opposite the angle by the length of the hypoteneuse (the sloping side of the triangle).

To make this simpler, imagine a ball on a string rotating around a center point, as shown in Figure 5-48. Ignore the force of gravity, the resistance of air, and other annoying variables. Just measure the vertical height of the ball and divide it by the length of the string, at regular instants of time, as the ball moves around the circular path at a constant speed. Plot the result as a graph, and there's your sine wave, shown in Figure 5-49. Note that when the ball circles below its horizontal starting line, we consider its distance negative, so the sine wave becomes negative, too.

Why should this particular curve turn up in so many places and so many ways in nature? There are reasons for this rooted in physics, but I'll leave you to dig into that topic if it interests you. Getting back to the subject of audio reproduction, what matters is this:

- Any sound can be broken down into a mixture of sine waves of varying frequency and amplitude.

Or, conversely:

- If you put together the right mix of audio sine waves, you can create *any sound at all*.

Suppose that there are two sounds playing simultaneously. Figure 5-50 shows one sound as a red curve, and the other as pale blue. When the two sounds travel either as pressure waves through air or as alternating electric currents through a wire, the amplitudes of the waves are added together to make the more complex curve, which is shown in black. Now try to imagine dozens or even hundreds of different frequencies being added together, and you have an idea of the complex waveform of a piece of music.

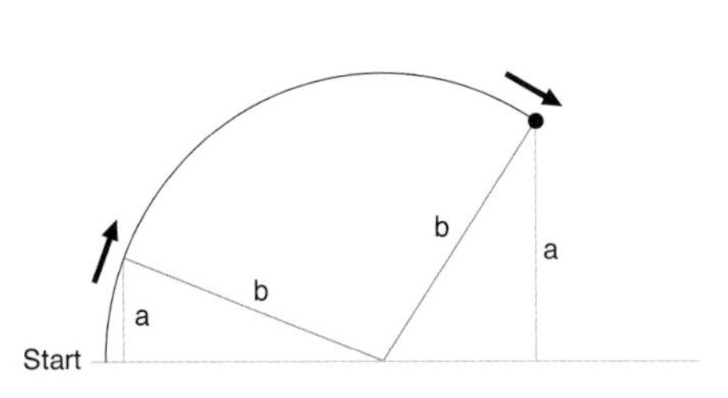

Figure 5-48. *If a weight on the end of a string (length b, in the diagram) follows a circular path at a steady speed, the distance of the weight from a horizontal center line (length a, in the diagram) can be plotted as a graph relative to time. The graph will be a sine wave, so called because in basic trigonometry, the ratio of a/b is the sine of the angle between line b and the horizontal baseline, measured at the center of rotation. Sinewaves occur naturally in the world around us, especially in audio reproduction and alternating current.*

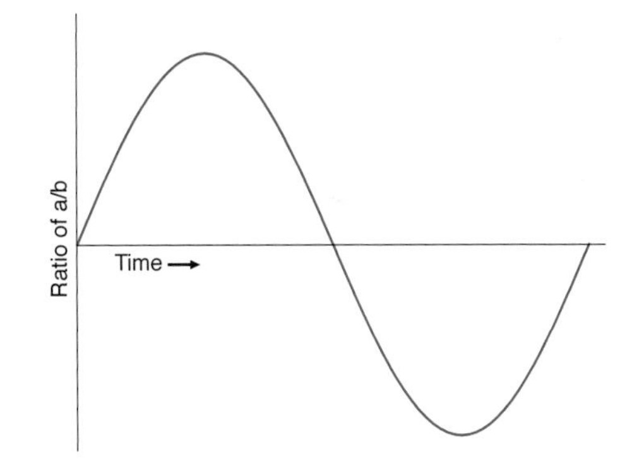

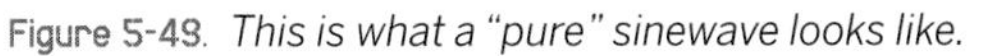

Figure 5-49. *This is what a "pure" sinewave looks like.*

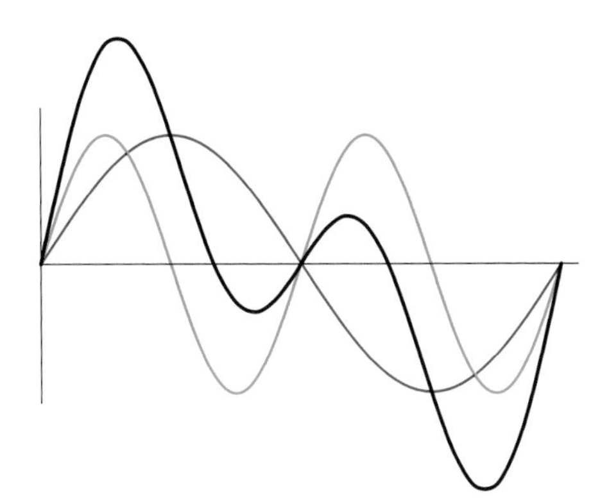

Figure 5-50. *When two sinewaves are generated at the same time (for instance, by two musicians, each playing a flute), the combined sound creates a compound curve. The blue sinewave is twice the frequency of the red sinewave. The compound curve (black line) is the sum of the distances of the sinewaves from the baseline of the graph.*

THEORY

Waveforms (continued)

You can create your own waveform as an input for your audio amplifier with the basic astable 555 timer circuit shown in Figure 5-51. You have to be careful, though, not to overload the amplifier input. Note the 680K series resistor on the output pin of the timer. Also note the 500Ω potentiometer.

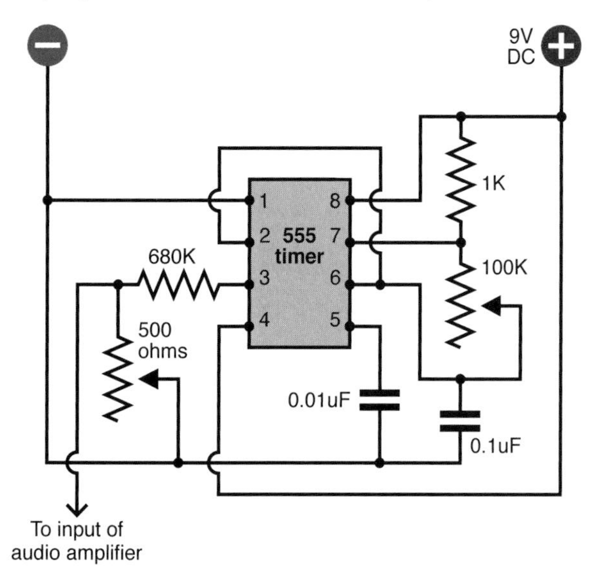

Figure 5-51. *A 555 timer is wired in astable mode using the component values shown here to generate a wide range of audible frequencies when the 100K potentiometer is adjusted. After the output is reduced in power, it can feed into the amplifier chip that was used previously.*

Disconnect your music player and connect the output from the 555 circuit to the input point (the 33K resistor) in the amplifier circuit shown earlier in Figure 5-41. You don't have to worry about a separate connection on the negative side as long as the 555 timer shares the same breadboard and the negative side of its power supply.

Make sure that the 500Ω potentiometer is turned all the way to short the output from the timer to the negative side of the power supply. This functions as your volume control. Also make sure the 100K potentiometer is in the middle of its range. Switch on the power and slowly turn up the 500Ω potentiometer until you hear a tone.

Now adjust the 100K potentiometer to create a low-pitched note. You'll find that it doesn't have a "pure" sound. There are some buzzing overtones. This is because the 555 timer is generating square waves such as those shown in Figure 5-52, not sine waves, and a square wave is actually a sum of many different sine waves, some of which have a high frequency. Your ear hears these harmonics, even though they are not obvious when you look at a square-shaped waveform.

Route one of the connections to your loudspeaker through your spool of hookup wire, and now you should hear a much purer tone, as the buzzing high frequencies are blocked by the self-inductance of the coil. Remove the coil and substitute the 10 μF capacitor, and now you hear more buzzing and less bass.

You've just taken a small step toward sound synthesis. If this subject interests you, you can go online and search for oscillator circuits. For a thorough understanding of the relationship between waveforms and the sounds you hear, you'll really need an oscilloscope, which will show you the shape of each waveform that you generate and modify.

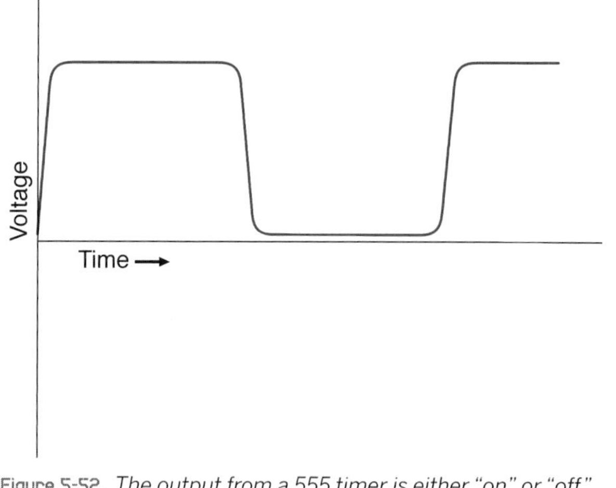

Figure 5-52. *The output from a 555 timer is either "on" or "off," with a very fast transition between those two states. The result is an almost perfect square wave. Theoretically, this can be disassembled into a complex set of sine waves that have many different frequencies. The human ear hears the high frequencies as harsh overtones.*

Experiment 30: Fuzz

Let's try one more variation on the circuit in Experiment 28. This will demonstrate another fundamental audio attribute: distortion.

You will need:

- One more 100K potentiometer.
- Generic NPN transistors: 2N2222 or similar. Quantity: 2.
- Various resistors and capacitors.

BACKGROUND

Clipping

In the early days of "hi-fi" sound, engineers labored mightily to perfect the process of sound reproduction. They wanted the waveform at the output end of the amplifier to look identical with the waveform at the input end, the only difference being that it should be bigger, so that it would be powerful enough to drive loudspeakers. Even a very slight distortion of the waveform was unacceptable.

Little did they realize that their beautifully designed tube amplifiers would be abused by a new generation of rock guitarists whose intention was to create as much distortion as possible.

The most common form of waveform abuse is technically known as "clipping." If you push a vacuum tube or a transistor to amplify a sine wave beyond the component's capabilities, it "clips" the top and bottom of the curve. This makes it look more like a square wave, and as I explained in the section on waveforms, a square wave has a harsh, buzzing quality. For rock guitarists trying to add an edge to their music, the harshness is actually a desirable feature.

Figure 5-53. *This Vox Wow-Fuzz pedal was one of the early stomp boxes, which deliberately induced the kind of distortion that audio engineers had been trying to get rid of for decades.*

The first gadget to offer this on a commercial basis was known as a "fuzz box," which deliberately clipped the input signal. An early fuzz box is shown in Figure 5-53. The clipping of a sine wave is shown in Figure 5-54.

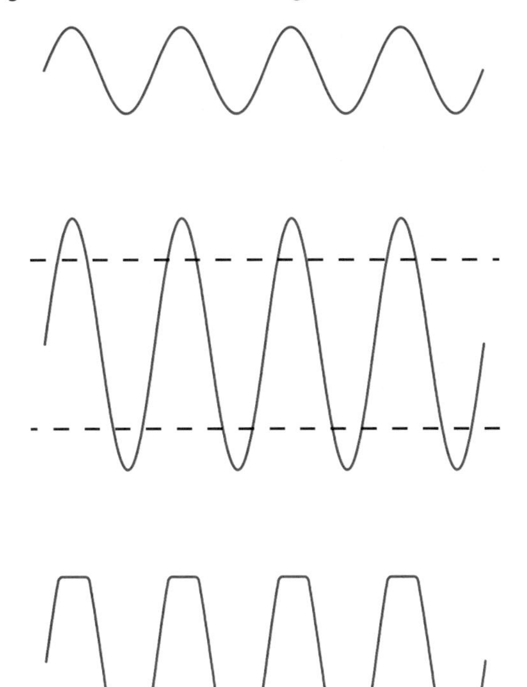

Figure 5-54. *When a sinewave (top) is passed through an amplifier which is turned up beyond the limit of its components (shown as dashed lines, center), the amplifier chops the wave (bottom) in a process known as "clipping." The result is close to a square wave and is the basic principle of a fuzz box commonly used to create a harsh guitar sound.*

Schematic

The output from the 555 timer is a square wave, so it already sounds quite "fuzzy," but we can make it more intense to demonstrate the clipping principle. I've redrawn the whole circuit in Figure 5-55, as several components have changed. The principal alteration is the addition of two NPN transistors.

If you assemble this circuit on your breadboard, note carefully that the 33K and 10K resistors at the bottom of the amplifier chip have been removed, and there's now just an 820Ω resistor in that location. The bottom of the adjacent 0.22 µF capacitor is still the input point for the amplifier, and if you follow the connection around to the middle of the schematic, you'll find it leading to a 100K potentiometer. This is your "fuzz adjuster."

The two NPN transistors are arranged so that the one on the left receives output from the 555 timer. This signal controls the flow of electricity through the transistor from a 33K resistor. This flow, in turn, controls the base of the righthand transistor, and the flow of current through that is what ultimately controls the amplifier.

When you power up the circuit, use the 100K potentiometer attached to the 555 timer to adjust the frequency (as before) and crank the "fuzz adjuster" potentiometer to hear how it adds increasing "bite" to the sound until ultimately it becomes pure noise.

The two transistors act as amplifiers. Of course, we didn't need them for that purpose—the input level for the amplifier chip was already more than adequate. The purpose of the lefthand transistor is simply to overload the righthand transistor, to create the "fuzz" effect. And when you turn up the output from the transistors with the "fuzz adjuster," eventually they overload the input of the amplifier chip, creating even more distortion.

If you want to tweak the output, try substituting different values for the 1K resistor and the 1 µF capacitor that are positioned between the emitter of the righthand transistor and the negative side of the power supply. A larger resistor should overload the transistor less. Different capacitor values should make the sound more or less harsh.

You can find literally thousands of schematics online for gadgets to modify guitar sound. The circuit I've included here is one of the most primitive. If you want something more versatile, you should search for "stomp box schematics" and see what you can find.

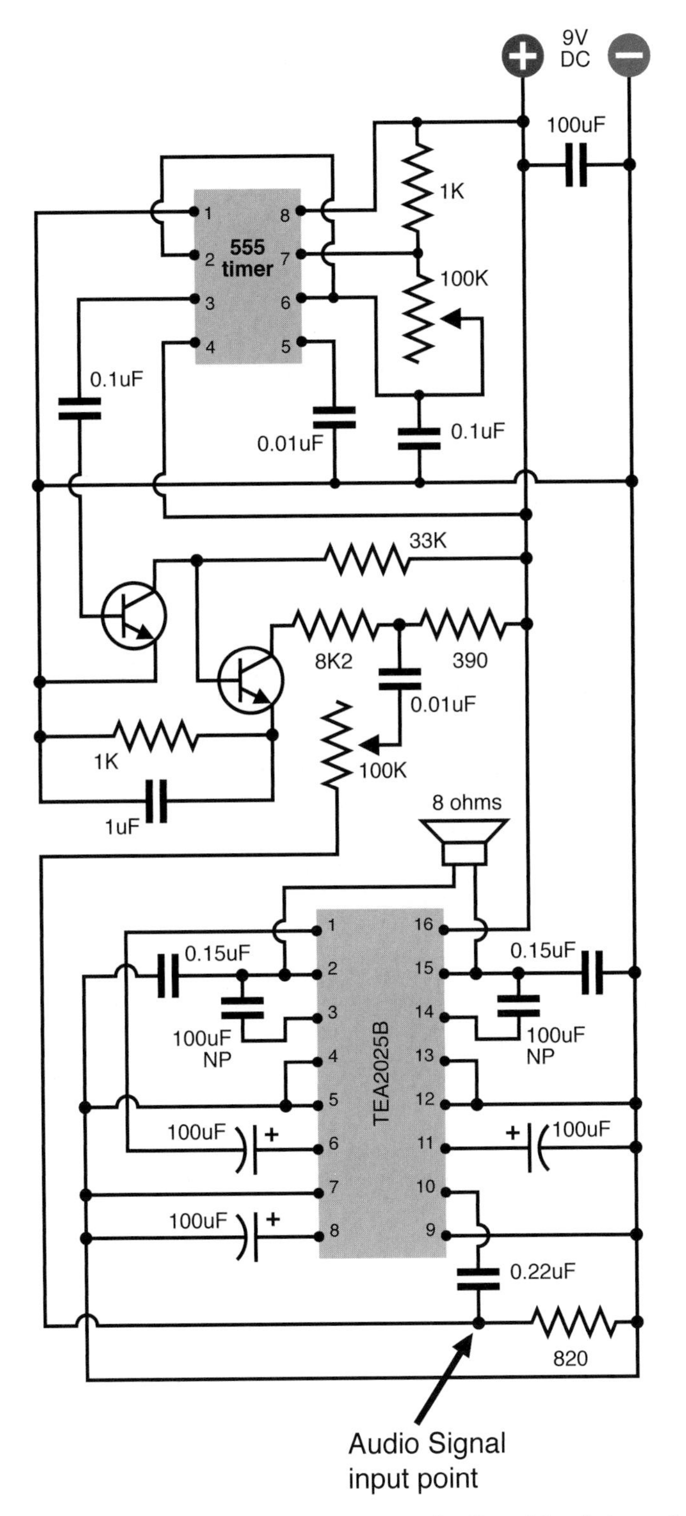

Figure 5-55. *For a quick demo of clipping, insert a couple of transistors between the output from the 555 timer and the input of the amplifier chip. One transistor overdrives the other, so that when you adjust the potentiometer at the center of the circuit, you hear an increasingly harsh, distorted sound.*

BACKGROUND

Stomp-box origins

The Ventures recorded the first single to use a fuzz box, titled "The 2,000 Pound Bee," in 1962. Truly one of the most awful instrumentals ever made, it used distortion merely as a gimmick and must have discouraged other musicians from taking the concept seriously.

Ray Davies of the Kinks was the first to embody distortion as an integral part of his music. Davies did it initially by plugging the output from one amp into the input of another, supposedly during the recording of his hit "You Really Got Me." This overloaded the input and created clipping—the basic fuzz concept. From there it was a short step to Keith Richards using a Gibson Maestro Fuzz-Tone when the Rolling Stones recorded "(I Can't Get No) Satisfaction" in 1965.

Today, you can find thousands of advocates promoting as many different mythologies about "ideal" distortion. In Figure 5-56, I've included a schematic from Flavio Dellepiane, a circuit designer in Italy who gives away his work (with a little help from Google AdSense) at *http://www.redcircuits.com*. Dellepiane is self-taught, having gained much of his knowledge from electronics magazines such as the British *Wireless World*. In his fuzz circuit, he uses a very high-gain amplifier consisting of three field-effect transistors (FETs), which closely imitate the rounded square-wave typical of an overdriven tube amp.

Dellepiane offers dozens more schematics on his site, developed and tested with a dual-trace oscilloscope, low-distortion sinewave oscillator (so that he can give audio devices a "clean" input), distortion meter, and precision audio volt meter. This last item, and the oscillator, were built from his own designs, and he gives away their schematics, too. Thus his site provides one-stop shopping for home-audio electronics hobbyists in search of a self-administered education.

Before fuzz, there was tremolo. A lot of people confuse this with vibrato, so let's clarify that distinction right now:

- Vibrato applied to a note makes the frequency waver up and down, as if a guitarist is bending the strings.
- Tremolo applied to a note makes its volume fluctuate, as if someone is turning the volume control of a guitar up and down very quickly.

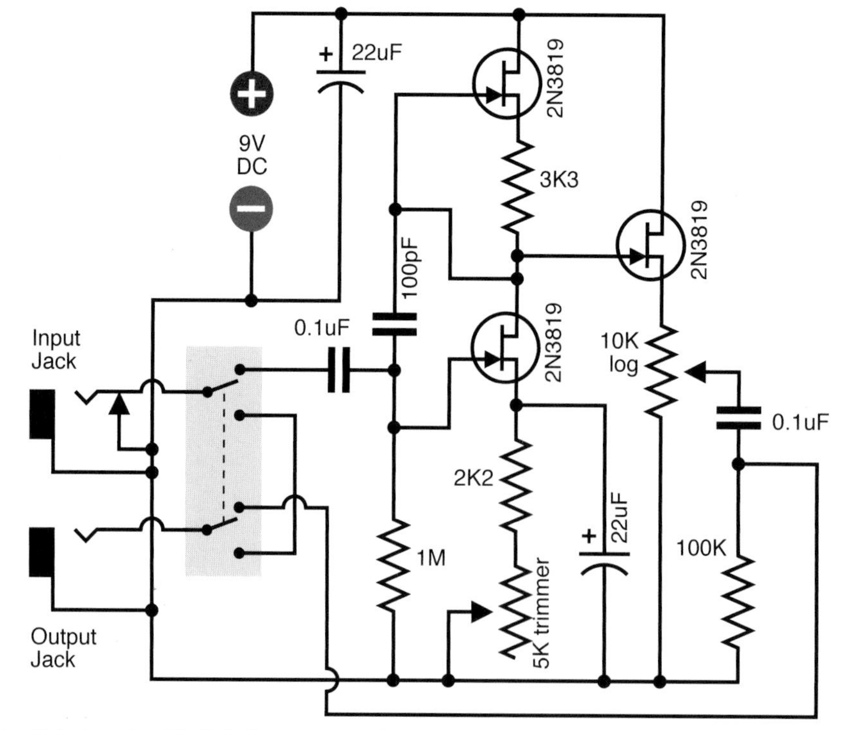

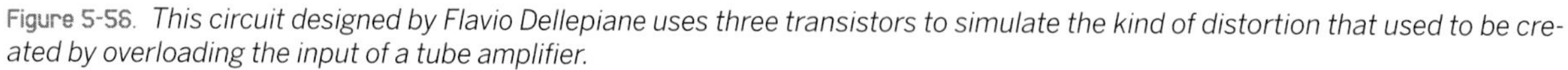
Figure 5-56. *This circuit designed by Flavio Dellepiane uses three transistors to simulate the kind of distortion that used to be created by overloading the input of a tube amplifier.*

BACKGROUND

Stomp-box origins (continued)

Harry DeArmond sold the first tremolo box, which he named the Trem-Trol. It looked like an antique portable radio, with two dials on the front and a carrying handle on top. Perhaps in an effort to cut costs, DeArmond didn't use any electronic components. His steam-punkish Trem-Trol contained a motor fitted with a tapered shaft, with a rubber wheel pressing against it. The speed of the wheel varied when you turned a knob to reposition the wheel up and down the shaft. The wheel, in turn, cranked a little capsule of "hydro-fluid," in which two wires were immersed, carrying the audio signal. As the capsule rocked to and fro, the fluid sloshed from side to side, and the resistance between the electrodes fluctuated. This modulated the audio output.

Today, Trem-Trols are an antique collectible. When industrial designer Dan Formosa acquired one, he put pictures online at *http://www.danformosa.com/dearmond.html*. And Johann Burkard has posted an MP3 of his DeArmond Trem-Trol so you can actually hear it: *http://johannburkard.de/blog/music/effects/DeArmond-Tremolo-Control-clip.html*.

The idea of a mechanical source for electronic sound mods didn't end there. The original Hammond organs derived their unique, rich sound from a set of toothed wheels turned by a motor. Each wheel created a fluctuating inductance in a sensor like the record head from a cassette player.

It's easy to think of other possibilities for motor-driven stomp boxes. Going back to tremolo: imagine a transparent disc masked with black paint, except for a circular stripe that tapers at each end. While the disc rotates, if you shine a bright LED through the transparent stripe toward a light-dependent resistor, you would have the basis for a tremolo device. You could even create never-before-heard tremolo effects by swapping discs with different stripe patterns. Figures 5-57 and 5-58 show the kind of thing I have in mind. For a real fabrication challenge, how about an automatic disc changer?

Today's guitarists can choose from a smorgasbord of effects, all of which can be home-built using plans available online. For reference, try these special-interest books:

- *Analog Man's Guide to Vintage Effects* by Tom Hughes (For Musicians Only Publishing, 2004). This is a guide to every vintage stomp box and pedal you can imagine.
- *How to Modify Effect Pedals for Guitar and Bass* by Brian Wampler (Custom Books Publishing, 2007). This is an extremely detailed guide for beginners with little or no prior knowledge. Currently it is available only by download, from sites such as *http://www.openlibrary.org*, but you may be able to find the previous printed edition from secondhand sellers, if you search for the title and the author.

Of course, you can always take a shortcut by laying down a couple hundred dollars for an off-the-shelf item such as a Boss ME-20, which uses digital processing to emulate distortion, metal, fuzz, chorus, phaser, flanger, tremolo, delay, reverb, and several more, all in one convenient multi-pedal package. Purists, of course, will claim that it "doesn't sound the same," but maybe that's not the point. Some of us simply can't get no satisfaction until we build our own stomp box and then tweak it, in search of a sound that doesn't come off-the-shelf and is wholly our own.

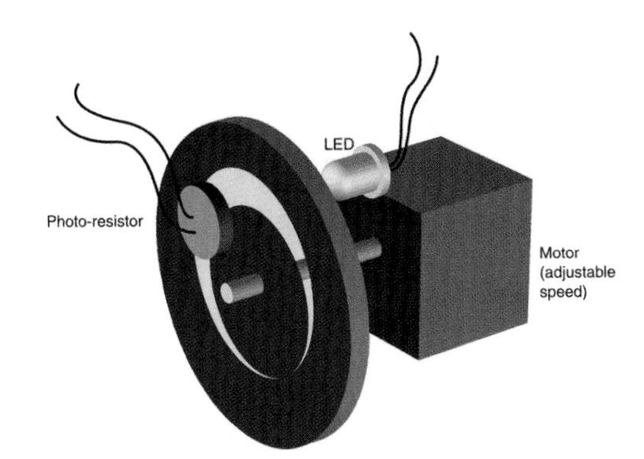

Figure 5-57. *Although electromechanical audio devices are obsolete now, some unexplored possibilities still exist. This design could create various tremolo effects, if anyone had the patience to build it.*

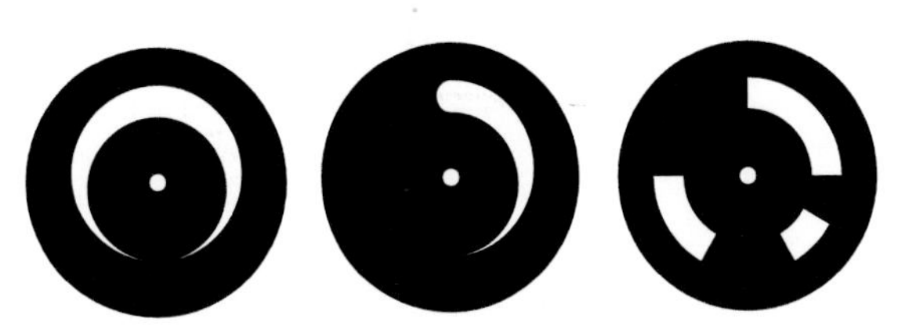

Figure 5-58. *Different stripe patterns could be used in conjunction with the imaginary electromechanical device in Figure 5-57 to create various tremolo effects.*

Experiment 31: One Radio, No Solder, No Power

Time now to go back one more time to inductance and capacitance, and demonstrate an application which also makes use of the way that waveforms can be added to each other. I want to show you how a simple circuit with no power supply at all can receive AM radio signals and make them audible. This is often known as a crystal radio, because the circuit includes a germanium diode, which has a crystal inside it. The idea dates back to the dawn of radio, but if you've never tried it, you've missed an experience that is truly magical.

Figure 5-59. *Just add wire and a coil, and this is all you need to receive AM radio signals. The black disc becomes the tuning dial, after it is screwed onto the variable capacitor (right). This is actually an optional extra. A germanium diode (center) rectifies the radio signal. The high-impedance earphone (top) creates a barely audible sound.*

You will need:

- Rigid cylindrical object, such as a vitamin bottle. Quantity: 1.
- 22-gauge hookup wire, solid-core. Quantity: 60 feet.
- 16-gauge wire, stranded. Quantity: 100 feet.
- Polypropylene rope ("poly rope") or nylon rope. Quantity: 10 feet.
- Germanium diode. Quantity: 1.
- High-impedance headphone. Quantity: 1.

The diode and headphone can be ordered from *http://www.scitoyscatalog.com*. You cannot use a modern headphone of the type you wear with an MP3 player.

Some of these items are shown in Figure 5-59.

First, you need to make a coil. It should be about 3 inches in diameter, and you can wind it around any empty glass or plastic container of that size, so long as it's rigid. A soda bottle or water bottle isn't suitable, because the cumulative squeezing force of the turns of wire can deform the bottle so that it isn't circular anymore.

I chose a vitamin bottle that just happened to be exactly the right size. To remove the label, I softened its adhesive with a heat gun (lightly, to avoid melting the bottle) and then just peeled it off. The adhesive left a residue, which I removed with Xylol (also known as Xylene). This is a handy solvent to have around, as it can remove "permanent" marker stains as well as sticky residues, but you should always use latex or nitrile gloves to avoid getting it on your skin, and minimize your exposure the fumes. Because Xylol will dissolve some plastics, clearly it's not good for your lungs.

Figure 5-60. *A large, 3-inch diameter empty vitamin bottle makes an ideal core for a crystal radio coil. The drilled holes will anchor wire wrapped around the bottle.*

After you prepare a clean, rigid bottle, drill two pairs of holes in it, as shown in Figure 5-60. You'll use them to anchor the ends of the coil.

Now you need about 60 feet of 22-gauge solid-core wire. If you use magnet wire, its thin insulation will allow the turns of the coil to be more closely spaced, and the coil may be slightly more efficient. But everyday vinyl-insulated wire will do the job, and is much easier to work with.

Begin by stripping the first 6 inches of insulation from the end of the wire. Now measure 50 inches along the insulated remainder and apply your wire strippers at that point, just enough to cut the insulation without cutting the wire. Use your two thumb nails to pull the insulation apart, revealing about a half-inch of bare wire, as shown in Figure 5-61. Bend it at the center point and twist it into a loop, as shown in Figure 5-62.

You just made a "tap," meaning a point where you will be able to tap into the coil after you wind it. You'll need another 11 of these taps, all of them spaced 50 inches apart. (If the diameter of the bottle that you'll be using as the core of your coil is not 3 inches, multiply its diameter by 16 to get the approximate desired spacing of taps.)

After you have made 12 taps, cut the wire and strip 6 inches off that end. Now bend the end into a U shape about a half-inch in diameter, so that you can hook it through the pair of holes that you drilled at one end of the bottle. Pull the wire through, then loop it around again to make a secure anchor point.

Now wind the rest of the wire around the bottle, pulling it tightly so that the coils stay close together. When you get to the end of the wire, thread it through the remaining pair of holes to anchor it as shown in Figure 5-63. The completed coil is shown in Figure 5-64.

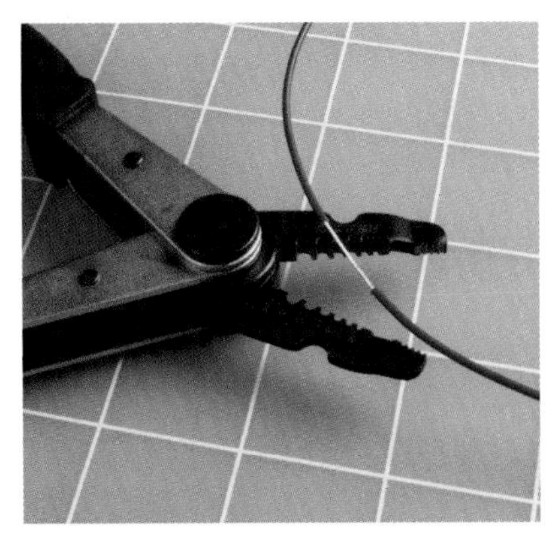

Figure 5-61. *Wire strippers expose the solid conductor at intervals along a 22-gauge wire.*

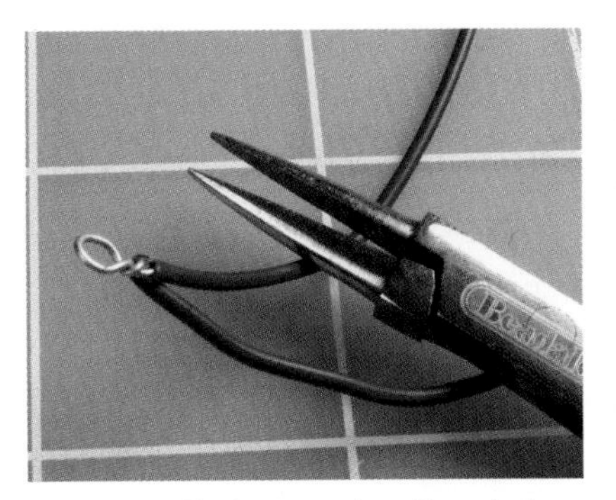

Figure 5-62. *Each exposed section of wire is twisted into a loop using sharp-nosed pliers.*

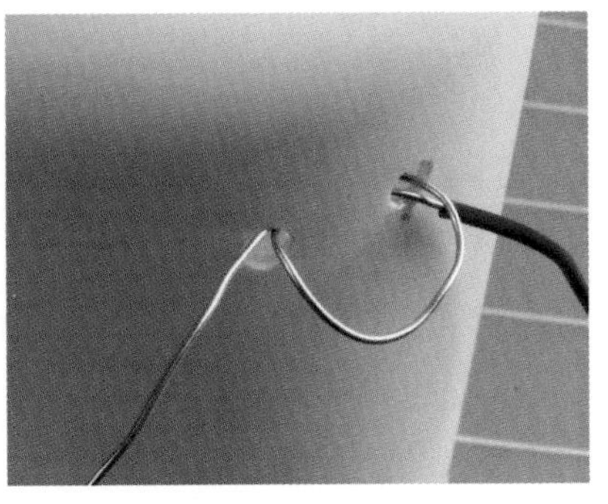

Figure 5-63. *The stripped end of the wire is secured through the holes drilled in the bottle.*

Figure 5-64. *The completed coil, wrapped tightly around the bottle.*

Your next step is to set up an antenna. If you live in a house with a yard outside, this is easy: just open a window, toss out a reel of 16-gauge wire while holding the free end, then go outside and string up your antenna by using polypropylene rope ("poly rope") or nylon rope, available from any hardware store, to hang the wire from any available trees, gutters, or poles. The total length of the wire should be about 100 feet. Where it comes in through the window, suspend it on another length of poly rope. The idea is to keep your antenna wire as far away from the ground or from any grounded objects as possible.

If you live in an apartment where you don't have access to a yard outside, you can try stringing your antenna around the room, hanging the wire from more pieces of poly rope. The antenna should still be about 100 feet long, but obviously it won't be in a straight line.

Hook the free end of your antenna to one end of your coil. At this point, you also need to add a germanium diode, which functions like a silicon-based diode but is better suited to the tiny voltages and currents that we'll be dealing with. The other end of the diode attaches to one of the wires leading to a high-impedance

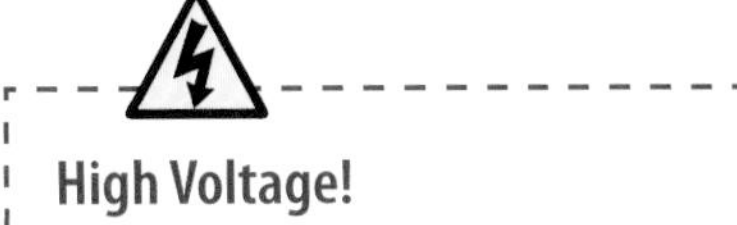

High Voltage!

The world around us is full of electricity. Normally we're unaware of it, but a thunderstorm is a sudden reminder that there's a huge electrical potential between the ground below and the clouds above.

If you put up an outdoor antenna, never use it if there is any chance of a lightning strike. This can be extremely dangerous. Disconnect the indoor end of your antenna, drag it outside, and push the end of the wire into the ground to make it safe.

earphone. *A normal modern earphone or headphone will not work in this circuit.* The return wire from the headphone is connected to a jumper wire, the other end of which can be clipped to any of the taps in your coil.

One last modification, and you'll be ready to tune in. You have to ground the jumper wire. By this I mean connect it to something that literally goes into the ground. A cold-water pipe is the most commonly mentioned option, but (duh!) this will work only if the pipe is made of metal. Because a lot of plumbing these days is plastic, check under the sink to see if you have copper pipes before you try using a faucet for your ground.

Another option is to attach the wire to the screw in the cover plate of an electrical outlet, as the electrical system in your house is ultimately grounded. But the sure-fire way to get a good ground connection is to go outside and hammer a 4-foot copper-clad grounding stake into reasonably moist earth. Any wholesale electrical supply house should be able to sell you a stake. They're commonly used to ground welding equipment.

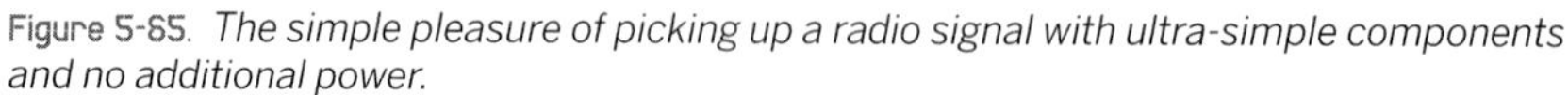
Figure 5-65. *The simple pleasure of picking up a radio signal with ultra-simple components and no additional power.*

Figures 5-66 and 5-67 show the completed radio.

If you've managed to follow these instructions (one way or another), it's time to tune your radio to the nearest station. Move the alligator clip at the end of your patch cord from one tap to another on your coil. Depending on where you live, you may pick up just one station, or several, some of them playing simultaneously.

It may seem that you're getting something for nothing here, as the earphone is making noise without any source of power. Really, though, there is a source of power: the transmitter located at the radio station. A large amplifier pumps power into the broadcasting tower, modulating a fixed frequency. When the combination of your coil and antenna resonates with that frequency, you're sucking in just enough voltage and current to energize a high-impedance headphone.

The reason you had to make a good ground connection is that the radio station broadcasts its signal at a voltage relative to ground. The earth completes the circuit between you and the transmitter. For more information on this and other concepts relating to radio, see the upcoming section "Theory: How radio works."

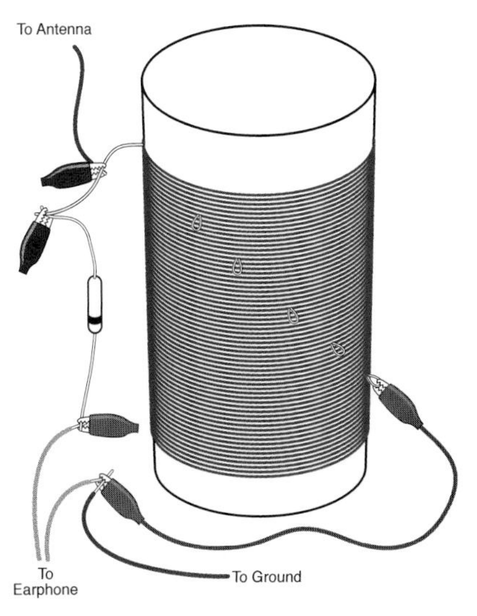

Figure 5-66. *A signal from the antenna can pass through the coil to ground. If the jumper wire is attached to an appropriate tap on the coil, it resonates with the radio signal, just powerfully enough to energize the earphone which is wired in series with a diode.*

Enhancements

The higher your antenna is, the better it should work. In my location, this is a major problem, as I live in a desert area without any trees. Still, just stringing the wire out of my window and tethering it (with rope) to the front bumper of my car enabled me to pick up a faint radio signal.

To improve the selectivity of your radio, you can add a variable capacitor, as shown in the following section. This allows you to "tune" the resonance of your circuit more precisely. Variable capacitors are uncommon today, but you can find one at the same specialty source that I recommended for the earphone and the germanium diode: the Scitoys Catalog (*http://www.scitoyscatalog.com*).

This source is affiliated with a smart man named Simon Quellan Field, whose site suggests many fun projects that you can pursue at home. One of his clever ideas is to remove the germanium diode from your radio circuit and substitute a low-power LED in series with a 1.5-volt battery. This didn't work for me, because I live 40 miles from the nearest AM broadcaster. If you're closer to a transmitter, you may be able to see the LED varying in intensity as the broadcast power runs through it.

Figure 5-67. *The real-life version of Figure 5-66.*

THEORY

How radio works

When electrical frequencies are very high, the radiation they create has enough energy to travel for miles. This is the principle of radio transmission: A high-frequency voltage is applied to a broadcasting antenna, relative to the ground.

When I say "ground" in this instance, I literally mean the planet beneath us. If you set up a receiving antenna, it can pick up a faint trace of the transmission relative to the ground—as if the earth is one huge conductor. Actually the earth is so large and contains so many electrons, it can function as a common sink, like a gigantic version of the file cabinet that I suggested you should touch to get rid of static electricity in your body before touching a CMOS logic chip.

To make a radio transmitter, I could use a 555 timer chip running at, say, 850 kHz (850,000 cycles per second), and pass this stream of pulses through an amplifier to a transmission tower; if you had some way to block out all the other electromagnetic activity in the air, you could detect my signal and reamplify it.

This was more or less what Marconi (shown in Figure 5-68) was doing in 1901, after he had purchased rights to Edison's wireless telegraphy patent, although he had to use a primitive spark gap, rather than a 555 timer, to create the oscillations. His transmissions were of limited use, because they had only two states: on or off. You could send Morse code messages, and that was all.

Figure 5-68. *Marconi, the great pioneer of radio (photograph from Wikimedia Commons).*

Five years later, the first true audio signal was transmitted by imposing lower audio frequencies on the high-frequency carrier wave. In other words, the audio signal was "added" to the carrier frequency, so that the power of the carrier varied with the peaks and valleys of the audio.

At the receiving end, a very simple combination of a capacitor and a coil detected the carrier frequency out of all the other noise in the electromagnetic spectrum. The values of the capacitor and the coil were chosen so that their circuit would "resonate" at the same frequency as the carrier wave. Figures 5-69 and 5-70 illustrate these concepts.

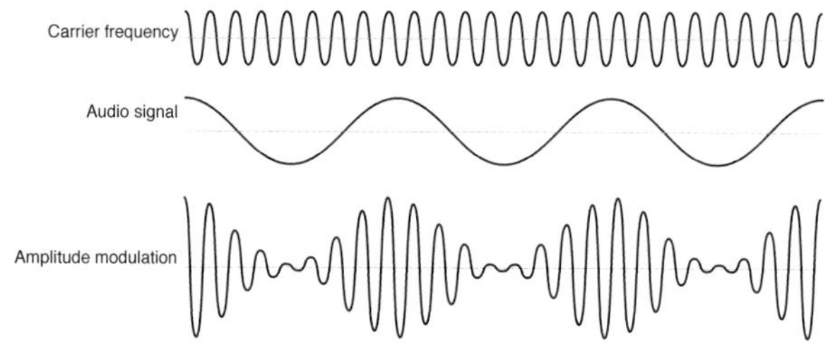

Figure 5-69. *When an audio signal (middle) is combined electronically with a high carrer frequency (top), the result looks something like the compound signal at the bottom. In actuality, the carrier frequency would be much higher compared with the audio frequency, by a radio of perhaps 1,000:1.*

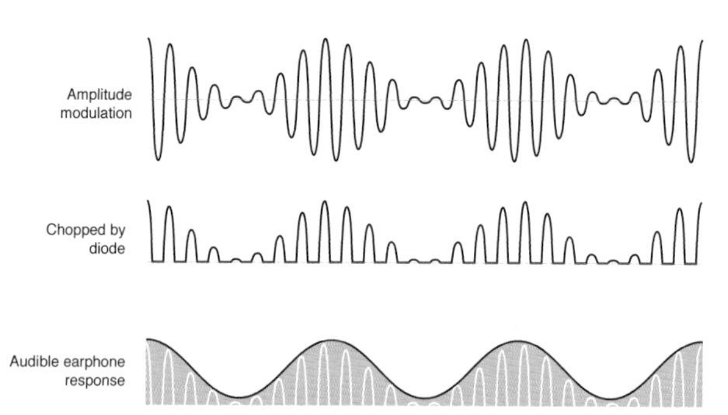

Figure 5-70. *When the compound signal is passed through a diode, only the upper half remains. An earphone cannot react fast enough to reproduce the high carrier frequency, so it "rides" the peaks and thus reproduces the audio frequency.*

The schematic in Figure 5-71 shows the simple circuit that you built by wrapping a coil around an empty vitamin bottle. When a positive pulse was received by the antenna, it resonated with the antenna and the coil, provided that the antenna was long enough and the coil was tapped at the appropriate number of turns.

THEORY

How radio works (continued)

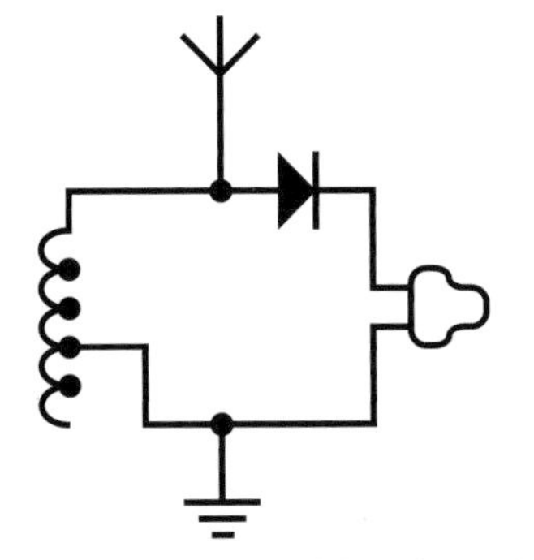

Figure 5-71. *An antenna at the top of the schematic picks up faint electromagnetic radiation from a distant transmitter. The coil at the left side is tapped at intervals so that its resonance can be adjusted to match the carrier frequency of the radio signal. Other frequencies are grounded (at the bottom of the schematic). The diode passes the "top half" of the signal to the earphone at the right, which is incapable of responding fast enough to reproduce the carrier frequency, and thus filters it out, leaving only the audio frequencies that were superimposed on it.*

By adding a capacitor, you can tune the circuit. Now an incoming pulse from the transmitter is initially blocked by the self-inductance of the coil, while it charges the capacitor. If an equally negative pulse is received after an interval that is properly synchronized with the values of the coil and the capacitor, it coincides with the capacitor discharging and the coil conducting. In this way, the right frequency of carrier wave makes the circuit resonate in sympathy. At the same time, audio-frequency fluctuations in the strength of the signal are translated into fluctuations in voltage in the circuit.

What happens to other frequencies pulled in by the antenna? The lower ones pass through the coil to ground; the higher ones pass through the capacitor to ground. They are just "thrown away."

The righthand half of the circuit samples the signal by passing it through a germanium diode and an earphone. The power from the transmitter is just sufficient to vibrate the diaphragm in the earphone, after the diode has subtracted the negative half of the signal.

Look back at the diagram of the amplitude-modulated signal. You'll see that it fluctuates up and down so rapidly, the earphone cannot possibly keep up with the positive-negative variations—hence the need for the diode. It will remain hesitating at the midpoint between the highs and lows, producing no sound at all. The diode solves this problem by subtracting the lower half of the "audio envelope," leaving just the positive spikes of voltage. Although these are still very small and rapid, they are now all pushing the diaphragm of the earphone in the same direction, so that it averages them out, approximately reconstructing the original sound wave.

Figure 5-72 shows how the circuit can be enhanced with a variable capacitor, to tune it without needing to tap the coil at intervals.

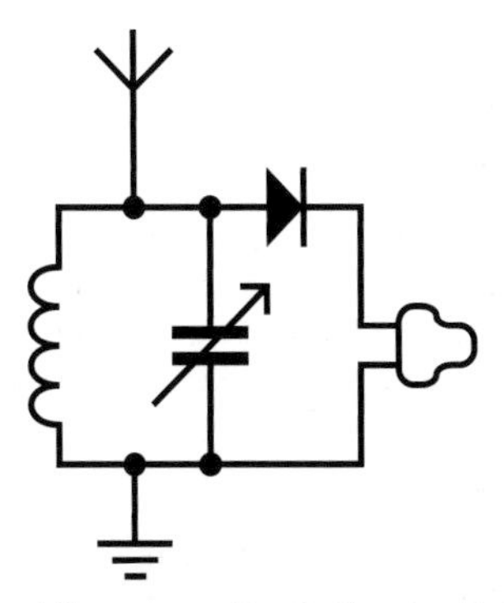

Figure 5-72. *By adding a capacitor to the circuit, its resonance can be tuned more precisely. The diagonal arrow indicates that a variable capacitor is used.*

The radio can pull in the stations on the AM (amplitude-modulated) waveband that happen to be most powerful in your area. The waveband ranges from 300 kHz to 3 MHz. If you find yourself interested in radio, your next step could be to build a powered radio using a couple of transistors. Alternatively you could build your own (legal) low-power AM transmitter. There's an ultra-simple kit available from *http://www.scitoys.com* consisting of just two principal components: a crystal oscillator, and a transformer, shown in Figure 5-73. That's all it takes.

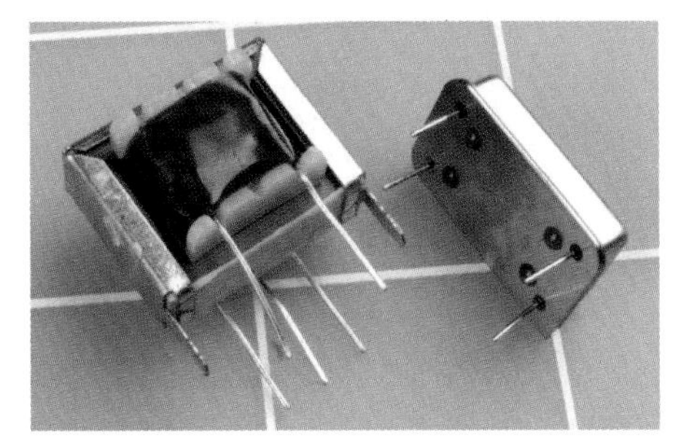

Figure 5-73. *An AM radio transmitter can be made from just two components: a transformer (left) and a crystal oscillator (right), available from http://www.scitoys.com.*

Experiment 32: A Little Robot Cart

Robotics is another application of electronics that deserves a book in itself—or several books. So, once again, I'm going to give you an introduction followed by some points that you can follow if you want to go further. As always, I will start with the simplest possible device, which in the world of robotics is a cart that finds its way around your living room.

You will need:

- SPST or SPDT microswitches requiring minimal pressure to activate them. A force between 0.02 and 0.1 newtons would be ideal. Quantity: 2. See Figure 5-74.
- DC gear-motor, rated for any voltage between 5 and 12, drawing a maximum of 100mA in its free-running state, output shaft turning between 30 and 60 RPM. Quantity: 1. A motor is shown in Figure 5-75.
- Disc or arm that fits securely onto your motor shaft. Quantity: 1.
- 555 timer. Quantity: 1.
- DPDT nonlatching relay rated for the same voltage as your motor. Quantity: 1.
- 1/4-inch plywood or plastic, one piece about 2 feet square.
- #4 sheet-metal screws, 5/8 inch or 3/4 inch long. Quantity: 2 dozen.
- #6 bolts, 3/4 inch long, with nylon-insert lock nuts. Quantity: 2 dozen.
- 1/4-inch bolts, 1 inch long, with nuts, to mount the wheels. Quantity: 4.

Figure 5-74. *A microswitch has a small button (at the front, righthand side in this picture) that is often actuated by a pivoted metal lever. The switch can respond to a very light pressure, but can handle relatively high currents.*

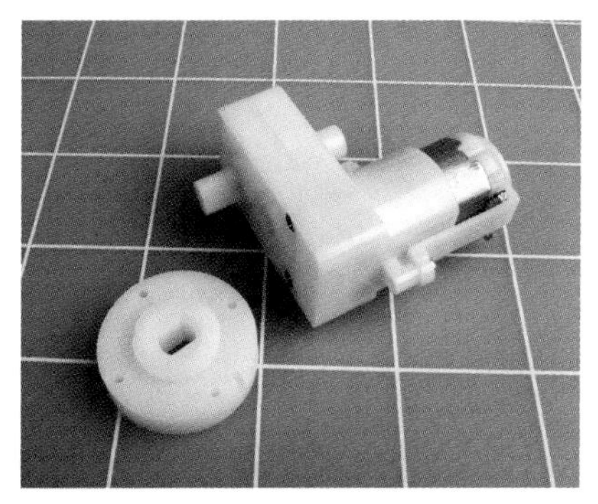

Figure 5-75. *For the Little Robot Cart, I found this 5-volt motor, which is supplied with a disc that fits its output shaft. The combination cost less than $10.*

I'm not specifying one particular motor, because if I did, it might not be available by the time you read this. Motors aren't like logic chips, which have retained their basic function throughout various improvements over a period of several decades. Motors come and go, and many that you may run across will be surplus parts that will never been seen again. Search online for "gear-motor" or "gearhead motor" and find one as close as possible to the specification that I have provided. The mechanical power output of the motor shouldn't be important, because we won't be requiring it to do much work.

The important consideration when you buy your motor is that you should also obtain something that fits onto its output shaft. Typically, this will be a disk or arm that can be screwed into place. To this you can then add a larger wheel of your own, which you can cut with a hole saw or make from the screw-on lid of a jar, or anything else circular that you may find in the house.

A larger wheel will make your cart move faster than a smaller wheel, but will reduce its torque, thus limiting its power to overcome obstacles.

This brings me to my next topic: fabrication. Although this is an electronics book, motors are electromechanical devices, and you have to be able to install them in some kind of a machine to get any interesting results. You can use plywood to complete the two little robotics projects here (ideally, the kind of thin, high-quality plywood sold by hobby stores) but I recommend something that looks better and is easier to work with: ABS plastic. Before you start on the robotic cart, you may want to check the section "Fundamentals: All about ABS."

FUNDAMENTALS

All about ABS

Unless you think the steampunk movement isn't going back far enough, you probably don't want your autonomous robot cart to resemble a relic from before the 1800s. Therefore, wood may not be the best construction material. Metal can look nice, but is not easy to work with. For quick results that have a twentieth-century look (maybe even a 21st-century look), plastic is the obvious choice, and I feel that ABS is the best type of plastic to use, because it provides such quick, easy results. ABS stands for "acrylonitrile butadiene styrene." Lego® blocks are made of ABS. Car-stereo installers and model-railroad buffs use it. You can use it, too. You can saw it, drill it, sand it, whittle it, and drive screws into it, and it won't warp, split, or splinter. It's washable, doesn't need to be painted, and will last almost forever.

Delrin is another type of plastic, but tends to cost more and is a little tougher to drill and cut. It's a matter of personal preference. ABS machines fairly well, but when you drill it, for example, it can "catch" on the bit and the piece will spin with the bit due to the way that plastic chips off with the bit. Delrin is self-lubricating and has better melting properties under the heat of machining, so it drills and cuts much more cleanly and easily than ABS.

Where to find ABS

Pieces of ABS a couple of feet square are available from online sources such as *http://hobbylinc.com* or estreetplastics (an eBay store), but you'll save money if you can truck on down to your nearest plastic supply house and buy it like plywood, in sheets measuring 4 by 8 feet. To discover whether you have a nearby plastic supply house, search for "plastic supply" in your yellow pages or Google Local.

Piedmont Regal Plastics has many supply centers around the nation, but you'll have to collect it yourself, and they may not be willing to cut small pieces. You can check online at *http://www.piedmontplastics.com* for their locations.

Stock colors of ABS include black, white, and "natural," which is beige. Sheets usually are textured on one side, which is the side that should face outward, as it is more scratch-resistant than the smooth side.

Because you won't be adding paint or other finishes, you'll have to be careful not to scuff the plastic or scratch it while working. Clean your bench thoroughly before you begin, taking special care to remove any metal particles, which tend to become embedded in the plastic. Use wooden shims in the jaws of your vise, and avoid resting the plastic accidentally on any sharp tools or screws. Working with ABS requires a clean environment and a very gentle touch.

Cut with Care

You can saw ABS, but if you use a table saw, the plastic will tend to melt and stick to the blade. These smears will get warm and sticky when you feed the next piece of plastic into the saw, and the result will be extremely unpleasant. The whirling blade will grab the plastic and hurl it at you powerfully enough to break bones. This is known as "kickback" and is a very serious risk when sawing plastic.

If you have extensive experience using a table saw, you are actually more vulnerable, because the reflexes and cautions you have developed while dealing with wood will not be adequate for working with plastic. Please take this warning seriously!

Your first and most obvious precaution is to use a plastic-cutting blade, which has a larger number of thicker teeth to absorb the heat. The blade I have used is a Freud 80T, but there are others. If you use a blade that is not suitable, you will see it starting to accumulate sticky smears. This is the only warning you will get. Clean that blade with a solvent such as acetone, and never use it for ABS again.

Regardless of other precautions, always wear gloves and eye protection when using a table saw, and stand to one side when feeding materials into it. Personally, after one episode of kickback that I thought had broken my arm, I prefer not to use a table saw on plastic at all.

For long, straight cuts, the alternatives include:

- *Panel saw (big and expensive, but safe and accurate).*
- *Miniature handheld circular saw with a blade around 4 inches in diameter, guided with a straight edge clamped to the sheet.*
- *Hand saw. This is my old-school preference. My favorite is a Japanese pull-to-cut saw, which makes very clean cuts: the Vaughan Extra-Fine Cross-Cut Bear Saw, 9-1/2 inches, 17 tpi (teeth per inch). If you use one of these, be careful to keep your free hand out of the way, as the saw can easily jump out of the cut. Because it is designed to cut hard materials such as wood, it has no difficulty cutting soft flesh. Gloves are strongly recommended.*

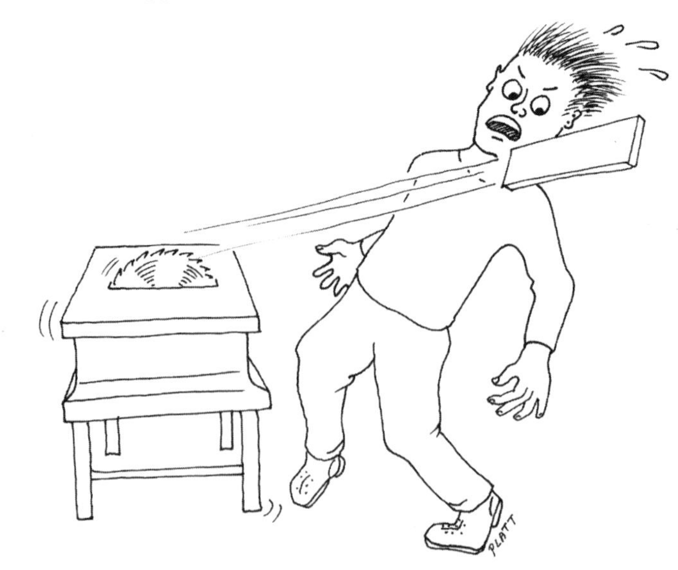

Figure 5-76. *The perils of kickback. Plastic easily sticks to the blade of a table saw, which will hurl it at you unexpectedly. Use other tools to cut plastic.*

Curving Cuts

Curving cuts involve relatively little danger, although eye protection and gloves are still advisable. My preferred tools:

- Band saw with a 3/8- or 1/4-inch blade designed for thin wood or plywood.
- Jigsaw. I have a special liking for the DeWalt XRP using Bosch blades that are designed for hardwood or plastic. This will cut complex curves in ABS as easily as scissors cutting paper.

No matter what type of saw you use, you'll have to clean ragged bits of plastic off the cut afterward, and the absolutely necessary item for this purpose is a deburring tool, available from *http://www.mcmaster.com* and most other online hardware sources. A belt sander or disc sander is ideal for rounding corners, and a metal file can be used to remove bumps from edges that are supposed to be straight.

Figures 5-77 through 5-80 show various cutting tools. Figure 5-81 shows a deburring tool, and Figure 5-82 shows a disc sander.

Figure 5-77. *A band saw is an ideal tool for cutting complex shapes out of ABS plastic. You can often find them secondhand for under $200.*

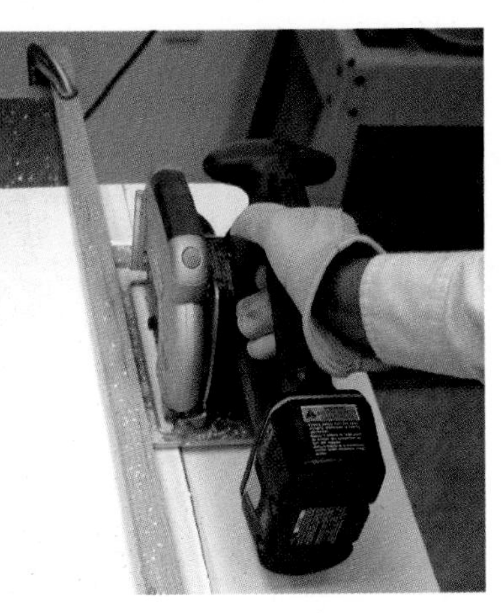

Figure 5-78. *A handheld circular saw, running along a straight-edge, is much safer than a table saw for cutting plastic, and can produce comparable results.*

Figure 5-79. *This Japanese-style saw cuts when you pull it, rather than when you push it. After some practice, you can use it to make very accurate cuts. Because ABS is so soft, minimal muscle-power is required.*

Figure 5-80. *This DeWalt jigsaw can run at very slow speeds, enabling precise and careful work with plastic.*

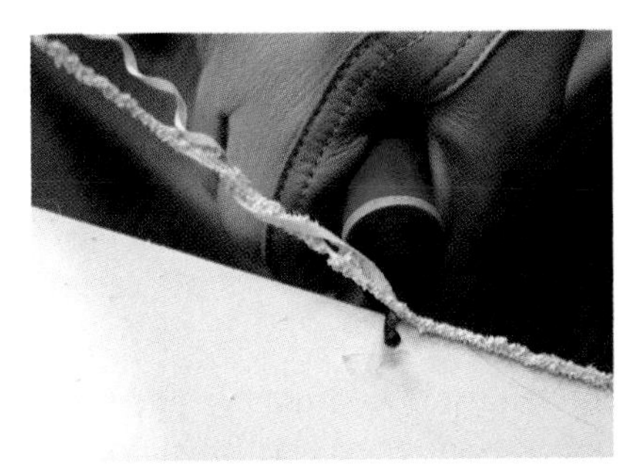

Figure 5-81. *A deburring tool will clean and bevel the sawn edge of a piece of plastic in just a couple of quick strokes.*

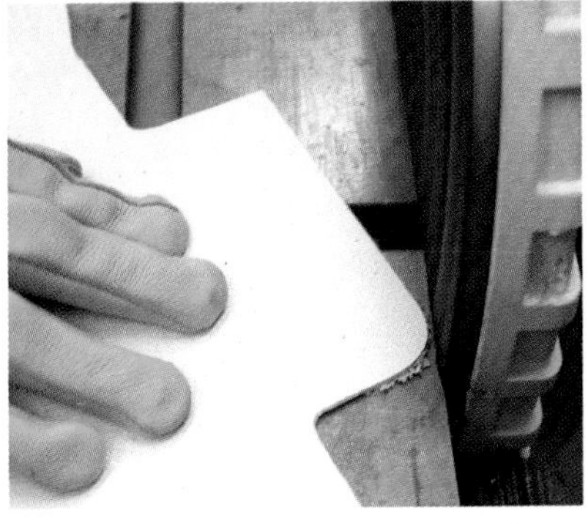

Figure 5-82. *A belt sander or disc sander is the ideal tool for rounding corners when working with ABS plastic.*

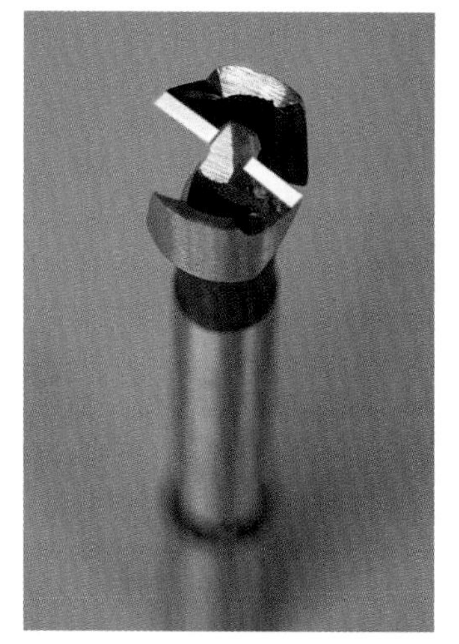

Figure 5-83. *A Forstner drill bit makes clean, precise holes; a large regular drill bit will chew up ABS plastic and make a mess.*

Figure 5-84. *By drilling holes at any location where two bends intersect, you reduce the risk of the plastic fissuring.*

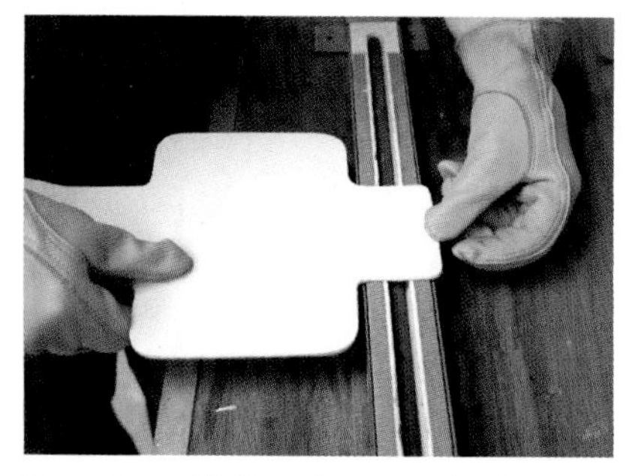

Figure 5-85. *Making clean, precise bends in ABS is simply a matter of resting the plastic over a bender that consists of an electric heating element.*

Making Plans

I like to use drawing software to create plans, and I try to print them at actual scale. I tape them to the smooth side of a piece of white or natural-color ABS, then use an awl to prick through the plan into the soft surface beneath. I remove the paper and connect the awl marks by drawing onto the plastic using a pencil or a fine-point water-soluble pen. Its lines can be wiped away later with a damp cloth. Don't use a permanent marker, as the solvents that you will need to clean it may dissolve the plastic.

Because ABS will tend to open a fissure when you bend it at any inside corner where you don't have a smooth radius, you need to drill holes at these locations, as shown in the cart plans in Figure 5-92 on page 275.

A regular half-inch drill bit is too aggressive; it will tend to jam itself into the plastic within one turn of the drill. Use Forstner bits (shown in Figures 5-83 and 5-84) to cut nice smooth circles.

Note that the heat from bending will tend to make any marks on the plastic permanent.

Bending It

A big advantage of plastic over wood is that you can make complex shapes by bending them, instead of cutting separate pieces and joining them with nails, screws, or glue. Unfortunately, bending does require an appropriate bender: an electric heating element mounted in a long, thin metal enclosure that you place on your workbench. The bender I use is made by FTM, a company that offers all kinds of neat gadgets for working with plastic. Their cheapest bender, shown in Figure 5-85, is just over $200 with a 2-foot element. You can get a 4-foot model for about $50 more. Check them out at *http://thefabricatorssource.com*.

Avoid Burns While Bending

A plastic bender will inflict serious burns if you happen to rest your hand on it accidentally, and because it has no warning light, you can easily forget that you have left it plugged in. Use gloves!

To bend plastic, lay it over the hot element of a plastic bender for a brief time (25 to 30 seconds for 1/8-inch ABS, 40 to 45 seconds for 3/16-inch, and up to a minute for 1/4-inch). If you overheat the plastic, you'll smell it, and when you turn it over you'll find it looks like brown melted cheese. Naturally you should learn to intervene before the plastic reaches that point.

ABS is ready to bend when it yields to gentle pressure. Take it off the bender and bend it *away* from the side that you heated. If you bend it toward the hot side, the softened plastic will bunch up inside the bend, which doesn't look nice.

You can work with it for about half a minute, and when you have it the way you want it, spray or sponge water onto it to make it set quickly. Alternatively, if you need more time, you can reheat it. The amount of force necessary to bend the sheet increases in proportion with the length of the bend, so a long bend can be difficult, and I usually insert it into a loose vise, push it a bit, move it along to the next spot, and push it again.

Because plastic bending is very similar to making shapes in origami, it's a good idea to model your projects in paper before you commit yourself to ABS.

If you decide that you don't want to spend money on a bender, don't abandon plastic just yet—you can use screws to assemble separate sections with greater ease and convenience than if you were working with wood.

Making 90-Degree Joints

Driving screws into the edge of a piece of plywood will almost always separate its layers, but ABS has no layers (or grain, either), and never splits or shatters. This means that you can easily join two pieces at 90° using small screws (#4 size, 5/8-inch long).

Figures 5-86 through 5-90 show the procedure for joining 1/8-inch (or thicker) ABS to 1/4-inch ABS, which I regard as the minimum thickness when you're inserting screws into its edge:

1. Mark a guideline on the thinner piece of plastic, 1/8 inch from its edge. For #4 screws, drill holes using a 7/64-inch bit. If you're using flat-headed screws, countersink the holes very gently.

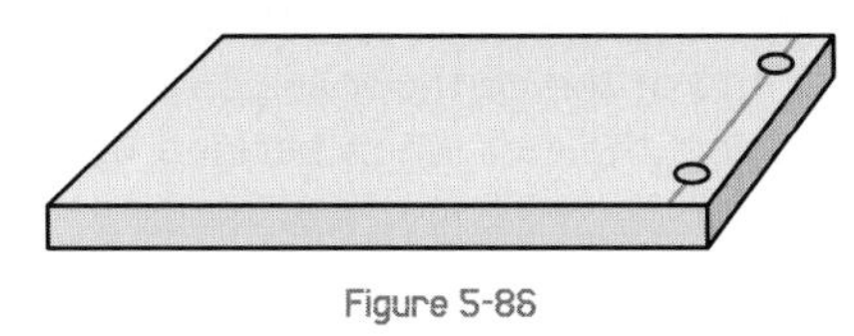

Figure 5-86

2. Hold or clamp the pieces in place and poke a pen or pencil through the holes to mark the edge of the 1/4-inch plastic beneath.
3. Remove the thin plastic, clamp the 1/4-inch plastic in a vise, and drill guide holes for the screws at each mark, centered within the thickness of the plastic. Because ABS does not compress like wood, the holes must be larger than you may expect; otherwise, the plastic will swell around the screw. A 3/32-inch bit is just right for a #4 screw.
4. Assemble the parts. Be careful not to overtighten the screws; it's easy to strip the threads that they cut in the soft plastic.

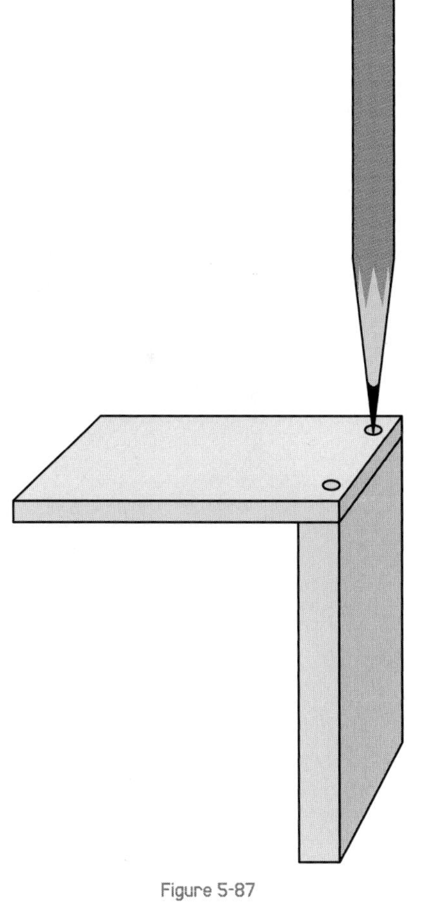

Figure 5-87

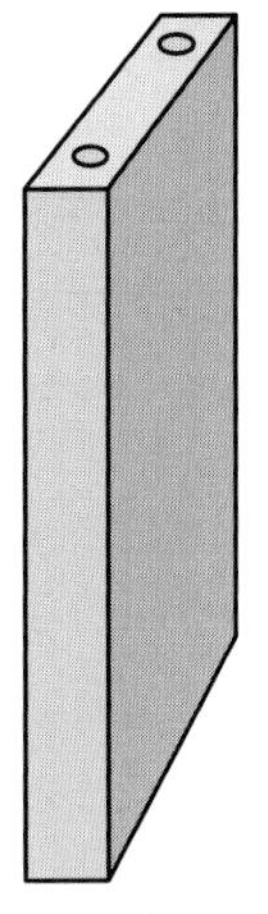

Figure 5-88

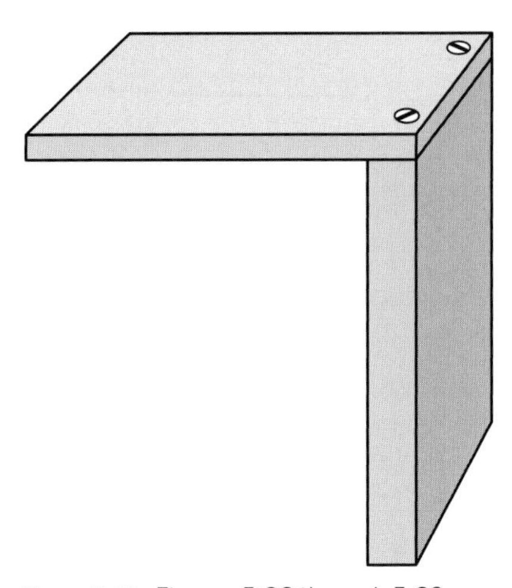

Figure 5-89. *Figures 5-86 through 5-89 illustrate four steps to join two pieces of ABS using #4 sheet-metal screws. Cut 7/64-inch holes on a line 1/8 inch from the edge of the first piece, then mark through the holes to the edge of the second piece. Drill 3/32-inch holes that are precisely centered in the edge, then screw the pieces together.*

Figure 5-90. *Three #4 screws driven into the edge of ABS, using a 1/16-inch guide hole, a 5/64-inch guide hole, and a 3/32-inch guide hole. respectively Because the first two guides holes were too small, the plastic swelled around the screw (but did not break).*

Figure 5-91. *If you have 3D rendering software, it can be a great way to test the feasibility of a construction project before you start cutting materials and trying to fit pieces together. This rendering was a proof-of-concept for the Little Robot Cart.*

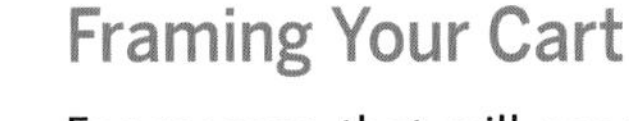

Framing Your Cart

For reasons that will soon be apparent, I've chosen an unusual diamond-shaped configuration of wheels. In the rendering shown in Figure 5-91, the front wheel (at the far end of the cart) applies power, the rear wheel (at the near end of the cart) steers the cart when backing up, and the side wheels prevent it from falling over.

Depending on the type of motor that you buy, you'll have to improvise a way to mount it in the front section of the cart. Don't be afraid to use kludges such as cable ties, duct tape, or even rubber bands to attach the motor to the frame. We're making a rough prototype, here, not a thing of beauty (although if you decide you like the cart, you can always rebuild it beautifully later).

The plan in Figure 5-92 shows the pieces that you will need. Part A is the body of the cart. If you're going to bend it from ABS, you should drill half-inch holes, with a forstner bit, at the four inside corners, so that these corners have rounded edges. If you simply saw the plastic to make sharp 90° corners, the plastic may develop fissures at the corners when you bend it. If you don't have a plastic bender and don't feel inclined to buy one, you can make Part A from three separate rectangles and then screw them together.

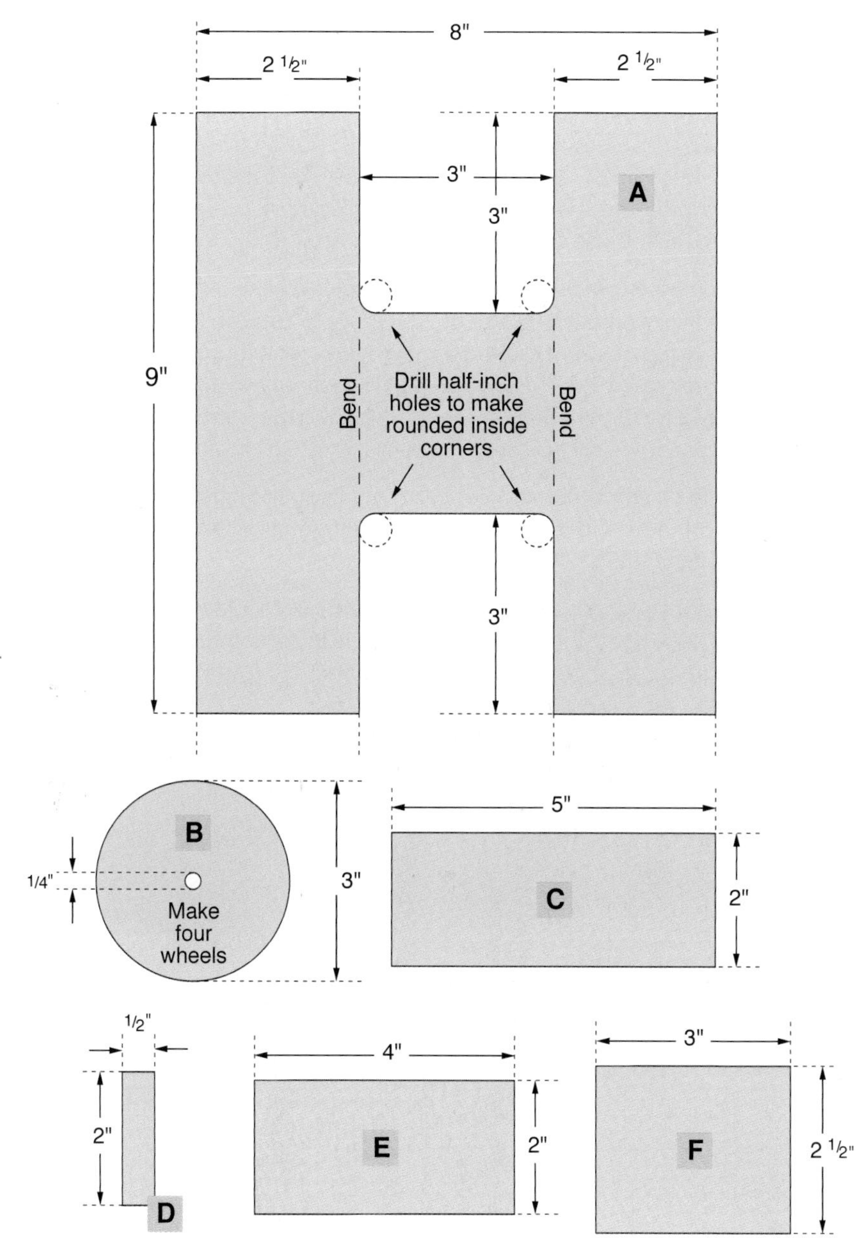

Figure 5-92. *These sections of 1/4-inch plastic can be assembled to create the simple cart described in Experiment 32.*

Part B is a wheel, of which you will need four. I cut them using a 3-inch hole saw. The front wheel is screwed to whatever disc or arm you obtained to mate with the shaft of your motor. See Figure 5-93.

Figure 5-93. *A 3-inch wheel is screwed to the disc that mates with the drive shaft of the motor.*

Parts C, D, and E assemble to form a yoke in which the rear wheel is mounted. I used a 2-inch hinge to pivot the yoke. The hinge is mounted on Part F, which is a partition located midway in the frame of the cart. The photographs in Figures

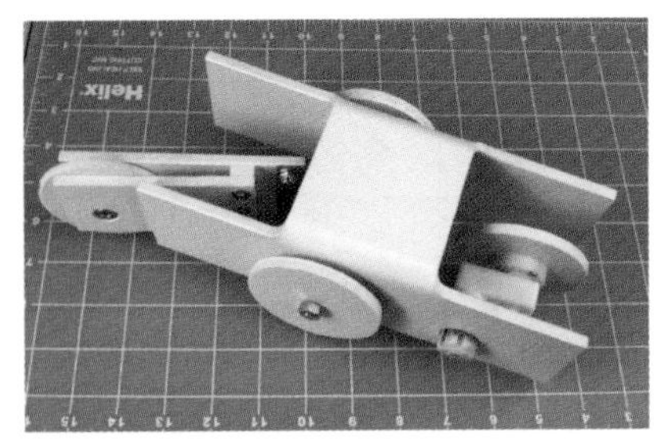

Figure 5-94. *The assembled body of the cart, before adding control electronics. The wheel at the righthand end will pull the cart from left to right. The hinged trailing wheel will allow the cart to move in a relatively straight line when it moves forward, but will tend to turn it when it backs up.*

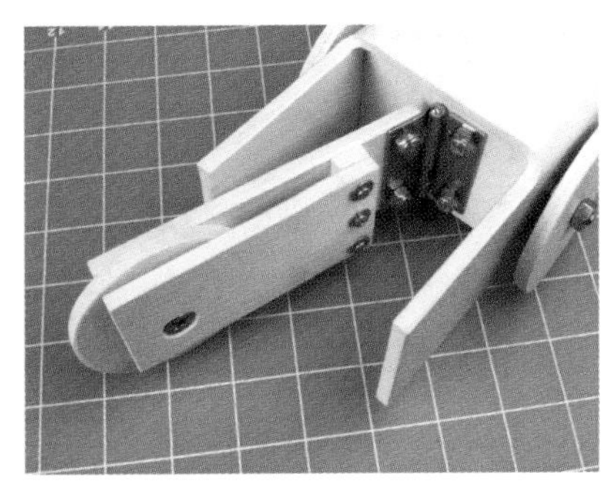

Figure 5-95. *A closeup of the hinged trailing wheel, which rotates freely and can flip from side to side with minimal friction.*

5-94 and 5-95 will help to make this clear. Initially, when you install Part F, use only two screws, one each side, so that you can adjust its angle a little. This will be necessary to optimize the contact of the wheels with the floor.

The side wheels and rear wheel must spin freely, but on the other hand, they shouldn't wobble. I simply tightened the nuts on the bolts that serve as axles for the wheels, until there was maybe half a millimeter of clearance. I added a drop of Loctite to stop the nuts from getting loose.

The plans don't show precisely where to drill holes for the axle bolts, because the location will depend on the size of your wheels. You can figure this out as you go along. Just make sure that the side wheels aren't mounted too low. We don't want them to lift the front wheel or the rear wheel off the floor. If the side wheels are a fraction higher off the ground than the front and rear wheels, that's good.

If you have tile or wood floors, your cart may acquire better traction if you wrap a thick rubber band around each disc that you use for the drive wheel and the steering wheel.

The most important aspect of the construction is to place microswitches where they'll be triggered when the cart runs into something. I placed mine at the front corners, as shown in Figures 5-96 and 5-97. And that brings me to the electronics.

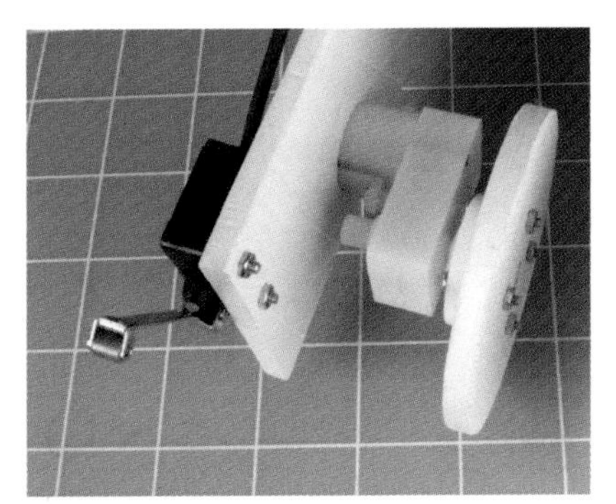

Figure 5-96

Figure 5-97. *Two microswitches with metal arms are mounted on each side of the cart, where they will sense any obstacle.*

The Circuit

The schematic is very, very simple, with only four principal components: two microswitches that sense obstacles in front of the cart, one relay, and one 555 timer. You will also need a small power switch, a battery or battery pack, and a resistor, and capacitors to go with the timer. A trimmer potentiometer will allow you to adjust the "on" time of the 555 timer, which will determine how long the cart takes to back up. See Figure 5-98.

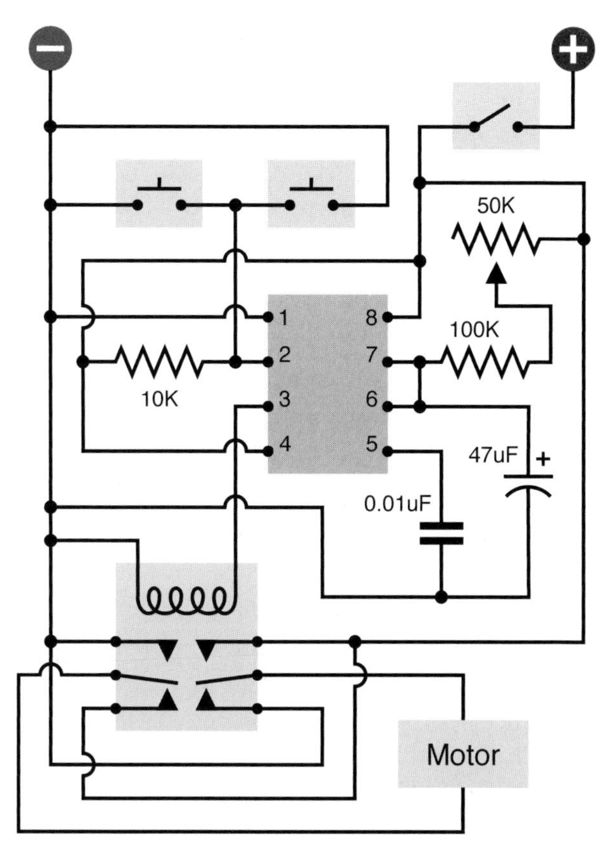

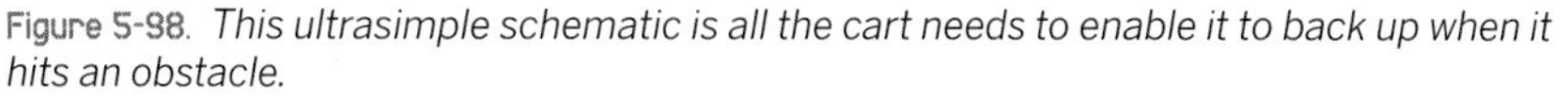
Figure 5-98. *This ultrasimple schematic is all the cart needs to enable it to back up when it hits an obstacle.*

The motor I chose requires 5 volts, so I had to use a voltage regulator with a 9-volt battery. If your motor uses 6 volts, you can wire four AA batteries to it directly. If you have a 12-volt motor, you can use two 9-volt batteries in series, supplying power through a 12-volt voltage regulator.

Assemble the components, mount them on the cart, and switch it on, and it should move forward slowly in a more-or-less straight line. If it moves backward, reverse your connection to the terminals on the motor.

When the cart bumps into something, either of the microswitches will connect negative voltage to the input pin of the 555 timer. This triggers the timer, which runs in monostable mode, generating a single pulse lasting about 5 seconds, which closes the relay, which is wired so that it reverses the voltage to the motor.

When the voltage is reversed to a simple DC motor, it runs backward. So the cart backs up. Because the rear wheel is mounted in a yoke that pivots, the yoke will tend to flip one way or the other, causing the cart to describe an arc as it moves backward. At the end of the timer cycle, the relay relaxes and the cart starts moving forward again. In forward mode, the rear wheel just follows along without applying any steering force, so the cart tends to follow a straight line—until it hits another obstacle, at which point it backs up, and tries another path.

FUNDAMENTALS

All about limit switches

The most obvious enhancement for your cart would be a better steering mechanism. You could use another motor to take care of this, with a pair of limit switches. Because limit switches are a basic, important idea in conjunction with motors, I'll explain them in detail.

Figure 5-99 shows three successive views of a motor with an arm attached to it, which can press either a lower pushbutton or an upper pushbutton. Both of the pushbuttons are normally closed, but will open when pressed by the motor arm. These buttons are the limit switches. Typically you would use microswitches for this purpose, just like the ones that I suggested as barrier-sensors at the front of the cart.

In addition, there's a DPDT relay that is activated by a simple on/off switch at the righthand side. On the cart, the 555 timer takes the place of the on/off switch, by feeding power to the relay.

Suppose that the motor begins with the arm pointing downward, as shown in the top view in Figure 5-99, and the motor is wired so that when it receives negative voltage at its lower terminal and positive at its upper terminal, it rotates counter-clockwise. This is what happens when the on/off switch closes and sends power to the DPDT relay. Positive voltage from the relay contacts cannot pass through the upper diode, but can pass through the upper limit switch, which is closed. Negative voltage cannot pass through the lower limit switch, because it's open, but can pass through the lower diode. So, the motor starts to turn counterclockwise. During the midpoint of its arc, it receives power through both of the limit switches.

Finally, the motor arm reaches the upper switch, and opens it. This prevents positive voltage from reaching the motor through that switch, and the positive voltage is also blocked by the upper diode. So, at this time, the motor stops.

Now suppose that the on/off switch is opened, as in the top view in Figure 5-100. The relay loses its power, so its contacts relax. The voltage to the motor is now reversed. Negative voltage passes through the upper diode, while positive voltage reaches the motor through the lower limit switch. The motor starts running clockwise, until its arm hits the lower switch, opening it and cutting off power to the motor.

Limit switches are necessary, because if you continue to apply voltage to a simple DC motor that is unable to turn, the motor sucks more current, gets hot, and may burn out.

You can easily see how this kind of system could be used to control the cart's steering. Even though the motor has only two positions, these are sufficient to make the cart turn when going backward, and proceed straight ahead when going forward.

To reduce power consumption, the DPDT relay could be replaced with a two-coil latching relay. The circuit would then have to be revised so that the relay is flipped to and fro by a pulse to each of its coils.

FUNDAMENTALS

All about limit switches (continued)

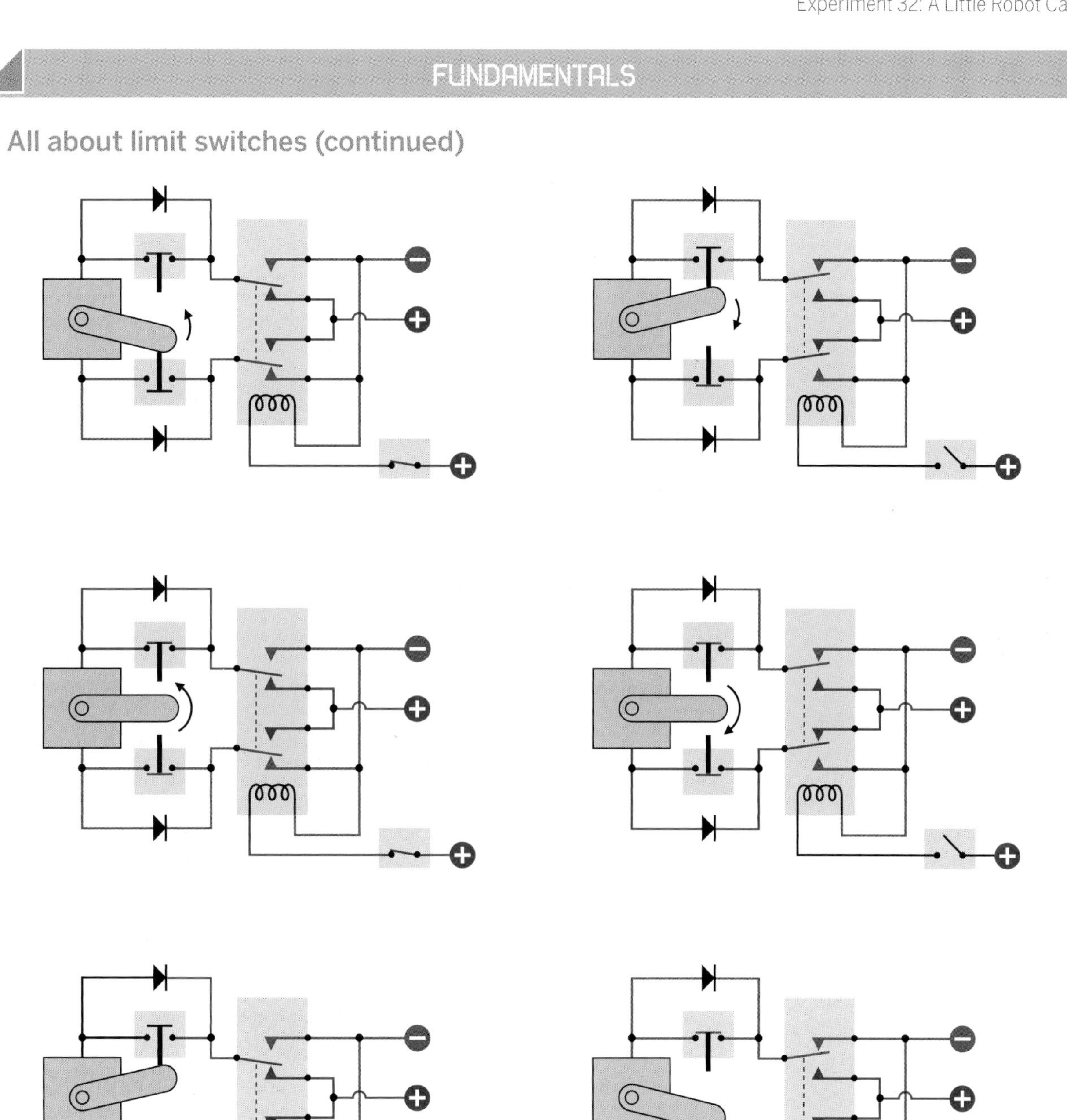

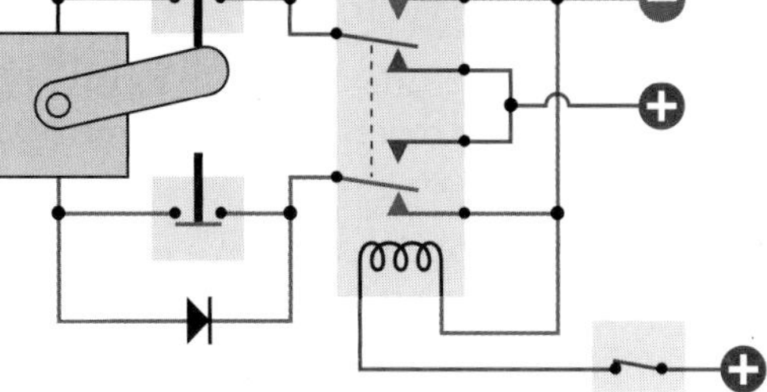

Figure 5-99. *The three diagrams, from top to bottom, show three snapshots of a motor controlled by a DPDT relay and two limit switches. When the on/off switch at bottom-right sends power to the relay, the lower relay contacts cause the motor to run counterclockwise until it stops itself as its arm opens the upper limit switch.*

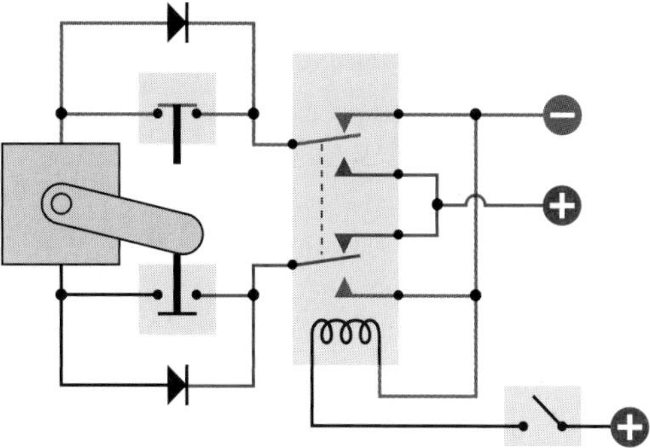

Figure 5-100. *When the on/off switch at bottom-right opens, the relay connects its upper contacts. This causes the motor to run clockwise until its arm opens the lower limit switch. Limit switches avoid the overheating and possible damage that are likely when power is delivered to a motor that is prevented from turning.*

FUNDAMENTALS

All about motors

Brushed DC motor

This is the oldest, simplest design for an electric motor, shown in very simplified form in Figure 5-101. Coils are attached to a shaft where they can interact with stationary magnets around them. The magnetic attraction turns the shaft a little, at which point the next coil on the shaft is energized to turn the shaft a little more, and then the next coil—and so on. To make this happen, electricity has to be fed into the coils by "brushes," often consisting of soft carbon pads that conduct power to a hub, known as a commutator, divided into sections, each of which is connected to a separate coil.

This basic design has several advantages if we want to build a small motorized gadget, such as a miniature robot or even a model airplane:

- Widely available
- Low cost
- Simple
- Reliable
- Will run in reverse when voltage reverses

In addition, brushed motors are often sold with reduction gearing built in. Such units are known as *gearhead motors* or *gear motors*. They free you from the need to use your own gears or belts to adjust the output speed yourself. You simply choose the motor that fits your specification.

DC stepper motor

This requires a controller, consisting of some electronics to tell the motor to rotate its shaft in small, discrete steps. The advantages of a stepper motor are:

- Precise positioning of the shaft
- Precise speed adjustment

Stepper motors are ideal for devices such as computer printers, where the paper has to roll up by a precise distance and the print head has to move laterally by an equally precise distance, but they are also useful in robots. If the motor is small enough to draw less than 200mA and will run on 12 volts or less, you can control it with pulses from a 555 timer. I'll describe stepper motors in more detail in Experiment 33.

Servo motor

This is generally used in conjunction with a programmable microcontroller, which sends instructors to rotate the motor shaft to a specific position and then hold it there. I'll mention servo motors when I introduce you to microcontrollers, but we won't be dealing with them in detail.

Other types of motors exist, including brushless DC motors (which require a different type of controller and are found in computer disk drives and CD players), and AC motors (including synchronous motors, which synchronize their rotation with the frequency of AC voltage, and were used extensively in clocks, before clocks mostly became digital).

In this book, I'll be talking mostly about brushed DC motors and DC stepper motors.

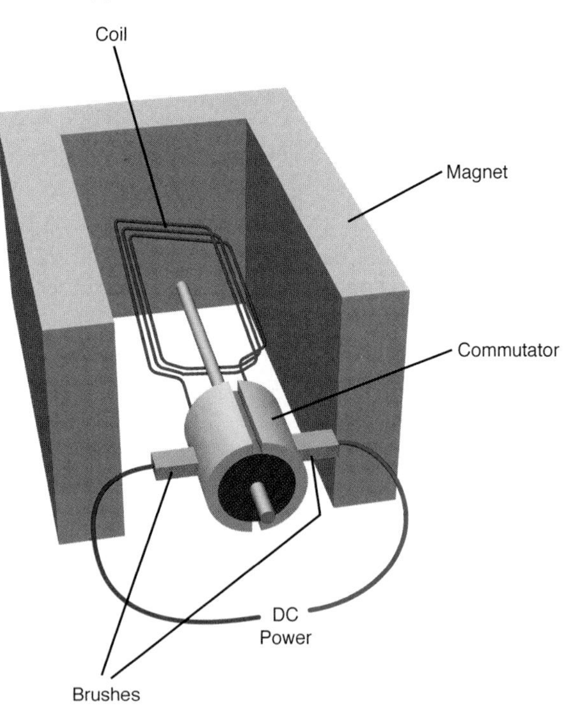

Figure 5-101. *The basic principle of a simple DC motor. The commutator passes electricity through a coil, creating a magnetic field that interacts with a magnet around the motor. The coil turns, and the commutator turns with it, until the electric field through the coil is reversed. This causes the process to repeat. In reality, a motor is likely to have a commutator formed from multiple segments, connected with multiple coils. The principle, however, remains the same.*

Take-home messages from this experiment include the following:

- You can buy simple DC motors with reduction gearing built in, providing your choice of RPM. Literally hundreds of websites will sell you small motors for robotics projects.
- When you reverse the voltage to a DC motor, the motor runs in reverse.
- A DPDT relay can be wired so that when it closes its contacts, it reverses a power supply to a motor.
- You can use two limit switches and a pair of diodes to stop a motor at two positions. In each of its stopping positions, the motor consumes no power and you won't have the risk of it burning out.

What other projects can you imagine using this simple set of techniques?

Mechanical Power

In the United States, the turning force, or torque, of a motor is usually measured in pound-feet or ounce-inches. In Europe, the metric system is used to measure torque in Newton-meters.

A pound-foot is easy to understand. Imagine a lever pivoted at one end, as shown in Figure 5-102. If the lever is one foot long, and you hang a one-pound weight at the end of it, the turning force is one pound-foot.

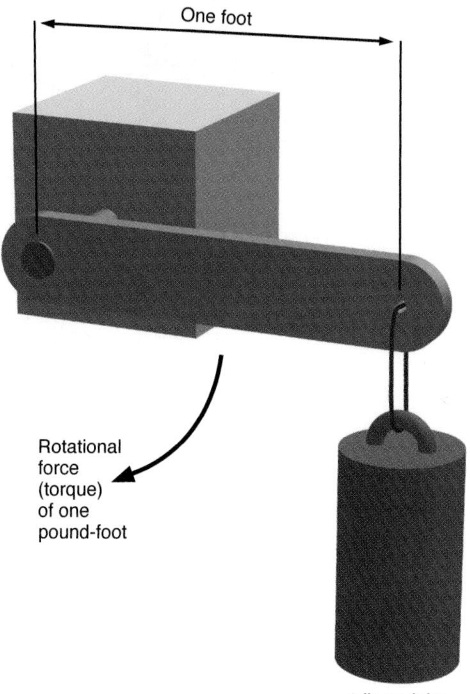

Figure 5-102. *The rotational force created by a motor is known as "torque," and in the United States it is measured in pound-feet (or ounce-inches, for small motors). In the metric system, torque is measured in dynes. Note that the torque created by a motor will vary according to the speed at which the motor is running.*

FUNDAMENTALS

Wire gauges

If you're going to power larger motors, or other components that take more current than LEDs or small relays, you really need to know about wire gauges. In particular, what's the relationship between wire thickness and AWG (American Wire Gauge)? And what gauge of wire should you use for any given current?

You can find numerous charts and tables if you go online, but many of these sources contradict each other, especially on the topic of how much current is safe to run through each gauge of wire.

After making several comparisons (and testing some wire samples myself), I've compiled the table in Figure 5-103, which I recommend as a compromise. Note the following:

- This table applies to solid-core copper wire.
- For stranded wire, or copper that has been tinned (giving it a silver appearance), the number of ohms per foot will *increase*, the number of feet per ohm will *decrease*, and the maximum amperage will *decrease*, probably by around 20%.

The maximum amperage assumes that the wire is insulated, preventing it from radiating heat as effectively as a bare conductor. I am also assuming that the wire is likely to be at least partially enclosed, inside a box or cabinet. At the amperages listed for each gauge of wire, you should expect the wire to become noticeably warm, and personally I would tend to use thicker wire instead of the maximums indicated in the table.

Most tables of this type only tell you the resistance of each gauge of wire in ohms per 1,000 feet. I have included that number but have also expressed the function the other way around, as the number of feet per ohm, as this doesn't require you to do so much arithmetic with decimals.

AWG		Diameter in inches	Ohms per 1,000 ft	Feet per ohm	Maximum amps (insulated)
0000	●	0.46	0.049	20,400	225
000	●	0.41	0.062	16,200	200
00	●	0.365	0.078	12,800	175
0	●	0.325	0.098	10,200	150
1	●	0.289	0.124	8,070	125
2	●	0.258	0.156	6,400	100
3	●	0.229	0.197	5,080	90
4	●	0.204	0.249	4,020	80
5	●	0.182	0.313	3,190	70
6	●	0.162	0.395	2,510	60
7	●	0.144	0.498	2,010	51
8	●	0.128	0.628	1,590	44
10	●	0.102	0.999	1,000	32
12	•	0.081	1.59	630	23
14	•	0.064	2.53	396	17
16	•	0.051	4.02	249	13
18	•	0.04	6.39	157	10
20	·	0.032	10.2	99	8
22	·	0.025	16.1	62	5
24	·	0.02	25.7	39	2.5
26	·	0.016	40.1	25	1.5
28	·	0.013	64.9	15	1.0
30	·	0.010	103.2	10	0.5

Figure 5-103. *American wire gauges (AWG) and their properties.*

THEORY

Calculating voltage drop

Another fact that you often need to know is how much of a voltage drop a particular length of wire will introduce in a circuit. If you want to get maximum power from a motor, you don't want to lose too much voltage in the wires that go to and from the motor.

Voltage drop is tricky, because it depends not only on the wire, but also on how heavily the circuit is loaded. Suppose that you are using 100 feet of 22-gauge wire, which has a resistance of about 1.5 ohms. If you attach it to a 12-volt battery and drive an LED and a series resistor offering a total effective resistance of about 1,200 ohms, the resistance of the wire is trivial by comparison. According to Ohm's Law:

amps = volts / ohms

so the current through the circuit is only about 10mA.

Again, by Ohm's Law:

volts = ohms × amps

so the wire with resistance of 1.5 ohms imposes a voltage drop of 1.5 × 0.01 = 0.015 volts.

Now suppose you're running a motor. The coils in the motor create impedance, rather than resistance, but still if we measure how much current is going through the circuit, we can establish its effective resistance. Suppose the current is 1 amp. Repeating the second calculation:

volts = ohms × amps

So the voltage drop in the wire is now 1.5 × 1 = 1.5 volts! This is illustrated in Figure 5-104.

Bearing these factors in mind, I have compiled a table for you. I've rounded the numbers to just two digits, as variations in the wire that you use make any pretense of greater accuracy unrealistic.

To use this table, you need to know how much current is passing through your circuit. You can calculate it (by adding up all the resistances and dividing it into the voltage that you are applying) or you can simply measure the current with a meter. Just make sure that your units are consistent (all in ohms, amps, and volts, or milliohms, milliamps, and millivolts).

In the table, I have arbitrarily assumed a length of 10 feet of wire. Naturally you will have to make allowances for the actual length of wire in your circuit. The shorter the wire, the less the loss will be. A circuit with only 5 feet of wire, and the same amperage and voltage, will suffer half of the percentage loss shown in the table. A circuit with 15 feet of wire, and the same amperage and voltage, will suffer 1.5 times the percentage loss. So, to use the table:

1. Divide your length of wire by 10. (Make sure that you measure the length in feet.)
2. Use the result to multiply the number in the table.

The table also arbitrarily assumes that you have a 12-volt supply. Again, you will have to make allowances if you are using a different voltage. So, to use the table:

1. Divide 12 by the actual voltage of your power supply.
2. Use the result to multiply the number in the table.

I can summarize those two steps like this:

Percent voltage lost = P × (12 / V) × (L / 10)

where P is the number from the table, V is your power-supply voltage, and L is the length of your wire.

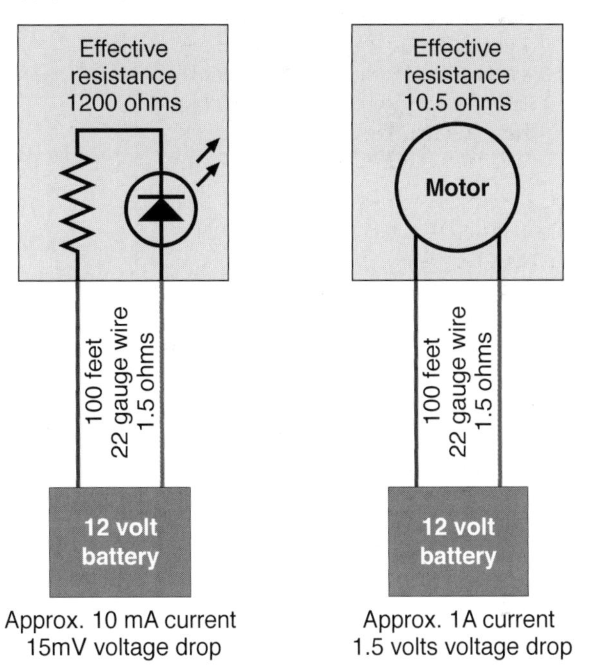

Figure 5-104. *The voltage drop imposed by wiring will depend on the current and the resistance in the circuit. The drop will be greatest when the resistance of the circuit is low and the amperage is high.*

THEORY

Calculating voltage drop (continued)

This table shows the percent voltage lost in a circuit with 10-foot wire at 12 volts.

Wire Gauge	Amperes 1	2	3	4	5	6	7	8	9	10
10	0.08	0.17	0.25	0.33	0.42	0.50	0.58	0.67	0.75	0.83
12	0.13	0.27	0.40	0.53	0.66	0.80	0.93	1.1	1.2	1.3
14	0.21	0.42	0.63	0.84	1.1	1.3	1.4	1.5	1.9	2.1
16	0.33	0.67	1.0	1.3	1.7	2.0	2.3	2.7	3.0	3.4
18	0.53	1.1	1.6	2.1	2.7	3.2	3.7	4.3	4.8	5.3
20	0.85	1.7	2.6	3.4	4.3	5.1	6.0	6.8	7.7	8.5
22	1.3	2.7	4.0	5.4	6.7	8.1	9.4	11	12	13
24	2.1	4.3	6.4	8.6	11	13	15	17	19	21
26	3.4	6.8	10	14	17	20	24	27	31	34
28	5.4	11	16	22	27	32	38	43	49	54
30	8.6	17	26	34	43	52	60	69	77	86

Remember, though, that the wire resistance will be higher if you are using stranded copper wire or tinned copper wire, and this will increase the percentage of voltage lost.

Experiment 33: Moving in Steps

Time now to build something more sophisticated: a cart that orients itself toward a light source. I'm going to tell you all you need to get started on this project, but this time I won't go all the way to the end in exhaustive detail. I want you to get into the habit of figuring out the details, improving on plans, and eventually inventing things for yourself.

You will need:

- 555 timers. Quantity: 8.
- Trimmer potentiometer, 2K linear. Quantity: 2.
- LEDs. Quantity: 4. If you get tired of using series resistors to protect LEDs in a 12-volt circuit, consider buying 12-volt LEDs such as Chicago Miniature 606-4302H1-12V, which contain their own resistors built in. However, the schematic in Figure 5-108 assumes that you will use regular 2V or 2.5V LEDs.
- Stepper motor: Unipolar, four-phase, 12-volt. Parallax 27964 or similar, consuming 100mA maximum. Quantity: 2.
- Photoresistors, ideally 500 to 3,000Ω range. Quantity: 2.

- ULN2001A or ULN2003A Darlington arrays by STMicroelectronics. Quantity: 2.
- CMOS octal or decade counter. Quantity: 2.
- Various resistors and capacitors.

Exploring Your Motor

I've specified a unipolar, four-phase, 12-volt motor because this is a very common type. A typical sample is shown in Figure 5-105. If you can't easily find the one that I've listed, you should feel safe in buying any other that has the same generic description. "Unipolar" means that you don't have to switch the power supply from positive to negative and back to positive again, to run the motor. Four-phase means that the pulses that run the motor must be applied in sequence to four separate wires. Because you will be running your motor directly from 555 timers, the lower its power consumption, the better.

Figure 5-105. *A typical stepper motor. The shaft rotates in steps when negative pulses are applied to four of the wires in sequence, the fifth wire being common-positive.*

First, though, we can apply voltage to the motor without using any other components at all. Most likely it will have five wires already attached, with the ends stripped and tinned, so that you can easily insert them into holes in a breadboard, as shown in Figure 5-106. Check the data sheet for your motor; you should find that four of the wires are used to energize the motor and turn it in steps, while the fifth is the common connection. In many cases, the common connection should be hooked to the positive side of your power supply, while you apply negative voltage to the other four wires in sequence, one step at a time.

The data sheet will tell you in what sequence to apply power to the wires. You can figure this out by trial and error if necessary. One thing to bear in mind: a stepper motor is very tolerant. As long as you apply the correct voltage to it, you can't burn it out.

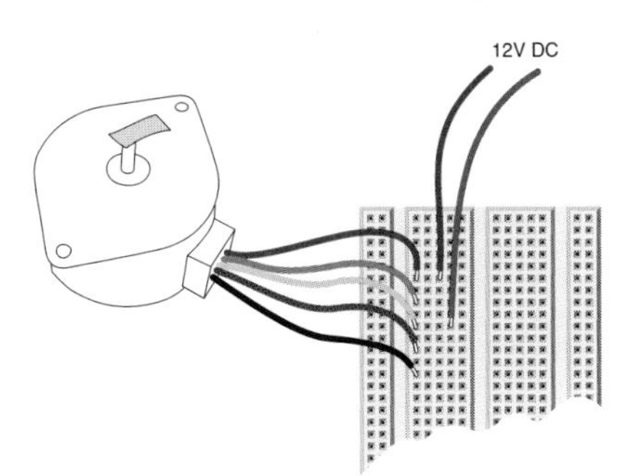

Figure 5-106. *The simplest test of a stepper motor is to apply voltage manually to each of its four control wires, while a piece of duct tape, attached to the output shaft, makes it easy to see how the motor responds.*

To see exactly what the motor is doing, stick a piece of duct tape to the end of the shaft. Then apply voltage to wires, one at a time, by moving your negative power connection from one to the next. You should see the shaft turning in little steps.

Inside the motor are coils and magnets, but they function differently from those in a DC motor. You can begin by imagining the configuration as being like the diagram in Figure 5-107. Each time you apply voltage to a different coil, the black quadrant of the shaft turns to face that coil. In reality, of course, the motor turns less than 90° from one coil to the next, but this simplified model is a good way to get a rough idea of what's happening. For a more precise explanation, see the upcoming section "Theory: Inside a stepper motor."

Bear in mind that as long as any of the wires of the motor are connected, it is constantly drawing power, even while sitting and doing nothing. Unlike a regular DC motor, a stepper motor is designed to do nothing for much of the time. When you apply voltage to a different wire, it steps to that position and then resumes doing nothing.

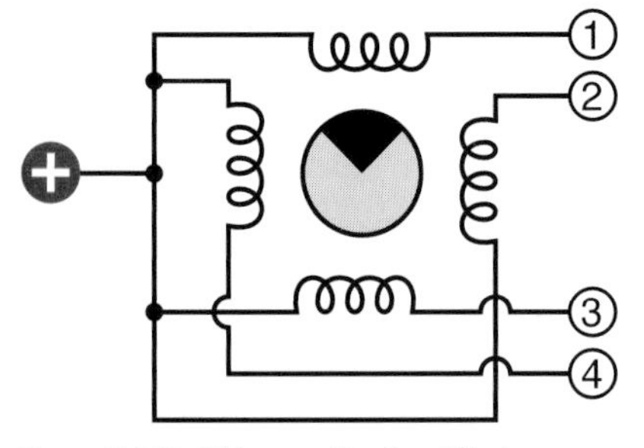

Figure 5-107. *This greatly simplified diagram helps in visualizing the way in which a stepper motor works. In reality, almost all motors rotate by less than 90° in response to each pulse.*

The coil inside the motor is holding the shaft in position, and the power that the motor draws will be dissipated as heat. It's quite normal for the motor to get warm while you're using it. The trouble is, if you use a battery to power it, and you forget that you have it connected, the battery will not hold its charge for long.

Quick Demo

Now that you've proved that your motor is functional, how can you actually run it? You need to send a pulse to each of the four wires in turn, in a rapidly repeating sequence. If you can also adjust the speed of the pulses, so much the better. I'm thinking that for a quick and simple demo, you can handle the challenge simply by using four 555 timers, all of them in monostable mode, with each one triggering the next.

The schematic in Figure 5-108 shows what I have in mind. It looks more complex than it really is. Each timer has the same pattern of components around it, so after you create the first module, you just make three copies of it.

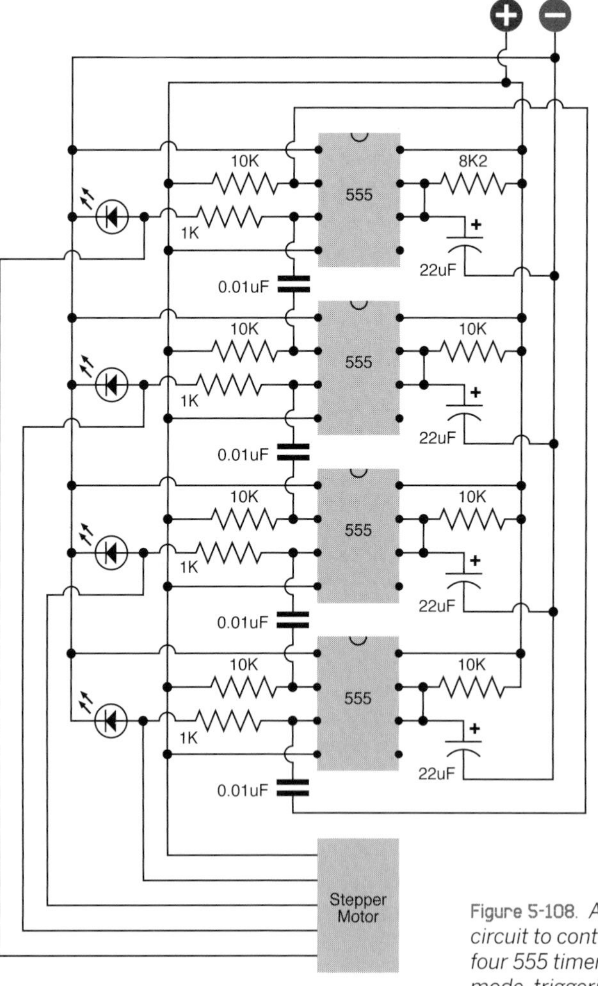

Figure 5-108. *A very quick and simple circuit to control a stepper motor uses four 555 timers, each in monostable mode, triggering each other in a repeating sequence.*

I've used a 10K resistor to pull up the input to each 555, so that the timers are naturally in their quiescent state. A 0.01 µF capacitor links the output from one timer to the input of the next so that they are electrically isolated from each other, and the capacitor just conveys a "spike" of voltage when one timer finishes its "on" cycle, and its output goes low, which triggers the next.

On the righthand side, I've used 10K resistors and 22 µF capacitors to generate a cycle of about a quarter of a second—except that the topmost timer has a 8K2 timing resistor. The reason for this is that when power is first applied, the timers will all be waiting for each other to begin, and timers 2 and 4 or 1 and 3 may fire together. By giving one timer a shorter cycle than the others, I minimize this problem.

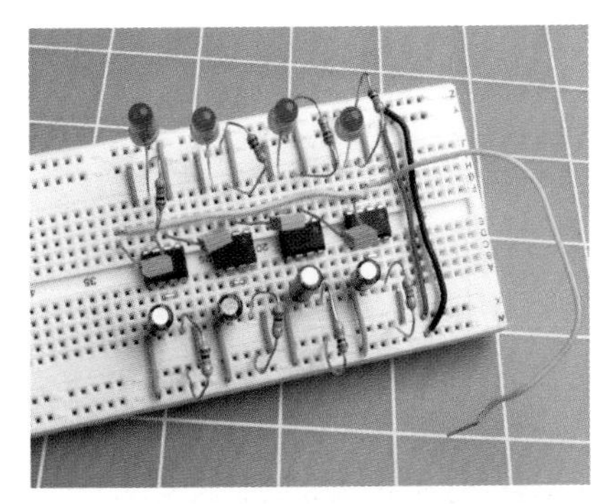

Figure 5-109. *To test the control circuit for errors, four LEDs show the outputs from the four 555 timers. The loose yellow wire at the righthand side connects to pin 2 of the first timer. Touch the free end of this wire to the positive side of the power supply to reset the timers, and then, if necessary, make a brief negative connection with the free end of the wire to restart their sequence.*

The LEDs are included just to give you some visual verification of what's happening. Without them, if you make a wiring error, the motor may turn to and fro erratically, and you won't know why. Initially you can run your circuit with only the LEDs connected, just to make sure it works. Figure 5-109 shows the breadboarded circuit before the motor is plugged in. Then add the motor by plugging its wires into the breadboard, where you'll make connection with the outputs (pins 3) of the timers. See Figure 5-110.

Figure 5-110. *After the circuit has been tested, the motor can be added by hooking its control wires to the outputs of the four 555 timers.*

Apply power, and you should see the motor turning in steps, in sequence with the LEDs. If the LED sequence isn't stable:

1. Connect a wire directly from the input (pin 2) of the topmost timer to the positive side of the voltage supply, and wait for the timers to calm down.
2. Restart the sequence by disconnecting the free end of this wire, or (if necessary) touch the free end of it briefly to the negative side of the supply, to trigger the first timer.
3. If you have trouble getting the motor to respond at all, try disconnecting the LEDs shown in Figure 5-108. Alternatively, try driving the four poles of the motor directly from the four outputs of the timers, instead of through 1K resistors. To verify that this is not overloading the timers, substitute a much higher value timing resistor for the 8K2 resistor, to slow the first timer to a 5-second pulse, and insert your meter, set to measure mA, between pin 3 of this timer and the load that it is driving. The maximum current must not exceed 200mA.

One thing you may have noticed, if you're paying very close attention: the common terminal of the motor is connected to positive. Therefore, when each timer flashes positive, that positive signal isn't actually powering the motor. The *low* outputs from the three timers that are *not* firing at any given moment are sinking current from the motor. It seems quite happy with this arrangement. You'll need some theory to understand why.

THEORY

Inside a stepper motor

If you check the Wikipedia entry for stepper motors, you may see a very nice 3D rendering showing a toothed rotor and four coils arrayed around it. Maybe stepper motors used to be manufactured like this once upon a time, but not anymore.

Imagine two horizontal rows of coils. In the space between them is a series of little magnets, like a freight train, that can move left or right, as shown in Figures 5-111 and 5-112. Each coil has two windings, in opposite directions, so that current through one winding will create an upward magnetic force while current through the other will create a downward force. Each row of windings is connected in parallel, so that they switch on and off simultaneously.

In Step 1, the negative connection energizes the upper windings of the upper coils, which creates an upward magnetic force. I've shown this force using blue-green arrows so that you won't mistake it for a flow of electricity. It so happens that this force attracts the north poles of the magnets and repels the south poles, so if the magnets begin in the position shown in Step 1, they will want to move one step to the right.

This brings them to the position shown in Step 2. Now the upper windings of the lower coils are energized, and again, this produces an upward force, which again attracts the north poles and repels the south poles.

This advances the magnets to their location in Step 3. Now the lower windings of the upper coils are energized, producing a downward force. This repels the north poles of the magnets and attracts their south poles. So the magnets keep moving.

They reach the position shown in Step 4. The lower windings of the lower coils are energized, producing a downward force which continues to attract the south poles while repelling the north poles. So the magnets move a final step to the right—which leaves them in the same orientation shown in Step 1. And the process can repeat all over again.

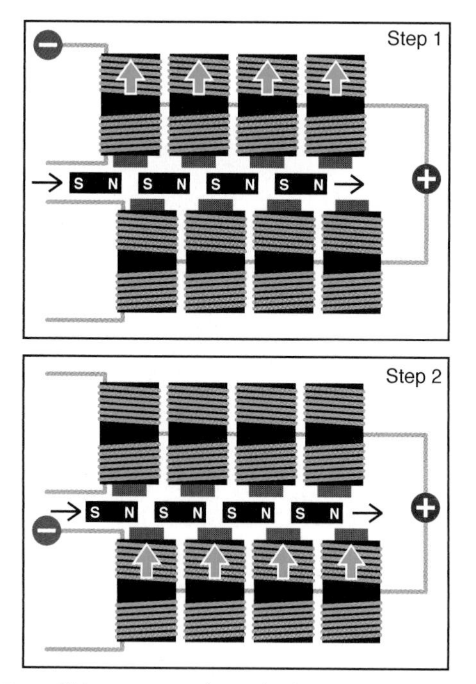

Figure 5-111. *This sequence shows the first two steps as the rotor of a stepper motor (shown as a series of north-south magnets) moves in response to pulses through electromagnets.*

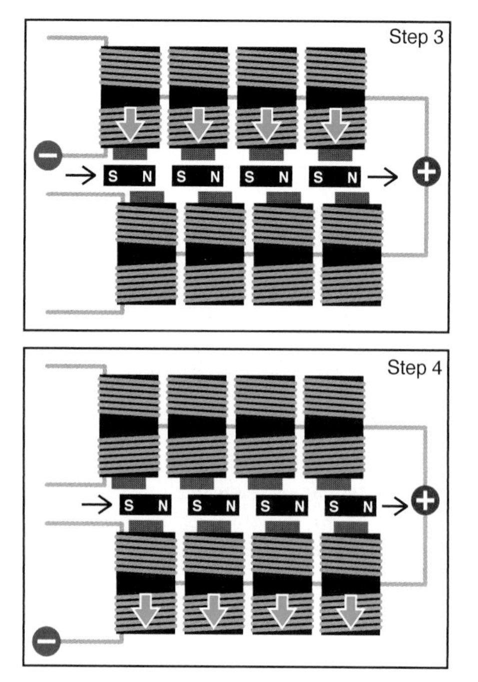

Figure 5-112. *After taking another two steps, the motor will be back where it started at Step 1 in Figure 5-111.*

THEORY

Inside a stepper motor (continued)

In reality, the magnets are not separate from each other. The edge of a rotor is magnetized in zones that alternate between south and north polarity. And instead of multiple coils, there are just four windings that go around all the magnetic cores. But the principle is exactly the same. The 3D rendering gives a general idea, and the photograph shows what I found when I cracked open a typical stepper motor.

Now bear in mind that when this device is driven by a set of 555 timers, we don't just connect negative to one wire at a time on the left, leaving the others floating. In reality, at any given moment, three of the timers have a negative output and the fourth has a positive output. The last diagram in Figure 5-112 shows this situation.

Suppose the top wire is positive while the other three are negative, as shown in Figure 5-113. The positive output does nothing, because it is balanced by the positive power on the other end of the coils. The two negatives attached to the bottom set of coils create equal and opposite forces that cancel each other out (while wasting some power). So the net result is the same as in Step 3.

In fact, you should find that you can disconnect the common wire completely while using the stepper motor with 555 timers, and the motor will still turn, because one of the timers is providing positive power while the others are negative. In fact, you'll be running them more efficiently this way.

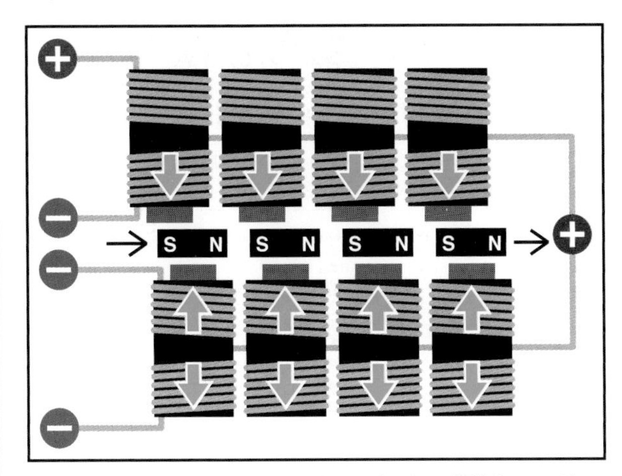

Figure 5-113. *When the motor is driven by four 555 timers, they are activating it by sinking positive voltage from it. The interior workings of the motor look something like this. It's not the most efficient way to do the job.*

Figures 5-114 and 5-115 may help to give you a clearer idea of what the motor actually looks like inside.

Figure 5-114. *This 3D rendering gives a better idea of what a typical stepper motor looks like inside. The copper coils and gray cylinders are stationary, while the black disc rotates between them.*

Figure 5-115. *When a stepper motor is broken open, this is what you're likely to find. On the left, the rotor of the motor, which has a magnetized band around its circumference, is still attached to the lower half of the casing. On the right, the upper half of the casing has been opened, and the coil has been removed (actually the winding you can see consists of two coils, wound in opposite directions). The spikes are the magnetic cores that exert force on the rotor.*

Speed Control

If you are a truly exceptionally observant, you may have noticed that I left pin 5 of each of the timers unconnected in the schematic for driving the stepper motor in Figure 5-108. Normally, pin 5 should be grounded through a capacitor to prevent it from picking up stray voltages which can affect the accuracy of the chip.

I left the pins unconnected because I had a plan for them. In fact, changing the timing of the chip is exactly what we want to do now, as a way to change the speed of the stepper motor.

If you tie pin 5 of all four timers together, as shown in Figure 5-116, and put a 2K trimmer potentiometer (shown in Figure 5-117) between them and the negative side of the power supply, you'll find that as you turn the trimmer to reduce its resistance, the timers start to run faster. Figure 5-118 shows the breadboard layout. Eventually, when the resistance goes below around 150 ohms, everything stops. The LEDs go dark, because you've reduced the voltage on pin 5 below the threshold level that the 555 timer finds acceptable.

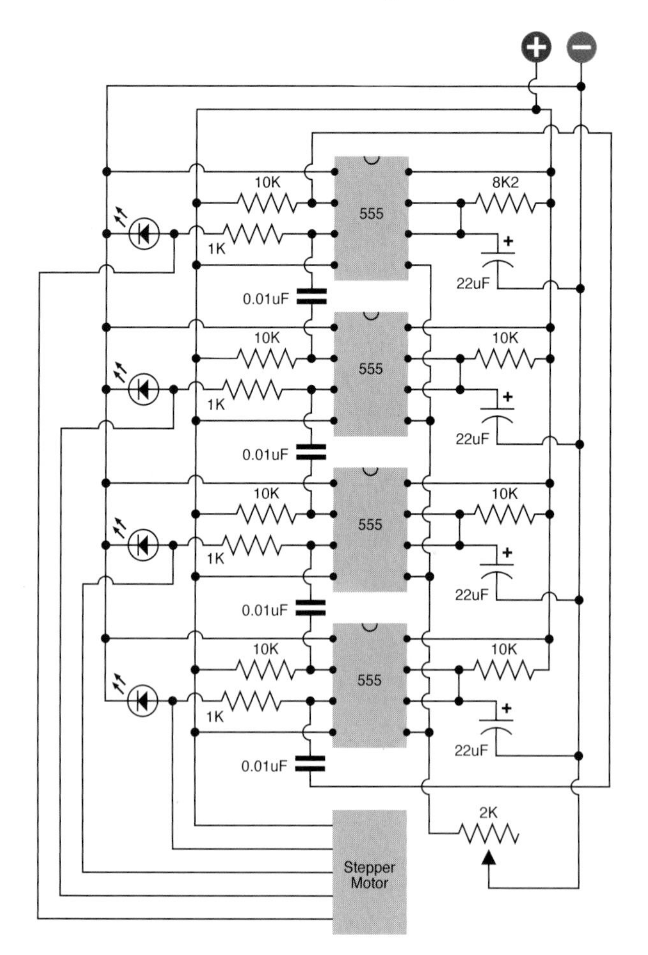

Figure 5-116. *To adjust the speed of the sequence of 555 timers, their control pins (pin 5 on each timer) are linked together and attached to a trimmer potentiometer that adjusts the resistance between the pins and the negative side of the power supply.*

Figure 5-117. *Close-up of a trimmer potentiometer with pins spaced at 1/10 inch for insertion in a breadboard or perforated board. The brass screw, at top-left, turns a worm gear inside the unit, allowing precise adjustment of internal resistance.*

Figure 5-118. *The trimmer potentiometer has been added to the circuit, allowing motor speed control.*

Initially I suggested a step time of 1/4 second just so that you could see what was happening. When you're actually using this circuit, you'll never need it to run as slowly as that. So you can increase the entire range of speeds. Remove the 22 μF timing capacitors and substitute, say, 4.7 μF capacitors, or smaller. Now when you adjust the potentiometer, you'll get a useful range of speed.

Adding Autonomy

Currently, the circuit simply does what you tell it to do. The next step is to make it autonomous—in other words, give it the illusion of making up its own mind. I'm thinking that instead of a trimmer potentiometer, we could substitute a photocell, properly known as a photoresistor. Typically, the resistance of a cadmium sulfide photo resistor is highest in the dark, and lowest when light shines on it.

One problem with photoresistors is that they're not as widely available as many other types of electronic components. If you search Mouser.com, for instance, you'll find virtually nothing. Partly this is because the online search function at Mouser is the weakest feature of the site, and partly it's because Mouser is not oriented toward hobbyists. What you need to do is conduct a "product search." Go to *http://www.google.com/products*, enter the search terms "CdS" and "photocell," and you'll find a bunch of cheap cadmium sulfide components from places you may never have heard of.

Because photoresistors seem to come and go as erratically as DC motors, I am not offering any part numbers. You can buy any product that has an appropriate minimum resistance (in bright light) and maximum resistance (in the dark). If you find a component that ranges from 500 to 3,000Ω, that would be a good choice. If the only ones you can find have a higher minimum than 500Ω, you could consider putting a couple of them in parallel.

Setting Up Your Light Seeking Robot

Why would you want to control the speed of a stepper motor by using a photo resistor? Because the original objective was to build a robot that is attracted to light.

The idea is simple enough: use two stepper motors, each controlling the speed of one wheel of the cart. Use two photoresistors, each controlling the speed of the opposite stepper motor. When the righthand photoresistor picks up more light, its resistance lowers, causing the lefthand set of timers to run faster, which will make the lefthand wheel run faster. Thus, the cart will turn toward the light. Figure 5-119 illustrates the concept.

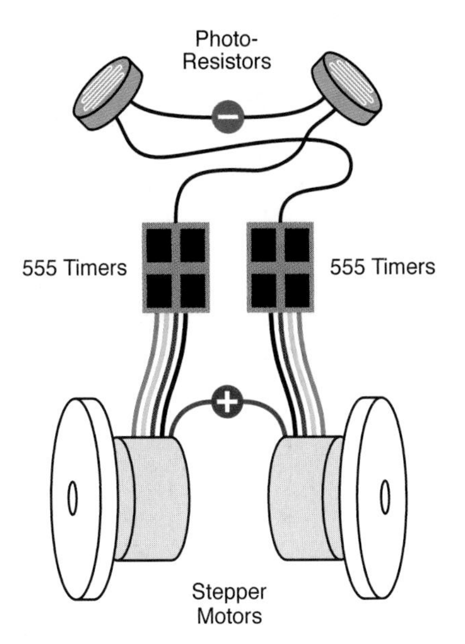

Figure 5-119. *If two photoresistors control the speed of two 555-timer arrays, the difference in speed between one wheel and the other can turn the cart toward a light source.*

Before you start wiring more 555 timers, though, you might consider doing the job with a more appropriate component. The ULN2001A and ULN2003A are chips containing Darlington amplifiers specifically designed to deliver current to inductive loads such as solenoids, relays, and (you guessed it) motors. Each chip has seven inputs that require very little current, and seven outputs that can deliver 500mA each. The inputs are TTL and CMOS compatible (the 2001A has a wider tolerance for voltages than the 2003A) and each channel of

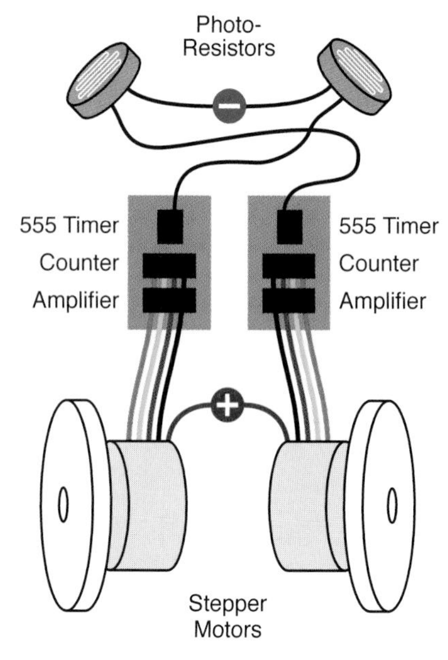

Figure 5-120. *A more efficient way to drive the motors is to use just one timer to set the speed of each, with a counter and amplifier (such as a Darlington array chip) sending the pulses down the wires. The principle is still the same, though.*

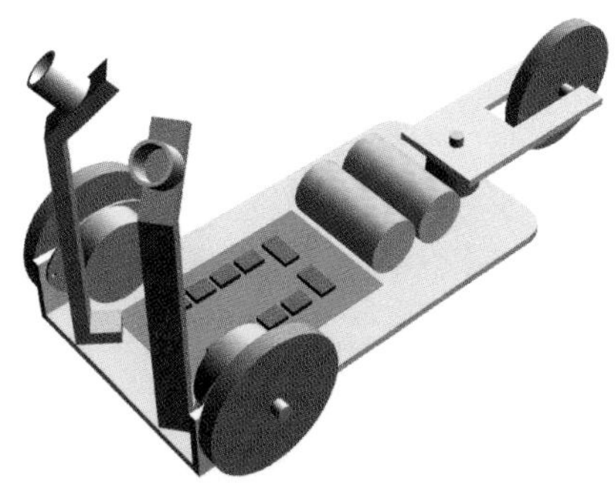

Figure 5-121. *This 3D rendering shows a possible configuration of the light-seeking cart, with two photoresistors enclosed in small tubes to restrict their response to light.*

the chip functions as an inverter, so that when the input goes high, the output goes low and sinks current. This is of course just what we need for our stepper motor that has a common positive connection.

The ULN2001A is only an amplification device, so you have to precede it with a counter that runs from 1 to 4 and then repeats. You can stick with your 555 timers, as you've already assembled them, or substitute almost any CMOS octal or decade counter that sends its output pulses to a series of pins. Just use the output from the fifth pin as the "carry" output to restart the counting sequence. I suggest a CMOS counter simply because it will run on 12 volts, so you can use the same power supply that suits your stepper motors.

If you switch to CMOS counters, you will still need a pair of 555 timers sending pulses to the counters. The timers will be free-running in astable mode, and your photoresistors will control their speed. Figure 5-120 shows the configuration.

One last item: you'll need a 12-volt battery. You can of course put eight AA cells together, but I think you should consider a rechargeable pack from a source such as *http://www.all-battery.com*, which has a section entirely devoted to "robot batteries."

If you put it all together, you should find that when you place your robot cart in a very dimly lit room, it will turn toward the beam from a bright, well-focused flashlight. To get reliable results, you may have to recess each of the photoresistors in little tubes, so that they receive much more light when they face your flashlight than when they face away from it. Figure 5-121 is a 3D rendering of the concept.

Another idea is to rewire your cart so that it actually runs away from the light. Can you imagine how this might be done?

Just one more thought: if you use infrared photoresistors, you can control your cart with beams from infrared LEDs, in normal room lighting. If you and a couple of friends all have infrared transmitters, you can get your cart to run from one of you to the next, like an obedient dog.

This takes us about as far as I'm going to go into robotics. I urge you to check out the sites online if you want to pursue the topic further. You can also buy a wide variety of robot kits, although of course I feel that it's more fun to invent or develop things for yourself.

All that's left now is to perform one last introduction: to a device that should make your life much easier, even though the device is much more complicated than anything we have dealt with so far.

Experiment 34: Hardware Meets Software

Throughout this book, in accordance with the goal of *learning by discovery*, I have asked you to do an experiment first, after which I've suggested the general principles and ideas that we can learn from it. I now have to change that policy, because the next experiment involves so much setup that it's only fair to tell you what to expect before you begin the preparations.

We are about to enter the realm of controller chips, often known as MCUs, which is an acronym for micro controller unit. An MCU contains some flash memory, which stores a program that you write yourself. The flash memory is like the memory in a portable media player, or the memory card that you use in a digital camera. It needs no electricity to power it. In addition, the chip has a processor which carries out the instructions in your program. It has RAM to store the values of variables on a temporary basis, and ROM, which tells it how to perform tasks such as sensing a varying voltage input and converting it into digital form for internal use. It also contains an accurate oscillator, so that it can keep track of time. Put it all together, and it's a tiny computer that you can buy for under $5.

Let's suppose that you have a greenhouse where the temperature must never fall below freezing. You set up a temperature sensor, and you have two different heaters. You want to switch on the first heater if the temperature falls below 38° Fahrenheit. But if, for some reason, that heater is broken, you want to switch on the second, backup heater when the temperature goes below 36° Fahrenheit.

Programming a MCU to take care of this can be very simple indeed. You could even add extra features, such as a second temperature sensor, just in case the first one fails, and you could tell the chip to use whichever sensor gives a lower reading.

Another application for a MCU would be in a fairly elaborate security system. The chip could monitor the status of various intrusion sensors, and can take various preprogrammed steps, depending on the sensors' status. You could include delay intervals, too.

Many MCUs have additional useful features built in, such as the ability to control servo motors that turn to a specific angle in response to a stream of pulses. Servos are widely used in radio-controlled model boats, airplanes, and hobby robotics.

Perhaps you are now wondering why, if MCUs can do all this, haven't we been using them all along? Why did I spend so much time describing the development of an alarm system using discrete components, if one chip could have done everything?

There are three answers:

1. MCUs cannot do everything. They need other components to help them interact with the world, such as transistors, relays, sensors, and amplifiers. You need to know how those things work, so that you can make intelligent use of them.
2. MCUs can introduce their own kinds of problems and errors, associated with using software in addition to hardware. I'll have more to say about this later.
3. MCUs have limits and restrictions, most obviously their requirement for a 5-volt regulated power supply, and their inability to source or sink much current from each pin. They also demand that you learn a programming language (which differs from one brand of MCU to the next). And to get the program into the chip, you have to be able to plug it into a computer and do a download, which is not always convenient.

In this experiment, you'll learn how to write a program for a small and simple MCU, and you'll transfer the program into it and see how it works.

BACKGROUND

Origins of programmable chips

In factories and laboratories, many procedures are repetitive. A flow sensor may have to control a heating element. A motion sensor may have to adjust the speed of a motor. Microcontrollers are perfect for this kind of routine task.

A company named General Instrument introduced an early line of MCUs in 1976, and called them PICs, meaning Programmable Intelligent Computer—or Programmable Interface Controller, depending which source you believe. General Instrument sold the brand to another company named Microchip Technology, which owns it today.

"PIC" is trademarked, but is sometimes used as if it's a generic term, like Scotch tape. In this book, I've chosen a range of controllers based on the PIC architecture. They are licensed by a British company named Revolution Education Ltd., which calls its range of chips the PICAXE, for no apparent logical reason other than that it sounds cool.

I like these microcontrollers because they were developed originally as an educational tool and because they are very easy to use. They're cheap, and some of them are quite powerful. Despite their odd name, I think they're the best way to get acquainted with the core concepts of MCUs.

After you play with the PICAXE, if you want to go farther into MCUs, I suggest the BASIC Stamp (which uses a very similar language, but with additional powerful commands) and the very popular Arduino (which is a more recent design, packed with powerful features, but requires you to learn a variant of C language to program it). I'll say more about these chips later.

If you search for "picaxe" on Wikipedia, you'll find an excellent introduction to all the various features. In fact, I think it's a clearer overview than you'll get from the PICAXE website.

Supplies

The PICAXE 08M was the lowest-price chip in the range for many years, but was superceded more recently by the 08M2, which was a major upgrade. The 08M is still being sold from many suppliers, and the steps described in this chapter will work if you have that version. But if you buy the newer 08M2, that should work too, and will allow you to tackle more ambitious projects in the future. Figure 5-122 shows the features of the two chips as specified by the vendor, while Figure 5-123 shows a closeup of the 08M with its legs safely embedded in a piece of conductive foam.

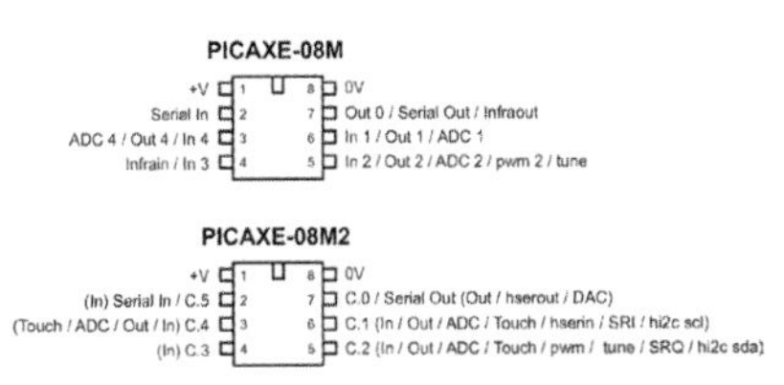

Figure 5-122. *A page from the PICAXE catalog lists only some of the chips that are available. What began as an educational aid has become a useful prototyping tool.*

In the United States, there are three distributors of this chip:

- *http://www.advancedmicrocircuits.com*
- *http://www.phanderson.com/picaxe*
- *http://www.sparkfun.com*

I like P. H. Anderson for its grass-roots hobbyist approach, and they have good prices if you want to buy multiple chips. But SparkFun Electronics offers other associated products that you may find interesting.

All the distributors will want to sell you "starter kits," such as the one in Figure 5-124, perhaps because the PICAXE itself is so cheap that it doesn't offer much of a profit margin. Still, for our purposes, you should buy the chip as a standalone item. And buy two of them, just in case you damage one (for example, by connecting voltage to it incorrectly).

Figure 5-123. *When supplied by one of its American distributors, a PICAXE 08M arrives embedded in a little square of conductive foam. The chip is the same size as a 555 timer but has the power of a tiny computer.*

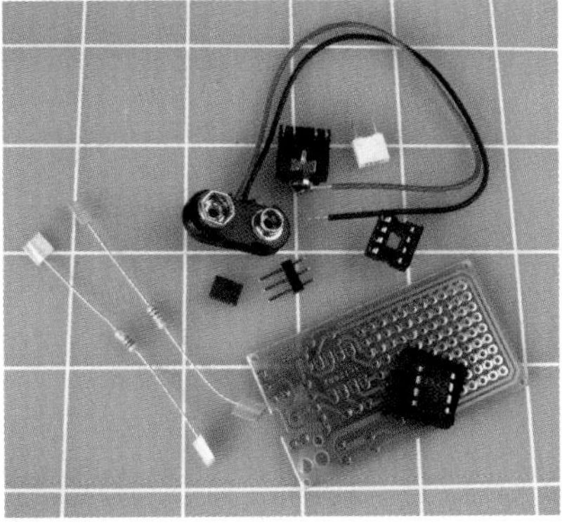

Figure 5-124. *A typical PICAXE kit includes a printed circuit board, which you may not really need, and some other not-entirely-essential items. But the 3.5-mm stereo jack socket (top, center) is absolutely necessary.*

To download your programming instructions into the chip, first you'll type the instructions on a computer, and then you'll feed them through a cable into the PICAXE memory. So you'll need to buy a cable, and you'll need software to help you to write the program.

Figure 5-125. *The USB download cable made for use with the PICAXE terminates in a 3.5-mm audio plug. This should not be inserted in any audio device. It establishes a serial connection with a computer, allowing program code to be downloaded into the chip.*

Figure 5-126. *Closeup of the 3.5-mm stereo socket that is used with the USB download cable.*

You can use the PICAXE with a serial cable, but I don't recommend it. The old RS-232 serial communications standard on PCs is pretty much obsolete, and PICAXE has recognized this by offering a USB cable (which contains a serial converter inside its plug). The USB cable is a little more expensive, but is also simpler and compatible with Apple computers. From any of the U.S. distributors, buy USB cable part AXE027, also sold as part PGM-08312 by *http://www.sparkfun.com* (quantity: 1). The cable is shown in Figure 5-125.

To write your software and send it down the wire to the chip, the PICAXE Programming Editor is the tool of choice. It comes in only a Windows version. For those who prefer Mac OS or Linux, you can get a free download of another piece of software known as AXEpad, which has fewer features, but will do the job. All the downloadable software is freely available from *http://www.rev-ed.co.uk/picaxe/software.htm*.

Finally, you need a 3.5-mm stereo audio socket with solder connections. The reason for this is that the manufacturers of the PICAXE have used a stereo audio plug on the free end of their USB cable, and you have to be able to plug it into something. The PICAXE breadboard adapter, SparkFun stock number DEV-08331, contains the necessary stereo socket in addition to a few other little items. Quantity: 1. See Figure 5-126.

Oddly enough, the USB cable is the most expensive item on the list, because of the electronics hidden inside it.

Software Installation and Setup

Now you have to go through a setup procedure. There is no way around this. Here is what you will be doing:

1. Install a driver so that your computer will recognize the special USB cable.
2. Install the Programming Editor software (or AXEpad for Mac/Linux) so that your computer will help you to write programs and then download them into the chip.
3. Mount the PICAXE on your breadboard and add the socket to receive downloads.

These steps are explained in the following sections.

The USB driver

Fair warning: If you go to the PICAXE website and try to use its search function, it probably won't find what you want. Search for "USB Driver," for instance, and it will pretend it has never heard of such a thing.

The PICAXE home page also has irritating drop-down menus that tend to disappear just when you're about to click on them, but at the time of writing, you can bypass these issues by going straight to the Software Downloads section at *http://www.rev-ed.co.uk/picaxe/software.htm*.

Scroll down past all the software until you get to Additional Resources. Look for the AXE027 PICAXE USB Download Cable. At first glance, it looks as if they want to sell you a cable, but in fact this is the list of drivers. Double-click the one appropriate to your computer, and choose a destination on your computer for the download—a place where you will find it easily, such as your desktop.

Be careful not to download the driver for the USB010 USB-Serial adapter by mistake. The USB-Serial adapter is something else entirely.

The download will leave you with a zipped file folder. You will have to unzip it. On Windows XP, right-click the folder and choose "Extract all." View the extracted files and you will find a PDF installation guide. Linux and Mac users can find instructions currently stashed at *http://www.rev-ed.co.uk/docs/AXE027.pdf.*

When installing the driver on a Windows platform, here are a few tips to minimize your exasperation level:

1. Remember, the special USB cable contains some electronics. It is not just a cable, but a device designed for interacting with a PICAXE chip. *Don't try to use it for anything else!*
2. You have to plug the cable into a USB port *before* you install the driver, because your computer will need to verify that the driver matches the cable.
3. You must not attach the PICAXE to the other end of the cable until *after* you have installed the driver.
4. Every USB port on your computer has a separate identity. Whichever one you choose when you first plug in the cable, you should use that port *every time in the future*. Otherwise, you will have to repeat the process of telling your computer what the cable is.
5. Bearing in mind Tip #4, you should avoid using the cable in a standalone USB hub.
6. The cable is fooling the PICAXE into thinking that it's talking to a serial port on your computer. Those "communication" ports are known as COM1, COM2, COM3, or COM4. When you install the driver, the installer will choose one of those COM ports for you, and later you will have to know which one it is. The PDF guide should help you through this procedure. Unfortunately, you cannot skip it.

The Programming Editor software

If you have come this far, you're ready for the next big step, which is much easier. You need the PICAXE Programming Editor, available for free on the Software Downloads web page where you found the USB driver. (If you are using a Mac or Linux, you will need AXEpad, which is on the same web page.)

Downloading and installing the Programming Editor should be simple and painless. Once you have done that, you should find that it has placed a shortcut on your desktop. Double-click it, go to View→Options, and in the window that opens (shown in Figure 5-127), click the Serial Port tab. You should see a dialog box like the one in Figure 5-128. Now make sure that the Programming Editor is looking at the same COM port that was chosen by the USB driver. Otherwise, the Programming Editor won't know where to find your PICAXE chip.

In the Programming Editor, go to View→Options and click the Mode tab, then click the button to select the 08M chip.

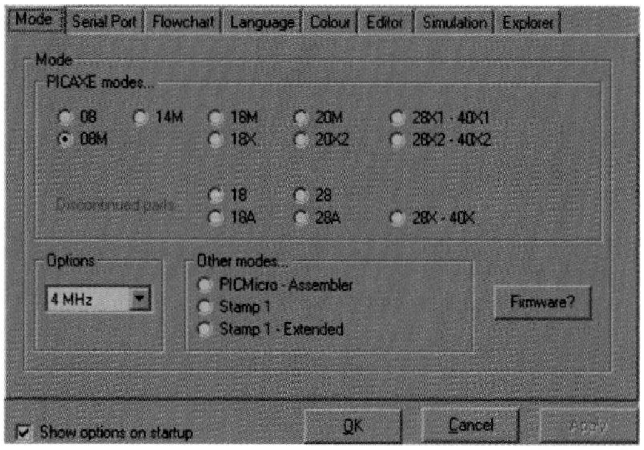

Figure 5-127. *This screenshot shows the options window of the PICAXE Program Editor, which you must use to select the type of chip that you intend to program (in our case, the 08M).*

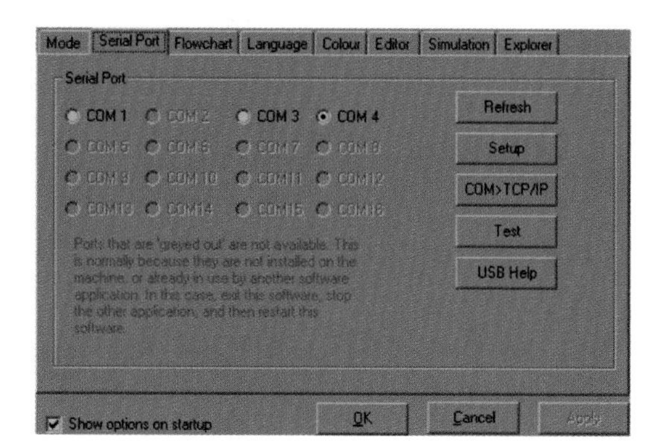

Figure 5-128. *Another screenshot of the options window shows the second essential choice that you must make: selection of the COM port that the installer chose on your computer.*

Are we having fun yet? Obviously not, but you're through with software hassles for the time being. The last step before you're ready to use the PICAXE is to mount it, and its socket, on your breadboard.

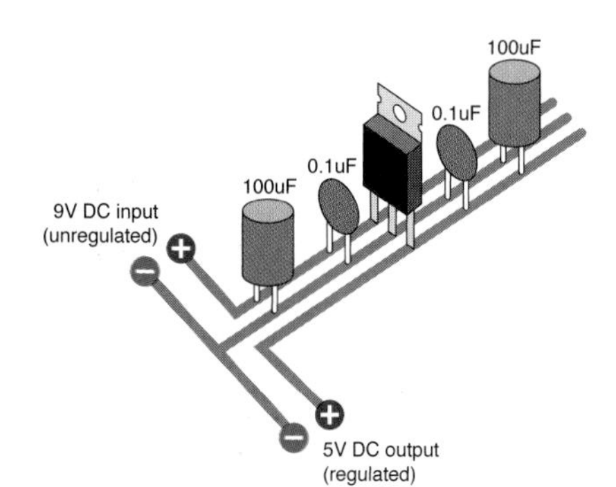

Figure 5-129. *PICAXE documentation specifies a 100 μF and 0.1 μF capacitor on the input side of a 5-volt regulator, and a similar pair of capacitors on its output side. On a breadboard, they can be arrayed like this.*

Setting up the hardware

The PICAXE 08M looks like a 555 timer. (Other chips in the PICAXE range have more pins and more features.) It requires a properly regulated 5 volts, just like the logic chips you dealt with previously. In fact, the PICAXE people are rather emphatic about protecting it from voltage spikes. They want you to use two capacitors (one 100 μF, one 0.1 μF) on either side of an LM7805 regulator. This seems like overkill, but the PICAXE is more inconvenient to replace than a 555 timer. You certainly can't run down to RadioShack to buy one. So let's do what the manufacturer says, just in case, and set up a breadboard as shown in Figures 5-129 and 5-130.

Now for the chip itself. Note that the pins for positive and negative power are *exactly opposite* to those for the 555 timer, so be careful!

Set up your breadboard following the schematic shown in Figure 5-131. Note that I am showing the stereo socket on its underside, because I think that's how you'll have to use it with the breadboard. If you try to stick its pins into the holes in the board, they will fit, but when you insert the plug into the socket, the thickness of the plug will tend to raise the socket up so that it loses contact. I really think the way to go is to solder wires to the pins on the socket and push the wires into the breadboard. See Figure 5-133.

Figure 5-130. *The actual components for power regulation, applied to a breadboard, delivering 5 volts (positive and negative) down each side.*

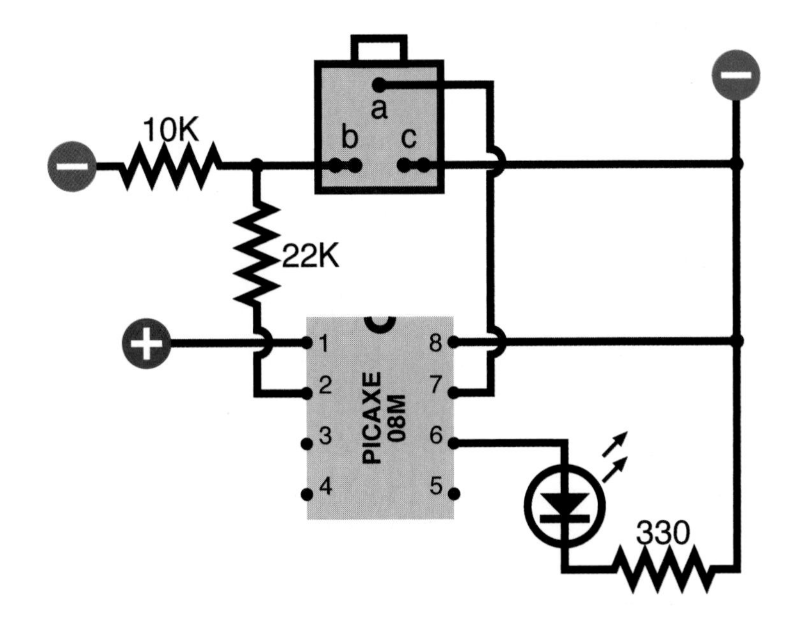

Figure 5-131. *The schematic of a test circuit for the PICAXE 08M shows the underside of the stereo socket, the essential 10K and 22K resistors on the input pin, and an LED to show an output from the chip.*

Be aware that the PICAXE manual shows things differently (although I have retained their labeling convention for the parts of the socket and the parts of the plug, identified as a, b, and c).

One little detail about the socket that is commonly supplied for use with the PICAXE: typically it has two *pairs* of contacts for the connections labeled b and c in the manual, and in my diagram. When you solder a connection, your solder joint should include *both* of the contacts in each pair, as shown in Figure 5-132.

Remember that the PICAXE must have 5 volts DC, and remember that your voltage regular will deliver this voltage reliably only if you give it extra voltage on its input side. If you provide it with 9 volts, that will provide a good amount of headroom.

The 22K and 10K resistors are essential for using the chip; see the following warning note for an explanation. My schematic also includes an LED and a 330Ω resistor, but they are needed only for the test that we'll be making momentarily.

Pin 2 Pull-Down

Always include the 22K resistor and the 10K resistor in the configuration shown in Figure 5-131. These resistors apply correct voltage to the serial connection, and when you're using the PICAXE on its own, they pull down the voltage on pin 2.

If pin 2 is left unconnected (floating), it may pick up random voltages, which the chip can misinterpret as a new program or other instructions, with unpredictable and undesirable results.

The 22K and 10K resistors should be regarded as permanent items accompanying your PICAXE regardless of whether you have it attached to your computer.

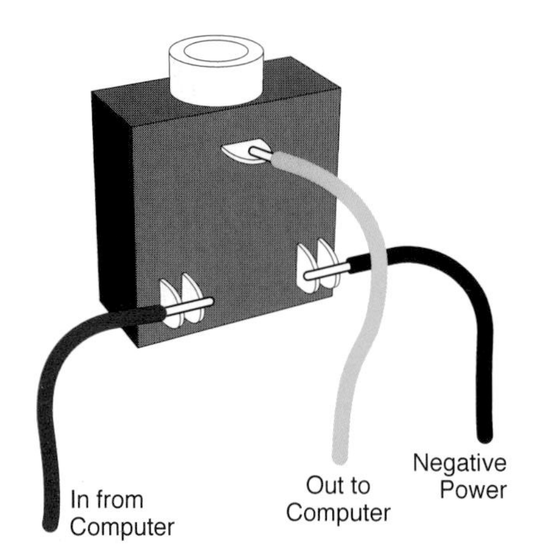

Figure 5-132. *Correct wiring of the socket is essential. When soldering wires to the lower terminals, make sure that you attach the wires to both of the terminals in each pair.*

Figure 5-133. *The breadboard version of the test schematic, with the plug of the USB download cable inserted in the socket on the board. The PICAXE chip can now receive a downloaded program, and will immediately start to execute it.*

Verifying the Connection

Follow these steps carefully every time you want to program or reprogram your PICAXE chip:

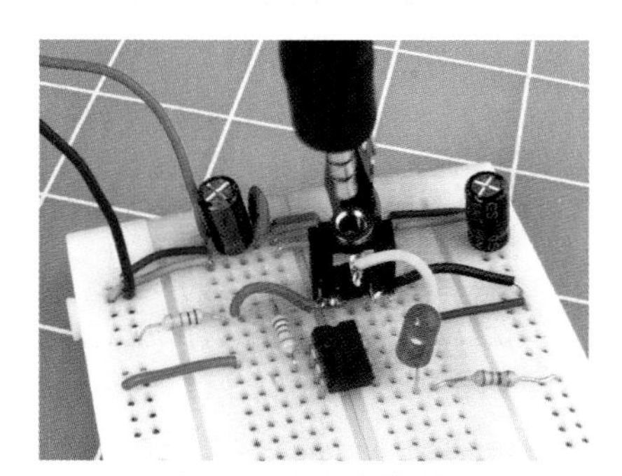

Figure 5-134. *After the program has been downloaded, the plug can be removed, and the program will continue to run, causing the LED to continue flashing.*

1. Insert the USB plug of your PICAXE cable into the same USB port that you used before.
2. Start the Programming Editor (or AXEpad if you are using a Mac OS or Linux).
3. In the Programming Editor, select View→Options to verify that the editor is using the right COM port and is expecting the 08M PICAXE chip.
4. Plug the stereo plug on the free end of the USB cable into the stereo socket that is now wired into your breadboard. See Figures 5-133 and 5-134.
5. Check your wiring, and then connect your power supply to the breadboard.
6. Click the button labeled "program" in the Program Editor window to tell the software to look for the PICAXE.

What If It Doesn't Work?

The first thing to do is pull out the plug of the USB cable from the PICAXE breadboard, leaving the other end of the cable attached to your computer. Set your multimeter to measure DC volts, and attach its probes to sections b and c of the plug. See Figure 5-135. Now click the "program" button again, and your meter should show 5 volts briefly coming out of your computer to the plug on the end of the cable.

If you detect the voltage, the software is installed and working properly. In that case, there's a problem on your breadboard, either in the chip or in the wiring around it.

If you cannot detect any voltage, the software probably wasn't installed properly, or is looking for the wrong serial port. Try uninstalling it and reinstalling it.

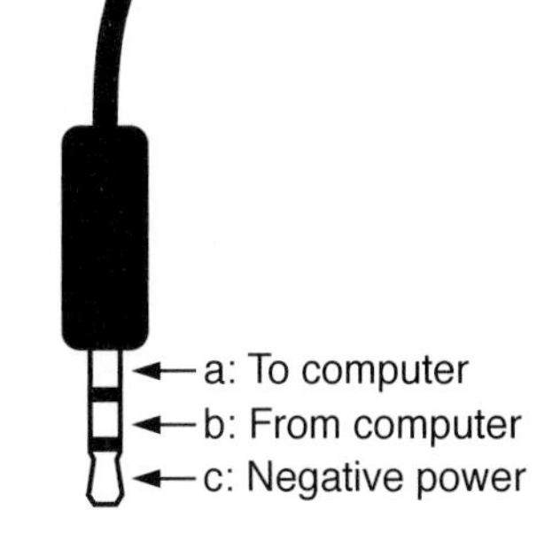

Figure 5-135. *The stereo plug on the end of the USB download cable can be used for fault tracing. A multimeter set to measure DC volts can be attached to sections b and c of the plug to establish whether the Programming Editor is sending data through the serial connection.*

Your First Program

Finally you're ready to create your first program. Type the following code into the Programming Editor window:

```
main:
    high 1
    pause 1000
    low 1
    pause 1000
    goto main
```

Be sure to include the colon after the word "main" on the first line. See Figure 5-136 for a screenshot. The indents are created by pressing the Tab key. Their only purpose is to make program listings more legible. The software ignores them.

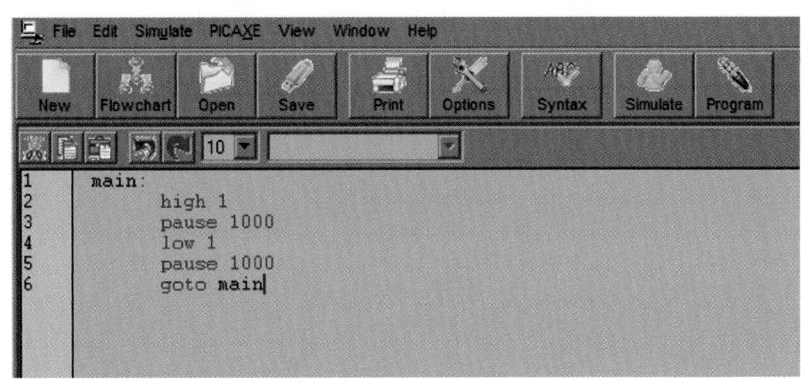

Figure 5-136. *This screenshot shows the first test program as it should be displayed by the Programming Editor (on a Windows computer).*

Click the Program button in the Programming Editor to download this program into the chip. As soon as the download is complete, the chip should start flashing the LED, lighting it for 1 second and then switching it off for 1 second. Figure 5-137 shows the steps that you should have followed to program the chip.

Now for the interesting part: disconnect the USB cable from the breadboard. The chip should continue flashing.

Disconnect the power supply from the breadboard and wait a minute or two for the capacitors to lose their charge. Reconnect the power, and the chip will start flashing again.

The program that you downloaded to the chip will remain in the memory inside the chip and will begin running every time power is applied to the chip.

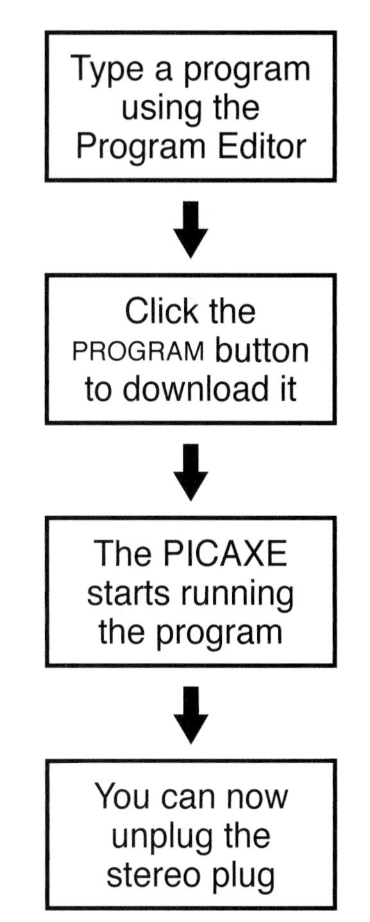

Figure 5-137. *Four steps to create and run a program on the PICAXE controller chip.*

Decoding the Code

The program doesn't use the usual pin numbers on the chip. It uses what I will call "logic pin numbers." Figure 5-138 shows how they are numbered. Figure 5-139 shows their multiple functions. I have put Logic Pin 0 in parentheses because its main purpose is to send data to the computer through the USB cable. It can do double duty as a digital output, but you have to disconnect it from the USB cable first. It's easy to forget to do this. It's a hassle that I prefer to avoid.

Let's take a look at the little program that you typed in. The first line identifies a section of the program. This program only has one section, and we're calling it "main." Any word with a colon after it is the name of a section of a program:

```
main:
```

The second line tells the chip to send a high output from Logic Pin 1:

```
high 1
```

The third line of the program tells the chip to wait for 1,000 milliseconds. This of course is the same as one second:

```
pause 1000
```

The fourth line tells the chip to change Logic Pin 1 back to its low state:

```
low 1
```

The fifth line tells the chip to wait for another 1,000 milliseconds:

```
pause 1000
```

The last line tells the chip to go back to the beginning of the "main" section:

```
goto main
```

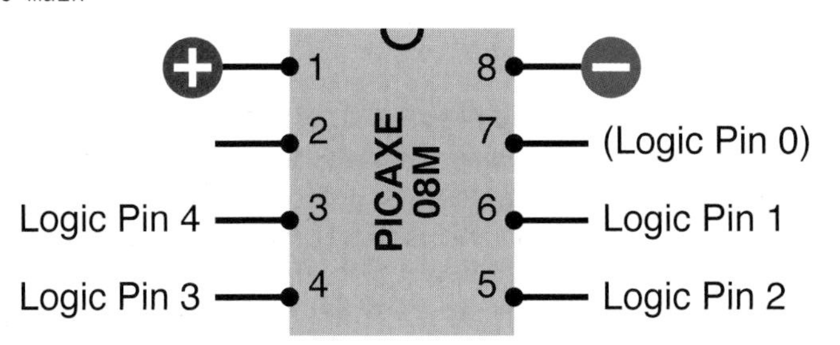

Figure 5-138. *The conventional pin numbers of the PICAXE chip are incompatible with the numbering system that is used in the PICAXE programming language. To minimize confusion, this guide refers to "Logic Pins" when using the numbering system that is required for programming the chip.*

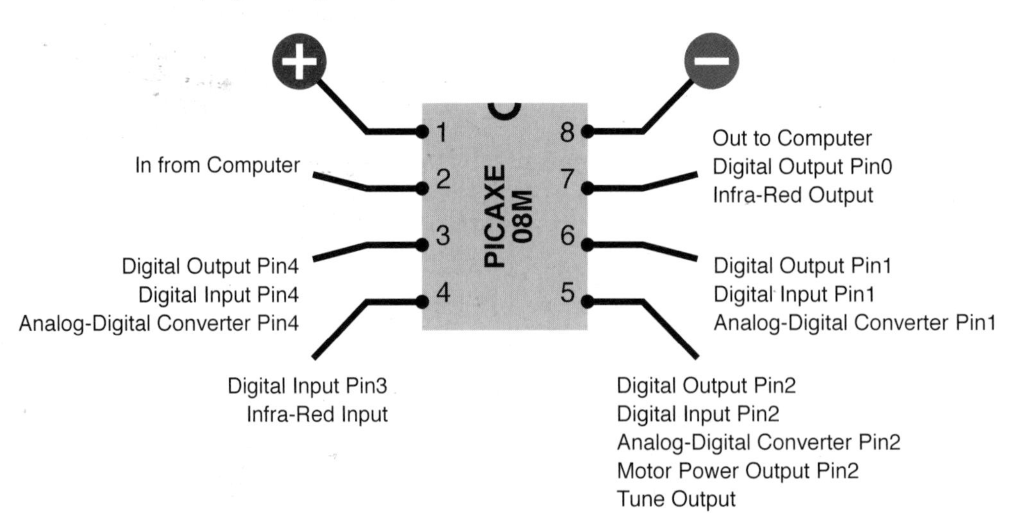

Figure 5-139. *Many of the pins on the PICAXE 08M have multiple functions, which can be selected by appropriate program instructions.*

Editing

What if you want to change the program? No problem! Use the Programming Editor to change one of the lines in the program. Substitute 100 instead of 1000 milliseconds, for instance. (The `pause` command can be followed by any number up to 65535.) In your program, don't use a thousands comma in any of the numbers that you specify.

Plug the USB cable into the breadboard again, hit the Program button on the screen, and the new version of the program will be automatically downloaded to the chip, overwriting the old version.

What if you want to save the program for future use? Just go to the File menu in the Programming Editor and save the program onto your computer's hard drive. Because the PICAXE uses a variant of the BASIC computer language, it adds a *.bas* filename extension.

Simulation

If you make a simple typing error, the Programming Editor will find it and stop you from downloading your program. It will leave you to figure out how to fix the line that contains the error.

Even if all the statements in your program are correctly typed, it's a good idea to run a simulation of what they'll do, before you download them. This is easily done: click the "simulate" button on the menu bar of the Programming Editor. A new window will open, displaying a diagrammatic view of the PICAXE chip and showing you the states of its pins. (Note that if you use very short `pause` commands, the simulation won't run fast enough to display the time accurately.) A simulation screenshot is shown in Figure 5-140.

You'll need to check the second part of the PICAXE documentation, which contains all the programming statements and their correct syntax. At the time of writing, this is stored at http://www.rev-ed.co.uk/docs/picaxe_manual2.pdf.

The >> button at the bottom-right corner of the simulation window will open up a list of all the variables in your program. So far, it doesn't have any variables, but it soon will. All the zeros on the righthand side are binary numbers, which you can ignore for now.

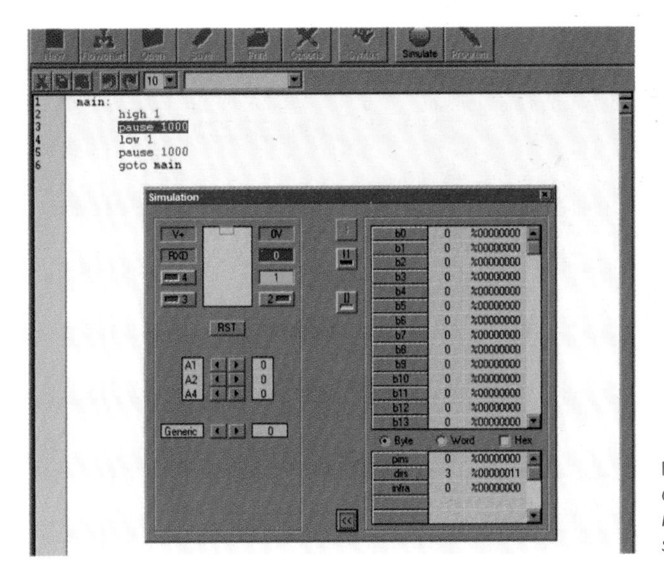

Figure 5-140. *This screenshot shows the simulation window that can be opened in the Program Editor to test program code before it is downloaded to the chip. The values of variables are shown in the section on the right. The pin states are shown on the left.*

Loops

Here's another thing I'd like you to try. Rewrite your program code as shown here and download it onto the PICAXE:

```
main:
    for b0 = 1 to 5
        high 1
        pause 200
        low 1
        pause 200
    next
    wait 2
    goto main
```

Note that b0 is letter b followed by a zero, not letter b followed by letter O. The extra indents once again are added to make the listing easier to understand. The four lines beginning "high 1" and ending "pause 200" will be executed repeatedly. It's helpful to see them as a block.

Watch the light and see what happens. It should flash five times quickly, then wait for two seconds, and then repeat. You just added a *loop* to your program. You can use a loop if you want something to happen more than once.

Figure 5-141. *To understand how a program works, visualize a variable as being like a "memory box" with its name on the outside and a number stored on the inside.*

b0 is known as a *variable*. Think of it as being like a little "memory box" with its name, b0, on a label on the outside. Figure 5-141 illustrates this concept. This particular memory box can contain any number from 0 through 255. The loop begins by telling the computer to put number 1 in the box, then process the remaining statements, until the word "next" sends the processor back to the first line, at which point it adds 1 to the contents of b0. If the value of b0 is 5 or less, the loop repeats. If the value is 6, the loop has run five times, so it's over, and the PICAXE skips down to the "wait 2" statement after "next." See Figure 5-142 for an annotated version of the program listing.

"Wait" is a PICAXE command that is measured in whole seconds, so "wait 2" waits for 2 seconds. Then "goto main" begins the procedure all over again.

If your flashing-light demo worked out as planned, it's time to take the next step and make the chip do something more useful.

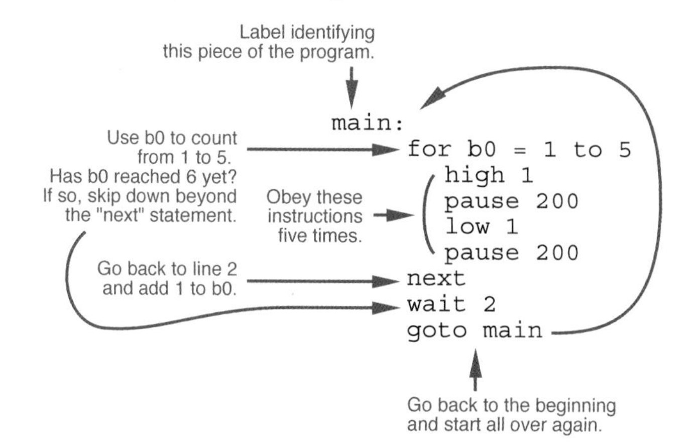

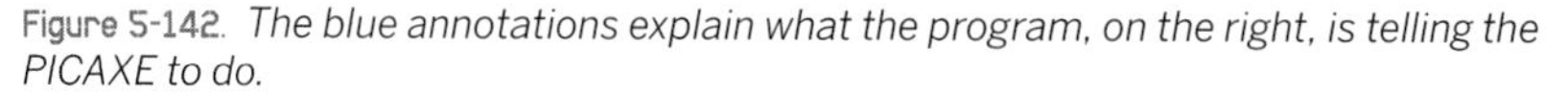

Figure 5-142. *The blue annotations explain what the program, on the right, is telling the PICAXE to do.*

FUNDAMENTALS

Basic PICAXE parameters

Here are some of the most useful parameters of the PICAXE:

- The PICAXE requires 5 volts DC, regulated.
- The inputs and outputs of the PICAXE are compatible with 5-volt logic chips. You can attach them directly.
- Each PICAXE pin can sink or source up to 20mA. The whole chip can deliver up to 90mA. This means that you can run LEDs directly from the pins, or a piezo noisemaker (which draws very little current), or a transistor.
- You can use a chip such as the ULN2001A Darlington array (mentioned in the previous experiment) to amplify the output from the PICAXE and drive something such as a relay or a motor.
- The chip executes each line of your program in about 0.1 milliseconds.
- The 08M chip has enough flash memory for about 80 lines of program code. Other PICAXE chips have more memory.
- The PICAXE provides 14 variables named b0 through b13. The "b" stands for "byte," as each variable occupies a single byte. Each can hold a value ranging from 0 through 255.
- No negative or fractional values are allowed in variables.
- You also have 7 double-byte variables, named w0 through w6. The "w" stands for "word." Each can hold a value ranging from 0 through 65535.
- The "b" variables share the same memory space as the "w" variables. Thus:
  - b0 and b1 use the same bytes as w0.
  - b2 and b3 use the same bytes as w1.
  - b4 and b5 use the same bytes as w2.
  - b6 and b7 use the same bytes as w3.
  - b8 and b9 use the same bytes as w4.
  - b10 and b11 use the same bytes as w5.
  - b12 and b13 use the same bytes as w6.
  - b14 and b15 use the same bytes as w7.

  Therefore, if you use w0 as a variable, do not use b0 or b1. If you use b6 as a variable, do not use w3, and so on.
- Variable values are stored in RAM, and disappear when the power is switched off.
- The program is stored in nonvolatile memory, and remains intact when the power is off.
- The manufacturer's specification claims that the nonvolatile memory is rewritable up to about 100,000 times.
- If you want to attach a switch or pushbutton to a pin and use it as an input, you should add a 10K pull-down resistor between the pin and the negative side of the power supply to hold the pin in a low state when the switch is open. Figure 5-143 shows how pull-down resistors should be used in conjunction with a SPST switch or a pushbutton.
- On the 08M chip, if you apply a varying resistance between Logic Pins 1, 2, or 4, and the negative side of the power supply, the chip can measure it and "decide" what to do. This is the "Analog-Digital Conversion" feature—which leads to our next experiment.

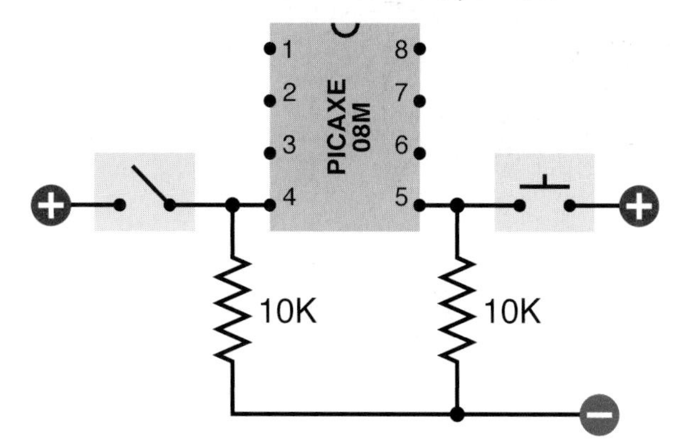

Figure 5-143. *The PICAXE can respond to the state of a switch or button attached to any of its input-capable pins. A 10K resistor must be used to pull down the state of the pin when the contact of the switch or button is open. Otherwise, you may get unpredictable results.*

Experiment 35: Checking the Real World

Often we want a microcontroller to measure something and respond in an appropriate way. For instance, it can measure a low temperature and sound an alarm, as I suggested in the example that I gave earlier.

The PICAXE has three analog-to-digital converters (ADCs) built in, accessible via logic pins 1, 2, and 4, as shown in Figure 5-139. The best way to use them is by applying a potential somewhere between 0 and 5 volts. In this experiment, I'll show you how to calibrate the response of the chip.

You will need:

- Trimmer potentiometer, 2K. Quantity: 1.
- PICAXE 08M chip and associated USB cable and socket. Quantity: 1 of each.

Procedure

Take the same trimmer potentiometer that you used in Experiment 32 and wire its center terminal to Logic Pin 2 of the PICAXE (which is hardware pin 5). The other two terminals of the 2K trimmer go to positive and to negative, respectively. So depending how you set the trimmer, the pin of the PICAXE is directly connected to positive (at one end of the scale), or directly connected to negative (at the other end of the scale), or somewhere in between. See Figure 5-144 for the revised schematic, and Figure 5-145 for a photograph of the breadboarded circuit.

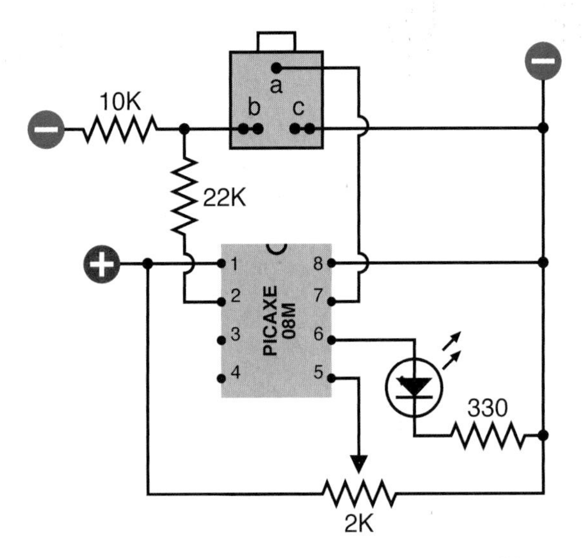

Figure 5-144. *This schematic, drawn in a layout suitable for breadboarding, shows how a 2K potentiometer can be used to apply a varying voltage to one of the pins of the PICAXE that is capable of converting an analog signal to a digital value.*

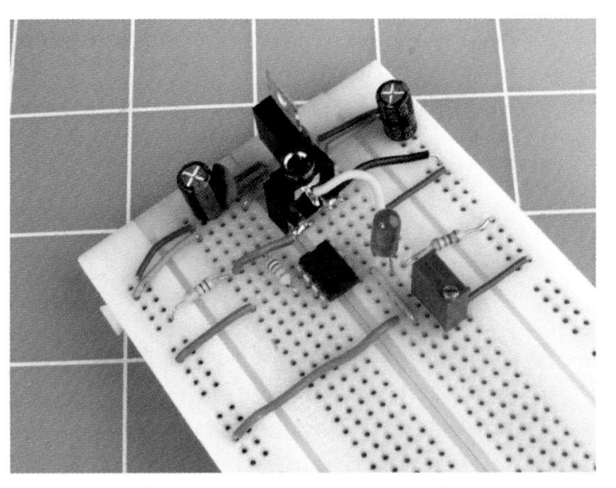

Figure 5-145. *The trimmer potentiometer added to the previously breadboarded circuit.*

Now we need a program to tell the chip what to do. Using the Programming Editor, start a new document. The code should look like this:

```
main:
    readadc 2,b0
    debug b0
    goto main
```

The command "readadc 2,b0" means "read the analog input on Logic Pin 2, convert from analog to digital, and store the result in b0."

The command "debug b0" tells the chip to go into program debugging mode, in which it uses its USB cable to tell the Programming Editor the values of all the variables while the program is running. The variables are displayed in a debugging window.

Download the program, and as the program starts to execute, the debugging window should open. Start adjusting the trimmer while looking at the value of b0, and you'll see b0 change its value.

You can make a table and draw a graph showing the relationship between the resistance between Logic Pin 2 and ground, and the value of b0. Just pull the trimmer off the breadboard, measure its resistance with a meter, then increase its resistance by, say, 200Ω, put it back into the breadboard, and look at the value of b0 again.

This is laborious, but calibrating equipment is always laborious—and in any case, I decided to do it for you. The graph is shown in Figure 5-146. You can also see the raw data numbers in the following table. I was pleased to find that the PICAXE gives a very precise, linear response to the input voltage. In other words, the graph is a straight line.

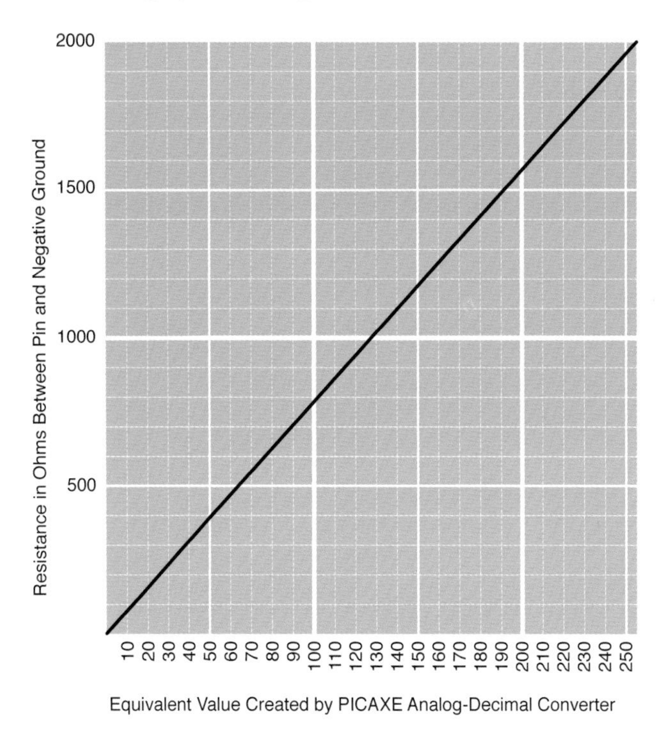

Figure 5-146. *When an ADC input pin is hooked up to a 2K potentiometer, which is connected across the same voltage that powers the chip, you should find that the resistance between the input pin and the negative side of the power supply generates the series of digital values shown on the graph. Note that the potentiometer must have a 2K value, and the power supply is assumed to be precisely 5 volts.*

This table shows measurements made with PICAXE 08M controller.

Resistance (in ohms) between the ADC pin and the negative supply	Equivalent digital value
2000	255
1900	243
1800	230
1700	218
1600	205
1500	192
1400	179
1300	166
1200	154
1100	141
1000	128
900	115
800	102
700	90
600	77
500	64
400	51
300	38
200	26
100	13
0	0

Now we can modify the program to make it do something with the information that it's taking in:

```
main:
    readadc 2,b0
    let w1 = 5 * b0
    high 1
    pause w1
    low 1
    pause w1
    goto main
```

Notice what's happening here. First we get a value in b0, and then on the next line, we do some arithmetic with it. The asterisk means "multiply." So the statement says, "Take whatever value is in b0, multiply by 5, and transfer it to another variable, w1." We have to use a w variable, because when we multiply the value of b0 by 5, we may get a number that is bigger than 255—too big to fit into a byte variable.

Finally, we take variable w1 and use it with a "pause" statement instead of a fixed number value. We're saying to the PICAXE, "pause for whatever number of milliseconds you get by checking the value of w1."

So the software checks a variable resistance, turns it into a number, and applies that number to adjust the flashing speed of the LED.

Think back to the need of the cart powered by stepper motors. It was supposed to check two photoresistors, and adjust the speed of each motor accordingly. Well, this PICAXE program is a step in that direction. It can measure voltage on a pin and change the output frequency on another pin. If you had two PICAXE chips, you could wire each of them to a photoresistor and a motor. Then you could adjust the behavior of your cart by editing the second line in the program, where it converts the value of b0 to the value of w1 which will be used in the "pause" command to determine the number of pulses per second. Instead of multiplying by 5 you could multiply by 7 or whatever number gives you the result you need. This leads to an important conclusion: *a big advantage of a programmable chip is that you can make adjustments in software.*

Because the PICAXE 08M actually has more than one ADC input, and has three pins that can be used for output, you might wonder whether you could use just the one chip to control both motors in response to inputs from two sensors. The problem is that the three output pins on the 08M also function as the three ADC input pins. You'd do better to buy one of the more advanced PICAXE chips, such as the 18M, which has more pins to choose from. It uses the same basic set of programming instructions, and doesn't cost much more money.

Also, you should read the PICAXE documentation and look up the "pwmout" command, which is short for "pulse-width modulation output," but you can think of as meaning "power motor output." This is specifically intended to run stepper motors. It establishes an output frequency of pulses that will continue while the chip obeys other instructions in its program.

FUNDAMENTALS

Extra features

A complete guide to the 08M would fill a book of its own, and of course such books already exist (just search the books section of Amazon.com for keyword "picaxe"). But I'll finish my introduction to the controller by listing some of its extra capabilities, leaving you to look them up and explore them. Then I'm going to suggest one last experiment.

Interrupts

The PICAXE 08M allows you to set one "interrupt." This feature tells the chip to make a mental note that if a particular event occurs—such as a switch applying voltage to one pin—it should stop doing whatever else it was doing, and respond to the interruption.

Infrared

One pin on the PICAXE 08M can be used to receive infrared signals from a TV-style remote that you can buy from the same suppliers that sell the PICAXE itself. With an infrared sensor attached to the chip, you can issue commands remotely. If you want to build a remote-controlled robot, the chip is specifically designed with this in mind.

Servo motors

Every PICAXE chip has at least one pin that can send a stream of pulses to control a typical servo motor. On the 08M chip, it's Logic Pin 2. The width of each pulse tells the motor how far to rotate from its center position before stopping. A 555 timer can send this stream, but the PICAXE makes it easier. You can search online for more information about servo motors, which are especially useful for applications such as steering model vehicles, adjusting the flaps on model airplanes, and actuating robots.

Music

The PICAXE has an onboard tone generator that can be programmed with a "tune" command to play tunes that you write using a simple code.

Alphanumeric input/output

The "kbin" programming command is available in the PICAXE models 20X2, 28X1 and 28X2, and 40X1 and 40X2. You can plug a standard computer keyboard into the chip, and it will read the keypresses. You can also attach alphanumeric displays, but these procedures are nontrivial. For instance, when you're trying to figure out which key someone has pressed on a keyboard, your program has to contain a list of the special hexadecimal codes that the keyboard creates.

Pseudorandom number generation

All PICAXE models can generate pseudorandom numbers using a built-in algorithm. If you initialize the number generator by asking the user to press a button, and you measure the arbitrary time that this takes, you can seed the pseudorandom number generator with the result, and the pseudorandom number generator will have a less repeatable sequence.

Visit *http://www.rev-ed.co.uk/docs/picaxe_manual1.pdf* to learn more.

Experiment 36: The Lock, Revisited

The combination lock that I described in Experiment 20 is especially appropriate for a microcontroller, because it requires a series of operations that resemble a computer program. I'm going to show how this project can be redesigned using a PICAXE 08M, and then leave it to you to consider how some of the other projects in this book could be converted.

You will need:

- The same type of keypad and relay recommended in Experiment 20.
- A transistor or Darlington array to amplify the output from the PICAXE so that it can drive the relay.

Getting the User Input

Any of the input pins on the PICAXE can sense a switch closing. The trouble is that we only have three pins capable of doing this, and you have to buy a much bigger PICAXE chip to get 10 input pins. So how can we attach a 10-key keypad to the 08M?

I have a suggestion: attach various resistors to the keypad, so that each key applies a different voltage to one of the ADC pins. Then use the ADC feature to convert the voltage to a number, and use a table of possible numbers to figure out which key is being pressed. This may not be the most elegant solution, but it works!

The keypad can be wired as shown in Figure 5-147. The asterisk key is still being used to supply power, as in the original experiment, while the pound key resets the relay at the end of your computing session, as before.

Current flows through a series of resistors, beginning with one that has a value of 500Ω. Because this is not a standard value, you will either have to make it by combining other resistors in series, or by presetting a trimmer potentiometer. After that, each button is separated from the next button by a 100Ω resistor. Finally, at the end of the chain, a 600Ω resistor separates the last button from the negative side of the power supply. Again, this is not a standard value, and you may have to use a trimmer.

Add up all the resistances and you have 2K, which is the range that the PICAXE wants us to use. When you press a button, you tap into the chain of resistances. Button 9 puts 600Ω between the PICAXE ADC pin and ground. Button 6 is 700Ω, button 3 is 800Ω, and so on. (You may prefer to lay out the buttons so that the resistance progresses in a more logical fashion. That's up to you. I chose to lay them out in the way that would be easiest to visualize on a keypad.)

Each time you press a button, you insert a different resistance between the PICAXE ADC pin and ground. You might think that pressing keypad number 3, for instance, would insert 600Ω. But, it's not so simple, because I had to add a 2K resistor (at the top of the schematic in Figure 5-147) so that the ADC inside the PICAXE will always have at least some voltage applied to it, even when no buttons are pressed. If I had not done this, the ADC input would have "floated," giving random values. Bearing this in mind, here's a table showing the resistance which I got by pressing each button. Since your power supply and resistors might vary in value, I also show an acceptable range.

Button	Resistance	Range
3	108	100-114
6	120	115-126
9	132	127-137
2	143	138-148
5	153	149-158
8	163	159-168
0	173	169-178
1	183	179-187
4	192	188-196
7	202	197-208

Suppose you attach the common pin of your keypad to ADC Logic Pin 2 of the PICAXE. You can now use the Program Editor to write a program that looks like this:

```
getkey:
    readadc 2,b0
    let b1 = 3
    if b0 < 115 then finish
    readadc 2,b0
    let b1 = 6
    if b0 < 127 then finish
    readadc 2,b0
    let b1 = 9
    if b0 < 138 then finish
    readadc 2,b0
    let b1 = 2
    if b0 < 149 then finish
    readadc 2,b0
    let b1 = 5
    if b0 < 159 then finish
    readadc 2,b0
    let b1 = 8
    if b0 < 169 then finish
    readadc 2,b0
    let b1 = 0
    if b0 < 179 then finish
    readadc 2,b0
    let b1 = 1
    if b0 < 188 then finish
    readadc 2,b0
    let b1 = 4
    if b0 < 197 then finish
    readadc 2,b0
    let b1 = 7
    if b0 < 210 then finish
    goto getkey

finish:
    readadc 2,b0
    if b0 < 250 then finish
    return
```

What does the word "return" mean at the end? I'll get to that in a second. I want to explain the rest of the routine first.

b0 receives the value supplied by the analog-digital converter when it looks at the keypad. After storing the number in b0, the routine has to figure out which keypad key it matches. The key identity (0 through 9) will be stored in another variable, b1.

The program starts by assigning value 3 to b1. Then it checks to see whether b0 < 115. This means "if b0 is less than 115." If it is, then it's in the acceptable range, so the routine says "finish," which means "jump to the finish: label." But if b0 is not less than 115, by default the PICAXE continues on to the next line, which makes a second attempt at guessing which key has been pressed. It assigns number 6 to b1. Now there's another if-then test, and so on. If the resistance value fails all the tests, it must mean that no key has been pressed, so the getkey: procedure repeats. Lastly the finish: routine repeats until the user lets go of the key.

If you're familiar with other dialects of BASIC, this may seem a bit laborious to you. You may wonder why we can't use a statement such as this:

```
if b0 > 114 and b0 < 127 then b1 = 6
```

The answer is that PICAXE BASIC isn't sufficiently sophisticated to allow this. An if-then statement has to result in a jump to another section of the program. That's the only permitted outcome.

If you don't have any prior programming experience, the routine may still seem laborious to you, and perhaps a bit puzzling, too. This is understandable, because you're getting a crash course in software design without any formal preparation. Still, the PICAXE Programming Editor can be a big help, because it has its simulation feature. Before you can use this, though, you have to precede the routine that I just supplied with a control routine that you must type above it. The screenshot in Figure 5-148 shows you how it should look.

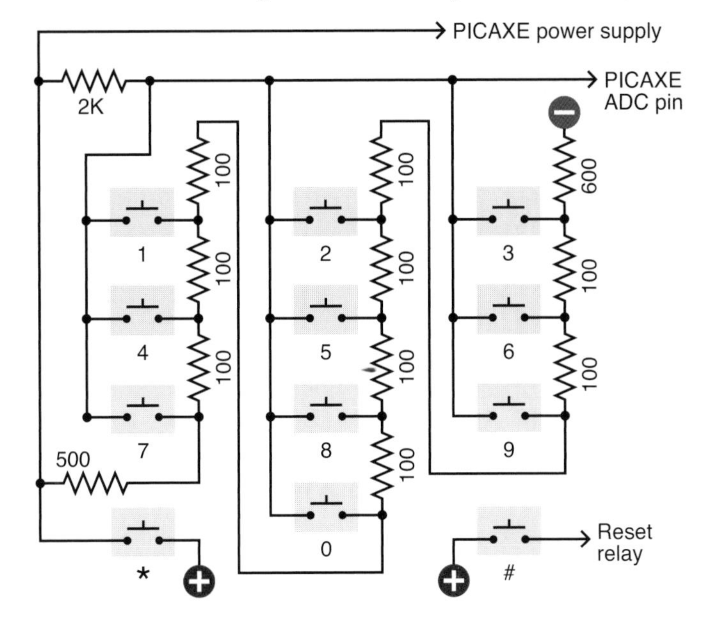

Figure 5-147. *A quick and simple way of attaching a keypad to provide numeric input to the PICAXE uses a chain of resistors totalling 2,000Ω. When a button is pressed, it connects the ADC input pin to a point in the chain. The resistance detected by the input pin can then be converted by the program in the chip to determine which key has been pressed.*

I have chosen an arbitrary combination of 7-4-1 for our combination lock. Using these numbers, the main section of the program looks like this:

```
main:
    low 1
    gosub getkey
    if b1<>7 then main
    gosub getkey
    if b1<>4 then main
    gosub getkey
    if b1<>1 then main
    high 1
    end
```

I should explain that the <> pair of symbols mean "is not equal to." So the fourth line of the program means, "if b1 is not equal to 7."

The value of b1 is supposed to be 7 if the user is putting in the correct combination. So if it's not 7, the user has entered the wrong value, and the if-then statement sends the PICAXE back to the beginning. In fact anytime the user inputs a number that is not in the correct 7-4-1 sequence, the program sends the PICAXE back to the beginning. This is the way the pure-hardware version of this experiment was set up.

But what is this word "gosub"? It means "go to a subroutine." A subroutine is any sequence of program statements that ends with the instruction to "return." So "gosub getkey" tells the PICAXE to mark its current place in the program while it skips to the `getkey:` section of code, which it obeys, until it finds the word "return," which returns it to the place from where it came.

The PICAXE continues in this fashion until it reaches the word "end." I had to insert the word "end" because otherwise the PICAXE will continue executing the program and will fall into the subroutine. "End" stops it from doing so. Figure 5-148 shows a screenshot of the complete listing.

So—is that all? Yes, that's it. If you enter the code into the Programming Editor exactly as I have supplied it, you should be able to run it in simulation mode, and in the simulation window, click the right-arrow beside Logical Pin A2 to increase its value in steps. Each time you pass one of the values in the `getkey:` subroutine, you should see the value for variable b1 change in the display.

This is really all you need to perform the functions of the combination lock. When the PICAXE runs this program, it waits for the correct combination. If it receives the combination, it sends the output from logical pin 1 high; otherwise, logical pin 1 stays low.

The only additional item you need is a transistor or CMOS gate between logical pin 1 and the relay that unlocks the computer, because the PICAXE cannot deliver enough current to operate the relay by itself.

Putting this procedure into a controller chip not only simplifies the circuit, but offers another advantage: you can change the combination simply by rewriting the program and downloading the new version into the chip.

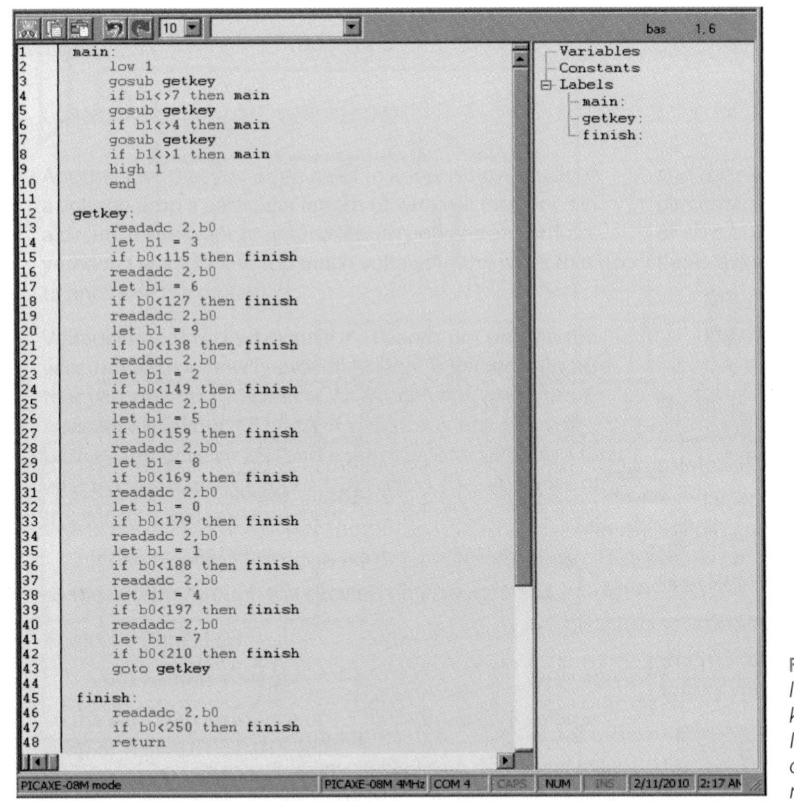

Figure 5-148. *This screenshot shows the complete listing of a program to read a sequence of three keypresses in conjunction with a combination lock. If the sequence is correct, the PICAXE sends a high output from one of its pins. If the sequence is incorrect, the program loops back to the beginning.*

FUNDAMENTALS

Limitations of MCUs

The PICAXE does have some disadvantages. Its voltage requirements alone restrict you from using it with the kind of freedom of a 555 timer.

Also, although I can get an instant result by plugging a 555 timer into a breadboard and adding a couple of resistors and a couple of capacitors, the PICAXE requires me to add a download socket, hook it up to my computer, write a program in the Programming Editor, and download the program.

Some people don't like writing software, or they have difficulty thinking in the relentlessly left-brain way that computer programming requires. They may prefer the hands-on process of assembling hardware.

Other people may have the opposite preference. This of course is a matter of taste, but one thing we know beyond all doubt is that computer programs often contain errors that may not reveal themselves until weeks or months later.

The PICAXE, for instance, doesn't protect you if a number is assigned to a variable that exceeds the limit for that type of variable. Suppose b1=200 and b2=60 and your program tells the PICAXE:

```
let b3 = b1 + b2
```

The result should be 260, but byte-size variables can only count up to 255. What happens? You will find that b3 acquires a value of 4, without any warning or explanation. This is known as an "overflow error," which can be very difficult to predict, because it happens at runtime, when external factors are in control. The code looks perfectly good; the Programming Editor doesn't find any syntax errors; the simulation behaves properly. But in the real world, days or even months later, an unexpected set of circumstances results in an input that causes the overflow, and because the code is residing inside the chip at this point, you may have a hard time figuring out what on earth went wrong.

Software has its problems. Hardware has its advantages.

FUNDAMENTALS

Unexplored territory

If you've taken the time to complete most of the projects in this book with your own hands, you have gained a very rapid introduction to the most fundamental areas of electronics.

What have you missed along the way? Here are some topics that remain wide open for you to explore. Naturally you should search online if they interest you.

The informal, learning-by-discovery approach that I have used in this book tends to be light on theory. I've avoided most of the math that you'd be expected to learn in a more rigorous course on the subject. If you have mathematical aptitude, you can use it to gain a much deeper insight into the way in which circuits work.

I didn't deal much with computer architecture, either. We didn't go very far into binary code, and you didn't build a half-adder, which is a great way to learn how computers function on the most fundamental level. Perhaps you should think about assembling one.

I avoided going deeply into the fascinating and mysterious properties of alternating current. Here again, some math is involved, but just the behavior of current at high frequencies is an interesting topic in itself.

For reasons already stated, I avoided surface-mount components—but you can still go into this area yourself for a relatively small investment, if you like the idea of creating fascinatingly tiny devices. This may be the future of hobby electronics, so if you stick with it, you'll probably end up in the world of surface-mount.

Vacuum tubes were not mentioned, because at this point, they are mainly of historical interest. But there's something very special and beautiful about tubes, especially if you can enclose them in fancy cabinetwork. In the hands of a skilled craftsperson, tube amplifiers and radios become art objects.

I didn't show you how to etch your own printed circuit boards. This is a task that appeals to only certain people, and the preparation for it requires you to make very neat drawings or use computer software for that purpose. If you happen to have those resources, you might want to do your own etching. It could be a first step toward mass-producing your own devices.

I didn't cover static electricity at all. High-voltage sparks don't have any practical applications, and they entail some safety issues—but they are stunningly impressive, and you can easily obtain the necessary information to build the equipment. Maybe you should try.

Other Controllers

If you want something more powerful, a BASIC Stamp is the logical next step after a PICAXE. the BASIC Stamp is so called because it originally looked like a postage stamp. The BASIC Stamp has a larger vocabulary of commands and a bigger range of add-on devices (including displays with graphical capability, and a little keyboard that is specifically designed for use with the controller). The BASIC Stamp is shown in Figure 5-149.

Figure 5-149. *The BASIC Stamp controller consists of surface-mounted components on a platform that has pins spaced at 1/10-inch intervals, for insertion in a breadboard or perforated board. This component uses a version of BASIC that is similar to the programming language of the PICAXE, but has many more extensions. The BASIC Stamp is available for use with a wide range of peripheral devices, including many alphanumeric dot-matrix displays.*

On the downside, you'll find that everything associated with the BASIC Stamp is a bit more expensive than in the PICAXE world, and the download procedure isn't quite as simple.

One of the more recent developments in the world of MCUs is the Arduino, which is both sophisticated and powerful. It does require programming in the C language. This language is a little more difficult to understand, and has only the vaguest similarity to the syntax that is used in the PICAXE and BASIC Stamp. On the other hand, because C dominates the larger world of computing, learning it might not be such a bad idea—and the Arduino offers some truly amazing capabilities. Because it is so popular, there are also many software tools, documentation, user forums, and many enthusatic hobbyists to help you. Two other Make: Books titles that I mentioned previously, *Getting Started with Arduino* and *Making Things Talk*, provide a great introduction.

In Closing

I believe that the purpose of an introductory book is to give you a taste of a wide range of possibilities, leaving you to decide for yourself what you want to explore next. Electronics is ideal for those of us who like to do things ourselves, because almost any application—from robotics, to radio-controlled aircraft, to telecommunications, to computing hardware—allows opportunities that we can explore at home, with limited resources.

As you delve deeper into the areas of electronics that interest you most, I trust you'll have a satisfying learning experience. But most of all, I hope you have lots of fun along the way.

Online Retail Sources and Manufacturers

A

This appendix contains URLs for companies mentioned as retail sources or manufacturers, along with the commonly used name of the source and the company name.

Colloquially used name	Actual corporate or company name	URL
3M	Minnesota Mining and Manufacturing Co.	*http://solutions.3m.com/en_US/*
Ace Hardware	Ace Hardware Corporation	*http://www.acehardware.com*
Advanced Micro Circuits	Advanced Micro Circuits Corp	*http://www.advancedmicrocircuits.com*
Akro-Mils	Myers Industries, Inc.	*http://www.akro-mils.com*
Alcoswitch	Division of Tyco Electronics Corporation	*http://www.tycoelectronics.com/catalog/menu/en/18025*
All Electronics	All Electronics Corporation	*http://www.allelectronics.com*
All Spectrum Electronics	All Spectrum Electronics	*http://www.allspectrum.com*
All-Battery.com	Tenergy Corporation	*http://www.all-battery.com*
Alpha potentiometers	Alpha Products Inc.	*http://www.alphapotentiometers.net*
ALPS pushbutton	ALPS Electric Co., Ltd.	*http://www.alps.com*
Amazon	Amazon.com, Inc.	*http://www.amazon.com*
Amprobe	Amprobe Test Tools	*http://www.amprobe.com*
Arduino	No corporate identity	*http://www.arduino.cc*
ArtCity	ArtCity	*http://www.artcity.com*
AutoZone	AutoZone, Inc.	*http://www.autozone.com*
Avago	Avago Technologies	*http://www.avagotech.com*
BASIC Stamp	Brand owned by Parallax, Inc.	*http://www.parallax.com*
BI Technologies	BI Technologies Corporation	*http://www.bitechnologies.com*
BK Precision	B&K Precision Corp.	*http://www.bkprecision.com*
Bussmann fuses	Cooper Bussman, Inc.	*http://www.cooperbussmann.com*
C&K switch	CoActive Technologies, Inc.	*http://www.ck-components.com*
Chicago lighting	CML Innovative Technologies	*http://www.cml-it.com*
CraftAmerica	Cardinal Enterprises	*http://www.craftamerica.com*
Darice	Darice Inc.	*http://www.darice.com*

Colloquially used name	Actual corporate or company name	URL
DeWalt	DeWalt Industrial Tool Company	*http://www.dewalt.com*
Digi-Key	Digi-Key Corporation	*http://www.digikey.com*
Directed switches	Directed Electronics Inc.	*http://www.directed.com*
Doctronics	Doctronics Educational Publishing	*http://www.doctronics.co.uk*
eBay	eBay Inc.	*http://www.ebay.com*
Elenco	Elenco Electronics Inc.	*http://www.elenco.com*
Everlight	Everlight Electronic Co. Ltd.	*http://www.everlight.com*
Extech	Extech Instruments Corporation	*http://www.extech.com*
Fairchild	Fairchild Semiconductor Incorporated	*http://www.fairchildsemi.com*
FTM	FTM Incorporated	*http://thefabricatorssource.com*
Fujitsu	Fujitsu America, Inc.	*http://www.fujitsu.com/us/*
GB wire strippers	Gardner Bender Inc.	*http://www.gardnerbender.com*
Hobbylinc	Hobbylinc Hobbies	*http://www.hobbylinc.com*
Home Depot	Homer TLC, Inc.	*http://www.homedepot.com*
Ideal wire strippers	Ideal Industries Inc.	*http://www.idealindustries.com*
Jameco	Jameco Electronics	*http://www.jameco.com*
K&J Magnetics	K&J Magnetics Inc.	*http://www.kjmagnetics.com*
Kingbright	Kingbright Corporation	*http://www.kingbrightusa.com*
Kobiconn	No web page found; use mouser.com	
KVM Tools	KVM Tools Inc.	*http://www.kvmtools.com*
Lowe's hardware	LF, LLC	*http://www.lowes.com*
Lumex	Lumex Inc.	*http://www.lumex.com*
McMaster-Carr	McMaster-Carr Supply Company	*http://www.mcmaster.com*
Megahobby	Megahobby.com	*http://www.megahobby.com*
Meter Superstore	Division of SRS Market Solutions Inc.	*http://www.metersuperstore.com*
Michaels craft stores	Michaels Stores, Inc.	*http://www.michaelscrafts.com*
Mill-Max	Mill-Max Manufacturing Corp.	*http://www.mill-max.com*
Mitutoyo	Mitutoyo America Corporation	*http://www.mitutoyo.com*
Motorola	Motorola, Inc.	*http://www.motorola.com/us*
Mouser electronics	Mouser Electronics, Inc.	*http://www.mouser.com*
Mueller alligator clip	Mueller Electric Company	*http://www.muellerelectric.com*
Newark	Subsidiary of Premier Farnell plc	*http://www.newark.com*
NKK switches	Nihon Kaiheiki Industry Co. Ltd.	*http://www.nkkswitches.com*
NXP semiconductors	NXP Semiconductors	*http://www.nxp.com*
Omron	Omron Corporation	*http://www.omron.com*
On Semiconductor	Semiconductor Components Industries, LLC	*http://www.onsemi.com*
Optek	Subsidiary of TT Electronics plc	*http://www.optekinc.com*
Panasonic	Panasonic Electric Works Corporation	*http://pewa.panasonic.com*

Colloquially used name	Actual corporate or company name	URL
PanaVise	Panavise Products, Inc.	*http://www.panavise.com*
Parallax	Parallax, Inc.	*http://www.parallax.com*
Pep Boys	Pep Boys-Manny, Moe and Jack	*http://www.pepboys.com*
Philips	Koninklijke Philips Electronics N.V.	*http://www.usa.philips.com*
PICAXE	Revolution Education Ltd.	*http://www.rev-ed.co.uk*
Piedmont Plastics	Piedmont Plastics, Inc.	*http://www.piedmontplastics.com*
Plano storages boxes	Plano Molding Company	*http://www.planomolding.com*
Pomona test equipment	Pomona Electronics Inc.	*http://www.pomonaelectronics.com*
RadioShack	RadioShack Corporation	*http://www.radioshack.com*
Sears	Sears Brands, LLC	*http://www.sears.com*
SparkFun Electronics	Sparkfun Electronics	*http://www.sparkfun.com*
Stanley tools	The Stanley Works	*http://www.stanleytools.com*
STMicroelectronics	STMicroelectronics Group	*http://www.st.com*
Texas Instruments	Texas Instruments Incorporated	*http://www.ti.com*
Tower Hobbies	Tower Hobbies	*http://www.towerhobbies.com*
Twin Industries	Twin Industries	*http://www.twinind.com*
Tyco	Tyco Electronics Corporation	*http://www.tycoelectronics.com*
Vaughan	Vaughan & Bushnell Mfg.	*http://hammernet.com/vaughan/*
Velleman keyboards	Velleman nv	*http://www.velleman.eu*
Vishay	Vishay Intertechnology Inc.	*http://www.vishay.com*
Wal-Mart	Wal-Mart Stores, Inc.	*http://www.walmart.com*
Weller	Division of Cooper Industries, LLC	*http://www.cooperhandtools.com/brands/weller/*
X-Acto	Division of Elmer's Products, Inc.	*http://www.xacto.com*
Xcelite	Division of Cooper Industries, LLC	*http://www.cooperhandtools.com/brands/xcelite/*
Xytronic	Xytronic Industries Ltd.	*http://www.xytronic-usa.com*

Acknowledgments

My association with MAKE magazine began when its editor, Mark Frauenfelder, asked me to write for it. I have always been very grateful to Mark for his support of my work. Through him I became acquainted with the exceptionally capable and motivated production staff at MAKE. Gareth Branwyn eventually suggested that I might like to write an introductory guide to electronics, so I am indebted to Gareth for initiating this project and supervising it as my editor. After I wrote an outline in which I described my idea for "Learning by Discovery" and the associated concept that cutting open components or burning them up can be an educational activity, MAKE's publisher, Dale Dougherty, uttered the memorable phrase, "I want this book!" Therefore I offer special thanks to Dale for his belief in my abilities. Dan Woods, the associate publisher, was also extremely supportive.

The production process was swift, competent, and painless. For this I thank my editor at O'Reilly, Brian Jepson; senior production editor Rachel Monaghan; copyeditor Nancy Kotary; proofreader Nancy Reinhardt; indexer Julie Hawks; designer Ron Bilodeau; and Robert Romano, who tweaked my illustrations. Most of all I am indebted to Bunnie Huang, my technical advisor, who reviewed the text in detail and knows a bunch of stuff that I don't know. Any residual errors are still my fault, even though I would prefer to blame them on Bunnie.

Thanks also to Matt Mets, Becky Stern, Collin Cunningham, Marc de Vinck, Phillip Torrone, Limor Fried, John Edgar Park, John Baichtal, and Jonathan Wolfe for helping out with some last-minute project testing.

Lastly I have to mention the genius of John Warnock and Charles Geschke, founders of Adobe Systems and creators of the very beautiful PostScript language, which revolutionized all of publishing. The horror of attempting to create this book using graphic-arts tools from...some other company...is almost unimaginable. In fact, without Illustrator, Photoshop, Acrobat, and InDesign, I doubt I would have attempted the task. I am also indebted to the Canon 1Ds with 100mm macro lens, which took many of the pictures in this book.

No free samples or other favors were received from any of the vendors mentioned herein, with the exception of two sample books from MAKE, which I read to ensure that I was not duplicating anything that had already been published.

Index

Numbers

A

B

C

D

E

F

G

H

I

O

P

R

S

T

U

V

W

X

Z

Colophon

The heading and cover font are BentonSans, the text font is Myriad Pro, and the code font is TheSansMonoCondensed.

About the Author

Charles Platt became interested in computers when he acquired an Ohio Scientific C4P in 1979. After writing and selling software by mail order, he taught classes in BASIC programming, MS-DOS, and subsequently Adobe Illustrator and Photoshop. He wrote five computer books during the 1980s.

He has also written science-fiction novels such as *The Silicon Man* (published originally by Bantam and later by Wired Books) and *Protektor* (from Avon Books). He stopped writing science fiction when he started contributing to *Wired* in 1993, and became one of its three senior writers a couple of years later.

Charles began contributing to MAKE magazine in its third issue and is currently a contributing editor. *Make: Electronics* is his first title for Make Books. Currently he is designing and building prototypes of medical equipment in his workshop in the northern Arizona wilderness.